Essential Maths for Students

Engineering Maths

Leslie Mustoe

Addison-Wesley

Harlow, England • Reading, Massachusetts • Menlo Park, California
New York • Don Mills, Ontario • Amsterdam • Bonn • Sydney
Singapore • Tokyo • Madrid • San Juan • Milan • Mexico City
Seoul • Taipei

Addison Wesley Longman Limited
Edinburgh Gate
Harlow
Essex
CM20 2JE
England

and Associated Companies throughout the world.

Typeset by 32

Printed and bound by Antony Rowe Ltd, Eastbourne

First printed 1997

ISBN 1-84479-326-5

British Library Cataloguing-in-Publication Data
A catalogue record for this book is available from the British Library

Library of Congress Cataloging-in-Publication Data is available

To Stephen, Jamie, Matthew, Stephen
and Paul

Contents

Preface

Mathematics is the language of engineering. If you have embarked on an engineering course in an institution of higher education then you will find that mathematics is present in all aspects of your course. You will need to acquire a stockpile of useful and applicable techniques and an understanding of the key concepts. You will also need to become confident in, and competent at, applying mathematics to the solution of engineering problems. *Engineering Maths* will give you a solid base in the mathematics that you will require in order to complete the first year of your course successfully.

An informal, user-friendly approach has been adopted and mathematical processes are described in everyday language. Mathematical ideas are developed by example rather than by formal proof. This reflects my experience that most students learn from example better than they do from an abstract approach. Where it is felt to be appropriate, the worked examples are developed with a great deal of detail so that you are not left to try and sort out how one part of the solution has been obtained from what went before.

In *Engineering Maths*, the objectives of each chapter are clearly stated at the beginning of that chapter. Key points and formulae are highlighted throughout the book. Self-assessment questions are provided at the end of most of the sections, the purpose being to test whether you understand the important ideas of that section; you can check your answers by referring back to the material in the section. These questions are followed by exercises and it is essential that you attempt them; the only way that you will develop your understanding and competence is by practice with examples of this kind. Detailed solutions to these exercises are given at the end of the book, but you should refer to them only when you have made a reasonable attempt at the exercise: reading a solution is no substitute for working it out yourself. A further set of test and assignment exercises is provided at the end of each chapter. They can be used either by your tutor as regular assignments or tests throughout your course or by you as material for revision. Solutions to these exercises are not provided in the book.

In order to keep the book to a reasonable length, only those topics which are believed to be the most important for a first-year course have been included. They are topics which are widely applicable and upon which work in subsequent years can build. The difficulties experienced by previous students have been taken into account when writing this book. However, in recent years many of the difficulties experienced by students have their origins in a lack of skill in the manipulation of algebraic expressions. Many of the students entering higher education to take engineering courses have not developed this skill to the level that is needed. Although an attempt is made in this book to explain the algebraic steps taken as carefully as space permits, you may need to revise or to acquire these skills in order to make progress and you may wish to refer to the book *Foundation Maths*, also in this series.

The best advice to offer you is to read through each section, paying careful attention to the worked examples and their solutions. Many of the examples contain important results which you will need to know for subsequent sections or chapters and it is a good idea to try each example yourself before looking at its solution. It is worth repeating the advice that only by *doing* mathematics rather than by reading what someone else has done will you really master the necessary techniques and really understand the underlying concepts. At the end of each section there are self-assessment questions for you to test your understanding. If you are unable to answer any of them, read the section again to discover the answers. Then you should attempt the exercises to reinforce your knowledge and skills; you should check your answers with those given at the back of the book.

If you have a graphic calculator, or access to a software package such as Derive, Maple or Mathematica, there will be occasions when you can enhance your learning of a topic by their use.

Finally, please remember that learning mathematics takes time, patience and perseverance; there are no short cuts. If you work through several exercises and experience a variety of problems then you will build up your own expertise and be more likely to tackle successfully unfamiliar and more challenging problems which arise in the other parts of your course. If you find *Engineering Maths* useful then it will have served its purpose. Good luck!

Leslie Mustoe

1 Discrete mathematics

Objectives

This chapter

- defines a set and simple set operations
- states the laws of set algebra
- explains the application of Boolean algebra to switching circuits
- shows how to design simple logic circuits
- applies the laws of statements to logical argument

1.1 Sets

A **set** is a collection of objects and membership of the set can be defined. Sets are denoted by capital letters, for example, $A,B,C,$ and its members, or elements, by lower-case letters, for example x,y,z. We write the statement 'x is a member of A' as $x \in A$ and the statement 'x is not a member of A' as $x \notin A$. A set which has a finite number of members is called a **finite set** whereas one with an infinite number of members is an **infinite set**. As an example, the set of the days of the week is a finite set and the set of all integers (whole numbers) is an infinite set.

Some commonly occurring sets have special notations. Four of these are:

$\mathbb{N}$ the set of natural numbers, that is, 0,1,2,3, ...
$\mathbb{Z}$ the set of integers, that is, $0, \pm 1, \pm 2, \pm 3, ...$
$\mathbb{Q}$ the set of rational numbers, that is, fractions
$\mathbb{R}$ the set of real numbers

Membership of a set can be defined in two ways. For a finite set we can list the membership, either in full or by implication. For example, the set A which comprises the numbers on the face of a die can be specified as:

$$A = \{1,2,3,4,5,6\}$$

where the braces { } enclose the members of the set. A second example is the set M of the months of the year:

$$M = \{\text{January, February, March, ..., December}\}$$

where the use of the symbol ... indicates that the months from April to November inclusive are also members of M.

We can specify the membership of any set by a rule. For example, our set A can be written:

$$A = \{x : x \in \mathbb{N} \text{ and } 1 \leqslant x \leqslant 6\}$$

We read the statement as 'A is the set of x such that x is a natural number and has a value between 1 and 6 inclusive'. The symbol : is read as 'such that'. Our second example can be written:

$$M = \{x : x \text{ is a month of the year}\}$$

An example of an infinite set is $B = \{x : x \in \mathbb{R} \text{ and } 0 \leqslant x \leqslant 2\}$ which is the set of real numbers between 0 and 2 inclusive. Two sets are **equal** if they have *exactly* the same membership. The order of listing the members is not important and any duplications are ignored.

If all the members of some set A are also members of a second set B then A is a **subset** of B. For example, if B is the set of parallelograms and A is the set of rectangles then A is a subset of B and we write $A \subseteq B$ (we could also write $B \supseteq A$). This definition of a subset allows the possibility that A and B are equal. If A is a subset of B and $A \neq B$ then A is a **proper subset** of B and we write $A \subset B$. Note that if $A \subseteq B$ and $B \subseteq A$ then A and B are equal.

Worked examples

1.1 Which of the following sets are equal?

(a) $\{1,3,5,7\}$, $\{7,3,1,5\}$, $\{3,3,7,1,1\}$ and $\{1,1,3,5,7,7\}$

(b) $A = \{x : x^2 - 4x + 3 = 0\}$, $B = \{x : x \in \mathbb{N} \text{ and } 0 \leqslant x \leqslant 2\}$, $C = \{1,3\}$

Solution (a) The third set does not contain 5; the other sets are equal.

(b) Sets A and C are equal; $B = \{0,1,2\}$.

1.2 Which of the following sets are subsets of $A = \{a,b,c,d,e\}$?

$B = \{a\}$, $C = \{b,c,d,e\}$, $D = \{a,e\}$, $E = \{a,b,c,d,e\}$, $F = \{a,c,c,e\}$, $G = \{a,f\}$

Solution All except G since $f \notin A$. Remember that duplication of members is ignored so that $F = \{a,c,e\}$. E is not a *proper* subset of A.

In most situations it is possible to define a **universal set** U to which all items under consideration belong; all sets mentioned subsequently will be subsets of U. For example, if we are dealing with quadrilaterals then U could be the set of all quadrilaterals; if we are discussing integers then U could be the set of all integers. Suppose that U is the set of all quadrilaterals and A is the set of parallelograms, then those members of U which do *not* belong to A form a set which is called the **complement** of A (in U); the set is denoted $\bar{A}$. In general, we can write formally:

$$\bar{A} = \{x : x \in U \text{ but } x \notin A\}$$

Relationships between sets can be pictured by the use of **Venn diagrams**. These are schematic drawings in which a set is represented by the interior of a circle or an oval shape. The sets are enclosed within a rectangular box whose interior represents the universal set, U. In Figure 1.1(a) we show the set A as a subset of the set B; the region representing A is entirely inside that representing B. Figure 1.1(b) illustrates the complement of a set: the shaded region represents the complement of A, that is, $\bar{A}$.

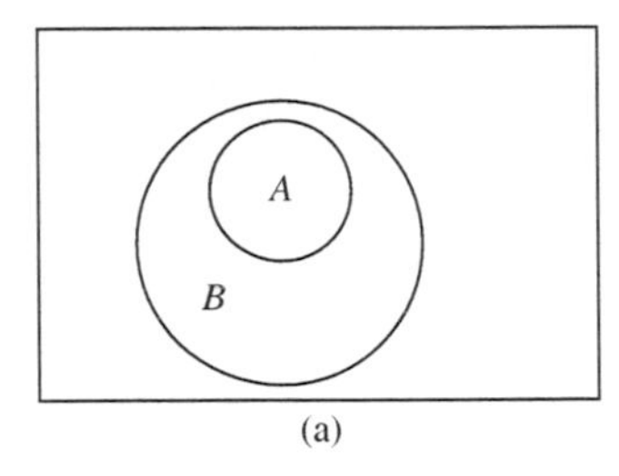

(a)

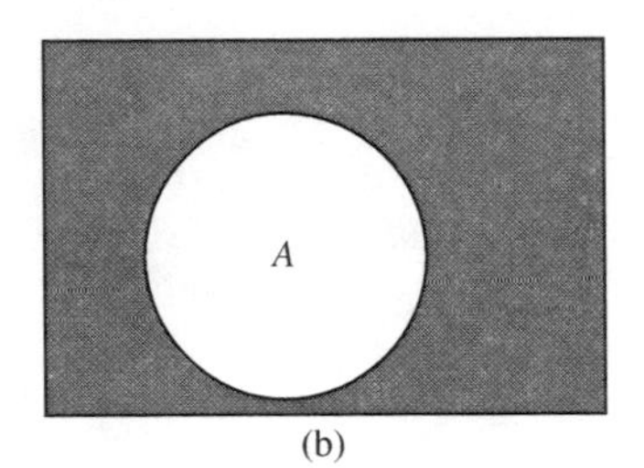

(b)

Figure 1.1.
(a) Subset
(b) Complement of a set

From two given sets A and B we can form a new set which contains those elements common to both A and B; this is called the **intersection** of A and B, written $A \cap B$ and read as 'A intersection B'. Formally,

$$A \cap B = \{x : x \in A \text{ and } x \in B\}$$

Note that $A \cap B = B \cap A$. If A and B have no elements in common then they are said to be **disjoint** and we write $A \cap B = \varnothing$ where $\varnothing$ is the **empty set**, that is, the set with no members. Note that $\varnothing$ is a subset of all sets.

In Figure 1.2(a) the intersection of A and B is shaded. In Figure 1.2(b) the sets D and E are disjoint; compare these diagrams with Figure 1.1(a) which depicts a subset.

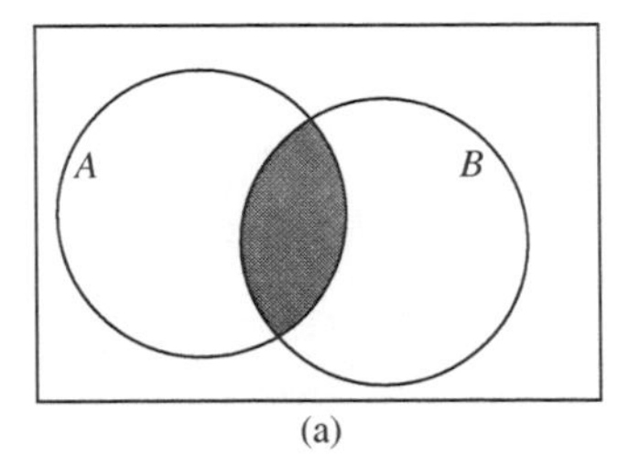

(a)

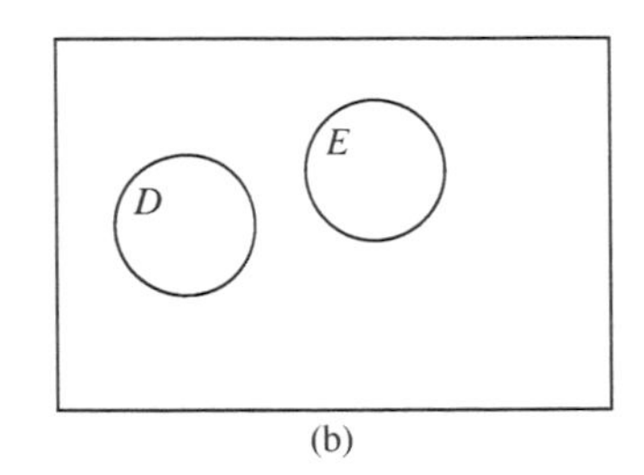

(b)

Figure 1.2.
(a) Intersection of two sets
(b) D and E are disjoint

From two given sets A and B we can form a second new set which contains all the members of A and all the members of B; this is called the **union** of A and B, written $A \cup B$ and read as 'A union B'. The formal definition is:

$$A \cup B = \{x : x \in A \text{ or } x \in B \text{ or both}\}$$

Some books omit 'or both' on the understanding that, in this context, 'or' is used in its *inclusive* sense, that is, it includes the idea of 'both'. Note that $A \cup B = B \cup A$. If A and B have members in common then these common members are counted once only. In Figure 1.3(a) the union of A and B is shaded: in part (b) the sets are disjoint.

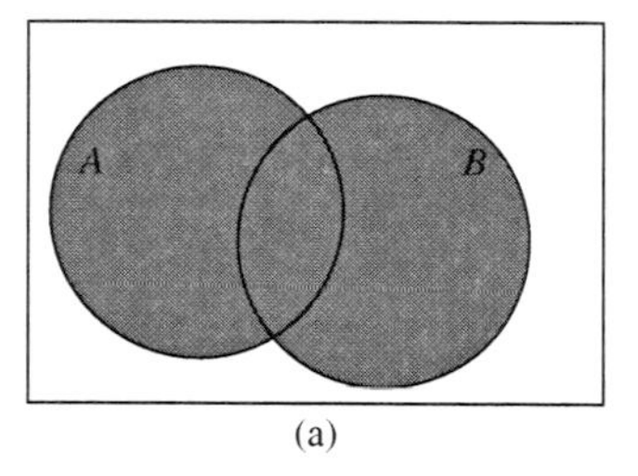

(a)

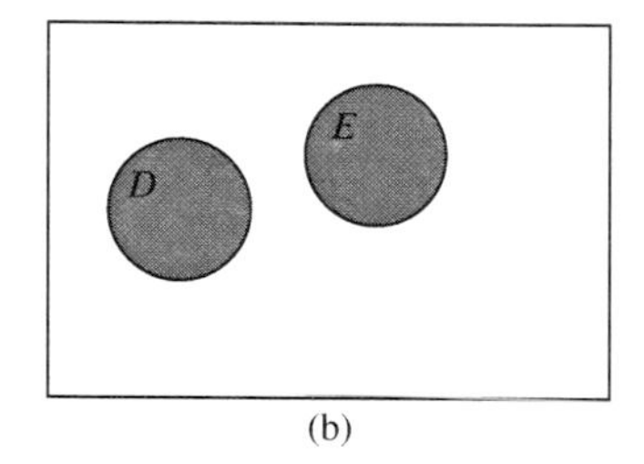

(b)

Figure 1.3. Union of two sets

KEY POINT

$A \cap B = \{x : x \in A \text{ and } x \in B\}$,

$A \cup B = \{x : x \in A \text{ or } x \in B \text{ or both}\}$

$\bar{A} = \{x : x \in U \text{ but } x \notin A\}$

Worked examples

1.3 If $A = \{1,2,3,4,5\}$, $B = \{2,4,6,8\}$ and $C = \{1,3,5\}$ find $A \cap B$, $B \cap C$ and $A \cap C$.

Solution $A \cap B = \{2,4\}$, $B \cap C = \varnothing$, and $A \cap C = \{1,3,5\}$. Note that, because C is a subset of A then $A \cap C = C$.

1.4 If $A = \{1,2,3,4,5\}$, $B = \{2,4,6,8\}$ and $C = \{1,3,5\}$ find $A \cup B$, $B \cup C$ and $A \cup C$.

Solution $A \cup B = \{1,2,3,4,5,6,8\}$, $B \cup C = \{1,2,3,4,5,6,8\}$, $A \cup C = \{1,2,3,4,5\}$. Note that, because C is a subset of A then $A \cup C = A$.

Venn diagrams can be helpful in solving some problems involving the numbers of items in sets.

Worked example

1.5 A survey was carried out into the reasons why 200 washing machines of a particular make had been reported as failing to pump out the water. The service engineers' reports showed that 69 machines had a blocked water outlet, 77 had a blocked filter, 80 had a faulty programmer, 14 had both water outlet and filter blocked, 11 had a blocked filter and a faulty programmer and 13 had a blocked water outlet and a faulty programmer; 4 had none of these defects.

How many machines had all three defects and how many had a blocked filter only?

Solution Let W be the set of machines with a blocked water outlet, F be the set of machines with a blocked filter and P be the set of machines with a faulty programmer. If $n(W)$ represents the number of items in the set W then the information can be translated as follows: $n(W) = 69$, $n(F) = 77$, $n(P) = 80$, $n(W \cap F) = 14$, $n(F \cap P) = 11$, $n(W \cap P) = 13$ and the number in the complement of the set $W \cup F \cup P$ is 4, that is, $n(\overline{W \cup F \cup P}) = 4$. Figure 1.4(a) illustrates the information derived from the data, starting with m as the number of items in the set $W \cap F \cap P$. Then the number of items in that part of the set $W \cap F$ which excludes $W \cap F \cap P$ must be $14 - m$; similarly we obtain the numbers $11 - m$ and $13 - m$. The number of elements in that part of the set W which has no elements in common with the sets F and P is given by $69 - (14 - m) - (13 - m) - m = 42 + m$; similarly we can insert the numbers $52 + m$ and $56 + m$. The information is shown in Figure 1.4(b).

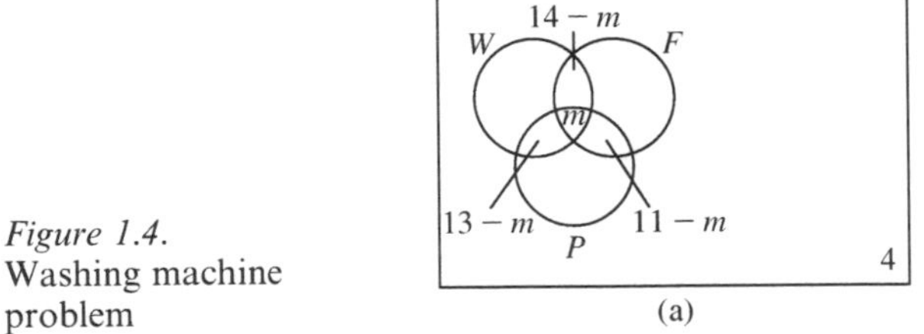

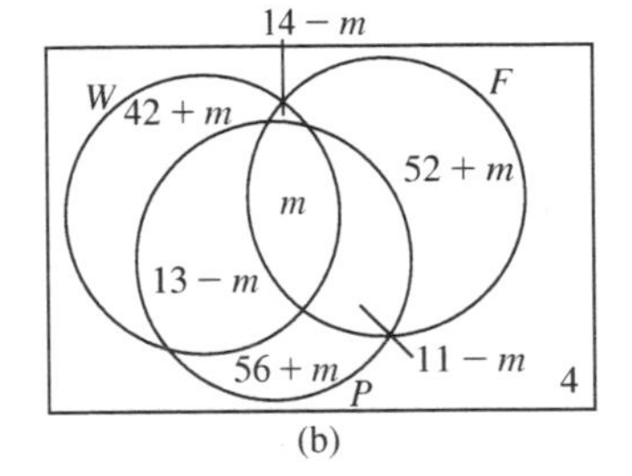

Figure 1.4. Washing machine problem

Now $n(W \cup F \cup P) = 200 - 4 = 196$ so that

$$\begin{aligned}(42 + m) + (52 + m) + (56 + m) + (14 - m)&\\ + (13 - m) + (11 - m) + m &= 196\\ 188 + m &= 196\\ m &= 8\end{aligned}$$

Therefore the number of machines which have all three defects is $m = 8$ and the number which have a blocked filter only is $52 + m = 60$.

We conclude this section with a statement of the laws of set algebra. They can be *illustrated* by Venn diagrams, but the formal proof requires a different approach, which we do not deal with here. In the following, A, B and C are any three sets belonging to the same universal set, U.

Commutative laws	$A \cup B = B \cup A$, $A \cap B = B \cap A$	1.1
Associative laws	$A \cup (B \cup C) = (A \cup B) \cup C$ $A \cap (B \cap C) = (A \cap B) \cap C$	1.2
Distributive laws	$A \cup (B \cap C) = (A \cup B) \cap (A \cup C)$ $A \cap (B \cup C) = (A \cap B) \cup (A \cap C)$	1.3
Identity laws	$A \cup \varnothing = A$, $A \cap U = A$ $A \cup U = U$, $A \cap \varnothing = \varnothing$	1.4
Complement laws	$A \cup \bar{A} = U$, $A \cap \bar{A} = \varnothing$	1.5
Idempotent laws	$A \cup A = A$, $A \cap A = A$	1.6

Other laws can be derived from the ones above. These laws are helpful when we want to simplify expressions involving union, intersection and complement.

Worked example

1.6 Simplify the following expressions:

(a) $A \cap (\bar{A} \cup (A \cap B))$

(b) $(A \cap B \cap C) \cup (A \cap B \cap \bar{C}) \cup (A \cap \bar{B} \cap C)$

Solution (a)

$$\begin{aligned} A \cap (\bar{A} \cup (A \cap B)) &= (A \cap \bar{A}) \cup (A \cap (A \cap B)) && \text{(distributive)} \\ &= \varnothing \cup (A \cap (A \cap B)) && \text{(complement)} \\ &= \varnothing \cup ((A \cap A) \cap B) && \text{(associative)} \\ &= \varnothing \cup (A \cap B) && \text{(idempotent)} \\ &= A \cap B && \text{(identity)} \end{aligned}$$

(b) $(A \cap B \cap C) \cup (A \cap B \cap \bar{C}) \cup (A \cap \bar{B} \cap C)$

$$\begin{aligned} &= (A \cap B \cap C) \cup (A \cap \bar{B} \cap C) \cup (A \cap B \cap \bar{C}) && \text{(commutative)} \\ &= (A \cap B \cap C) \cup (A \cap C \cap \bar{B}) \cup (A \cap B \cap \bar{C}) && \text{(commutative)} \\ &= ((A \cap C) \cap (B \cup \bar{B})) \cup (A \cap B \cap \bar{C}) && \text{(distributive)} \\ &= (A \cap C) \cup (A \cap B \cap \bar{C}) && \text{(complement identity)} \\ &= A \cap (C \cup (B \cap \bar{C})) && \text{(distributive)} \\ &= A \cap ((C \cup B) \cap (C \cup \bar{C})) && \text{(distributive)} \\ &= A \cap (C \cup B) \cap U && \text{(complement)} \\ &= A \cap (C \cup B) && \text{(identity)} \end{aligned}$$

Self-assessment questions 1.1

1. Explain the terms complement, union and intersection in relation to sets.
2. Explain the use of Venn diagrams to represent relationships between sets.

Exercise 1.1

1. List the membership of the following sets:

   $A = \{x : x \in \mathbb{N}, x < 12\}$
   $B = \{x : x^2 = 25\}$
   $C = \{x : 0 < x < 20, x \text{ is a multiple of } 3\}$
   $D = \{x : x \in Z, x \text{ is even}, 3 < x < 11\}$

   List the membership of the sets $A \cup C$, $A \cup D$, $C \cup D$, $A \cup C \cup D$, $A \cap B$, $A \cap C$, $A \cap C \cap D$, $A \cap (C \cup D)$. One of these sets is a subset of another; specify the relationship.

2. If the universal set is defined to be the set of natural numbers less than or equal to 18, find $\bar{A}$, $\bar{C}$, $\bar{D}$, $\overline{A \cup C}$, $\overline{A \cup D}$, $\overline{A \cap D}$, $\overline{A \cap C \cap D}$ for the sets of Question 1.

3. **De Morgan's laws** are $\overline{A \cup B} = \bar{A} \cap \bar{B}$, $\overline{A \cap B} = \bar{A} \cup \bar{B}$. Verify these laws for $A = \{1,2,3,4,5\}$, $B = \{1,3,5,7,9,11\}$, $U = \{x : x \in N, 0 < x < 13\}$.

4. Simplify the following using the laws of set algebra:

   (a) $A \cup (A \cap B)$ (b) $A \cap (A \cup B)$
   (c) $\bar{A} \cup (A \cap \bar{B})$ (d) $(A \cup B) \cup (\bar{A} \cap \bar{B})$
   (e) $(A \cup B \cup C) \cap (\bar{A} \cup B \cup C) \cap (\bar{B} \cup C)$
   (f) $(A \cup B \cup C) \cap ((A \cap B) \cup C)$

5. Of 200 vehicles which failed the MoT test 86 had faulty tyres, 71 had faulty brakes, 19 had faulty tyres and brakes, 23 had faulty brakes and exhausts, 26 had faulty tyres and exhausts, and 8 had all three defects. How many cars had faulty exhausts and how many had faulty brakes only?

1.2 Boolean algebra and switching circuits

Set algebra, which is governed by the laws (1.1) to (1.6), is an example of a general logic structure known as a **Boolean algebra**. A **Boolean variable** can take two values only, for example, true or false, on or off. A **Boolean expression** is a combination of variables using connectives corresponding to union and intersection and the operation corresponding to complement. In this section we consider the application to switching circuits.

A switch is either open (off) or closed (on). We denote the open state by 0 and the closed state by 1; the switch is denoted by a lower-case letter such as p; the variable p takes one of the values 0 and 1, and is therefore a Boolean variable. For the switch shown in Figure 1.5(a), $p = 0$ means that the switch is open and no current can flow through it, whereas $p = 1$ means that the switch is closed and current can flow.

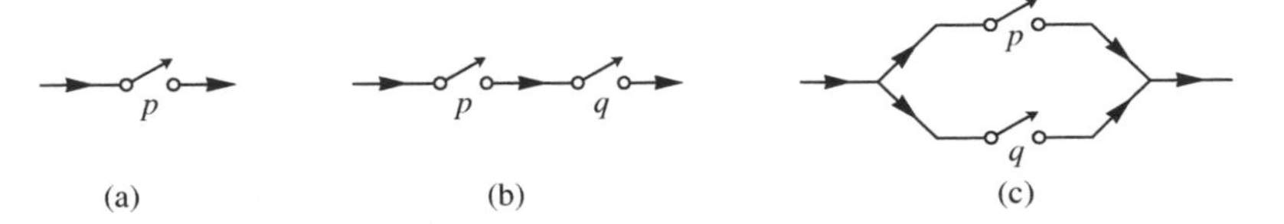

Figure 1.5. Switches and switching circuits

A **switching circuit** comprises an input, an output and a number of switches, connected in some way.

A series connection between two switches p and q is shown in Figure 1.5(b), and a parallel connection is shown in Figure 1.5(c). Each circuit has an associated Boolean function, denoted f, which takes the value 0 if no current flows through the circuit and the value 1 if current can flow. The value that f takes depends upon the states of the switches in the circuit. A summary of the behaviour of a circuit is provided by a **truth table**; Table 1.1(a) is for the series connection and Table 1.1(b) for the parallel connection. Since there are two switches in each case and each switch can be open or closed, there are four combinations to consider and each table has four rows of information.

For example, the only time that f takes the value 1 in Table 1.1(a) is when both p and q have the value 1, that is, only when both switches are closed will current flow through the series connection. Conversely, the only time that f takes the value 0 in Table 1.1(b) is when both p and q are 0, that is, unless both switches are open current will flow through the parallel connection. A series connection corresponds to the intersection of two sets (and); a parallel connection corresponds to the union (or). It is possible to connect two switches to each other so that they open or close together; in this case both switches are given the same letter. If the switches are connected so that when one opens the other closes, then we can denote one by p and the other by $\bar{p}$. Thus, when $p = 0$, $\bar{p} = 1$ and when $p = 1$, $\bar{p} = 0$. In what follows we replace the symbols $\cap$ and $\cup$ in set algebra by the symbols . and $+$, respectively.

To test whether two expressions are **logically equivalent** we compare their truth tables. If these tables are identical in all respects then the expressions take equal values for each combination of the inputs and the expressions are therefore logically equivalent.

KEY POINT

If two expressions have the same truth table then they are logically equivalent.

Worked examples

1.7 Show that the expressions $\overline{p.q}$ and $\bar{p} + \bar{q}$ are logically equivalent, that is, $\overline{p.q} = \bar{p} + \bar{q}$. Show also that $\overline{p+q} = \bar{p}.\bar{q}$.

Solution Table 1.2 shows the truth table for $\overline{p.q}$, $\bar{p} + \bar{q}$, $\overline{p+q}$ and $\bar{p}.\bar{q}$. Columns six and seven are identical, showing that $\overline{p.q} = \bar{p} + \bar{q}$. The last two columns are identical, showing that $\overline{p+q} = \bar{p}.\bar{q}$. These two results are De Morgan's laws in this algebra.

Table 1.1. Truth tables

(a)

p	q	$f = p.q$
1	1	1
1	0	0
0	1	0
0	0	0

(b)

p	q	$f = p + q$
1	1	1
1	0	1
0	1	1
0	0	0

Table 1.2. Truth table for Example 1.7

p	q	$\bar{p}$	$\bar{q}$	$p.q$	$\overline{p.q}$	$\bar{p}+\bar{q}$	$p+q$	$\overline{p+q}$	$\bar{p}.\bar{q}$
1	1	0	0	1	0	0	1	0	0
1	0	0	1	0	1	1	1	0	0
0	1	1	0	0	1	1	1	0	0
0	0	1	1	0	1	1	0	1	1

The laws of the algebra of switching circuits can be stated in a similar way to those of set algebra. In the following, p, q and r are any three switches, 1 replaces the universal set and 0 replaces the empty set.

Commutative laws	$p + q = q + p$, $p.q = q.p$	1.7
Associative laws	$p + (q + r) = (p + q) + r$ $p.(q.r) = (p.q).r$	1.8
Distributive laws	$p + (q.r) = (p + q).(p + r)$ $p.(q + r) = (p.q) + (p.r)$	1.9
Identity laws	$p + 0 = p$, $p.1 = p$, $p + 1 = 1$, $p.0 = 0$	1.10
Complement laws	$p + \bar{p} = 1$, $p.\bar{p} = 0$	1.11
Idempotent laws	$p + p = p$, $p.p = p$	1.12

Note that in some books $p.q$ is written simply as pq. Further, the operation . takes precedence over + so that, for example, $p + (q.r)$ can be written as $p + q.r$.

As with set algebra, Boolean expressions can be simplified using these laws.

Worked examples

1.8 Simplify the expression $p.q + \bar{q}.p + q.r + p.q.r$.

Solution First, note that $\bar{q}.p = p.\bar{q}$ (commutative). Then

$$\begin{aligned} p.q + p.\bar{q} + q.r + p.q.r &= p.(q + \bar{q}) + q.r + p.q.r && \text{(distributive)} \\ &= p.1 + q.r + p.q.r && \text{(complement)} \\ &= p + q.r + p.q.r && \text{(identity)} \\ &= p + 1.q.r + p.q.r && \text{(identity)} \\ &= p + (1 + p).(q.r) && \text{(distributive)} \\ &= p + 1.(q.r) && \text{(identity)} \\ &= p + q.r && \text{(identity)} \end{aligned}$$

1.9 Design a circuit comprising three switches p, q and r which will fail if all three switches are open or if p is open but q is closed.

Solution First we construct a truth table, Table 1.3, for the Boolean function f which takes the value 1 when the circuit allows current to pass and 0 when it does not. The last column shows the situation for each of the eight possibilities; for example, row three corresponds to p and r being closed and q being open, which we summarize as $p.\bar{q}.r$. Using the last two columns we see that the circuit allows current through ($f = 1$) in one of five states.

We can therefore write

$$f = p.q.r + p.q.\bar{r} + p.\bar{q}.r + p.\bar{q}.\bar{r} + \bar{p}.\bar{q}.r$$

Now we simplify this expression using laws (1.7) to (1.12):

$$\begin{aligned} f &= p.q.r + p.\bar{q}.r + p.q.\bar{r} + p.\bar{q}.\bar{r} + \bar{p}.\bar{q}.r && \text{(commutative)} \\ &= p.(q + \bar{q}).r + p.(q + \bar{q}).\bar{r} + \bar{p}.\bar{q}.r && \text{(distributive)} \\ &= p.1.r + p.1.\bar{r} + \bar{p}.\bar{q}.r && \text{(complement)} \\ &= p.r + p.\bar{r} + \bar{p}.\bar{q}.r && \text{(identity)} \\ &= p.(r + \bar{r}) + \bar{p}.\bar{q}.r && \text{(distributive)} \\ &= p.1 + \bar{p}.\bar{q}.r && \text{(complement)} \\ &= p + \bar{p}.\bar{q}.r && \text{(identity)} \end{aligned}$$

In fact, we can simplify the expression further, to $p + \bar{q}.r$, as is required by Exercise 1.2, Question 1. The circuit is shown in Figure 1.6.

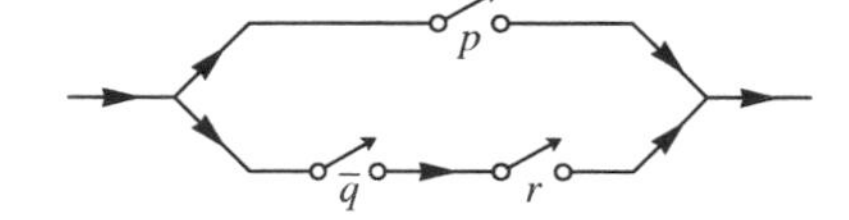

Figure 1.6. Switching circuit for Example 1.9

Table 1.3.
Truth table for function f

p	q	r	f	State
1	1	1	1	$p.q.r$
1	1	0	1	$p.q.\bar{r}$
1	0	1	1	$p.\bar{q}.r$
1	0	0	1	$p.\bar{q}.\bar{r}$
0	1	1	0	$\bar{p}.q.r$
0	1	0	0	$\bar{p}.q.\bar{r}$
0	0	1	1	$\bar{p}.\bar{q}.r$
0	0	0	0	$\bar{p}.\bar{q}.\bar{r}$

Self-assessment questions 1.2

1. Explain the use of truth tables in switching circuits.
2. Write down the truth tables for (a) a series connection, (b) a parallel connection between two switches.
3. Compare the laws of set algebra and the algebra of switching circuits.

Exercise 1.2

1. Show that $p + \bar{p}.\bar{q}.r = p + \bar{q}.r$.
2. By comparing appropriate truth tables verify the distributive laws (1.9).
3. Simplify the following Boolean expressions:

   (a) $p.(p.q + \bar{q})$ (b) $p.q + p + q + \bar{r}$
   (c) $p.\overline{(\bar{p}.\bar{q} + r)}$ (d) $p.q.r + p.q.\bar{r} + \bar{p}.q.r$
4. Draw the switching circuits represented by the Boolean expressions of Question 3(a), (b) and (d) and their simplifications.
5. Design a circuit which will turn a lamp on when the majority of three independent switches are closed.
6. Write down the Boolean expressions for the switching circuits shown in Figure 1.7.

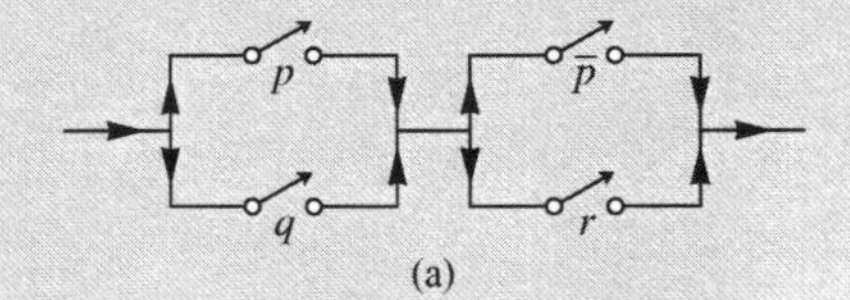

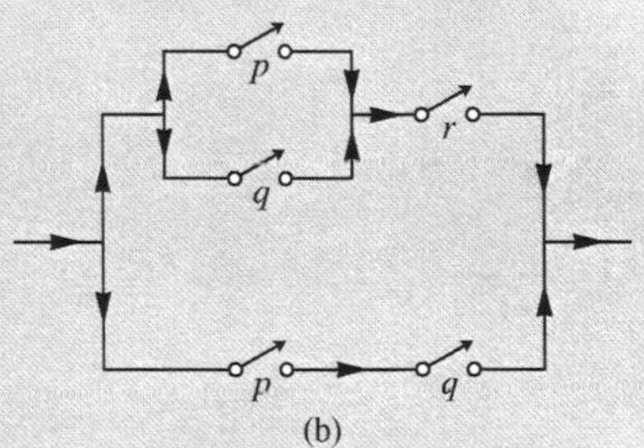

Figure 1.7
Switching circuits for Question 6

1.3 Logic circuits

The algebra of switching circuits can be adapted to deal with any system which is composed of devices that can exist in two states. In this section we consider **logic circuits** which consist of components called **logic gates** connected to each other. The gates correspond to standard

Boolean expressions and exist in states denoted 0 and 1, corresponding (in positive logic) to *low* and *high* voltage, respectively.

The **AND gate** is shown schematically in Figure 1.8. If both inputs p and q are in state 1 (high) then the output is also 1. For all other inputs the output is 0. The truth table is that of Table 1.4. It corresponds to the Boolean expression $f = p.q$.

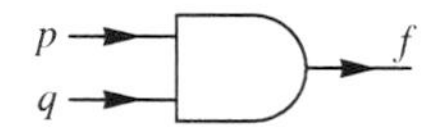

Figure 1.8.
The AND gate

The **OR gate** is shown schematically in Figure 1.9. If both inputs p and q are in state 0 (low) then the output is also 0. For all other inputs the output is 1. The truth table is that of Table 1.5. It corresponds to the Boolean expression $f = p + q$.

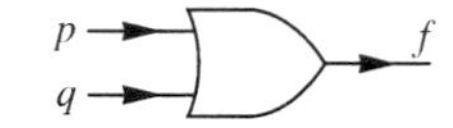

Figure 1.9.
The OR gate

The **NOT gate** is shown schematically in Figure 1.10. When the input p is in state 1 (high) then the output is in state 0 (low). When the input p is in state 0 then the output is in state 1. The truth table is that of Table 1.6. It corresponds to the Boolean expression $f = \bar{p}$.

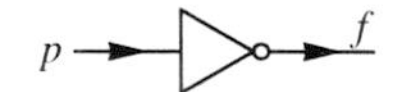

Figure 1.10.
The NOT gate

Worked example

1.10 Design a logic circuit to represent the Boolean expression $f = p + \bar{q}.r$ where p, q and r are inputs and f is the output.

Solution First we use a NOT gate to obtain $\bar{q}$ from the input q, then an AND gate to produce $\bar{q}.r$ and finally an OR gate to generate f. The circuit is shown in Figure 1.11.

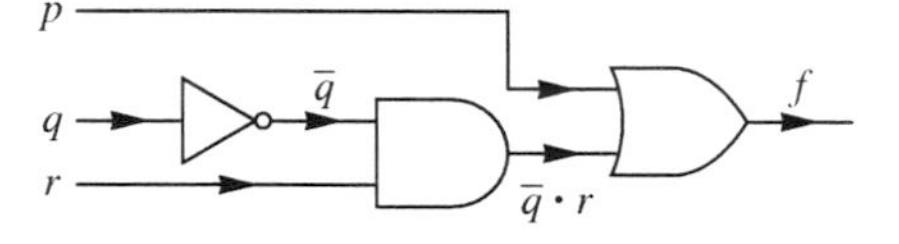

Figure 1.11.
Circuit for Example 1.10

Two further gates which are used frequently are the NOR gate and the NAND gate. The **NOR gate**, or 'not or' gate, corresponds to the Boolean expression $f = \overline{p + q}$ and is shown in Figure 1.12; its truth table is Table 1.7 and the equivalent circuit is shown in Figure 1.13. Using De Morgan's laws we see that $f = \overline{p + q} = \bar{p}.\bar{q}$.

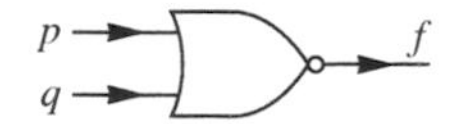

Figure 1.12.
The NOR gate

Table 1.4.
Truth table for the AND gate

p	q	$f = p.q$
1	1	1
1	0	0
0	1	0
0	0	0

Table 1.5.
Truth table for the OR gate

p	q	$f = p + q$
1	1	1
1	0	1
0	1	1
0	0	0

Table 1.6.
Truth table for the NOT gate

p	$f = \bar{p}$
1	0
0	1

Table 1.7.
Truth table for the NOR gate

p	q	$f = \overline{p + q}$
1	1	0
1	0	0
0	1	0
0	0	1

Table 1.8.
Truth table for the NAND gate

p	q	$f = \overline{p.q}$
1	1	0
1	0	1
0	1	1
0	0	1

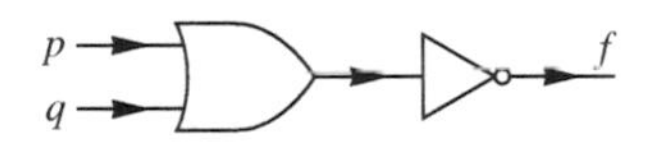

Figure 1.13.
Equivalent circuit for the NOR gate

The **NAND gate**, or 'not and' gate, corresponds to the Boolean expression $f = \overline{p.q}$ and is shown in Figure 1.14; its truth table is Table 1.8 and the equivalent circuit is shown in Figure 1.15. Using De Morgan's laws we see that $f = \overline{p.q} = \bar{p} + \bar{q}$.

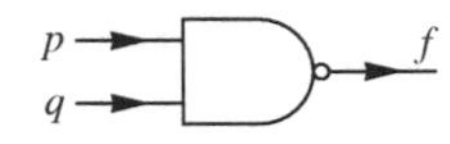

Figure 1.14.
The NAND gate

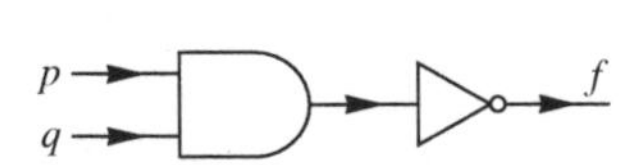

Figure 1.15.
Equivalent circuit for the NAND gate

It can be shown that *any* Boolean expression can be represented by a logic circuit consisting entirely of NOR gates, or by one consisting entirely of NAND gates.

Finally, we define a gate of practical significance – the **EXCLUSIVE OR gate**. This is shown in Figure 1.16 and corresponds to the Boolean expression $f = \bar{p}.q + p.\bar{q}$. When the inputs are in different states then $f = 1$, otherwise $f = 0$. The OR gate is strictly an 'inclusive or' gate.

Figure 1.16.
The EXCLUSIVE OR gate

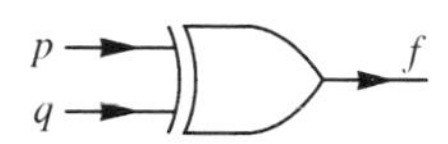

Self-assessment questions 1.3

1. Write down the truth tables for OR, AND and NOT gates.
2. Show that the NAND gate can be represented by the Boolean expression $f = \bar{p} + \bar{q}$ and that the NOR gate can be represented by the Boolean expression $f = \bar{p}.\bar{q}$.

Exercise 1.3

1. Produce the truth table for the EXCLUSIVE OR gate.
2. Find a Boolean function for the logic circuit in Figure 1.17.

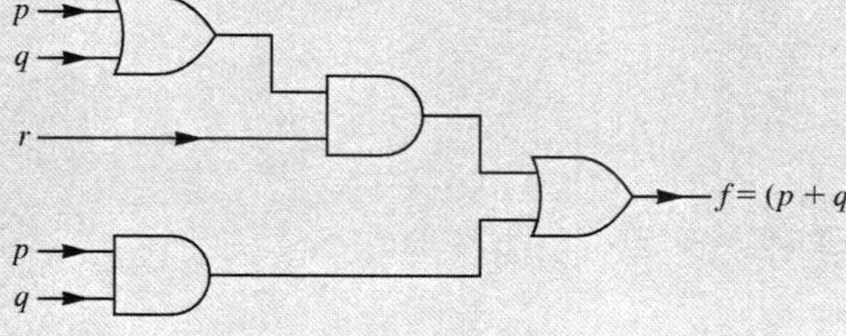

Figure 1.17.
Diagram for Question 2

3. Show how the Boolean expressions $p + q$ and $p.q$ can be implemented using only NAND gates.
4. Draw the logic circuit for the Boolean expressions $p.q + \bar{r}$ and $(p + \bar{q}).(p + r)$.

1.4 Implication and proof

A **proposition** is a statement which has a value of true or false. The truth or falseness of the statement can be determined immediately. For example, the statement that 1992 was an Olympic year is a proposition which is easily determined to be true, whereas the statement $x + y = 0$ is not a proposition since we do not know the values of x or y. In this case, however, once we know the values of x and y then we can decide whether the statement is true or false (for example, it is true if $x = 2$ and $y = -2$, and is false if $x = -3$ and $y = 5$). The statement is an example of a **predicate**; more generally, a predicate is an expression containing variables which becomes a proposition when actual values are substituted for the variables.

The notation we use is of the form p : The number 343 is divisible by 7, which we read as 'p is the statement that 343 is divisible by 7'. To each statement there is a related statement known as the **negation**; the negation of the statement above is $\bar{p}$: The number 343 is not divisible by 7. In general when p is true $\bar{p}$ is false and vice versa. The notation $\tilde{p}$ is also used.

Worked examples

1.11 Which of the following are propositions?

(a) 12 is an even number
(b) 26 is an odd number
(c) the next number will be prime
(d) the switch is open

Solution (a) and (b) are propositions; (c) and (d) are not propositions since we do not have enough information to decide whether they are true.

1.12 Negate the statements in Example 1.11.

Solution

(a) 12 is not an even number
(b) 26 is not an odd number
(c) the next number will not be prime
(d) the switch is closed.

Two propositions can be combined in ways similar to the intersection and union of sets. The **conjunction** of the propositions p and q is written $p \wedge q$ and corresponds to 'and'. The **disjunction** of the propositions p and q is written $p \vee q$ and corresponds to 'or'. The truth tables are shown as Table 1.9.

We can generate compound propositions using the logical connectives $\wedge$ and $\vee$ together with negation. For example, we can verify that the propositions $p \vee (q \wedge r)$ and $(p \vee q) \wedge (p \vee r)$ are **logically equivalent** by showing that they have identical truth tables (compare with equivalent Boolean expressions in Section 1.2). This is done in

Table 1.9. Truth table for conjunction and disjunction

p	q	$p \wedge q$	$p \vee q$
1	1	1	1
1	0	0	1
0	1	0	1
0	0	0	0

Table 1.10. Logically equivalent propositions

p	q	r	$(q \wedge r)$	$p \vee (q \wedge r)$	$(p \vee q)$	$(p \vee r)$	$(p \vee q) \wedge (p \vee r)$
1	1	1	1	1	1	1	1
1	1	0	0	1	1	1	1
1	0	1	0	1	1	1	1
1	0	0	0	1	1	1	1
0	1	1	1	1	1	1	1
0	1	0	0	0	1	0	0
0	0	1	0	0	0	1	0
0	0	0	0	0	0	0	0

Table 1.10 where we see that the fifth and eighth columns are identical. We can then write $p \vee (q \wedge r) \equiv (p \vee q) \wedge (p \vee r)$ where the symbol $\equiv$ is read 'is (logically) equivalent to'.

Worked examples

1.13 Given the propositions p : the switch is closed, q : the lamp is on, write in words the propositions (a) $p \wedge q$, (b) $p \vee q$, (c) $\bar{p} \vee q$, (d) $p \wedge \bar{q}$, (e) $\bar{p} \wedge \bar{q}$, (f) $\overline{p \vee q}$.

Solution
(a) the switch is closed and the lamp is on
(b) the switch is closed or the lamp is on
(c) the switch is not closed or the lamp is on
(d) the switch is closed and the lamp is not on
(e) the switch is not closed and the lamp is not on

It is difficult to express (f) in words as it stands, but if we construct appropriate truth tables then we see that it is equivalent to (e).

1.14 Given the statements p : the sum of two numbers is 7, q : the first number is 4, write in symbols the statements (a) the sum of two numbers is 7 and the first number is 4, (b) the sum of two numbers is 7 or the first number is not 4, (c) the sum of two numbers is not 7 and the first number is not 4.

Solution (a) $p \wedge q$, (b) $p \vee \bar{q}$, (c) $\bar{p} \wedge \bar{q}$.

The laws of statements are similar to those for set algebra and for switching circuits. Four groups are exactly similar.

Commutative laws	$p \vee q \equiv q \vee p$, $p \wedge q \equiv q \wedge p$	1.13
Associative laws	$p \vee (q \vee r) \equiv (p \vee q) \vee r$ $p \wedge (q \wedge r) \equiv (p \wedge q) \wedge r$	1.14
Distributive laws	$p \vee (q \wedge r) \equiv (p \vee q) \wedge (p \vee r)$ $p \wedge (q \vee r) \equiv (p \wedge q) \vee (p \wedge r)$	1.15
Idempotent laws	$p \vee p \equiv p$, $p \wedge p \equiv p$	1.16

In order to complete the picture we need to introduce two concepts which correspond to U and $\varnothing$ in set algebra. A **tautology** is a statement which is true in any circumstances, for example, $t \equiv q \vee \bar{q}$. A **contradiction** is a statement which is false in all circumstances, for example $c \equiv q \wedge \bar{q}$. Then we have:

Identity laws	$p \vee c \equiv p$, $p \wedge t \equiv p$, $p \vee t \equiv t$, $p \wedge c \equiv c$	1.17
Complement laws	$p \vee \bar{p} = t$, $p \wedge \bar{p} = c$	1.18

Implication is a cornerstone of mathematical proof. If p and q are two statements then the compound statement 'if p then q' or 'p implies q' means that the fact that p is true ensures that q is true. The symbolic form is $p \rightarrow q$ and the truth table is shown as Table 1.11. The common alternative form is $p \Rightarrow q$. What does it mean to say that $p \rightarrow q$ is true when p is false? Suppose we have the statements p: I service the car regularly and q: it will not break down. Then $p \rightarrow q$ reads 'If I service the car regularly then it will not break down'. If I do not service the car regularly then the implication statement is not false, whether or not the car breaks down, and hence the implication is true.

Worked examples

1.15 Given the statements p : switch A is closed, q : switch B is closed, r : the lamp is lit, write in symbolic form (a) if switch A is closed then the lamp

Table 1.11. Truth table for implication

p	q	$p \to q$
1	1	1
1	0	0
0	1	1
0	0	1

is lit, (b) if switch A is open and switch B is closed then the lamp is not lit, (c) if neither switch is closed then the lamp is not lit.

Solution (a) $p \to r$, (b) $\bar{p} \wedge q \to \bar{r}$, (c) $\bar{p} \wedge \bar{q} \to \bar{r}$.

1.16 Given the statements of Example 1.15, write in words the following:
(a) $q \to r$ (b) $p \wedge q \to r$ (c) $p \vee q \to r$
(d) $r \to p$ (e) $\bar{p} \to \bar{r}$ (f) $\bar{r} \to \bar{p}$

Solution
(a) if switch B is closed then the lamp is lit
(b) if both switches are closed then the lamp is lit
(c) if either switch is closed then the lamp is lit
(d) if the lamp is lit then switch A is closed
(e) if switch A is not closed then the lamp is not lit
(f) if the lamp is not lit then switch A is not closed

The statement $q \to p$ is called the **converse** of the statement $p \to q$. If one of these statements is true it does not follow that the other one is also true. For example, if we have statements p : current flows in the circuit and q : the switch is closed, then $p \to q$ is 'If current flows in the circuit then the switch is closed' and $q \to p$ is 'If the switch is closed then current flows in the circuit'. The first implication is true providing that the switch is an essential part of the circuit, but the second one may not be if there is a fault in the circuit.

If $p \to q$ then we say that p is a **sufficient condition** for q. If $q \to p$ then p is a **necessary condition** for q (sometimes expressed as 'q only if p'). If both $p \to q$ and $q \to p$ then p is a **necessary and sufficient condition** for q, written $p \leftrightarrow q$ (or $p \Leftrightarrow q$) and read 'p if and only if q'. We can make the definition $p \Leftrightarrow q \equiv (p \to q) \wedge (q \to p)$.

Worked example

1.17 In each of the following state whether p is a necessary condition for q, a sufficient condition, both or neither.
(a) $p : n$ is odd $q : n^2$ is odd
(b) $p : n$ is odd $q : 2n + 1$ is odd
(c) $p : n$ is odd $q : n$ is prime
(d) $p : n \geqslant 2$ $q : n + 2 \geqslant 4$
(e) $p : 2n$ is even $q : n$ is even

Solution (a) necessary and sufficient
(b) sufficient but not necessary, $n = 2$ is a suitable example
(c) neither, 9 is odd but not prime and 2 is prime but not odd
(d) necessary and sufficient
(e) necessary but not sufficient, $n = 3$ is a simple example

---

A **theorem** links a **hypothesis** p with a **conclusion** q by the implication $p \to q$. A sequence of valid propositions that ends with the conclusion is a **proof**. The main methods of proof are direct, indirect and by induction. An example of direct proof follows.

Theorem: If n is a positive integer then $n(n+1)(n+2)$ is divisible by 6. The hypothesis is that n is a positive integer and the conclusion is that the expression $n(n+1)(n+2)$ is divisible by 6.

Proof: If n is even then it is divisible by 2 and if n is odd then $(n+1)$ is even; in either case the given expression is divisible by 2. One and only one of the consecutive numbers n, $(n+1)$ and $(n+2)$ must be divisible by 3. Since the given expression is divisible by both 2 and 3 it must be divisible by 6.

Self-assessment questions 1.4

1. Explain the difference between a proposition and a statement.
2. Explain the terms contradiction, tautology and converse.
3. Explain what are meant by a necessary condition and a sufficient condition.

Exercise 1.4

1. Construct a truth table for the proposition $p \wedge (p \vee q)$.

2. Show that $p \vee (p \to q)$ is a tautology and that $\bar{q} \wedge p \wedge (p \to q)$ is a contradiction.

3. Which of the following are (i) necessary (ii) sufficient for the real numbers x and y to be equal?

   (a) $x^2 = y^2$,

   (b) $x^2 + y^2 = 2xy$,

   (c) $x + \dfrac{1}{y} = y + \dfrac{1}{x}$.

4. The **contrapositive** form of $p \to q$ is $\bar{q} \to \bar{p}$. Write down the contrapositive and the converse of

   (a) if a triangle is equilateral then it is isosceles,

   (b) if $x^2 + y^2 \leqslant 4$ then $x + y \leqslant 2$.

5. Show that the statement $x^2 - 3x + 2 \leqslant 0$ *for all* x is false by finding a **counter-example**, that is, one instance for which it is false.

6. Prove that the sum of four consecutive odd integers is divisible by 8.

Test and assignment exercises 1

1. If the universal set is the set of natural numbers less than or equal to 20 and $A = \{n : 2 < n < 9\}$, $B = \{n : n^2 < 150\}$, $C = \{n : n \text{ is odd}, n < 16\}$, find the following sets:

   $\bar{A}$, $A \cup \bar{B}$, $B \cap C$, $A \cup (\bar{B} \cap \bar{C})$, $\overline{A \cup (B \cap C)}$

2. Which of the following are equivalent to $A \subset B$ and which are equivalent to $A = B$?

   (a) $A \cap B = A$ (b) $A \cup B = B$
   (c) $A \cup B = A \cap B$ (d) $A \cap \bar{B} = \varnothing$
   (e) $(A \cap \bar{B}) \cup (B \cap \bar{A}) = \varnothing$

3. Draw switching circuits which correspond to the Boolean expressions:

   (a) $p.q + \bar{p}.\bar{q}$ (b) $p.q + p.\bar{q} + \bar{p}.q$

4. Simplify the following Boolean expressions:

   (a) $p.\overline{(q + \bar{p}.\bar{r})}$ (b) $\overline{(\bar{p}.q + p.\bar{q})}$

5. Construct truth tables for the Boolean expressions $(p + q).(\bar{p} + q).(q + r)$ and $p.q + \bar{p}.q + \bar{q}.r$. What do you conclude?

6. Draw logic circuits for the following Boolean expressions:

   (a) $(p + \bar{q}).r$ (b) $\overline{(p + r)}.q + \bar{q}$

7. Show that the following propositions are equivalent to $p \to q$:

   (a) $\bar{p} \vee q$ (b) $\bar{q} \to \bar{p}$ (c) $\overline{p \wedge \bar{q}}$

8. Show that the product of four consecutive integers is divisible by 24.

2 Sequences and series

Objectives

This chapter

- defines a sequence and a series
- shows how to specify the terms in a sequence
- defines arithmetic and geometric sequences
- explains the idea of the sum of an infinite series
- finds the sum of an arithmetic series and of a geometric series
- states the binomial theorem and uses it to obtain binomial expansions

2.1 Sequences

A **sequence** is a set of numbers written in a particular order. Hence the sequences $\{1,2,3,4,5,6\}$ and $\{2,5,1,6,4,3\}$ are different. Each item in the sequence is called a **term**. In our second sequence, the first term is 2 and the fifth term is 4. If a sequence has a finite number of terms it is called a **finite sequence**, otherwise it is an **infinite sequence**.

A finite sequence can be specified by a full listing or by a partial listing which indicates the missing terms. An infinite sequence can be specified by a partial listing.

Worked examples

2.1 How many terms are there in the following sequences?

(a) $\{1,3,5,\ldots,45\}$ (b) $\{1,\frac{1}{2},\frac{1}{4},\ldots\}$

Solution

(a) The use of the symbols ... in this example indicates that the pattern of the first three terms will continue. If you write out the list in full you will see that there are 23 terms.

(b) Here the use of ... at the end of the given terms indicates that the sequence is infinite. The next three terms are $\frac{1}{8}$, $\frac{1}{16}$ and $\frac{1}{32}$.

There are several notations for sequences in current usage. The one employed in this chapter is to denote the terms by u_1, u_2, u_3, and so on. Hence in sequence (a) in Example 2.1, $u_1 = 1$, $u_2 = 3$, $u_{23} = 45$.

We could specify both sequences in Example 2.1 by a **rule** which allows us to write down any of the terms. For example, in sequence (a) the rule is $u_k = 2k - 1$ (as you can verify), where k takes the values 1 to 23 inclusive. Sequence (b) can be specified by the rule

$$u_k = \frac{1}{2^{k-1}}$$

(Remember that $2^0 = 1$ so that $u_1 = 1/2^0 = 1/1 = 1$.)

A third method of specifying a sequence is by a **recurrence relation**. In the case of sequence (a) this is $u_{k+1} = u_k + 2$ and for sequence (b) it is $u_{k+1} = \frac{1}{2}u_k$. (We could equally well have written $u_k = u_{k-1} + 2$ and $u_k = \frac{1}{2}u_{k-1}$, respectively.)

We must start each recurrence relation with one or more initial terms so that in both (a) and (b) we specify $u_1 = 1$.

Sometimes it is straightforward to specify a sequence by either of these two methods.

Worked examples

2.2 Write out the first five terms of the sequences specified by:
(a) $u_k = k^2 + k$ (b) $u_k = 3u_{k-1} + 1$, $u_1 = 1$

In each case k takes the values 1,2,3,... .

Solution (a) $u_1 = 1 + 1 = 2$, $u_2 = 4 + 2 = 6$, $u_3 = 12$, $u_4 = 20$, $u_5 = 30$.
(b) $u_1 = 1$, $u_2 = 3 \times 1 + 1 = 4$, $u_3 = 3 \times 4 + 1 = 13$, $u_4 = 40$, $u_5 = 121$.

2.3 Specify the sequence (a) $u_k = 2u_{k-1}$, $u_1 = 3$ by a rule and the sequence (b) $u_k = 1/(3^k)$ (k takes the values 1,2,3,...) by a recurrence relation. List the first five terms of each sequence.

Solution (a) $u_1 = 3$, $u_2 = 6$, $u_3 = 12$, $u_4 = 24$, $u_5 = 48$.
Hence $u_k = 3 \times 2^{k-1}$.

(b) $u_1 = \frac{1}{3}$, $u_2 = \frac{1}{9}$, $u_3 = \frac{1}{27}$, $u_4 = \frac{1}{81}$, $u_5 = \frac{1}{243}$.

Hence $u_k = \frac{1}{3}u_{k-1}$, with the initial condition $u_1 = \frac{1}{3}$.

We can picture the behaviour of a sequence as k increases by plotting its graph as in Figure 2.1.

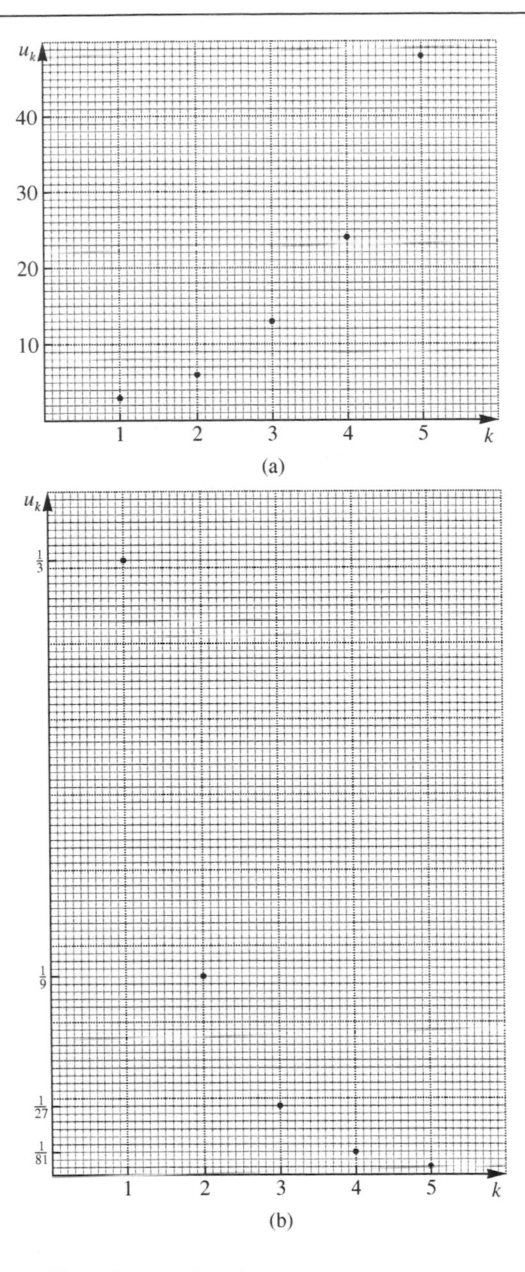

Figure 2.1. The sequences of Example 2.3

Two important sequences

An **arithmetic sequence** or **arithmetic progression** (AP) is formed when each term is obtained by adding the same fixed value, the **common difference**, to its predecessor. The sequence $\{1,3,5,...,45\}$ is an example with a **first term** of 1 and a common difference of 2.

In general the terms of an AP are written as:

KEY POINT

$$a, \quad a+d, \quad a+2d, \quad a+3d, \ ...$$

where a is the first term and d is the common difference. (Note that d could be negative, or even zero.)

The second term is $a + 1d$, the third term is $a + 2d$, the fourth term is $a + 3d$ and hence:

KEY POINT

> the nth term of an AP is $a + (n - 1)d$

Worked examples

2.4 Write down the first five terms of the following APs:
(a) $a = 2, d = 3$ (b) $a = -2, d = 3$ (c) $a = 2, d = -3$
(d) $a = -2, d = -3$ (e) $a = 2, d = 0$

Solution (a) 2,5,8,11,14 (b) -2,1,4,7,10 (c) 2,-1,-4,-7,-10
(d) -2,-5,-8,-11,-14 (e) 2,2,2,2,2

Sequence (e) is an example of a **constant sequence**.

2.5 Write down the first term and common difference of the following APs:
(a) $\frac{1}{3}, \frac{5}{3}, \frac{9}{3}, \frac{13}{3}, \ldots$ (b) $-0.1, -0.45, -0.8, -1.15$

Solution (a) The first term is $\frac{1}{3}$ and each term is $\frac{4}{3}$ more than the previous one. Hence $a = \frac{1}{3}, d = \frac{4}{3}$.

(b) Here $a = -0.1$, and to obtain the next term we subtract 0.35; hence $d = -0.35$.

2.6 If the fourth term of an AP is 11 and the tenth term is 23 find the first term and the common difference.

Solution We know that $a + 3d = 11$ and that $a + 9d = 23$. Subtracting the second equation from the first gives $6d = 12$. Hence $d = 2$ and using the first equation we have $a + 6 = 11$ so that $a = 5$.

A second important sequence is the **geometric sequence** or **geometric progression** (GP). Each term is obtained from its predecessor by multiplying by a fixed value, the **common ratio**. The sequence $\{2,6,18,54\}$ is an example, where the first term is 2 and the common ratio is 3. In general

KEY POINT

> a geometric progression can be written as $a, ar, ar^2, ar^3, \ldots$

where the first term is a, and the common ratio is r.

The second term is ar^1, the third term is ar^2, the fourth term is ar^3 so that, in general,

KEY POINT

> the nth term of a geometric progression is ar^{n-1}

Worked examples

2.7 Write down the common ratio of the following GPs:

(a) $1, -\frac{1}{2}, \frac{1}{4}, -\frac{1}{8}$ (b) $2, 8, 32, 128, \ldots$

Solution (a) Each term is multiplied by $-\frac{1}{2}$ to get the next term; hence $r = -\frac{1}{2}$.

(b) The first term is 2 and each term is multiplied by 4 to get the next term; hence $r = 4$.

2.8 Write down the first five terms of the following GPs:

(a) $a = -\frac{1}{2}, r = 3$ (b) $a = 3, r = -\frac{1}{2}$

Solution (a) $-\frac{1}{2}, -\frac{3}{2}, -\frac{9}{2}, -\frac{27}{2}, -\frac{81}{2}$ (b) $3, -\frac{3}{2}, \frac{3}{4}, -\frac{3}{8}, \frac{3}{16}$

2.9 If the second term of a GP is 0.1 and the fifth term is 0.0125, find the first term and the common ratio.

Solution We know that $ar = 0.1$ and that $ar^4 = 0.0125$ so that $r^3 = ar^4 \div ar = 0.0125 \div 0.1 = 0.125$. Hence $r = 0.5$ and $a \times 0.5 = 0.1$, giving $a = 0.2$.

The terms of the sequence $1, \frac{1}{2}, \frac{1}{4}, \frac{1}{8}, \ldots$ are getting closer to zero. They never become negative but we can find a term as close to zero as we wish by going far enough down the sequence. We say that the **limit of the sequence** is zero or that the sequence **converges** to zero.

In the general notation we write

$$\lim_{k \to \infty} \{u_k\} = 0$$

This is read as 'the limit as k tends to infinity of the sequence u_k is zero'.

Worked examples

2.10 Describe the behaviour of the following sequences as $k \to \infty$ and illustrate that behaviour by plotting the first five terms of each sequence. In each case the formula for the general term, u_k, is given.

(a) $3 + \dfrac{1}{k}$ (b) $3 - \dfrac{1}{k}$ (c) 2^k (d) $(-1)^k$

(e) $-k$ (f) $(-1)^k . \dfrac{1}{k}$ (g) $(-2)^k$

Solution Refer to Figure 2.2

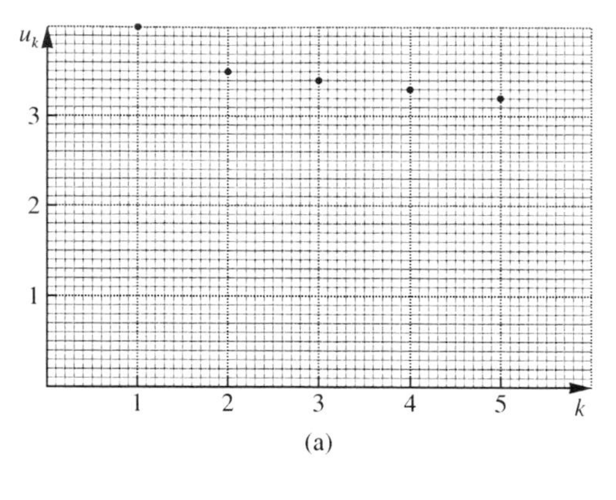
u_k
3
2
1
1 2 3 4 5 k
(a)

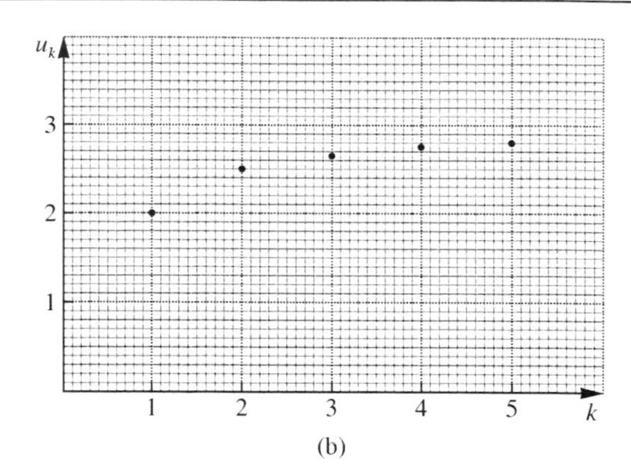
u_k
3
2
1
1 2 3 4 5 k
(b)

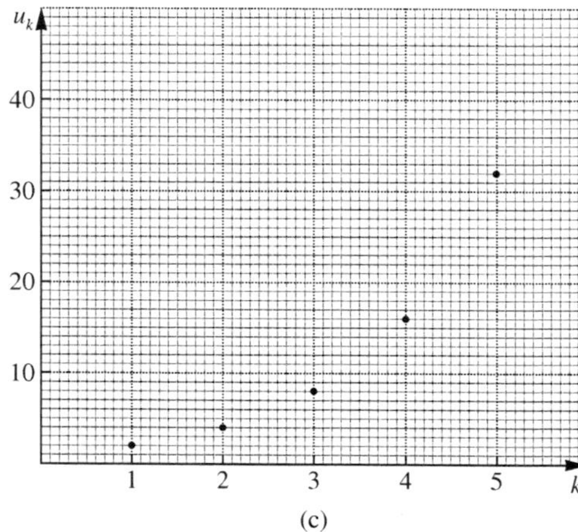
u_k
40
30
20
10
1 2 3 4 5 k
(c)

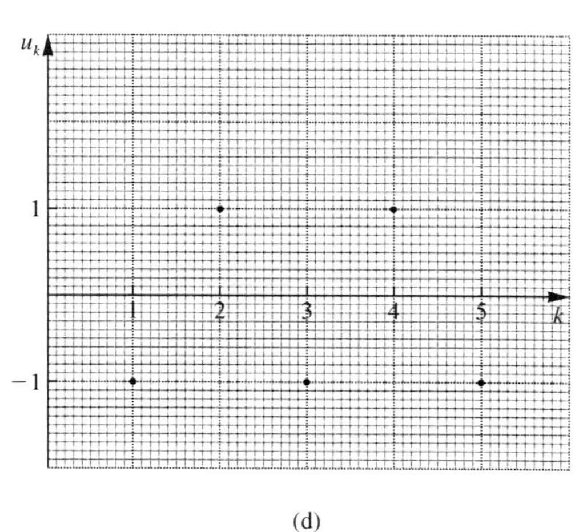
u_k
1
−1
1 2 3 4 5 k
(d)

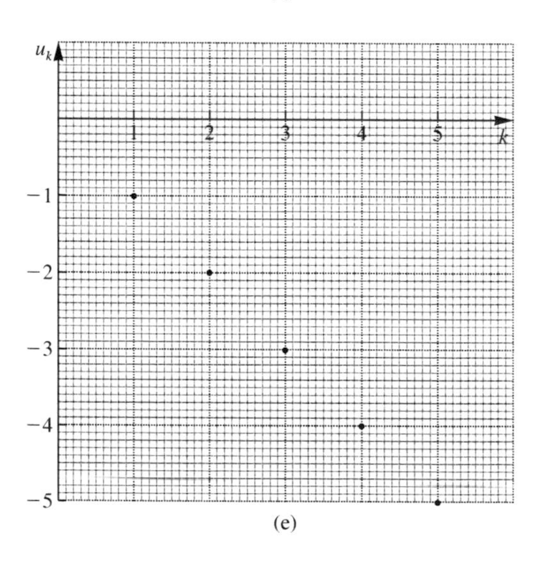
u_k
1 2 3 4 5 k
−1
−2
−3
−4
−5
(e)

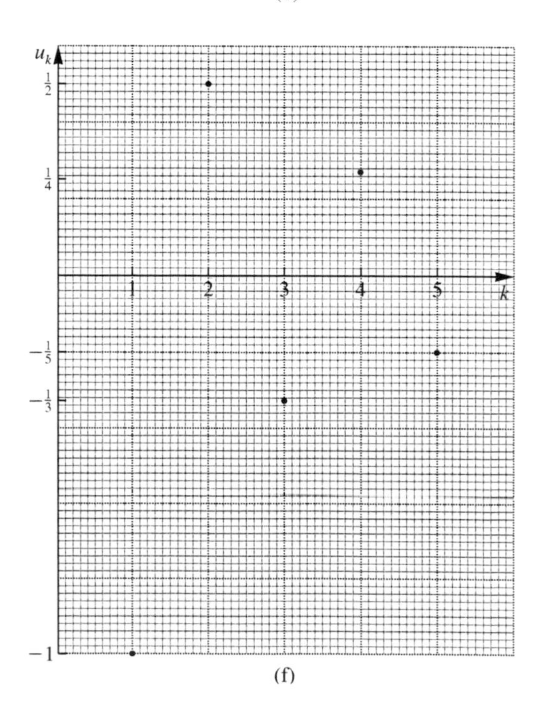
u_k
$\frac{1}{2}$
$\frac{1}{4}$
1 2 3 4 5 k
$-\frac{1}{5}$
$-\frac{1}{3}$
−1
(f)

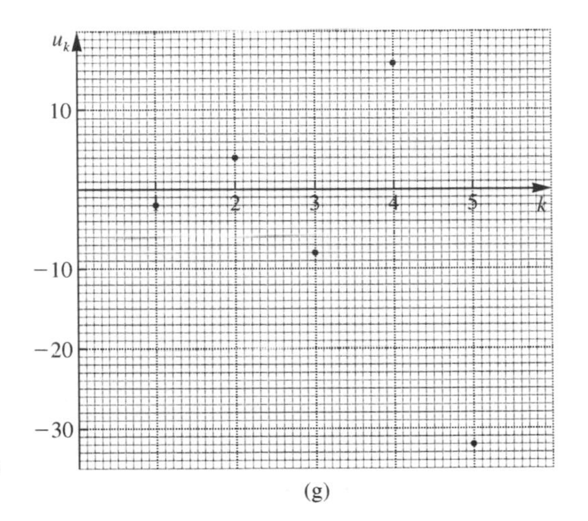

Figure 2.2.
Graphs of the sequences of Example 2.10

(a) $u_1 = 3 + 1 = 4, u_2 = 3 + \frac{1}{2} = 3\frac{1}{2}, u_3 = 3\frac{1}{3}, u_4 = 3\frac{1}{4}, u_5 = 3\frac{1}{5}$.

As k gets larger, $\frac{1}{k}$ gets closer to zero and u_k approaches the value 3 ever more closely, from above. Hence the limit of the sequence is 3.

(b) $u_1 = 3 - 1 = 2, u_2 = 3 - \frac{1}{2} = 2\frac{1}{2}, u_3 = 2\frac{2}{3}, u_4 = 2\frac{3}{4}, u_5 = 2\frac{4}{5}, \ldots$.

As k gets larger, $\frac{1}{k}$ gets closer to zero and u_k approaches the value 3 ever more closely, from below. Hence the limit of the sequence is again 3.

(c) $u_1 = 2, u_2 = 4, u_3 = 8, u_4 = 16, u_5 = 32$.

The terms increase towards infinity. The sequence does not have a limit. It is said to **diverge** (to infinity).

(d) $u_1 = -1, u_2 = 1, u_3 = -1, u_4 = 1, u_5 = -1$.

The sequence **oscillates**; there is no limit, but the terms do not diverge and the behaviour is called a finite oscillation.

(e) $u_1 = -1, u_2 = -2, u_3 = -3, u_4 = -4, u_5 = -5$.

The terms become progressively more negative and the sequence is said to diverge to minus infinity.

(f) $u_1 = -1, u_2 = \frac{1}{2}, u_3 = -\frac{1}{3}, u_4 = \frac{1}{4}, u_5 = -\frac{1}{5}$.

The terms are oscillating (in sign) but their magnitudes are becoming progressively smaller. The sequence converges (to zero).

(g) $u_1 = -2, u_2 = 4, u_3 = -8, u_4 = 16, u_5 = -32$.

The sequence oscillates and diverges.

Self-assessment questions 2.1

1. Explain what is meant by the statement 'the limit of a sequence is zero'.
2. What are the key features of an arithmetic sequence and a geometric sequence?

Exercise 2.1

1. A sequence is obtained by sampling the continuous function $f(t) = t^2 + 2t$ at $t = 0,1,2,3,4,\ldots$. Write down the first five terms of the sequence.

2. Write down the first six terms of the sequences defined by:

   (a) $u_k = \dfrac{1}{k}$ (b) $u_k = k^2 + \dfrac{1}{k^2}$

   (c) $u_{k+1} = 1 - u_k$, $u_1 = 10$

   (d) $u_{k+1} = \frac{1}{4}u_k$, $u_1 = 1024$.

   In all cases, k takes the values $1,2,3,\ldots$.

3. Find (a) a rule, (b) a recurrence relation for the sequence $\{2, \frac{2}{3}, \frac{2}{9}, \frac{2}{27}, \ldots\}$.

4. Write down the 9th and the 20th terms of the following arithmetic progressions:

   (a) 7,10,13,... (b) 18,15,12,...

5. The 10th term of an arithmetic progression is -62 and the 20th term is -142. Find the first term and the common difference.

6. The 10th term of a geometric progression is 1536 and the 20th term is 1 572 864. Find the first term and the common ratio.

7. Describe the behaviour as $k \to \infty$ of the sequences whose general term is

   (a) $1 + \dfrac{2}{k}$ (b) $1 - \dfrac{2}{k}$ (c) $\left(\frac{1}{4}\right)^k$

   (d) $\left(-\frac{1}{4}\right)^k$ (e) $(-3)^k$ (f) $\dfrac{1}{k^2}$

   Confirm your conclusions by calculating and plotting the first 50 terms of each sequence. If the sequence has a limit, how far along the sequence must you go until the terms are within 0.001 of the limit?

2.2 Series

A **series** is formed when the terms of a sequence are added. Hence

$$1 + 3 + 5 + 7 + 9$$

is a series.

If the series has a finite number of terms then its sum can be found by simple addition. The series above has a sum of 25. In calculating the overall sum we can form the **partial sums** 1, $1 + 3$, $1 + 3 + 5$, $1 + 3 + 5 + 7$ and, finally, $1 + 3 + 5 + 7 + 9$, that is, 1,4,9,16,25.

Is it possible to find the sum of an infinite series? If the sequence of partial sums converges then the limit of that sequence is defined to be the sum of the series.

Worked examples

2.11 Find, where possible, the sum of each of the following series:

(a) $1 + 3 + 5 + ... + 45$ (b) $1 + 1 + 1 + 1 + ...$

(c) $1 + 2 + 3 + 4 + ...$ (d) $1 + \frac{1}{2} + \frac{1}{4} + \frac{1}{8} + ...$

Solution

(a) It is straightforward, although tedious, to calculate the sum as 529.

(b) The sequence of partial sums is 1,2,3,4,... which clearly diverges; the series has no sum.

(c) The sequence of partial sums is 1,3,6,10,... which clearly diverges; the series has no sum.

(d) The sequence of partial sums is $1, \frac{3}{2}, \frac{7}{4}, \frac{15}{8}, ...$. It can be shown that the kth term in the sequence is

$$u_k = 2\left(1 - \left(\frac{1}{2}\right)^k\right)$$

As k increases this expression tends to the value $2(1) = 2$. Hence the sum of the series is 2.

To visualize what this last result means think of a rectangular sheet of paper 2 m by 1 m. Cut it in half and give one piece to a friend. Then cut the other part in half and give one piece to the friend. As you repeat this process again and again the total area of paper given to your friend increases and approaches, but never reaches, the value $2\,\text{m}^2$. It is $1 + \frac{1}{2} + \frac{1}{4} + \frac{1}{8} + ...\,\text{m}^2$.

We now quote two useful results.

KEY POINT

The sum of the first n natural numbers:

$$1 + 2 + 3 + ... + n = \frac{1}{2}n(n+1) \qquad 2.1$$

The sum of the squares of the first n natural numbers:

$$1^2 + 2^2 + 3^2 + ... + n^2 = \frac{1}{6}n(n+1)(2n+1) \qquad 2.2$$

Worked examples

2.12 Find the following sums:

(a) $1 + 2 + 3 + ... + 40$ (b) $1^2 + 2^2 + 3^2 + ... + (40)^2$

Solution In both cases $n = 40$.

(a) From formula (2.1) the sum is $\frac{1}{2} \times 40 \times 41 = 820$.

(b) From formula (2.2) the sum is $\frac{1}{6} \times 40 \times 41 \times 81 = 22\,140$.

You could, if you wished, verify these results by direct addition.

2.13 For $n = 1,2,3,4$ and 5 in turn, calculate the sums $1 + 2 + ... + n$ and $1^3 + 2^3 + ... + n^3$. What can you infer about the sum $1^3 + 2^3 + ... + n^3$ in general?

Solution We denote $1 + 2 + ... + n$ by S_1 and $1^3 + 2^3 + ... + n^3$ by S_3. The results are as follows:

n	1	2	3	4	5
S_1	1	3	6	10	15
S_3	1	9	36	100	225

It *appears* to be the case that $S_3 = (S_1)^2$. (In fact, $S_3 = \frac{1}{4}n^2(n + 1)^2$ for all natural numbers n but the proof is not given here.)

Two important series

An **arithmetic series** is formed by adding the terms of an arithmetic sequence. Hence

$$1 + 3 + 5 + ... + 45$$

is an arithmetic series. In general, such a series can be written

$$a + (a + d) + (a + 2d) + ... + (a + (n - 1)d)$$

Sometimes we denote the **last term** in the series by l so that

KEY POINT

$$l = a + (n - 1)d \qquad 2.3$$

Let the sum of the series be S then

$$S = a + (a + d) + ... + (a + (n - 2)d) + (a + (n - 1)d)$$

Alternatively, we can write the terms on the right-hand side in the reverse order, so that

$$S = (a + (n - 1)d) + (a + (n - 2)d) + ... + (a + d) + a$$

Adding these two formulae for S we find that

$$\begin{aligned} 2S &= (2a + (n - 1)d) + (2a(n - 1)d) + ... \\ &\quad + (2a + (n - 1)d) + (2a + (n - 1)d) \\ &= n(2a + (n - 1)d) \end{aligned}$$

since we are adding n terms with the same value. Hence

KEY POINT

$$S = \frac{n}{2}(2a + (n - 1)d) \qquad 2.4$$

Now $a + l = a + a + (n - 1)d = 2a + (n - 1)d$. Hence the sum of an arithmetic series is

KEY POINT

$$S = \frac{n}{2}(a + l) \qquad 2.5$$

If we know *three* of the five quantities a, d, l, n and S then we can find the other two from formulae (2.3) and (2.4) or from (2.5).

KEY POINT

For an arithmetic series, using standard notation,

$$l = a + (n - 1)d \qquad \text{and} \qquad S = \frac{n}{2}(a + l)$$

Worked examples

2.14 Find the sums of the following arithmetic series:
(a) $2 + 5 + 8 + \dots + 26$ (b) $81 + 76 + 71 + \dots + 26$

Solution (a) Here $a = 2, d = 3, l = 26$. From (2.3)

$$26 = 2 + (n - 1)3$$

so that:

$$24 = 3(n - 1)$$

that is, $n - 1 = 8$, giving $n = 9$. Therefore, from (2.5)

$$S = \tfrac{9}{2}(2 + 26) = \tfrac{9}{2} \times 28 = 126$$

(b) Here $a = 81, d = -5, l = 26$. In the same way as for Example 2.14(a) we find that $n = 12$ and hence:

$$S = \frac{12}{2}(81 + 26) = 6 \times 107 = 642$$

2.15 The sum of an arithmetic series is 420. If the first term is 60 and the fifth term is 44 find how many terms the series has.

Solution The first term is $a = 60$ and the fifth term is $a + 4d = 44$. Hence $d = -4$. From (2.4)

$$420 = \frac{n}{2}(120 - 4(n - 1))$$

that is

$$840 = n(124 - 4n) = 124n - 4n^2$$

(multiplying both sides of the equation by 2 and regrouping the terms on the right-hand side). Then

$$4n^2 - 124n + 840 = 0$$

so that

$$n^2 - 31n + 210 = 0$$

that is

$$(n - 10)(n - 21) = 0$$

Hence $n = 10$ or $n = 21$.

The series has the same sum, 420, after 10 terms are added, and again after 21 terms are added. (Write out the first 21 terms and check this result directly.)

A **geometric series** is formed by adding the terms of a geometric sequence. Hence $2 + 6 + 18 + 54$ is a geometric series. In general, such a series can be written

$$a + ar + \ldots + ar^{n-2} + ar^{n-1}$$

Let the sum of the series be S then

$$S = a + ar + \ldots + ar^{n-2} + ar^{n-1}$$

Multiplying by r we obtain

$$rS = ar + ar^2 + \ldots + ar^{n-1} + ar^n$$

Subtracting the second formula from the first, and noting that many terms cancel in pairs:

$$S - rS = (1 - r)S = a - ar^n = a(1 - r^n)$$

so that

KEY POINT

$$S = \frac{a(1 - r^n)}{(1 - r)} \qquad r \neq 1 \qquad 2.6$$

It is necessary to specify $r \neq 1$ since we cannot divide by zero and get a meaningful result. If $r = 1$ then the series is $a + a + a + \ldots + a$ which has sum na.

Worked examples

2.16 Find the sum of the following series:

(a) $2 + 6 + 18 + \ldots + 1458$

(b) $1 - \frac{1}{2} + \frac{1}{4} - \frac{1}{8} + \ldots + \frac{1}{1024}$

Solution (a) Here $a = 2$ and $r = 3$. If n is the number of terms then $1458 = 2 \times 3^{n-1}$. Noting that

$$1458 = 2 \times 729 = 2 \times 3^6$$

we see that $n - 1 = 6$ and therefore $n = 7$. Hence the sum is

$$S = \frac{2(1 - 3^7)}{1 - 3} = \frac{2(3^7 - 1)}{3 - 1} = \frac{2(2187 - 1)}{2} = 2186$$

(b) Here $a = 1$ and $r = -\frac{1}{2}$. The last term is

$$\frac{1}{1024} = \frac{1}{2^{10}} = ar^{10}$$

Hence $n = 11$ and the sum is

$$S = \frac{1\left(1 - \left(-\frac{1}{2}\right)^{11}\right)}{1 - \left(-\frac{1}{2}\right)} = \frac{1 + \frac{1}{2^{11}}}{1 + \frac{1}{2}} = \frac{2^{11} + 1}{2^{11} + 2^{10}}$$

$$= \frac{2049}{3072} \quad (= 0.666\,992 \quad 6\,\text{d.p.})$$

Suppose the geometric series is an infinite series. Does it have a sum? In formula (2.6) let n increase; then *provided* $-1 < r < 1$, r^n will decrease progressively in size and $1 - r^n \simeq 1$. In the limit as $n \to \infty$ then

KEY POINT

$$S = \frac{a}{1 - r} \qquad -1 < r < 1 \qquad 2.7$$

For the series

$$1 + \frac{1}{2} + \frac{1}{4} + \frac{1}{8} + \ldots$$

$a = 1$ and $r = 1/2$, which satisfies $-1 < r < 1$. Then

$$S = \frac{1}{1 - \frac{1}{2}} = \frac{1}{\frac{1}{2}} = 2$$

as we suggested earlier.

KEY POINT

The infinite geometric series

$$a + ar + \ldots + ar^{n-2} + ar^{n-1} + \ldots$$

has a sum given by

$$S = \frac{a}{1 - r}$$

provided that $-1 < r < 1$.

Worked examples

2.17 Find, where appropriate, the sum of the following geometric series:

(a) $1 - \frac{1}{2} + \frac{1}{4} - \frac{1}{8} + \dots$ (b) $1 + 2 + 4 + 8 + \dots$

Solution (a) Here $a = 1$ and $r = -\frac{1}{2}$, which satisfies $-1 < r < 1$; therefore

$$S = \frac{1}{1 + \frac{1}{2}} = \frac{1}{\frac{3}{2}} = \frac{2}{3}.$$

(b) Here $r = 2$ and therefore the sum does not exist.

Self-assessment questions 2.2

1. Explain what is meant by the statement that an infinite series has a sum.
2. When does an infinite geometric series have a sum?

Exercise 2.2

1. Find the following sums:

   (a) $2 + 3 + 4 + \dots + 41$

   (b) $20^2 + 21^2 + 22^2 + \dots + 40^2$

2. Find the sum of the first 10 terms of the following arithmetic series:

   (a) $a = 3, d = 4$ (b) $a = 3, d = -4$

   (c) $a = -3, d = -4$ (d) $a = -3, d = 4$

3. The sum of the first 10 terms of an arithmetic series is 50 and its common difference is -1.5. Find its first term.

4. The sum of the first 10 terms of an arithmetic series is 165 and the sum of the next 10 terms is 445. Find the sum of terms 21 to 30.

5. The sum to infinity of a geometric series is 1.125 times the sum of the first two terms. Find the common ratio between the terms.

6. One hundred and thirty circular cylinders are stacked in rows in such a way that the number in a row is one less than in the row below. The bottom row contains 16 cylinders. How many rows are there? How many cylinders are there in the top row?

7. A particle oscillates in a straight line. It starts at a point A, then reaches a point B, 10 m from A. It reverses direction and travels back towards A, but reaches a point C only 7.5 m from B. It reverses direction again and travels back towards B and so on. Each leg of the journey is three-quarters of the length of the previous leg. How far does the particle travel in total by the time it comes to rest and where does it finally stop?

2.3 Binomial series

An expression such as $(a + b)^3$ or $(2 - 3x)^6$ is called a **binomial expression**: binomial means 'two terms' and refers to the two terms in the brackets (parentheses). By multiplying out, the first expression can be shown to be $a^3 + 3a^2b + 3ab^2 + b^3$; this expansion is known as a **binomial expansion**. To achieve the expansion we proceed as follows:

$$\begin{aligned}(a+b)^2 &\equiv (a+b)(a+b) \equiv a^2 + ab + ba + b^2 \equiv a^2 + 2ab + b^2\\ (a+b)^3 &\equiv (a+b)^2(a+b) \equiv (a^2 + 2ab + b^2)(a+b)\\ &\equiv a^3 + 2a^2b + ab^2 + a^2b + 2ab^2 + b^3\\ &\equiv a^3 + 3a^2b + 3ab^2 + b^3\end{aligned}$$

It would clearly be tedious to expand $(a+b)^n$ for $n > 3$. Notice that in the expansion of $(a+b)^3$ each term consists of the product of three quantities (a^3, a^2b, ab^2 and b^3) multiplied by a coefficient (1, 3, 3, 1); as we move from one term to the next, one of the as is replaced by a b. The coefficients can be obtained from a pattern of numbers known as **Pascal's triangle**, the first seven rows of which are shown in Figure 2.3. Each row starts and ends with the number 1. The other entries are found by adding the numbers above left and above right in the previous row (for example, 5 and 10 in row six produce 15 in row seven).

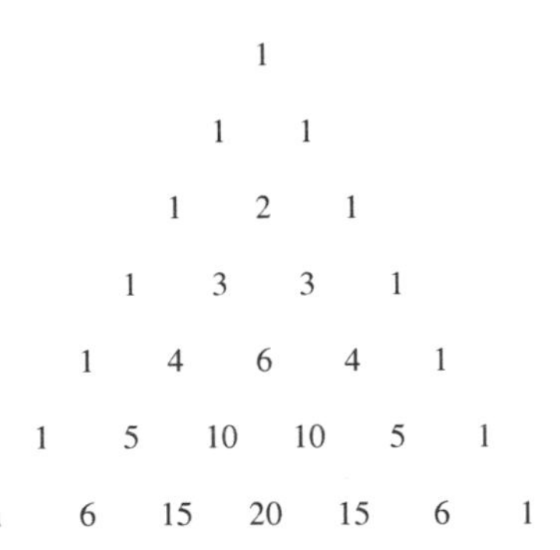

Figure 2.3.
The first seven rows of Pascal's triangle

Note that the numbers in row four, namely 1, 3, 3 and 1, are the coefficients in the expansion of $(a+b)^3$. Also note the symmetry of the numbers in a row about the middle value(s).

Worked examples

2.18 Expand (a) $(a+b)^5$ (b) $(2-3x)^4$

Solution (a) The first term will contain a^5 and successive terms will replace one 'a' by one 'b', that is, $a^4b, a^3b^2, a^2b^3, ab^4$ and b^5 (the sum of the indices is always 5). The coefficients come from row six and the expansion is therefore

$$1 \times a^5 + 5 \times a^4b + 10 \times a^3b^2 + 10 \times a^2b^3 + 5 \times ab^4 + 1 \times b^5$$

or

$$a^5 + 5a^4b + 10a^3b^2 + 10a^2b^3 + 5ab^4 + b^5$$

(b) We take the coefficients from row five. Now

$$2 - 3x \equiv 2 + (-3x)$$

and identifying 2 with a and $-3x$ with b we obtain

$$\begin{aligned}(2-3x)^4 &\equiv 1 \times 2^4 + 4 \times 2^3(-3x)^1 + 6 \times 2^2 \times (-3x)^2 \\ &\quad + 4 \times 2 \times (-3x)^3 + 1 \times (-3x)^4 \\ &\equiv 16 + 4 \times 8(-3x) + 6 \times 4 \times 9x^2 + 4 \times 2 \times (-27x^3) \\ &\quad + 1 \times 81x^4 \\ &\equiv 16 - 96x + 216x^2 - 216x^3 + 81x^4\end{aligned}$$

Such a process is relatively straightforward for cases where the power is a small positive integer. It would be awkward for large values. In order to deal more effectively in these situations we need a different approach. We quote the **binomial theorem** which states that for n, a positive integer

KEY POINT

$$(a+b)^n = a^n + \frac{n}{1}a^{n-1}b + \frac{n(n-1)}{1 \times 2}a^{n-2}b^2 + \frac{n(n-1)(n-2)}{1 \times 2 \times 3}a^{n-3}b^3 + \ldots + b^n \qquad 2.8$$

Many books make use of the **factorial** notation: the product $1 \times 2 \times 3 \times \ldots \times n$ is called 'n factorial' or 'factorial n' and written $n!$. Hence

$$1! = 1, \quad 2! = 1 \times 2 = 2, \quad 3! = 1 \times 2 \times 3 = 6, \ldots$$

With this notation (2.8) becomes

Note that $0! = 1$.

$$(a+b)^n = a^n + \frac{n}{1!}a^{n-1}b + \frac{n(n-1)}{2!}a^{n-2}b^2 + \frac{n(n-1)(n-2)}{3!}a^{n-3}b^3 + \ldots + b^n \qquad 2.9$$

An important special case is found by taking $a = 1$ and $b = x$. Then the theorem is restated:

$$(1+x)^n = 1 + \frac{n}{1}x + \frac{n(n-1)}{1 \times 2}x^2 + \frac{n(n-1)(n-2)}{1 \times 2 \times 3}x^3 + \ldots + x^n \qquad 2.10$$

Worked example

2.19 Use the binomial expansion to expand $(1+x)^5$.

Solution Using (2.10) with $n = 5$ we obtain:

$$(1+x)^5 = 1 + \frac{5}{1}x + \frac{5\times 4}{1\times 2}x^2 + \frac{5\times 4\times 3}{1\times 2\times 3}x^3 + \frac{5\times 4\times 3\times 2}{1\times 2\times 3\times 4}x^4 + x^5$$

Cancelling down the fractions gives

$$(1+x)^5 = 1 + 5x + 10x^2 + 10x^3 + 5x^4 + x^5$$

This last result *could* have been obtained more quickly using Pascal's triangle. Where the theorem is most useful is when n is not a positive integer. In these cases the series expansion (2.10) is infinite and convergence to a sum is guaranteed only for $-1 < x < 1$.

KEY POINT

For n not a positive integer then, provided that $-1 < x < 1$,

$$(1+x)^n = 1 + \frac{n}{1}x + \frac{n(n-1)}{1\times 2}x^2 + \frac{n(n-1)(n-2)}{1\times 2\times 3}x^3 + \ldots \qquad 2.11$$

Worked examples

2.20 Find the expansion as far as the term in x^4 of

(a) $(1+x)^{\frac{1}{2}}$ (b) $(1+x)^{-2}$ (c) $(1+x)^{-\frac{1}{3}}$

Solution Note that the next term on the right-hand side of (2.11) is

$$\frac{n(n-1)(n-2)(n-3)}{1\times 2\times 3\times 4}x^4$$

In each case assume that $-1 < x < 1$.

(a) Here $n = \dfrac{1}{2}$.

$$(1+x)^{\frac{1}{2}} \equiv 1 + \frac{\frac{1}{2}}{1}x + \frac{\frac{1}{2}\left(-\frac{1}{2}\right)}{1\times 2}x^2 + \frac{\frac{1}{2}\left(-\frac{1}{2}\right)\left(-\frac{3}{2}\right)}{1\times 2\times 3}x^3$$

$$+ \frac{\frac{1}{2}\left(-\frac{1}{2}\right)\left(-\frac{3}{2}\right)\left(-\frac{5}{2}\right)}{1\times 2\times 3\times 4}x^4 + \ldots$$

$$\equiv 1 + \frac{1}{2}x - \frac{1}{8}x^2 + \frac{1}{16}x^3 - \frac{5}{128}x^4 + \ldots$$

(b) Here $n = -2$ and therefore:

$$(1+x)^{-2} \equiv 1 + \frac{(-2)}{1}x + \frac{(-2)(-3)}{1 \times 2}x^2 + \frac{(-2)(-3)(-4)}{1 \times 2 \times 3}x^3$$

$$+ \frac{(-2)(-3)(-4)(-5)}{1 \times 2 \times 3 \times 4}x^4 + \ldots$$

$$\equiv 1 - 2x + 3x^2 - 4x^3 + 5x^4 + \ldots$$

(c) Here $n = -\dfrac{1}{3}$ and

$$(1+x)^{-\frac{1}{3}} \equiv 1 + \frac{\left(-\frac{1}{3}\right)}{1}x + \frac{\left(-\frac{1}{3}\right)\left(-\frac{4}{3}\right)}{1 \times 2}x^2 + \frac{\left(-\frac{1}{3}\right)\left(-\frac{4}{3}\right)\left(-\frac{7}{3}\right)}{1 \times 2 \times 3}x^3$$

$$+ \frac{\left(-\frac{1}{3}\right)\left(-\frac{4}{3}\right)\left(-\frac{7}{3}\right)\left(-\frac{10}{3}\right)}{1 \times 2 \times 3 \times 4}x^4 + \ldots$$

$$\equiv 1 - \frac{1}{3}x + \frac{2}{9}x^2 - \frac{14}{81}x^3 + \frac{35}{243}x^4 + \ldots$$

2.21 For which values of x would the following expansions be valid?

(a) $(3-x)^7$

(b) $(4+3x)^{\frac{1}{2}}$

(c) $(4+x^2)^{-1}$

(d) $(4-x^2)^{\frac{1}{2}}$

Solution (a) Since n is a positive integer the expansion will be valid for all x.

(b) $4 + 3x \equiv 4\left(1 + \frac{3}{4}x\right)$ so that $(4+3x)^{\frac{1}{2}} = 4^{\frac{1}{2}}\left(1 + \frac{3}{4}x\right)^{\frac{1}{2}}$. We require for convergence that $-1 < \frac{3}{4}x < 1$ and by multiplying by $\frac{4}{3}$ we find that $-\frac{4}{3} < x < \frac{4}{3}$.

Remember that when an inequality is multiplied by a negative number it reverses direction. Thus when $5 > 2$ is multiplied by -3 it becomes $-15 < -6$ or $-6 > -15$.

(c) $4 + x^2 \equiv 4(1 + x^2/4)$ and we require $-1 < x^2/4 < 1$ for convergence. Hence $-4 < x^2 < 4$ and since x^2 is never negative this reduces to $x^2 < 4$ or $-2 < x < 2$.

(d) $4 - x^2 \equiv 4(1 - x^2/4)$ and for convergence we take $-1 < -x^2/4 < 1$; multiplying by -1 we obtain $1 > x^2/4 > -1$ which is the same condition as in (c).

Hence $-2 < x < 2$, again.

2.22 Find the expansion as far as the term in x^3 of

$$\frac{3-2x}{4+x}$$

For what values of x is the infinite series expansion valid?

Solution

$$\frac{(3-2x)}{4+x} \equiv \frac{(3-2x)}{4\left(1+\frac{x}{4}\right)} \equiv \frac{1}{4}(3-2x) \times \left(1+\frac{x}{4}\right)^{-1}$$

$$= \frac{1}{4}(3-2x)\left(1+\frac{(-1)}{1}\left(\frac{x}{4}\right)+\frac{(-1)(-2)}{1\times 2}\left(\frac{x}{4}\right)^2\right.$$

$$\left.+\frac{(-1)(-2)(-3)}{1\times 2\times 3}\left(\frac{x}{4}\right)^3+\ldots\right) \qquad \text{Note 1}$$

$$\simeq \frac{1}{4}(3-2x)\left(1-\frac{x}{4}+\frac{x^2}{16}-\frac{x^3}{64}\right) \qquad \text{Note 2}$$

$$\simeq \frac{1}{4}\left(3-\frac{3}{4}x+\frac{3x^2}{16}-\frac{3x^3}{64}-2x+\frac{2x^2}{4}-\frac{2x^3}{16}\right) \qquad \text{Note 3}$$

$$= \frac{1}{4}\left(3-\frac{11}{4}x+\frac{11}{16}x^2-\frac{11}{64}x^3\right) \qquad \text{Note 4}$$

$$= \frac{3}{4}-\frac{11}{16}x+\frac{11}{64}x^2-\frac{11}{256}x^3$$

Notes

1. We stopped the expansion of $(1+x/4)^{-1}$ at the term in x^3 since all terms up to then might be needed but none afterwards would be.
2. We have now *ignored* terms after the one in x^3 and hence we have made an approximation.
3. We did not calculate the product $(-2x)(-x^3/64)$ since it would have given an unwanted term; hence we made another approximation.
4. This expression is *exactly equal* to the line above. The expansion is valid for $-1 < x/4 < 1$, that is, $-4 < x < 4$.

2.23 Expand $(1-2x)^{\frac{1}{2}}$ as far as the term in x^3. Put $x = \frac{1}{10}$ to obtain an approximation to $\sqrt{5}$ correct to 3 d.p.

Solution

$$(1-2x)^{\frac{1}{2}} = 1+\frac{\frac{1}{2}}{1}(-2x)+\frac{\left(\frac{1}{2}\right)\left(-\frac{1}{2}\right)}{1\times 2}(-2x)^2$$

$$+\frac{\left(\frac{1}{2}\right)\left(-\frac{1}{2}\right)\left(-\frac{3}{2}\right)}{1\times 2\times 3}(-2x)^3+\ldots$$

$$\simeq 1-x-\frac{1}{2}x^2-\frac{1}{2}x^3$$

With $x = 1/10$,

$$(1-2x)^{\frac{1}{2}} \simeq 1 - 0.1 - 0.005 - 0.0005 = 0.8945$$

But

$$(1-2x)^{\frac{1}{2}} = (1-0.2)^{\frac{1}{2}} = \left(\frac{4}{5}\right)^{\frac{1}{2}} = \frac{2}{\sqrt{5}} = \frac{2\sqrt{5}}{5}$$

Therefore

$$\sqrt{5} = \frac{5}{2} \times 0.8945 = 2.236 \qquad \text{(3 d.p.)}$$

You can verify with your calculator that this is the value of $\sqrt{5}$ correct to 3 d.p.

Self-assessment questions 2.3

1. Under what conditions does the binomial expansion of $(a+b)^n$ have a finite number of terms?
2. What is the requirement for an infinite binomial expansion to converge?

Exercise 2.3

1. Expand (a) $(a+b)^8$ (b) $(3-5x)^6$
2. Find the expansion as far as the term in x^5 of
   (a) $(1-x)^{\frac{1}{4}}$ (b) $(1-x)^{-3}$
   (c) $(1-2x)^{-\frac{1}{2}}$
3. For which values of x would the following expansions be valid?
   (a) $\left(4-\frac{1}{2}x\right)^6$ (b) $(2-5x)^{\frac{1}{3}}$
   (c) $(2+x^2)^{-2}$ (d) $(9-x^2)^{\frac{1}{3}}$
4. Find the expansion as far as the term in x^3 of $2+3x/(1-x)$. For which values of x is the expansion valid?
5. Expand $(1-4x)^{\frac{1}{2}}$ as far as the term in x^4. By putting $x=-1/5$ and rounding your answer to 2 d.p. find an approximation to $\sqrt{5}$ and compare this with the calculator value for $\sqrt{5}$.
6. Expand $\left(1+\frac{1}{x}\right)^{\frac{1}{2}}$ in descending powers of x. For which values of x is the expansion valid? Write the expression as $x^{-\frac{1}{2}}(1+x)^{\frac{1}{2}}$ and find the first four terms in the expansion in ascending powers of x. For which values of x is this second expansion valid?

Test and assignment exercises 2

1. Write down the first five terms of the sequences defined by the following:
   (a) $u_k = \dfrac{1}{k^2+1}$
   (b) $u_k = k^2 - \dfrac{1}{k^2}$
   (c) $u_{k+1} = 2u_k + 3,\ u_1 = 1$
   (d) $u_{k+1} = u_k + u_{k-1},\ u_1 = 1, u_2 = 1$
2. The fifth term of an arithmetic progression is -5 and the tenth term is -15. Find the first term and the common difference.

3. The fifth term of a geometric progression is 48 and the ninth term is 768. Find the first term and the common ratio.

4. Describe the behaviour as $k \to \infty$ of the sequences whose general term is:

   (a) $5 - \dfrac{1}{k^2}$ (b) $\left(\dfrac{-1}{5}\right)^k$

   (c) $5 + \dfrac{1}{k} - \dfrac{1}{k^2}$

5. Find the sums of the following series:

   (a) $7 + 8 + 9 + \ldots + 28$
   (b) $2 + 5 + 8 + \ldots + 32$
   (c) $1 + 3 + 5 + \ldots + 31$
   (d) $(37)^2 + (36)^2 + (35)^2 + \ldots + (8)^2$

6. The sum of the first 20 terms of an arithmetic series is -530. The sum of the next eight terms is -548. Find the first term and the common difference.

7. If the sum to infinity of a geometric series is $\frac{4}{3}$ times the sum of the first two terms, find the common ratio.

8. Expand

   (a) $(2 + 3x)^4$ (b) $(1 - x)^5$

9. Expand the following as far as the term in x^3:

   (a) $(2 - x)^{\frac{1}{5}}$ (b) $(2 + x)^{-2}$
   (c) $(1 + x)^{-\frac{1}{5}}$

   For which values of x is each expansion valid?

10. Expand $(1 - 8x)^{\frac{1}{2}}$ as far as the term in x^2. Put $x = -\frac{1}{10}$ to obtain an approximation to $\sqrt{5}$. Find the next term in the expansion and hence find two values between which $\sqrt{5}$ lies.

3 Functions I – basic ideas

Objectives

This chapter

- defines a function, its domain and range
- emphasizes the importance of graphing a function
- considers the composition of two functions
- develops the idea of transforming a function
- defines the inverse function

3.1 Definitions and graphs

A **function** is a relationship between two sets of variables which associates each member of the first set with a *unique* member of the second set. Examples of functions are 'add 3' and 'square'; simple engineering examples are that the voltage across a resistor is a function of the current through the resistor, and the extension of a loaded spring is a function of the applied load. We can represent the action of a function by a **box diagram**. Figure 3.1(a) shows the box diagram for the first function with a sample input and a general input and diagram (b) does likewise for the second function.

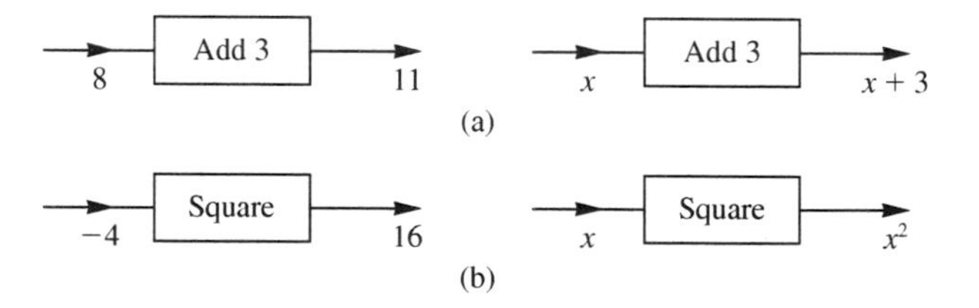

Figure 3.1. Box diagrams for the functions 'add 3' and 'square'

Note that each time we input 8 to the box 'add 3' we get one value, namely 11; this is the essence of a function. Symbolically we give the function a single-letter name. If there is only one function the letter chosen is usually f. We write $f: x \mapsto x+3$ which is read as 'the function f maps x onto $x+3$'. If we label our second function g, we can write $g: x \mapsto x^2$. Alternatively, we can write $f(x) = x+3$ and

$g(x) = x^2$; in this notation we can state $f(8) = 11$ and $g(-4) = 16$ as examples of the action of the function. The complete set of possible values of the input x to a function f is called the **domain** of the function. The set of values $f(x)$ which the function can assume is called the **range** of the function. In the case of 'add 3' both the domain and the range are the set of all real numbers. In the case of 'square' the domain is the set of all real numbers but the range is all non-negative real numbers: it is impossible to obtain a negative number by squaring a number.

KEY POINT

> A **function** is a relationship between two sets of variables which associates each member of the first set with a *unique* member of the second set.

Worked example

3.1 Represent each of the following functions $f(x)$ by a box diagram:

(a) 'halve' (b) 'subtract 2'

(c) 'cube' (d) 'take square root'

Use the symbolic form to evaluate $f(3)$, $f(0)$, $f(-5.1)$ for each function and state its domain and range.

Solution Figure 3.2 shows box diagrams for each of the functions.

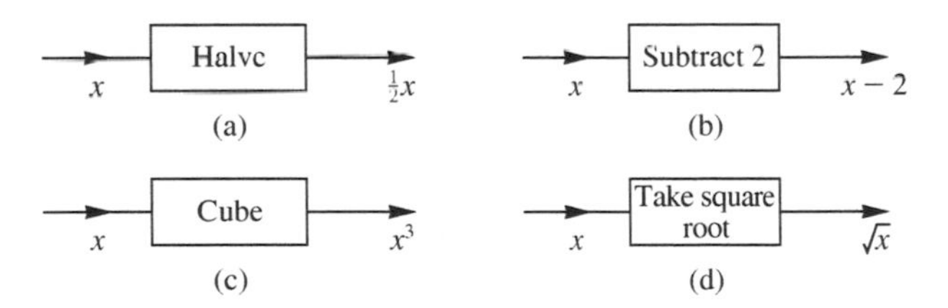

Figure 3.2. Box diagrams for the functions (a) 'halve' (b) 'subtract 2' (c) 'cube' (d) 'take square root'

(a) If $f(x) = \frac{1}{2}x$, $f(3) = 1.5$, $f(0) = 0$, $f(-5.1) = -2.55$

(b) If $f(x) = x - 2$, $f(3) = 1$, $f(0) = -2$, $f(-5.1) = -7.1$

(c) If $f(x) = x^3$, $f(3) = 27$, $f(0) = 0$, $f(-5.1) = -132.651$

(d) If $f(x) = \sqrt{x}$ or $x^{\frac{1}{2}}$ it is understood that the *positive* square root is to be taken where applicable. $f(3) = 1.732$ (3 d.p.), $f(0) = 0$; $f(-5.1)$ does not exist.

The domain of functions (a), (b) and (c) is the set of all real numbers, as is the range. Function (d) has as domain the set of real numbers greater than or equal to zero; the same set is its range.

Graphs of functions

A pictorial method of representing a function is its **graph**. The values of the domain lie along the horizontal or x-axis and the values of the range lie along the vertical or y-axis. A value of x is chosen and the function is used to calculate the corresponding value $f(x)$. The point with coordinates $(x, f(x))$ is plotted. This procedure is repeated for other values of x and a smooth curve is drawn through the points which have been plotted. Any line on a graph is called a curve, even if it is a straight line. The graphs of the functions 'add 3' and 'square' are shown in Figure 3.3. Note that in this representation we plot the graphs of $y = f(x)$. Here x is called the **independent variable** and y is called the **dependent variable**.

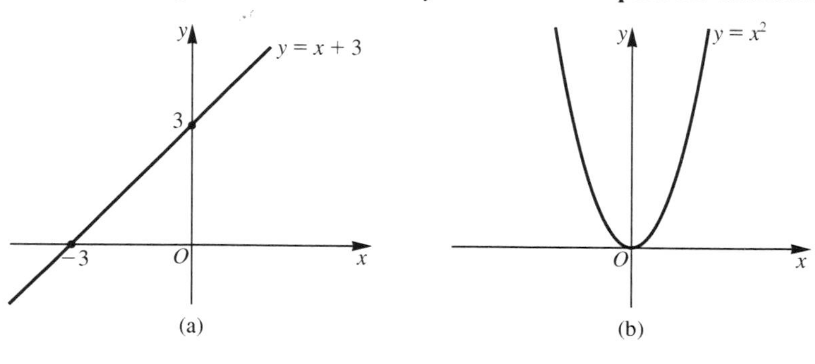

Figure 3.3.
Graphs of the functions 'add 3' and 'square'

Worked examples

3.2 Sketch the graphs of the functions in Example 3.1.

Solution Figure 3.4 shows the graphs.

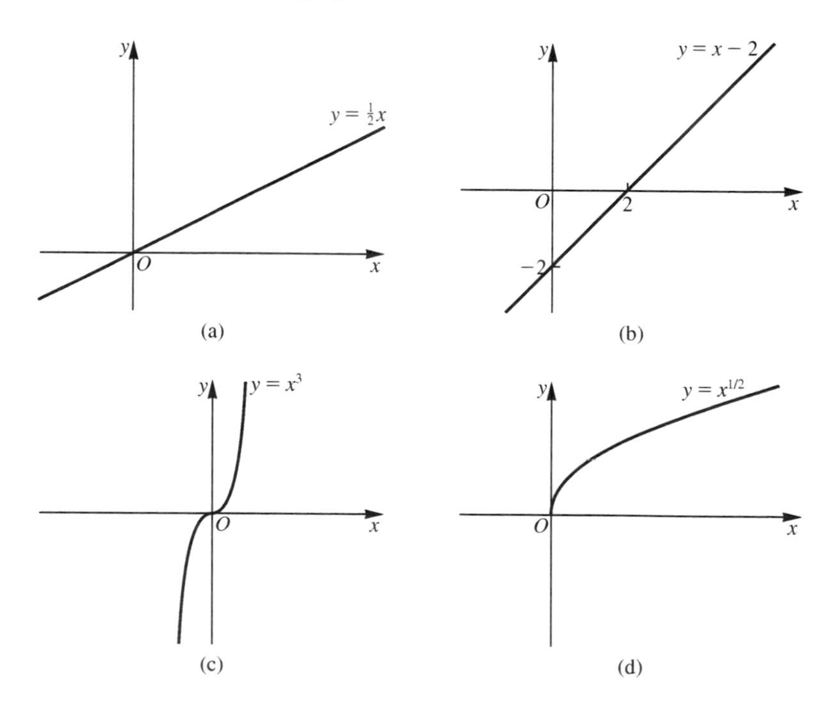

Figure 3.4.
Graphs of the functions of Example 3.2

3.3 The **absolute value** or **modulus function** is denoted $f(x) = |x|$. It can be defined by the statements $f(x) = x$ if $x \geqslant 0$ and $f(x) = -x$ if $x < 0$. Evaluate $f(3)$, $f(0)$ and $f(-2)$, state the domain and range of the function and draw its graph.

Solution $f(3) = 3$, $f(0) = 0$, $f(-2) = 2$. The domain of the function is the set of all real numbers and the range is the set of all non-negative real numbers. Figure 3.5(a) depicts its graph.

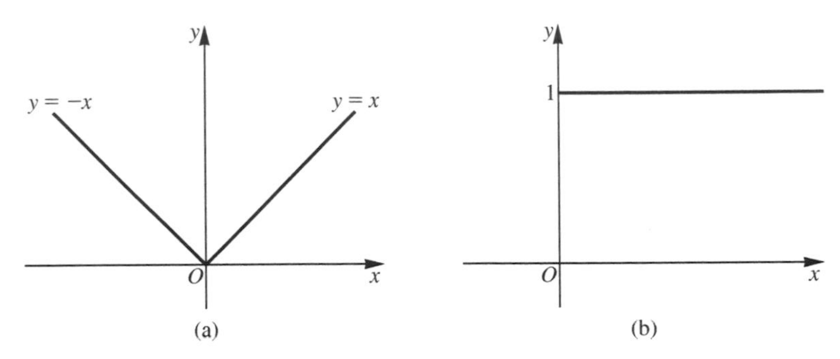

Figure 3.5. Graphs of (a) the absolute value function and (b) the unit step function

3.4 The **unit step function** is depicted in Figure 3.5(b), with $f(0) = 1$. Write defining statements and evaluate $f(3)$ and $f(-2)$. What is the range of the function?

Solution $f(x) = 1$ if $x \geqslant 0$ and $f(x) = 0$ if $x < 0$. Alternatively

$$f(x) = \begin{cases} 1 & x \geqslant 0 \\ 0 & x < 0 \end{cases}$$

Further, $f(3) = 1$, $f(-2) = 0$. The range is the set $\{0, 1\}$.

An example of a situation which can be modelled by the step function is where a switch is closed and a constant supply voltage is brought into play.

(Note that some authors use the term **co-domain** to denote the set in which the range is located. In this last example we say that f maps the set of real numbers (domain) into the set of real numbers (co-domain); the range is the subset of the co-domain which constitutes the set of actual values $f(x)$.)

Self-assessment questions 3.1

1. Explain the terms function, domain and range.
2. Explain the use of box diagrams to illustrate the action of a function.

Exercise 3.1

1. Represent the action of each of the following functions by a box diagram and state the domain and range:

   (a) 'divide by 3' (b) 'multiply by 5'
   (c) 'raise to the fifth power'
   (d) 'take fourth root'

2. Evaluate $f\left(\frac{1}{2}\right), f(-2), f(6.1)$ for the following functions:

   (a) $f(x) = x^3$ (b) $f(x) = \frac{1}{x}$
   (c) unit step function

3. Produce the graph of each of the following functions. State the domain and range.

   (a) $f(x) = \begin{cases} 0 & x < 0 \\ x & x \geqslant 0 \end{cases}$

   (b) $f(x) = \begin{cases} x+1 & x < 2 \\ 5-x & x \geqslant 2 \end{cases}$

   Function (a) is an example of a **ramp function**.

4. The speed of a vehicle is defined by the formula

   $$v(t) = \begin{cases} \frac{1}{2}t & 0 \leqslant t \leqslant 20 \\ 10 & 20 < t < 50 \\ 60 - t & 50 \leqslant t \leqslant 60 \end{cases}$$

   State the domain and range of the function $v(t)$ and sketch its graph.

5. Which of the following can be rearranged to give y as a function of x?

   (a) $x^3 + y = 0$
   (b) $x^3 + y^3 = 0$
   (c) $x = y^2$

3.2 Combining functions

Two or more functions can be combined to give new functions. We shall look at a few examples. The **sum** of two functions, $f+g$, is defined by $(f+g): x \mapsto f(x) + g(x)$; its domain is the intersection of the domains of f and g. The **product** of two functions $f \cdot g$ is defined by $(f \cdot g): x \mapsto f(x) \cdot g(x)$.

Note that the function $f+g$ is identical to the function $g+f$ and that the function $f \cdot g$ is identical to the function $g \cdot f$.

Worked example

3.5 Find the sum and product of each of the following pairs of functions, stating the domain and range in each case.

(a) $f(x) = x^2, g(x) = x^3$ (b) $f(x) = x + 3, g(x) = x - 3$

(c) $f(x) = 2x, g(x) = \frac{1}{2}x$ (d) $f(x) = x^2, g(x) = \dfrac{1}{x^2}$

Solution (a) $(f+g): x \mapsto x^2 + x^3$ $(f \cdot g): x \mapsto x^2 \cdot x^3 = x^5$
In both cases domain and range are the set of all real numbers.

(b) $(f+g): x \mapsto x+3+x-3 = 2x$ $(f \cdot g): x \mapsto x^2 - 9$
The domain of both functions is the set of all real numbers and this is also the range of $f+g$. The range of $f \cdot g$ is the set of all real numbers greater than or equal to -9.

(c) $(f+g): x \mapsto \frac{5}{2}x$ $(f \cdot g): x \mapsto 2x \cdot \frac{1}{2}x = x^2$

The domain of both sum and product is the set of real numbers, which is also the range of the sum. The range of the product is all real numbers greater than or equal to 0.

(d) $(f+g): x \mapsto x^2 + \frac{1}{x^2}$, which has domain all real numbers $\neq 0$ and range the set of all real numbers greater than 0. (We cannot achieve a value of 0 or a negative value.)

$(f \cdot g)(x) = 1$ has domain the set of real numbers and range $\{1\}$. (Note that $g(x)$ is not defined when $x = 0$.)

Composition of two functions

When two functions are applied in succession the result is known as a **composite function**. Suppose that the functions 'square' and 'add 3' are applied in turn. Figure 3.6(a) shows the box diagrams for the process. We can denote the function 'add 3' by f and the function 'square' by g.

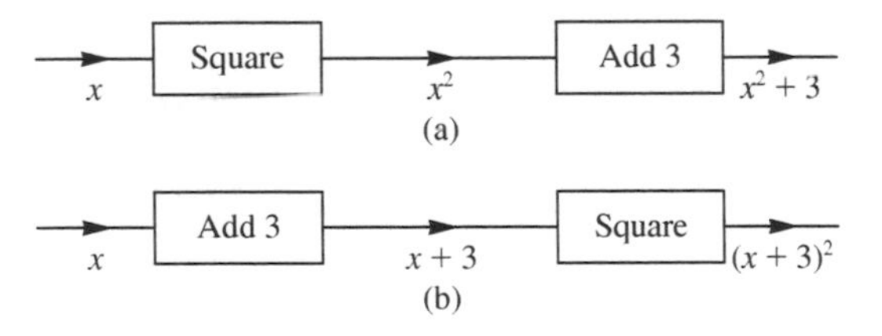

Figure 3.6.
Composite functions

Then, successively, $g(x) = x^2$ and $f(g(x)) = x^2 + 3$. Note the role of x. We wrote $g(x) = x^2$ but we could equally write $g(u) = u^2, g(v) = v^2$, and so on. It is the **process** of squaring which the function describes and x (or u or v) is the independent variable which helps explain the process.

Figure 3.6(b) depicts the composition $g(f(x)) = (x+3)^2$ which illustrates that, in general, $g(f(x))$ is *not* the same as $f(g(x))$.

The functions 'square' and 'add 3' both have domain all real numbers so that both $f(g(x))$ and $g(f(x))$ have the same domain. The range of g is real numbers greater than or equal to zero and therefore the range of $f(g(x))$ is real numbers greater than or equal to 3. The range of f is all real numbers and hence the range of $g(f(x))$ is all real numbers greater than or equal to zero.

Worked example

3.6 Given the functions $f(x) = x^{\frac{1}{2}}$ and $g(x) = x - 2$ determine $f(g(x))$ and $g(f(x))$, illustrate by box diagrams and state their domain and range.

Solution Figure 3.7(a) is the box diagram for $f(g(x))$ and Figure 3.7(b) shows $g(f(x))$.

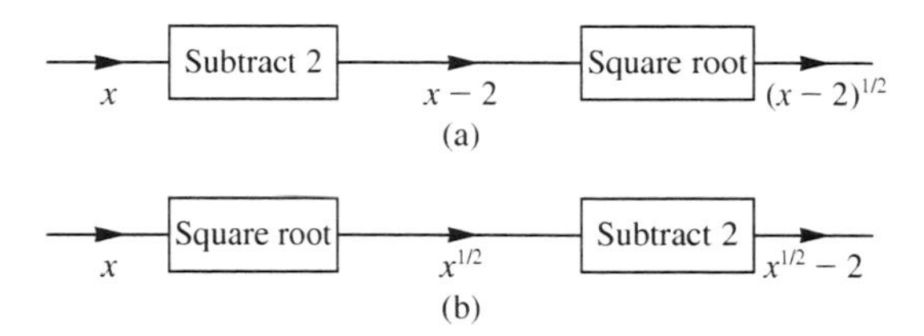

Figure 3.7. The functions of Example 3.6

Note that $f(g(x)) = (x - 2)^{\frac{1}{2}}$ and $g(f(x)) = x^{\frac{1}{2}} - 2$.

The domain of f is the set of real numbers greater than or equal to zero and hence the input $x - 2$ must be greater than or equal to zero so that $x \geqslant 2$. Hence the domain of $f(g(x))$ is real numbers $x \geqslant 2$ and the range is real numbers greater than or equal to zero.

The domain of $g(f(x))$ is real numbers greater than or equal to zero since the domain of f is real numbers greater than or equal to zero.

The range of f is also real numbers greater than or equal to zero and therefore the range $g(f(x))$ is real numbers greater than or equal to -2.

Self-assessment questions 3.2

1. Define the sum and product of two functions and in each case state the domain and range in relation to those of the two original functions.
2. Explain the term composite function and distinguish between $f(g(x))$ and $g(f(x))$.

Exercise 3.2

1. Find the sum and product of each of the following pairs of functions, stating the domain:

   (a) $f(x) = x^4,\ g(x) = x$

   (b) $f(x) = x^{\frac{1}{2}},\ g(x) = x^{\frac{1}{3}}$

   (c) $f(x) = \dfrac{1}{x},\ g(x) = \dfrac{1}{x^3}$

   (d) $f(x) = x^{\frac{1}{2}},\ g(x) = x^2$

   Obtain plots of each pair of functions and of their sum and product.

2. Under what circumstances is each of the following true?

   (a) $f(x) + g(x) \equiv f(x)$

   (b) $f(x) \cdot g(x) \equiv f(x)$

   (Note that in this context the use of the symbol $\equiv$ means that the functions on either side of it have exactly the same domain and range and the same effect on each member of the domain.)

3. Give examples of functions $f(x)$ and $g(x)$ for which

   (a) $f(g(x)) \equiv f(x)$

   (b) $f(g(x)) \equiv g(x)$

4. For the functions $f(x)$ and $g(x)$ of Question 1 find $f(g(x))$ and $g(f(x))$, stating the domain and range in each case.

   Obtain plots of the composite functions and compare them in each case.

5. Evaluate, where possible, $f(g(1))$, $g(f(1))$, $f(g(0))$ and $g(f(0))$ for the functions of Question 1(a), (b) and (d).

6. The product of a function $f(t)$ and the unit step function $u(t)$ has the effect of suppressing that part of $f(t)$ for which $t < 0$. Demonstrate this by drawing the graphs of $f(t)$ and $u(t)f(t)$ for

   (a) $f(t) = t + 2$

   (b) $f(t) = t^2$

3.3 Transformation of functions

In this section we consider how the effect of translating the axes, altering the horizontal and vertical scales of a graph and reflecting the graph in either axis can be represented in terms of related functions.

Translations

Figure 3.8 shows the graphs of $y = x^2 - 2$, $y = x^2$ and $y = x^2 + 1$. Each graph can be translated into the others by moving it parallel to the x-axis. This illustrates the general result:

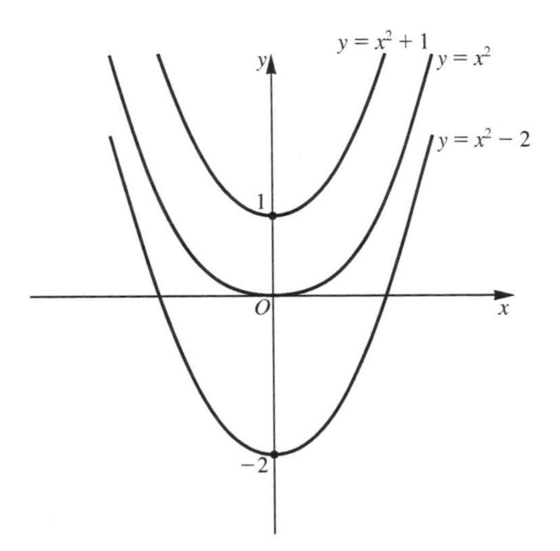

Figure 3.8.
Vertical translation

KEY POINT

The graph of the function $f(x) + a$ is the graph of $f(x)$ translated parallel to the y-axis, upwards if $a > 0$, downwards if $a < 0$.

Figure 3.9 shows the graphs of $y = x^2$, $y = (x - 2)^2$ and $y = (x + 1)^2$. Each graph can be translated into the others by moving it parallel to the y-axis. This illustrates the general result:

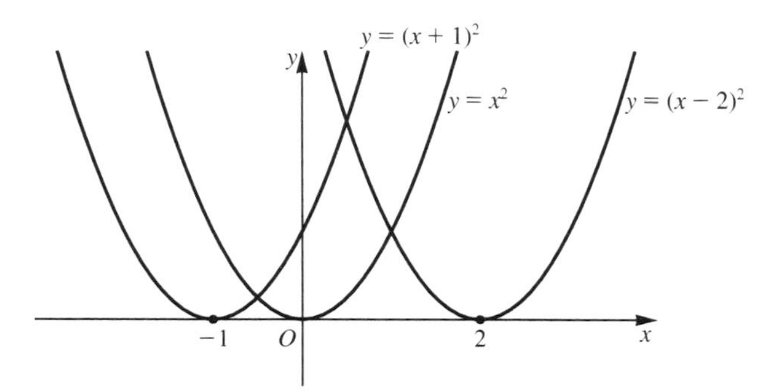

Figure 3.9.
Horizontal translation

KEY POINT

The graph of the function $f(x + a)$ is the graph of $f(x)$ translated parallel to the x-axis, to the left if $a > 0$, to the right if $a < 0$.

The direction of translation might be surprising since we normally think of positive as being to the right; therefore we should take care.

Worked examples

3.7 Sketch the graphs of $y = |x|$, $y = |x| + 2$, $y = |x + 2|$.

Solution Figure 3.10 shows the graphs.

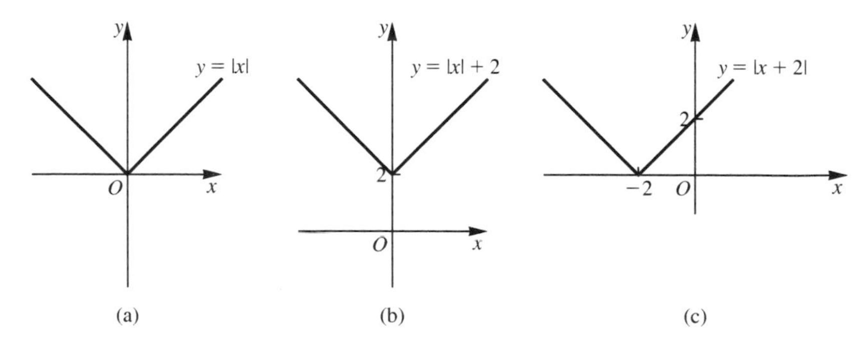

Figure 3.10.
Graphs of the functions of Example 3.7

3.8 Show that $f(x) = x^2 + 4x - 5 \equiv (x + 2)^2 - 9$ and hence sketch the graph of $y = f(x)$.

Solution $(x+2)^2 - 9 \equiv x^2 + 4x + 4 - 9 \equiv x^2 + 4x - 5$, as required. Starting with the graph of $y = x^2$, Figure 3.11(a), we obtain the graph of $y = (x+2)^2$ by translating 2 units to the *left* (Figure 3.11b), then we translate this graph downwards by 9 (Figure 3.11c).

Note that the order of translations could be reversed, that is, we could go via $y = x^2 - 9$.

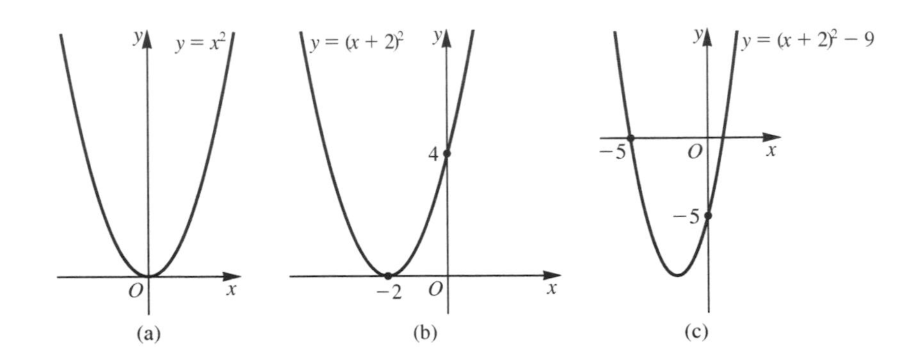

Figure 3.11.
Graphs of the functions of Example 3.8

Stretching the axes

Suppose $f(x) = 1 - x^2$. Then $2f(x) = 2 - 2x^2$ and $\frac{1}{2}f(x) = \frac{1}{2} - \frac{1}{2}x^2$. The graphs of these three functions are shown in Figure 3.12 (a)–(c).

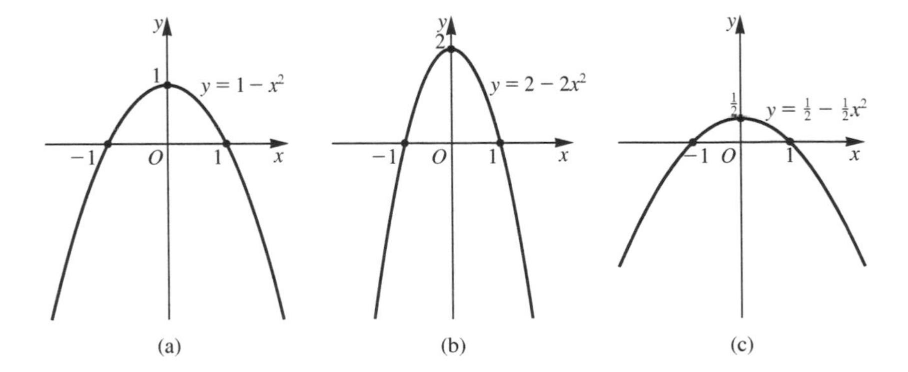

Figure 3.12.
Change of vertical scale:
(a) $f(x) = 1 - x^2$;
(b) $2f(x) = 2 - 2x^2$;
(c) $\frac{1}{2}f(x) = \frac{1}{2} - \frac{1}{2}x^2$

The transformations lead to a change of scale in the y-direction and illustrate the general rule:

KEY POINT

> The graph of $y = af(x)$ is obtained from that of $y = f(x)$ by changing the scale in the y-direction, stretching by a if $a > 1$ and compressing by a factor $\frac{1}{a}$ if $0 < a < 1$.

Figure 3.13 depicts the graphs of $f(x) = 1 - x^2$, $f(2x) = 1 - 4x^2$ and $f\left(\frac{1}{2}x\right) = 1 - \frac{1}{4}x^2$.

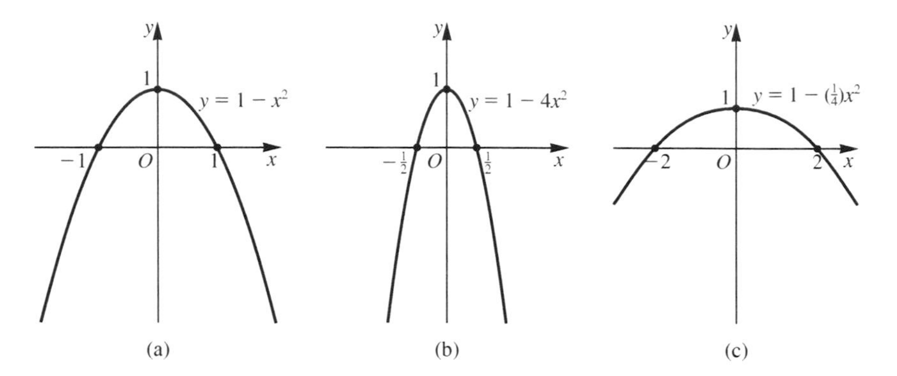

Figure 3.13. Change of horizontal scale:
(a) $f(x) = 1 - x^2$;
(b) $f(2x) = 1 - 4x^2$;
(c) $f\left(\frac{1}{2}x\right) = 1 - \frac{1}{4}x^2$

The transformations lead to a change of scale in the x-direction and illustrate the general rule:

KEY POINT

> The graph of $y = f(ax)$ is obtained from that of $y = f(x)$ by a change of scale in the x-direction, compressing by a factor a if $a > 1$ and stretching by a factor $\frac{1}{a}$ if $0 < a < 1$.

Note that the change in the x-scale is opposite to what our intuition might have suggested.

Worked example

3.9 Illustrate on the same axes the graphs of the functions $f(x) = x - 1$, $f(2x)$ and $f\left(\frac{1}{2}x\right)$.

Solution Figure 3.14 shows the graphs of $f(x) = x - 1$, $f(2x) = 2x - 1$ and $f\left(\frac{1}{2}x\right) = \frac{1}{2}x - 1$.

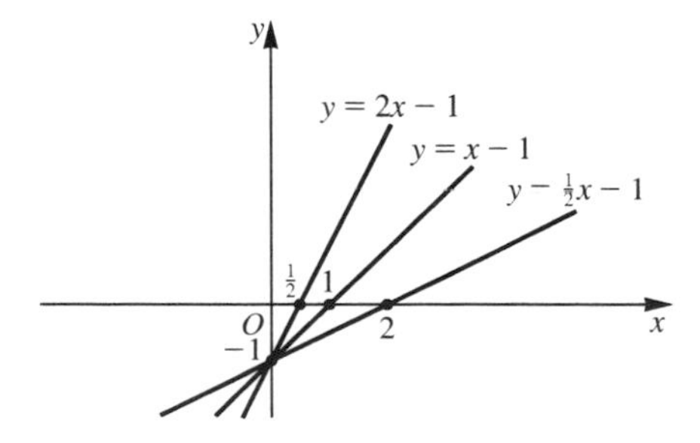

Figure 3.14. Graphs of the functions of Example 3.9

Reflections

Figure 3.15 depicts the graphs of $f(x) = 1 - x$, $g(x) = -f(x) = x - 1$ and $f(-x) = 1 + x$. The graph of $y = x - 1$ is the reflection of $y = 1 - x$ in the x-axis and the graph of $y = 1 + x$ is the reflection of $y = 1 - x$ in the y-axis.

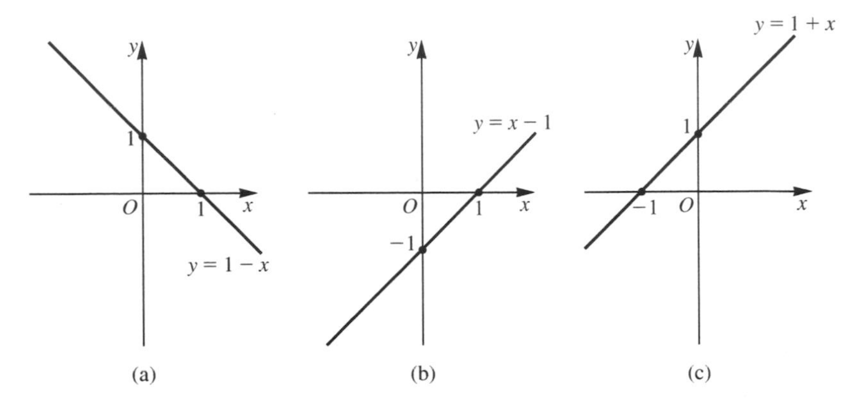

Figure 3.15. Reflection

KEY POINT

> The reflection of the graph of $f(x)$ in the x-axis is that of $-f(x)$ and the reflection of the graph of $f(x)$ in the y-axis is that of $f(-x)$.

Worked examples

3.10 For $f(x) = x^2 + 1$ show that the graphs of $f(-x)$ and $f(x)$ are identical. What does this imply about the graph of $f(x)$?

Solution $f(-x) = (-x)^2 + 1 = x^2 + 1 = f(x)$. Hence the graphs of $f(x)$ and $f(-x)$ are identical. Since the graph of $f(-x)$ is the reflection of the graph of $f(x)$ in the y-axis it follows that this graph is symmetrical about the y-axis. In such a case we say that $f(x)$ is an **even function**.

3.11 For $f(x) = x^3$ show that the graphs of $f(-x)$ and $-f(x)$ are identical.

Solution $f(-x) = (-x)^3 = -x^3 = -f(x)$. In such a case we say that $f(x)$ is an **odd function.**

3.12 Given that

$$\frac{1 - 2x}{x + 1} \equiv \frac{3}{x + 1} - 2$$

hence sketch the graph of

$$y = \frac{1 - 2x}{x + 1}$$

Solution We start with the graph of $f(x) = 1/x$ (Figure 3.16a) then displace it 1 unit to the left to obtain the graph of $f(x+1) = 1/(x+1)$ (Figure 3.16b); next, we stretch by a factor of 3 in the y-direction to obtain the graph of $3f(x+1) = 3/(x+1)$ (Figure 3.16c). Finally, we translate downwards by 2 units to obtain the graph of $3f(x+1) - 2$ (Figure 3.16d).

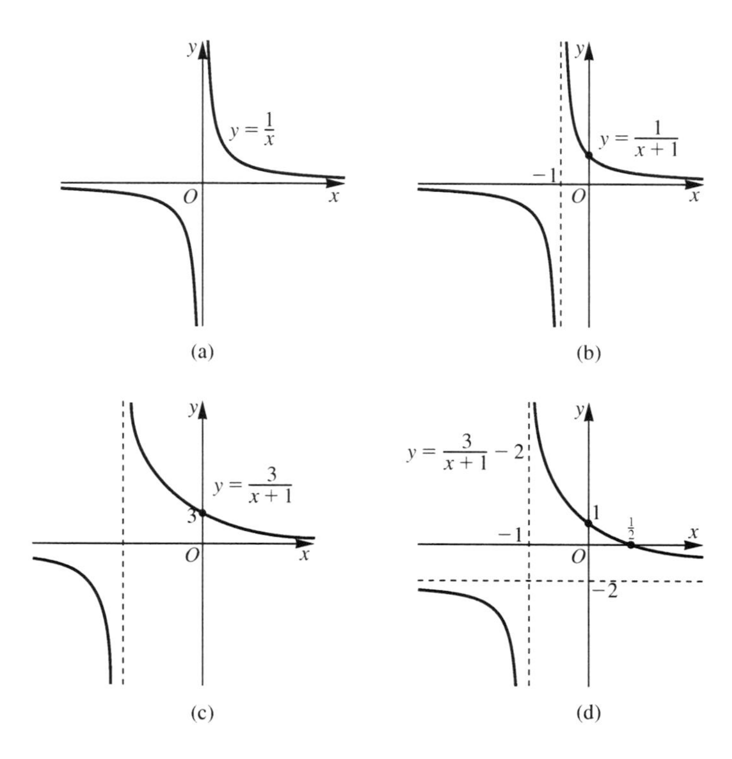

Figure 3.16.
Graphs for Example 3.12

Self-assessment questions 3.3

1. Explain the relationship between the graphs of $f(x) + a$, $f(x + a)$ and $f(x)$, distinguishing the cases $a > 0$ and $a < 0$.
2. Explain the relationship between the graphs of $af(x)$, $f(ax)$ and $f(x)$, distinguishing the cases $a > 1$, $0 < a < 1$, $-1 < a < 0$ and $a = -1$.

Exercise 3.3

In these exercises we use t as the independent variable. This is usually done when we deal with functions dependent on time. This is merely a change of symbol. Hence if $f(x) = x + 3$ then $f(t) = t + 3$ and vice versa.

1. The unit step function is denoted $u(t)$ (sometimes $H(t)$). Sketch the following functions:

   (a) $u(t-2)$ (b) $u(t+1)$
   (c) $3u(t)$ (d) $u(t) - 2$
   (e) $u(t) + 1$ (f) $2 - u(t)$
   (g) $u(-t)$ (h) $u(2-t)$

2. Sketch the following functions:

   (a) $u(t) - u(t-3)$
   (b) $u(t) - u(t-3) + u(t+6)$

3. Sketch the function

$$r(t) = \begin{cases} \frac{1}{h} & -\frac{h}{2} < t < \frac{h}{2} \\ 0 & \text{elsewhere} \end{cases}$$

   Then sketch the function $r(t-3)$.

4. The **impulse function**, $\delta(t)$, is formed by taking the limit of the function of Question 3 as $h \to 0$. Its graph is shown in Figure 3.17. Indicate on a diagram $\delta(t) + 2\delta(t-1)$.

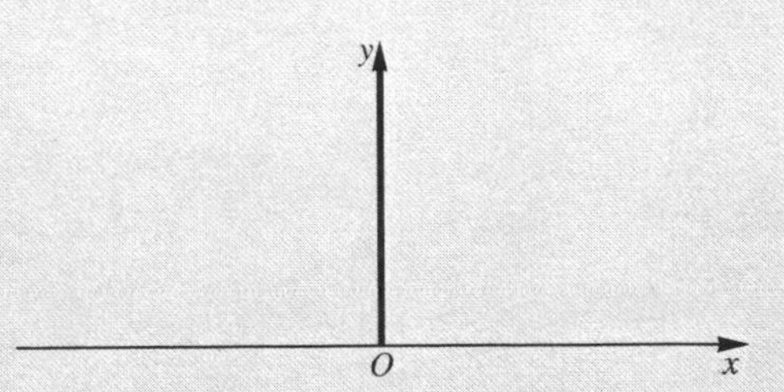

Figure 3.17.
The impulse function

5. Find an alternative way of defining the function $f(t) = [u(t-t_1) - u(t-t_2)]t^2$, where $t_2 > t_1$.

3.4 Inverse of a function

Suppose that we wish to reverse or 'undo' the effect of applying the function 'add 3' (depicted in Figure 3.18a). By reversing the direction of travel through the box the effect is to 'subtract 3' as shown in Figure 3.18(b). Algebraically, we write $y = x + 3$ and rearrange this to $x = y - 3$. Since we are used to describing functions in the form $y = f(x)$ we interchange x and y to obtain $y = x - 3$.

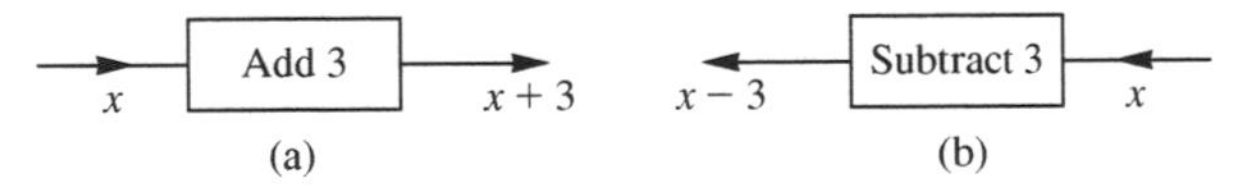

Figure 3.18.
The inverse of the function 'add 3'

The effect of the function 'multiply by 2' can be reversed by applying the function 'divide by 2'; see Figure 3.19. Algebraically, we write $y = 2x$ and rearrange this to $x = y/2$. Interchanging x and y we obtain $y = x/2$.

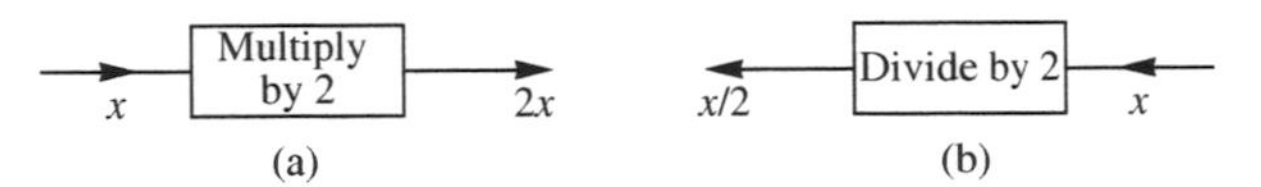

Figure 3.19.
The inverse of the function 'multiply by 2'

Worked example

3.13 Find algebraically the inverse of the following functions:

(a) $f(x) = x^3$ (b) $f(x) = \dfrac{1}{x}$, $x \neq 0$

Illustrate by box diagrams.

Solution (a) If $y = x^3$ then $y^{\frac{1}{3}} = x$ or $x = y^{\frac{1}{3}}$. Interchanging x and y gives $y = x^{\frac{1}{3}}$. Figure 3.20 shows the box diagrams.

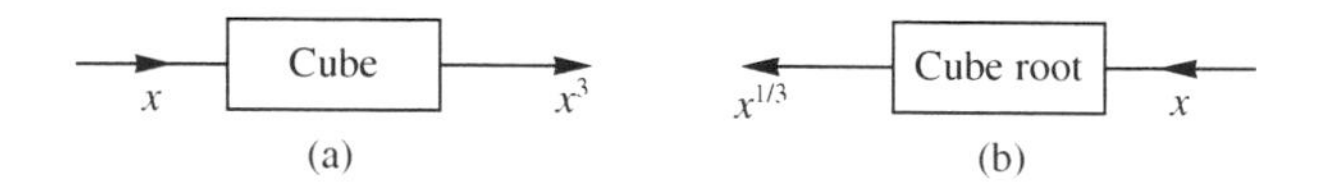

Figure 3.20. The inverse of the function 'cube'

(b) If $y = 1/x$ then $x = 1/y$. Interchanging x and y gives $y = 1/x$, again. The function is its own inverse. Note that $x \neq 0$ in both cases. Figure 3.21 shows the box diagrams.

Figure 3.21. The inverse of the function 'take reciprocal'

Note that for each of the four functions we have considered in this section the inverse operation is also a function, known as the **inverse function**. If the original function is denoted $f(x)$ then the inverse function is denoted $f^{-1}(x)$.

KEY POINT

> The inverse of a given function reverses the operations carried out by the function. If the inverse is also a function it is known as the inverse function.

Please note: f^{-1} does *not* mean $1/f$.

Hence:

if $f(x) = x + 3$ then $f^{-1}(x) = x - 3$

if $f(x) = 2x$ then $f^{-1}(x) = \frac{1}{2}x$

if $f(x) = x^3$ then $f^{-1}(x) = x^{\frac{1}{3}}$

if $f(x) = \frac{1}{x}$, $x \neq 0$ then $f^{-1}(x) = \frac{1}{x}$, $x \neq 0$

Of course, if we start with $f(x) = x - 3$ then $f^{-1}(x) = x + 3$. So each function in a pair is the inverse of the other.

In general, subject to any restrictions on the domain

$$f^{-1}\{f(x)\} = x \quad \text{and} \quad f\{f^{-1}(x)\} = x$$

We can obtain the graph of the inverse function by a simple procedure. Figure 3.22(a) shows the graphs of $y = x + 3$, $y = x - 3$ and $y = x$ drawn on the same axes.

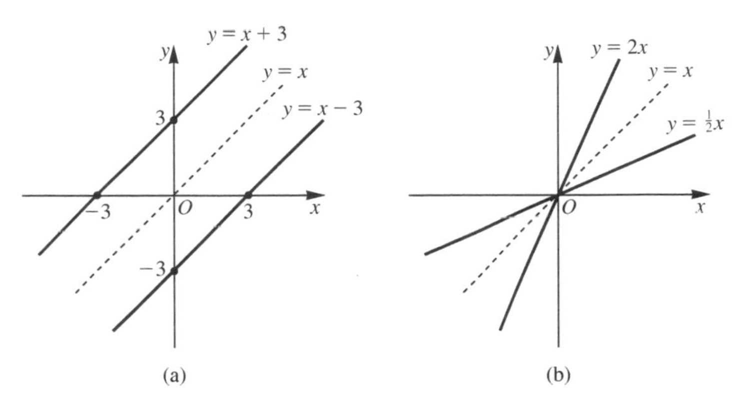

Figure 3.22.
Obtaining the inverse of a function graphically

Note that the graph of $y = x - 3$ is the reflection in the line $y = x$ of the graph of $y = x + 3$ (and vice versa). Note also that where one graph cuts the x-axis the other cuts the y-axis. Figure 3.22(b) shows the graphs of $y = 2x$, $y = \frac{1}{2}x$ and $y = x$. The graph of $y = \frac{1}{2}x$ is the reflection in the line $y = x$ of the graph of $y = 2x$. Note also that where the graphs of $y = 2x$ and $y = \frac{1}{2}x$ meet they do so on the line $y = x$. Therefore, if we are given the graph of $y = f(x)$ then we can obtain the graph of $y = f^{-1}(x)$ by reflecting the first graph in the line $y = x$.

Worked example

3.14 Using the line $y = x$ obtain the graphs of the inverse functions of

(a) $f(x) = x^3$ (b) $f(x) = \dfrac{1}{x}$, $x \neq 0$

Solution (a) Figure 3.23(a) shows the graphs of $y = x^3$, $y = x$ and $y = x^{\frac{1}{3}}$. The last graph has been obtained by reflecting that of $y = x^3$ in the line $y = x$. Note that where the graph of $y = x^3$ meets the line $y = x$ the inverse function also passes through this point.

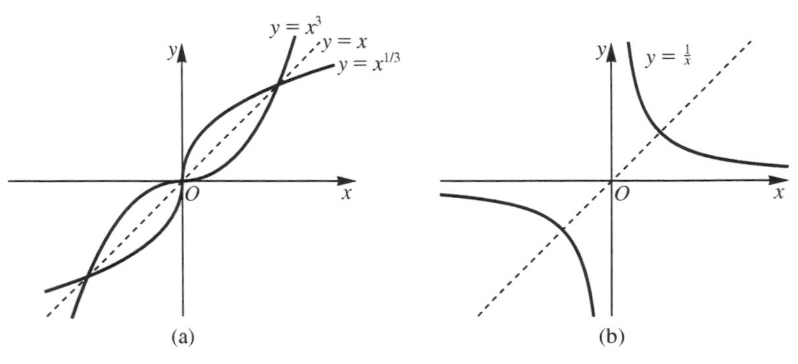

Figure 3.23.
Finding the inverse of the functions of Example 3.14

(b) In Figure 3.23(b) the graph of $y = 1/x$ is symmetrical about the line $y = x$, hence the graph of the inverse function is identical to that of $f(x)$. They are the same function, of course.

Note that in each case the domain of f^{-1} is the range of f, and vice versa.

Inverse of composite functions

To invert a composite function we can travel through the boxes in the reverse order as well as the reverse direction.

Consider the composite function $f(x) = x^3 + 1$ in Figure 3.24(a). Figure 3.24(b) shows that the inverse function is $f^{-1}(x) = (x - 1)^{\frac{1}{3}}$. Algebraically, if $y = x^3 + 1$ then $y - 1 = x^3$ or $x^3 = y - 1$ so that $x = (y - 1)^{\frac{1}{3}}$. Interchanging x and y gives $y = (x - 1)^{\frac{1}{3}}$.

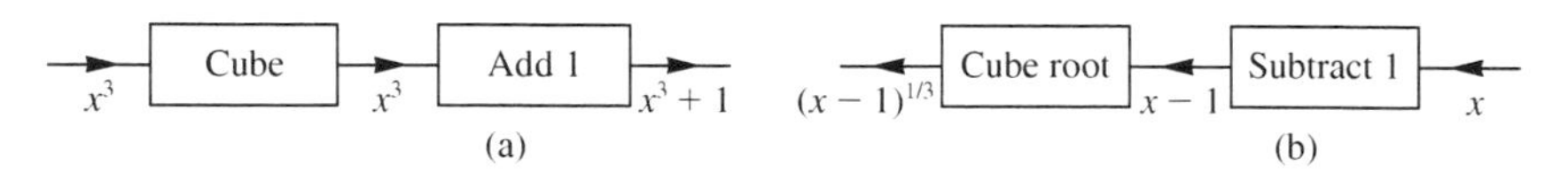

Figure 3.24.
The inverse of a composite function

Worked example

3.15 Invert the following composite functions:

(a) $f(x) = 3x + 1$ (b) $f(x) = \dfrac{1}{x + 2}$, $x \neq -2$

State the restrictions on domain and range in case (b).

Solution

(a) If $y = 3x + 1$ then $3x = y - 1$ and $x = \frac{1}{3}(y - 1)$. Interchanging x and y gives $y = \frac{1}{3}(x - 1)$. We write $f^{-1}(x) = \frac{1}{3}(x - 1)$. See Figure 3.25.

(b) If $y = 1/(x + 2)$ then $x + 2 = 1/y$ and $x = (1/y) - 2$. Hence, interchanging x and y, $y = (1/x) - 2$ or $f^{-1}(x) = (1/x) - 2$. Here $x \neq 0$. Hence the domain of $f(x)$ is all real numbers $x \neq -2$ and the range all non-zero real numbers. The domain of $f^{-1}(x)$ is all non-zero real numbers and the range is all real numbers except -2.

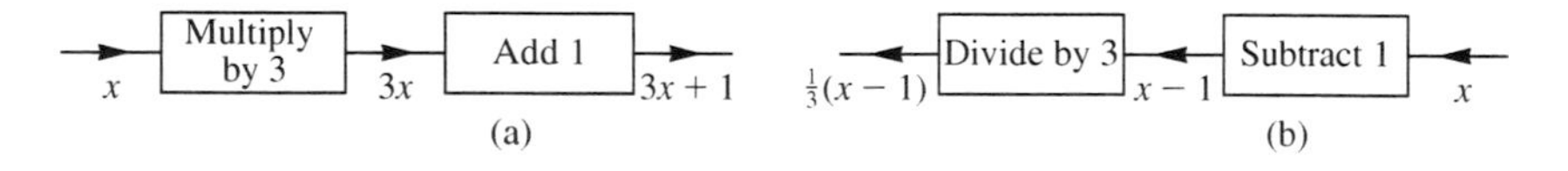

Figure 3.25.
Inverting the function of Example 3.15(a)

Not all functions have an inverse which is a function. As an example, if the graph of the function $f(x) = x^2$ is reflected in the line $y = x$ we obtain the graph with equation $y^2 = x$ shown in Figure 3.26. This does not represent a function because each value of $x > 0$ corresponds to two values of y.

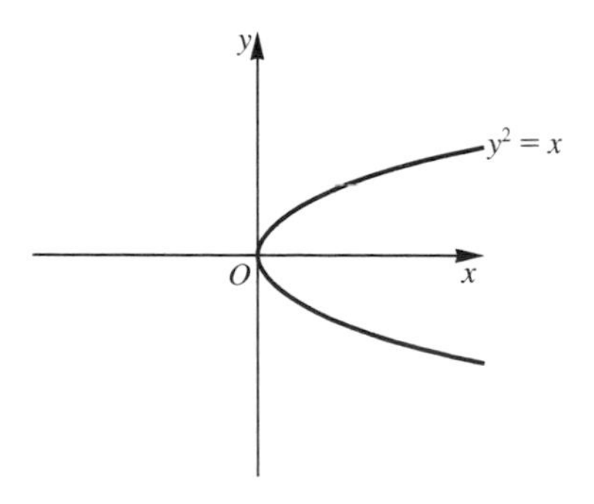

Figure 3.26.
An inverse which is not a function

Calculators have a square root button which gives the positive square root corresponding to that part of the curve above the x-axis. For example, if we input the number -3 and square it we obtain 9; pressing the square root button gives $+3$: we cannot recover the number -3. By convention, when we write $\sqrt{x}$ or $x^{\frac{1}{2}}$ we mean the positive square root.

Worked example

3.16 By restricting the domain of $f(x) = x^2$ show that it can have an inverse which is a function.

Solution The function $g(x) = x^{\frac{1}{2}}$ has domain $x \geqslant 0$ and range real numbers $\geqslant 0$. Hence if we restrict the domain of $f(x)$ to real numbers $\geqslant 0$ $f(x)$ has an inverse, namely $f^{-1}(x) = x^{\frac{1}{2}}$.

Self-assessment questions 3.4

1. Explain what is meant by the term inverse function.
2. Explain how to find the inverse of the function $f(g(x))$.

Exercise 3.4

1. Find the inverse of each of the following functions:

   (a) $f(x) = x + 2$ (b) $f(x) = 2 - x$

   (c) $f(t) = 4t$ (d) $g(t) = \dfrac{1}{t^3}$, $t \neq 0$

   (e) $\dfrac{x+1}{x}$, $x \neq 0$ (f) $\dfrac{x}{x-1}$, $x \neq 1$

2. Which of the following functions $f(x)$ has an inverse which is a function?

   (a) $6x^3 + 2$ (b) $6x^4 + 2$

   (c) $\sqrt[3]{1-x}$ (d) $\sqrt[3]{1-x^2}$

3. What restriction on a function $f(x)$ will guarantee that $f^{-1}(x)$ is also a function?

4. Find the inverse of $h(t)$ where $h(t) = f(g(t))$ and

   (a) $f(t) = 3t + 2$, $g(t) = 2t$
   (b) $f(t) = t^3$, $g(t) = t + 1$

Test and assignment exercises 3

1. What is the largest domain for each of the following functions?

   (a) $f(x) = x^3 - 3$
   (b) $f(x) = x + \sqrt{x-1}$
   (c) $f(x) = x + \sqrt{1-x^2}$
   (d) $f(x) = x^2 - \sqrt{1-x^2}$

2. Sketch the graphs of the following functions for $-2 \leqslant x \leqslant 2$.

   (a) $f(x) = x + |x|$
   (b) $f(x) = x - |x|$
   (c) $f(x) = |x| - x$
   (d) $f(x) = x|x|$
   (e) $f(x) = |x| \div x$

3. State the domain of the following functions, evaluate $f(2)$, $f(-2)$ and $f(0)$ and determine which are odd and which are even:

   (a) $f(x) = \dfrac{x+1}{x-1}$ (b) $f(x) = \dfrac{x-1}{x+1}$

   (c) $f(x) = \dfrac{x}{x^2+1}$ (d) $f(x) = \dfrac{x^2+1}{x^2}$

   (e) $f(x) = x^2 + \dfrac{1}{x^2}$

   (f) $f(x) = x^2 + \dfrac{1}{x^2+1}$

4. A function is defined by

$$f(x) = \begin{cases} 0 & x < -2 \\ x+2 & -2 \leqslant x \leqslant 0 \\ 2-x & 0 \leqslant x \leqslant 2 \\ 0 & x > 2 \end{cases}$$

   Sketch the graphs of $f(x)$, $f(x)+2$, $f(x)-1$, $f(x+1)$, $f(x-1)$, $f(x+2)$, $f(x-2)$ and $f(x-4)$.

5. For the function of Question 4, sketch the graphs of $2f(x)$, $\frac{1}{2}f(x)$, $f(2x)$, $f(\frac{1}{2}x)$, $-f(x)$, $f(-x)$ and $f(-2x)$.

6. For the function of Question 4, sketch the graphs of $f(2-x)$, $3f(x)+2$, $f(x-2)+2$ and $f(2x)-2$.

7. If $f(t) = 2t$ and $g(t) = 4$, sketch the graphs of

   (a) $f(t) + g(t).u(t)$
   (b) $f(t) + g(t).u(t-2)$.

8. For the following functions find $f(g(x))$ and $g(f(x))$. State the domains of the composite functions.

   (a) $f(x) = \dfrac{1}{x}$, $g(x) = \dfrac{x}{x-1}$

   (b) $f(x) = \sqrt{x+1}$, $g(x) = \dfrac{1}{x-1}$

9. Find the inverse of the functions $f(x)$, $g(x)$, $f(g(x))$ and $g(f(x))$ for the functions in Question 8. State the domain of each inverse function.

4 Functions II – standard engineering functions

Objectives

This chapter

- looks at polynomial functions
- studies the main features of rational functions
- explains how to find partial fractions of a rational function
- shows how to model decay and growth processes
- uses the exponential and logarithmic functions

4.1 Polynomial functions

A **polynomial function**, $P(x)$, has the form

$$P(x) = a_n x^n + a_{n-1}x^{n-1} + \dots + a_1 x + a_0 \qquad 4.1$$

where $a_0, a_1, \dots, a_n$ are constants, known as coefficients, and n is an integer. The expression on the right-hand side of (4.1) is called a **polynomial**.

The usual ways to write a polynomial are with powers either increasing (left to right) or, as in the example above, decreasing. The **degree** of a polynomial is the value of the highest power. Note that the only coefficient which cannot be zero is a_n.

Table 4.1 shows some low-degree polynomials; a, b, c and d are constants.

Two polynomials of the same degree are identical if and only if all corresponding coefficients are equal, that is

$$a_n x^n + a_{n-1}x^{n-1} + \dots + a_1 x + a_0 \equiv b_n x^n + b_{n-1}x^{n-1} + \dots + b_1 x + b_0$$

if and only if

$$a_n = b_n,\ a_{n-1} = b_{n-1},\ \dots,\ a_1 = b_1 \text{ and } a_0 = b_0$$

An accurate way of evaluating a polynomial, especially if it is of high degree, is by the use of **nested multiplication**. As a simple example, we can write the cubic polynomial

Table 4.1. Low-degree polynomials

Degree	General form	Name
0	0	Constant (polynomial)
1	$ax + b$	Linear
2	$ax^2 + bx + c$	Quadratic
3	$ax^3 + bx^2 + cx + d$	Cubic
4	$ax^4 + bx^3 + cx^2 + dx + e$	Quartic

$$P(x) = 3x^3 - 4x^2 - x + 2$$

as

$$P(x) = [(3x - 4)x - 1]x + 2$$

The evaluation is carried out in steps as follows:

- We multiply 3 by x and subtract 4.
- Then we multiply by x and subtract 1.
- Finally we multiply by x and add 2.

Worked examples

4.1 What is the degree of each of the following?
(a) $3x^4 + 2$ (b) $3x^4 + 2x^2$ (c) $3x^4$

(d) $2x^{\frac{1}{2}} + x^2$ (e) $x + \dfrac{1}{x}$

Solution (a) 4 (b) 4 (c) 4
(Only the term with the highest power of x is important in determining the degree.)
(d) and (e) are not polynomials since they include a power of x which is not an integer $\geqslant 0$.

4.2 For which values of A, B, and C are the following true?
(a) $A(x + 3) + B(x - 2) + C(x^2 + 4) \equiv x^2 - 3x$
(b) $A(x + 1)(x - 2) + B(x - 2)(x - 3) + C(x + 1)(x - 3) \equiv x^2$

Solution (a) Writing the first polynomial in decreasing order we obtain

$$Cx^2 + (A + B)x + 3A - 2B + 4C$$

Equating in turn coefficients of x^2, x and the constants gives

$$C = 1, \quad A + B = -3, \quad 3A - 2B + 4C = 0$$

The solutions are

$$A = -2, B = -1, C = 1$$

(b) We *could* use the same method as for (a). In this case, however, it is simpler to substitute certain values of x. Since the identity holds for all values of x we can choose values which will simplify the left-hand side.

Put $x = 3$ then $x - 3 = 0$ and $A(4)(1) + 0 + 0 = 9$, that is, $4A = 9$ and so $A = 9/4$.

Put $x = 2$ then $x - 2 = 0$ and $0 + 0 + C(3)(-1) = 4$, that is, $-3C = 4$ and so $C = -4/3$.

Put $x = -1$ then $x + 1 = 0$ and $0 + B(-3)(-4) = 1$, that is, $12B = 1$ and $B = 1/12$.

The solutions are $A = 9/4$, $B = 1/12$, $C = -4/3$. (In some problems a combination of the two methods is needed.)

4.3 Express in nested form the polynomial

$$P(x) = 2x^4 + 3x^3 + 5x^2 - x + 4$$

Compare the number of arithmetical operations in both nested form and original form.

Solution We start with $2x^4 + 3x^3 \equiv (2x + 3)x^3$. Then

$$2x^4 + 3x^3 + 5x^2 \equiv \{(2x + 3)x + 5\}x^2$$

Next,

$$2x^4 + 3x^3 + 5x^2 - x \equiv [\{(2x + 3)x + 5\}x - 1]x$$

Finally,

$$P(x) \equiv [\{(2x + 3)x + 5\}x - 1]x + 4$$

The nested form requires four multiplications; $2 \times x$ initially and three further multiplications by x. There are also four additions or subtractions. The original form could be written as

$$2 \times x \times x \times x \times x + 3 \times x \times x \times x + 5 \times x \times x - x + 4$$

and requires nine multiplications and four additions or subtractions. We have saved five multiplications leading to a more speedy and more accurate evaluation (less round-off error).

Zeros and roots

A polynomial with real coefficients can be expressed as a product of linear factors or irreducible quadratic factors or a mixture of both. An **irreducible quadratic expression** is one which cannot be expressed as the product of two linear factors with real coefficients. The process of obtaining these factors is called **factorization**. It is not always easy to find these factors and for polynomials of high degree a software package is probably necessary, but the following result helps.

If $P(a) = 0$ then $x = a$ is a **zero** of the polynomial $P(x)$. Note that $x = a$ is a zero of $P(x)$ if and only if $x - a$ is a factor of $P(x)$. Also if $x = a$ is a zero of the polynomial $P(x)$ then we say that a is a **root** of the polynomial equation $P(x) = 0$.

If $(x - a)^2$ is a factor of $P(x)$ then $x = a$ is a **double root** of $P(x)$. In general, if $(x - a)^n$ is a factor of $P(x)$ where $n > 1$ then $x = a$ is a **repeated root** of $P(x)$.

At a simple root the graph of the polynomial crosses the horizontal axis; at a double root the graph touches the axis.

KEY POINT

> If $P(a) = 0$ then $x = a$ is a zero of the polynomial $P(x)$; a is a root of the polynomial equation $P(x) = 0$.

A polynomial of odd degree always has at least one real zero. In general, a polynomial of odd degree n has either n real roots or $n - 2$ real roots or $n - 4$ or , ... , or 1 real root. A polynomial of even degree n has either n real roots or $(n - 2)$ or , ... , or 2 or 0 real roots.

If you have access to a graphics calculator or to a computer software graphics package you should plot the graphs of several polynomial functions to develop a 'feel' for them. Suitable examples include:

$$\pm x^2, \pm x^3, \pm x^4, \pm x^5, \pm x^2, \pm 2, \pm x^3,$$
$$x^3 - x^2, x^3 - x, (x^4 + 3)^4, (x - 3)^4, x^4 + 3, x^4 - 3, x^2 - x^4$$

Worked examples

4.4 Factorize the following polynomials where possible:
(a) $x^2 - 5x + 6$ (b) $x^2 - 6x + 10$
(c) $x^4 - 81$ (d) $x^3 - 3x^2 + 4x - 12$

Solution (a) If $P(x) = x^2 - 5x + 6$ then $P(2) = 4 - 10 + 6 = 0$. Since $P(x)$ is a quadratic it has either two real zeros or no real zeros. It must therefore have a second real zero. Trial and error with suitable values of x show that $P(3) = 0$. Hence $x^2 - 5x + 6 \equiv (x - 2)(x - 3)$.

(b) $x^2 - 6x + 10 \equiv (x - 3)^2 + 1$ and can never be zero. It is an irreducible quadratic. The process we just carried out is called **completing the square**. We took the coefficient of x, (-6), halved it (-3) and tried $(x - 3)^2$; this is $x^2 - 6x + 9$ and we need to add 1 to recover the original quadratic.

(Note that this process applied to the polynomial in part (a) gives

$$x^2 - 5x + 6 \equiv \left(x - \frac{5}{2}\right)^2 - \frac{1}{4}$$

so that the polynomial is zero if

$$\left(x - \frac{5}{2}\right)^2 = \frac{1}{4} = \left(\frac{1}{2}\right)^2$$

and hence

$$x - \frac{5}{2} = \pm\frac{1}{2}$$

giving $x = \frac{5}{2} \pm \frac{1}{2}$, that is, $x = 3$ or 2.)

(c) $x^4 - 81 \equiv (x^2 - 9)(x^2 + 9) \equiv (x - 3)(x + 3)(x^2 + 9)$.

(d) Trying some simple values of x we find that $P(3) = 0$ so that $P(x) \equiv x^3 - 3x^2 + 4x - 12 \equiv (x - 3) \times$ (quadratic). In order to obtain $x^3 \equiv x \times x^2$, the quadratic must have a leading term of x^2, that is:

$$\begin{aligned} P(x) &\equiv (x - 3)(x^2 + \alpha x + \beta) \qquad \text{for real constants } \alpha \text{ and } \beta \\ &\equiv x^3 - 3x^2 + \alpha x^2 - 3\alpha x + \beta x - 3\beta \end{aligned}$$

Considering the terms in x^2 we must take $\alpha = 0$. Then $\beta x \equiv 4x$ so that $\beta = 4$ and $-3\beta = -12$, as required.
Therefore $x^3 - 3x^2 + 4x - 12 \equiv (x - 3)(x^2 + 4)$.

4.5 Given that $x = 1$ is a double root of $x^4 - 2x^3 - 3x^2 + 8x - 4 = 0$, factorize the polynomial fully.

Solution If $x = 1$ is a double root of $P(x) = 0$ then $(x - 1)^2 \equiv x^2 - 2x + 1$ is a factor of $P(x)$.
Hence

$$\begin{aligned} x^4 - 2x^3 - 3x^2 + 8x - 4 &\equiv (x^2 - 2x + 1) \times \text{(quadratic)} \\ &\equiv (x^2 - 2x + 1)(x^2 + \alpha x + \beta) \end{aligned}$$

But

$$\begin{aligned} (x^2 - 2x + 1)(x^2 + \alpha x + \beta) \equiv x^4 &+ (\alpha - 2)x^3 + (\beta - 2\alpha + 1)x^2 \\ &+ (\alpha - 2\beta)x + \beta \end{aligned}$$

Comparing coefficients of x^3 gives $\alpha = 0$.

Comparing coefficients of x^2 gives $\beta + 1 = -3$, therefore $\beta = -4$.

(Check that the coefficients of x and the constant terms agree.)

Then $P(x) \equiv (x - 1)^2 (x^2 - 4) \equiv (x - 1)^2(x - 2)(x + 2)$.

4.6 What are the options for factorizing a quartic with real coefficients?

Solution A quartic equation has 4 real roots or 2 real roots or 0 real roots. Hence a quartic (polynomial) has 4 linear factors or 2 linear factors or 0 linear factors, any shortfall being made up by irreducible quadratics. The options for a quartic where the coefficient of x^4 is 1 are of the form

$$(x-a)(x-b)(x-c)(x-d)$$
$$(x-a)(x-b)(x^2+cx+d)$$
$$(x^2+ax+b)(x^2+cx+d)$$

where a, b, c and d are real coefficients (they can be positive, zero or negative).

Self-assessment questions 4.1

1. Explain the terms degree and zero in relation to polynomial functions.
2. Explain the connection between a zero and a root with reference to a polynomial.
3. What are the advantages of nested multiplication?

Exercise 4.1

1. Find the values of A, B, C and D for which the polynomials

$$P(x) = A(x+2) + B(x-2) + C(x^2-4) + D(x^3+x^2)$$

and

$$Q(x) = 2x^3 + 3x^2 - 2x + 4$$

are identical.

2. Express in nested form:

(a) $P(x) = 5x^4 - 2x^3 + x^2 - 4x + 2$
(b) $Q(x) = 5x^5 - 3x^3 + 2x$

3. Factorize as far as possible the following polynomials:

(a) $x^4 - 256$
(b) $x^4 + 81$
(c) $x^2 + 5x - 6$
(d) $x^3 + 8$
(e) $x^3 + x - 2$
(f) $x^3 + 6x^2 + 12x + 8$

4. Given that $x = 1$ is a multiple zero of the expression $x^4 - x^3 - 3x^2 + 5x - 2$, factorize the polynomial fully.

4.2 Rational functions

A **rational function** has the form

$$f(x) = \frac{P(x)}{Q(x)}$$

where $P(x)$ and $Q(x)$ are polynomials. If the degree of $P(x)$ is less than that of $Q(x)$ the rational function is said to be **proper**.

Figure 4.1(a, b) shows the graphs of the rational functions $f(x) = 1/x$ and $f(x) = (x + 2)/(x - 1)$, respectively.

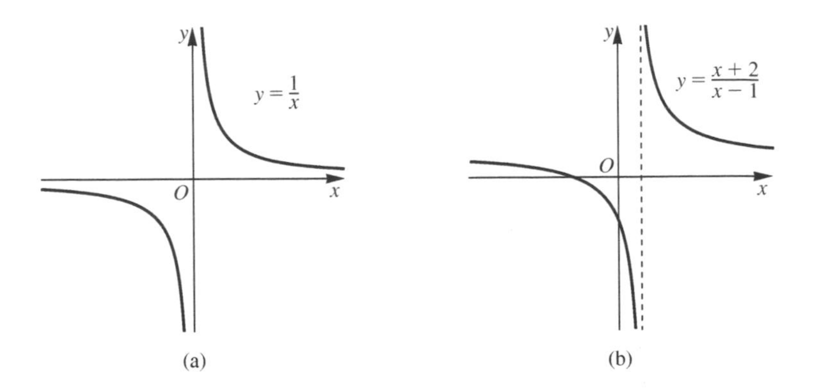

Figure 4.1.
Two rational functions

In part (a) as $x \to \pm\infty$, $y \to 0$. The x-axis ($y = 0$) is a **horizontal asymptote**; the graph of $f(x) = 1/x$ approaches it but never crosses or even meets it, although it gets ever closer as $x \to \pm\infty$. In the same way the y-axis is a **vertical asymptote**. In part (b) the asymptotes are the lines $x = 1$ and $y = 1$. We can write the expression

$$\frac{x+2}{x-1} \equiv 1 + \frac{3}{x-1}$$

As $x \to \pm\infty$,

$$\frac{1}{x-1} \to 0 \text{ and } f(x) \to 1$$

The zeros of the polynomial $P(x)$ are called the **zeros** of the rational function. The zeros of the polynomial $Q(x)$ are called the **poles** of the rational function and have a very important role, for instance, in control engineering, because their location is crucial in determining the stability of a system.

Worked example

4.7 Determine the zeros, poles and asymptotes of the following functions and sketch their graphs:

(a) $\dfrac{2x+1}{x}$ (b) $\dfrac{x-2}{x+3}$ (c) $\dfrac{x^2+x-1}{x-1}$ (d) $\dfrac{x}{x^2-x-2}$

Solution (a) Zero at $x = -1/2$, pole at $x = 0$. Vertical asymptote is $x = 0$. Because $(2x + 1)/x \equiv 2 + (1/x)$, the horizontal asymptote is $y = 2$.

(b) Zero at $x = 2$, pole at $x = -3$. The vertical asymptote is $x = -3$. Since

$$\frac{x-2}{x+3} \equiv \frac{x+3-5}{x+3} \equiv 1 - \frac{5}{x+3}$$

the horizontal asymptote is $y = 1$.

(c) Zero where $x^2 + x - 1 = 0$, that is, $x = (-1 \pm \sqrt{5})/2$. Pole at $x = 1$. Vertical asymptote $x = 1$. You can verify, by combining the terms on the right-hand side, that

$$\frac{x^2+x-1}{x-1} \equiv x + 2 + \frac{1}{x-1}$$

As $x \to \pm\infty$,

$$\frac{x^2+x-1}{x-1} \simeq x + 2$$

and there is an oblique asymptote of $y = x + 2$.

(d) Zero at $x = 0$, poles where $x^2 - x - 2 = 0$, that is, $(x-2)(x+1) = 0$; hence poles at $x = 2$ and $x = -1$. Vertical asymptotes are $x = 2$ and $x = -1$. As $x \to \pm\infty$,

$$\frac{x}{(x-2)(x+1)} \simeq \frac{x}{x.x} = \frac{1}{x}$$

Hence horizontal asymptote is $y = 0$.

Sketch graphs of the functions are shown in Figure 4.2.

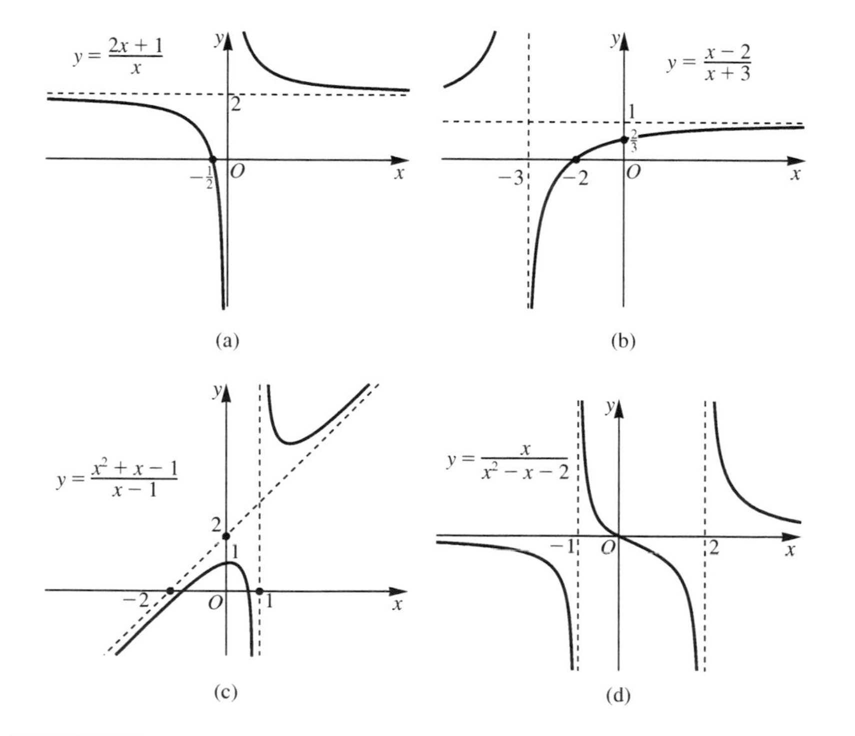

Figure 4.2.
Graphs of the functions of Example 4.7

Partial fractions

Any proper rational function can be expressed as a sum of simpler rational functions whose denominators are linear functions, powers of linear functions or irreducible quadratic functions. These simpler functions are called **partial fractions**.

To find the partial fractions of a given rational function the following points are helpful:

1. Factorize the denominator as far as possible.
2. Any linear factor of the form $ax + b$ leads to a partial fraction

$$\frac{A}{ax+b}$$

where A is constant.

3. A quadratic factor of the form $ax^2 + bx + c$ leads to a partial fraction

$$\frac{Ax+B}{ax^2+bx+c}$$

where A and B are constants.

4. A repeated linear factor gives rise to more than one partial fraction; for example, $(ax + b)^2$ leads to partial fractions

$$\frac{A}{ax+b} \text{ and } \frac{B}{(ax+b)^2}$$

and $(ax + b)^3$ leads to these fractions plus

$$\frac{C}{(ax+b)^3}$$

where A, B and C are constants.

Worked example

4.8 Express the following rational fractions as the sum of partial fractions:

(a) $\dfrac{x-7}{x^2+x-2}$ (b) $\dfrac{2x^2-x-7}{(x^2-1)(x+2)}$

(c) $\dfrac{2x^2-9x+1}{(x-1)^2(x+2)}$ (d) $\dfrac{3x^2-6x+8}{(x-2)(x^2+x+2)}$

Solution (a) Let

$$\frac{x-7}{x^2+x-2} \equiv \frac{x-7}{(x+2)(x-1)} \equiv \frac{A}{x+2} + \frac{B}{x-1}$$

$$\equiv \frac{A(x-1)+B(x+2)}{(x+2)(x-1)}$$

Then $A(x-1)+B(x+2) \equiv x-7$.
Put $x=1$ then $0+B.(3)=-6$ therefore $B=-2$.
Put $x=-2$ then $A(-3)+0=-9$ therefore $A=3$.
Hence

$$\frac{x-7}{(x+2)(x-1)} \equiv \frac{3}{x+2} - \frac{2}{x-1}$$

(b) Let $\dfrac{2x^2-x-7}{(x^2-1)(x+2)}$

$$\equiv \frac{2x^2-x-7}{(x-1)(x+1)(x+2)}$$

$$\equiv \frac{A}{x-1} + \frac{B}{x+1} + \frac{C}{x+2}$$

$$\equiv \frac{A(x+1)(x+2)+B(x-1)(x+2)+C(x-1)(x+1)}{(x-1)(x+1)(x+2)}$$

Then
$A(x+1)(x+2)+B(x-1)(x+2)+C(x-1)(x+1) \equiv 2x^2-x-7$.
Put $x=1$ then $A(2)(3)=-6$ so that $A=-1$.
Put $x=-1$ then $B(-2)(1)=-4$ so that $B=2$.
Put $x=-2$ then $C(-3)(-1)=3$ so that $C=1$.
Hence

$$\frac{2x^2-x-7}{(x^2-1)(x+2)} \equiv \frac{-1}{x-1} + \frac{2}{x+1} + \frac{1}{x+2}$$

(c) Let

$$\frac{2x^2-9x+1}{(x-1)^2(x+2)} \equiv \frac{A}{x-1} + \frac{B}{(x-1)^2} + \frac{C}{x+2}$$

$$\equiv \frac{A(x-1)(x+2)+B(x+2)+C(x-1)^2}{(x-1)^2(x+2)}$$

Then $A(x-1)(x+2)+B(x+2)+C(x-1)^2 \equiv 2x^2-9x+1$.
Put $x=1$ then $B(3)=-6$ so that $B=-2$.
Put $x=-2$ then $C(-3)^2=27$ so that $C=3$.
To find A we compare coefficients of x^2 on both sides of the identity $A+C=2$ so that $A=-1$.
Hence

$$\frac{2x^2-9x+1}{(x-1)^2(x+2)} \equiv \frac{-1}{x-1} - \frac{2}{(x-1)^2} + \frac{3}{x+2}$$

(d) Let

$$\frac{3x^2-6x+8}{(x-2)(x^2+x+2)} \equiv \frac{A}{x-2} + \frac{Bx+C}{x^2+x+2}$$

$$\equiv \frac{A(x^2+x+2)+(Bx+C)(x-2)}{(x-2)(x^2+x-2)}$$

Then $A(x^2+x+2)+(Bx+C)(x-2)\equiv 3x^2-6x+8$.
Put $x=2$ then $8A=8$ so that $A=1$.
Coefficients of x^2: $A+B=3$ so that $B=2$.
Constants: $2A-2C=8$ so that $C=-3$.

Hence

$$\frac{3x^2-6x+8}{(x-2)(x^2+x+2)}\equiv\frac{1}{x-2}+\frac{2x-3}{x^2+x+2}$$

(Note that in these examples A, B and C had simple integer values for clarity of understanding the method. In real life we cannot expect to be so lucky.)

Self-assessment questions 4.2

1. Define a rational function and explain the terms pole, zero and asymptote in that connection.
2. Explain the concept of a partial fraction of a rational function and discuss the cases to be considered in finding partial fractions.

Exercise 4.2

1. A circuit contains two resistors R_1 and R_2 in parallel. Their equivalent resistance, R, is given by
$$\frac{1}{R}=\frac{1}{R_1}+\frac{1}{R_2}$$
Express R as a function of R_1.
If R_2 is fixed, sketch the graph of R against R_1.
2. Sketch the following rational functions, indicating the asymptotes:
(a) $\frac{1}{x^2}$ (b) $\frac{1}{x^3}$
(c) $\frac{1}{x^4}$ (d) $\frac{1}{(x-1)^2}$
(e) $2+\frac{1}{(x-2)^2}$ (f) $\frac{1}{(x+2)^2}-3$
3. Find the poles, zeros and asymptotes of the following rational functions. Sketch the graphs of each function.
(a) $\frac{x+1}{x}$ (b) $\frac{x}{x+1}$
(c) $\frac{x}{x^2+1}$ (d) $\frac{x^2+1}{x}$
(e) $\frac{x-1}{(x+2)(x-3)}$ (f) $\frac{x+2}{(x-1)(x-3)}$
(g) $\frac{x-3}{(x+2)(x-1)}$
4. Express the following in partial fractions:
(a) $\frac{x}{(2x-3)(x+1)}$ (b) $\frac{6x+4}{x^2-5x+6}$
(c) $\frac{6x+4}{x^2+5x-6}$ (d) $\frac{x^2}{x^2+4}$
(e) $\frac{x^3}{x^2+4}$ (f) $\frac{2x^2+2x+4}{(x-2)^2(x+1)}$
(g) $\frac{x^2+x-1}{(x^2-1)^2}$ (h) $\frac{1}{x^3+1}$
(*Hint*: in (d) write x^2 as x^2+4-4 and in (e) write x^3 as $x(x^2+4)-4x$.)

4.3 Power laws: growth and decay

The quietest sound that can be heard by an average human has a level of 10^{-12} W/m^2. The power of a sound signal in W/m^2 is given by

$$P = 10^{-12}(10^{S/10})$$

where S is the intensity of sound in decibels (dB).

Functions of the form $f(x) = a^x$ are known as **power laws**. If $a > 1$ the function represents a **growth process** and if $0 < a < 1$ the function represents a **decay process**. In our example the power of the sound signal increases with the intensity of sound according to a power law.

Figure 4.3 shows the graphs of $y = 2^x$ and $y = \left(\frac{1}{2}\right)^x$. Note that each graph is a reflection in the y-axis of the other. This follows because $\left(\frac{1}{2}\right)^x = 1/(2^x) = 2^{-x}$ and $y = f(-x)$ is a reflection of $y = f(x)$ in the y-axis. Note that since $a^0 = 1$ all curves $y = a^x$ pass through the y-axis at $y = 1$. One special value of a is most frequently used to describe growth processes. This value is denoted by e and its value to 3 d.p. is 2.718; in truth, it cannot be expressed by a finite decimal. The importance of e is that the gradient in the curve $y = e^x$ at any x is e^x; in other words the rate of change of y is equal to y itself. The function $y = e^x$ is known as the **exponential function**. e^x is also written as $\exp(x)$.

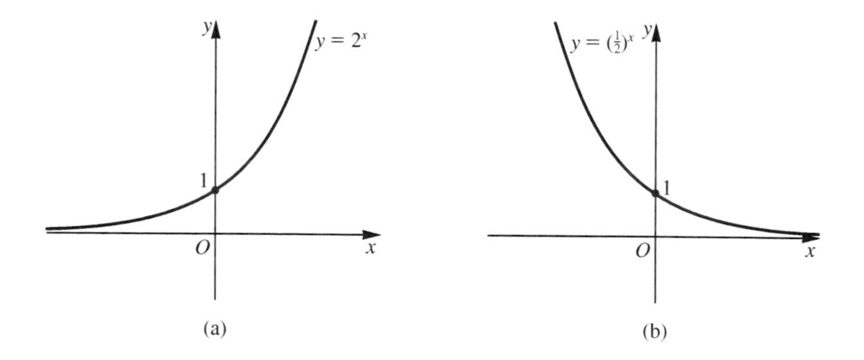

Figure 4.3. Decay and growth processes

More generally, if the rate of change of some quantity y is proportional to the current amount of that quantity y, then the functional relationship between x and y is

$$y = ae^{kx}$$

where a and k are constants.

If $k > 0$ then the process is one of growth, if $k < 0$ the process is one of decay and if $k = 0$ the process is one of **steady state.** Figure 4.4 shows the graphs of $y = e^{kx}$ for $k > 0$ and $k < 0$.

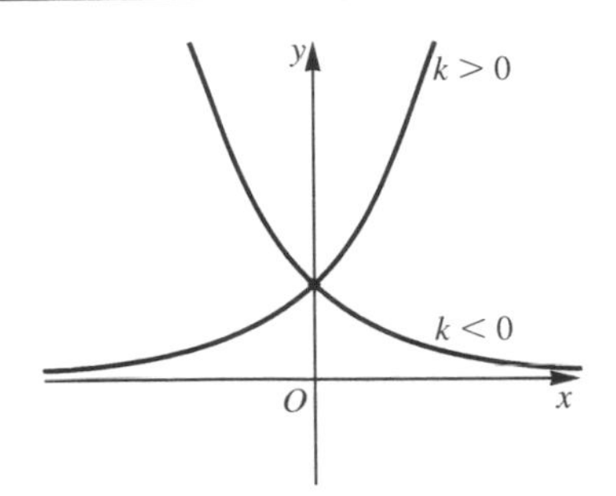

Figure 4.4.
Exponential growth and exponential decay

The domain of the exponential function is all real numbers and the range is real numbers greater than or equal to zero.

KEY POINT

The exponential function $f(x) = e^x$ is a power law relationship in which the rate of change of the function is equal to its value at every stage.

Worked examples

4.9 (a) The power received in conversation at normal levels is about $10^{-6}\,\mathrm{W\,m^{-2}}$. What is the corresponding sound level in dB?
(b) What is the power received by a listener at a rock festival where the sound level is 115 dB?

Solution (a) $10^{-6} = 10^{-12}(10^{S/10})$
Multiply by 10^{12} to obtain $10^6 = 10^{S/10}$ therefore $S/10 = 6$ and hence $S = 60\,\mathrm{dB}$.
(b) $P = 10^{-12}(10^{11.5}) = 10^{-0.5} = 0.316\,\mathrm{W\,m^{-2}}$ (3 d.p.)
(Note that to evaluate 10^y on a calculator we can use the x^y button with $x = 10$.)

4.10 Sketch on the same axes the graphs of $y = a^x$ for $a = 1, 2, 3, 4$.

Solution Figure 4.5 shows the graphs.

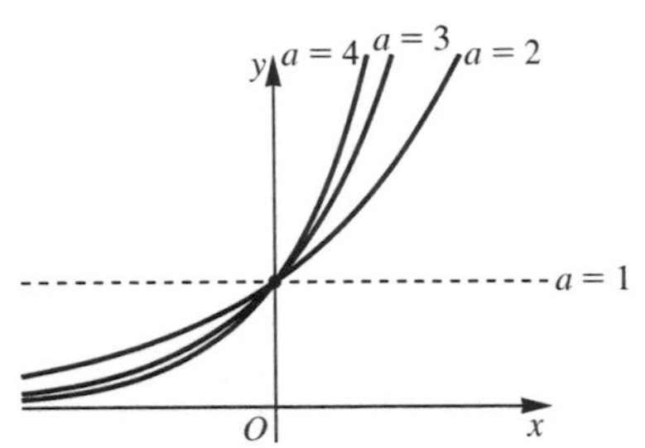

Figure 4.5.
Graphs of the functions of Example 4.10

4.11 The voltage across a discharging capacitor decays according to the law $V = V_0 e^{-t/\tau}$ where the quantity τ is known as the **time constant** of the circuit and t is the time elapsed since the switch was closed. Find the value of the voltage when $t = 5$, 10 and 20 if $\tau = 10$. What is the effect of a larger value of τ?

Solution When $t = 5$, $V = V_0 e^{-\frac{1}{2}} \simeq 0.607V_0$
When $t = 10$, $V = V_0 e^{-1} \simeq 0.368V_0$
When $t = 20$, $V = V_0 e^{-2} \simeq 0.135V_0$
Note that as τ decreases the decay is more rapid; see Figure 4.6.

Note that to evaluate the function e^x on some calculators it is necessary to input x and then press the INV and LN buttons.

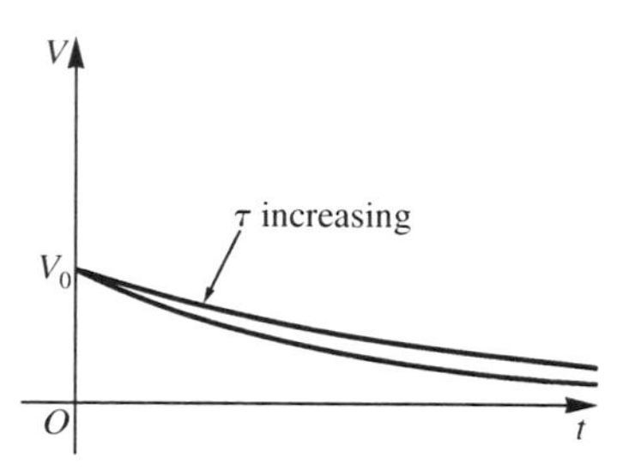

Figure 4.6.
Different decay rates

4.12 Atmospheric pressure p (kilopascals) decreases with increasing height h (km) according to the law

$$p = p_0 e^{-0.144h}$$

Find the pressure at heights 16 km, 32 km, 10 km and 26 km if p_0, the pressure at sea level, is 100 kPa and deduce a rule of thumb for the decrease of pressure with height.

Solution When $h = 16$, $p = 100e^{-0.144 \times 16} \simeq 9.98$ kPa.
When $h = 32$, $p = 100e^{-0.144 \times 32} \simeq 0.997$ kPa.
When $h = 10$, $p \simeq 23.7$ kPa.
When $h = 26$, $p \simeq 2.37$ kPa.
Roughly speaking, the pressure decreases by a factor of 10 for every 16 km of height gained.

Self-assessment questions 4.3

1. Explain the terms growth process and decay process.
2. What is the special property of the exponential function?

Exercise 4.3

1. Sketch on the same axes the graphs of
   (a) $e^x + 1$ (b) $1 + e^{-x}$
   (c) $1 - e^x$ (d) $1 - e^{-x}$
2. Sketch on the same axes the graphs of
   (a) e^{-x} (b) e^{-2x} (c) e^{-3x}
3. Find the sound level in dB corresponding to a power of 10^{-3} W m^{-2} and the power corresponding to a sound level of 100 dB.
4. Find the atmospheric pressure at heights of 12 km, 20 km and 28 km.
5. The luminous intensity at a depth of x m in a liquid is given by $I = I_0 e^{-3x}$. What is the interpretation of I_0? Find the percentage decrease in intensity when descending to depths of (a) 0.5 m, (b) 1 m, (c) 1.5 m.
6. The mass, m, of pollutant in a reservoir decreases according to the law $m = m_0 e^{-0.1t}$ where t is time in days and m_0 is the initial mass. After 100 days by how much has the mass decreased?

4.4 The logarithmic function

If $x = a^z$ for $a > 0$ then we say that z is the **logarithm** of x to the base a, written $z = \log_a x$. For example, $1000 = 10^3$ so that $3 = \log_{10} 1000$. The most common bases are 10 and e, appearing on most calculators as log and ln respectively; hence $\log_e x$ can be, and often is, written as $\ln x$. (ln is known as 'natural' logarithm.) If $y = e^x$ then $x = \ln y$ so that the natural logarithmic function is the inverse of the exponential function, and vice versa. For comparison, the graphs of $y = e^x$ and $y = \ln x$ are shown in Figure 4.7.

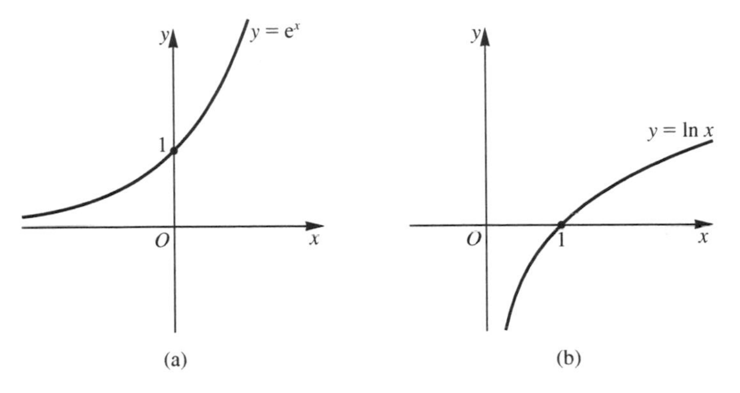

Figure 4.7. Exponential and logarithmic functions

The domain of the natural logarithmic function is real numbers > 0 and the range is all real numbers. Both functions increase from left to right. Notice that, whereas the gradient of the exponential function increases from left to right, that of the natural logarithmic function decreases from left to right.

KEY POINT

> The natural logarithmic function is the inverse of the exponential function.

The **laws of logarithmic functions** are stated below with the proviso that the results are valid *only* if all the logarithms exist.

$$\log_a(xy) = \log_a x + \log_a y$$ (the log of a product is the sum of the logs)

$$\log_a\left(\frac{x}{y}\right) = \log_a x - \log_a y$$ (the log of a quotient is the difference between the logs)

$$\log_a(x^n) = n\log_a x$$ (the log of a power is that power times the log)

These rules allow us to simplify expressions in logarithms.

Worked examples

4.13 Prove the product rule starting from the definitions of logarithms.

Solution Let $\log_a x = z$ and $\log_a y = w$ so that $x = a^z$ and $y = a^w$.
Then $xy = a^z . a^w = a^{z+w}$.
Hence $\log_a(xy) = z + w = \log_a x + \log_a y$.

4.14 Let a and b be positive numbers. Show that

$$\log_a x = \log_b x \div \log_b a \qquad \textbf{(change of base rule)}$$

What can be deduced if $x = b$?

Solution Let $\log_a x = z$ then $x = a^z$.
Taking logarithms to base b gives

$$\log_b x = \log_b(a^z) = z\log_b a \qquad \text{(power rule)}$$

that is

$$\log_b x = \log_a x . \log_b a$$

and the result follows by dividing both sides by $\log_b a$. If $x = b$ then $\log_b b = 1$, so that

$$\log_a b = 1/\log_b a$$

4.15 Simplify the following expressions:

(a) $\log_a x + \log_a x^4$ (b) $5\log_a x + \log_a\left(\frac{1}{x^2}\right)$

(c) $\log_a(xy^2) + \log_a x - 3\log_a y$ (d) $\log_3 81$

Solution (a) $\log_a x^4 = 4\log_a x$. The expression is $5\log_a x$ or $\log_a x^5$.
(b) $5\log_a x = \log_a x^5$; $\log_a x^5 + \log_a(1/x^2) = \log_a(x^5/x^2) = \log_a x^3$. Alternatively, $\log_a(1/x^2) = \log_a(x^{-2}) = -2\log_a x$ so that the expression is $5\log_a x - 2\log_a x = 3\log_a x$.
(c) $3\log_a y = \log_a y^3$ so that the expression is

$$\log_a(xy^2 \times x \div y^3) = \log_a\left(\frac{x^2y^2}{y^3}\right) = \log_a\left(\frac{x^2}{y}\right)$$
$$= 2\log_a x - \log_a y$$

(d) $81 = 3^4$ so that $\log_3 81 = 4$.

Equations involving exponential functions can be solved by taking logarithms and vice versa.

4.16 Solve the following equations:

(a) $e^x - e^{-x} = 4$ (b) $e^x + e^{-x} = 1$ (c) $\ln x^5 - \ln x^3 = 2.4$
(d) $6^x = 40$

Solution (a) Multiply by e^x, noting that $e^x.e^{-x} = e^0 = 1$, to obtain $(e^x)^2 - 1 = 4e^x$ or $(e^x)^2 - 4e^x - 1 = 0$. This is a quadratic equation in e^x with solution

$$e^x = \frac{4 \pm \sqrt{4^2 - 4 \times (-1)}}{2} = 2 \pm \sqrt{5}$$

However, $e^x > 0$ for all x so we reject $2 - \sqrt{5}$, leaving $e^x = 2 + \sqrt{5}$. Taking natural logarithms gives $x = \ln(2 + \sqrt{5})$.

(b) Repeating the procedure in (a) we obtain:

$$(e^x)^2 + 1 = e^x \qquad \text{or} \qquad (e^x)^2 - e^x + 1 = 0$$

The solution is given by:

$$e^x = \frac{1 \pm \sqrt{1 - 4}}{2}$$

and it is impossible to find any solution.

(c) *Method (i)* The left-hand side is $5 \ln x - 3 \ln x = 2 \ln x$. Hence $2 \ln x = 2.4$ and $\ln x = 1.2$. Taking exponentials we obtain $x = e^{1.2}$.
Method (ii) Using the quotient rule, the left-hand side is $\ln x^2$. Hence $\ln x^2 = 2.4$ and taking exponentials gives $x^2 = e^{2.4}$ so that $x = \pm(e^{2.4})^{\frac{1}{2}} = \pm e^{1.2}$. However, the negative answer is inadmissible since if $x < 0$ then $x^3 < 0$ and $x^5 < 0$. Since we cannot take the log of a negative number (or zero) neither $\ln x^3$ nor $\ln x^5$ can exist and the equation is meaningless.
(This is a warning to use the laws of logarithms with care.)

(d) Note first that, since $6^2 = 36$ and $6^3 = 216$, the value of x is between 2 and 3 and closer to 2. Taking logarithms (any base will do) we obtain:

$$\log 6^x = x \log 6 = \log 40$$

giving:

$$x = \log 40 \div \log 6 \simeq 2.054$$

4.17 A steel bar is cooled by running water. Its temperature (in °C) is given by:

$$\theta = 16 + 1200e^{-0.2t}$$

where t is measured in minutes. After how long will the temperature be a 'safe' 38 °C?

Solution Rearranging the formula we obtain, progressively,

$$1200e^{-0.2t} = \theta - 16$$

$$e^{-0.2t} = \frac{\theta - 16}{1200}$$

$$e^{0.2t} = \frac{1}{e^{-0.2t}} = \frac{1200}{\theta - 16}$$

$$0.2t = \ln\left(\frac{1200}{\theta - 16}\right)$$

$$t = 5 \ln\left(\frac{1200}{\theta - 16}\right)$$

Substituting the given value for θ,

$$t = 5\ln\left(\frac{1200}{22}\right) \simeq 20.0 \qquad (3 \text{ s.f.})$$

4.18 A substance decays according to the law $m = m_0 e^{-kt}$ where m is the mass at the time t, m_0 is the initial mass and k is a constant (the decay constant). The **half-life** is the interval of time over which the substance loses half its mass. For radioactive carbon the decay constant is 0.000 124 years^{-1}. Find its half-life and the times taken for the amount of the substance to decay to $\frac{1}{4}$ then $\frac{1}{8}$ of the original amount.

Solution If $m = \frac{1}{2}m_0$ then $\frac{1}{2}m_0 = m_0 e^{-kt}$ so that $\frac{1}{2} = 1/e^{kt}$ and $e^{kt} = 2$, which gives $kt = \ln 2$.

Finally

$$t = \frac{1}{k}\ln 2 = 5590 \text{ years} \qquad \text{(nearest year)}$$

If $m_0 = \alpha m_0$, then $\alpha = e^{-kt}$ so that $\ln \alpha = -kt$ and $t = -\ln \alpha / k$.

$$\text{When } \alpha = \frac{1}{4}, \quad t = 11180 \text{ years} \qquad \text{(twice the half-life)}$$

$$\text{When } \alpha = \frac{1}{8}, \quad t = 16770 \text{ years} \qquad \text{(three times the half-life)}$$

Modelling with power laws

If we suspect that two variables x and y are connected by a law $y = ax^m$ where a and m are constants then we can estimate a and m from experimental measurements. It is better to fit straight lines to data than to fit curves, and we linearize the equation by taking logarithms. The law is rewritten

$$\ln y = \ln a + \ln x^m = \ln a + m \ln x$$

If the values of $\ln y$ are plotted against $\ln x$ then the straight line has a gradient of m and an intercept of $\ln a$.

Special paper called log–log paper allows us to plot values directly onto the paper but we must be careful when estimating m and a. Another special paper called log–linear paper has a logarithmic scale vertically and a linear scale horizontally; an example of its use is given in Example 4.19.

Table 4.2.
Data for Example 4.19

t	5	10	15	20
θ	441.5	162.4	59.7	22.0

Worked example

4.19 The results in Table 4.2 show the temperature of a metal bar as measured in an experiment.

The law connecting the temperature θ (°C) with time t (minutes) is believed to be of the form

$$\theta = \theta_0 \mathrm{e}^{-kt}$$

Plot a suitable graph to estimate k and θ_0.

Solution Taking logarithms of the terms in the formula we obtain

$$\ln \theta = \ln \theta_0 + \ln(\mathrm{e}^{-kt}) = \ln \theta_0 - kt$$

We can plot $\ln \theta$ against t or use log–linear paper. Figure 4.8 shows the plot. We can estimate k and θ_0 by taking two values from the graph. Hence, for example,

$$\ln 441.5 = \ln \theta_0 - 5k$$

$$\ln 22.0 = \ln \theta_0 - 20k$$

Solving for k and θ_0 we obtain:

$$k = 0.2 \qquad \text{and} \qquad \theta_0 = 1200$$

so that the relationship is $\theta = 1200\,\mathrm{e}^{-0.2t}$.

Self-assessment questions 4.4

1. State the laws of logarithms.
2. Explain the link between the exponential function and the natural logarithmic function.

Exercise 4.4

1. Derive the power law for logarithms starting from the definition of a logarithm.

2. Simplify the following expressions:

   (a) $\log_a x^4 - \log_a x^2$

   (b) $3\log_a x - 2\log_a\left(\frac{1}{x}\right)$

   (c) $\log_a(x^3y) + \log_a(xy^2) - 2\log_a x$

   (d) $\log_2(0.25)$

3. Solve the following equations:

   (a) $\mathrm{e}^x - \mathrm{e}^{-x} = 5$ (b) $\mathrm{e}^x + \mathrm{e}^{-x} = 3$

   (c) $\ln(2x^2) + \ln(x^4) = 16$

   (d) $6^{-x} = 20$

4. A substance decays according to the law $m = m_0\mathrm{e}^{-kt}$ where t is measured in days. In 100 days the original mass m_0 has been reduced by a factor of 10. How many more days will it be before the initial mass has been reduced by a factor of 15?

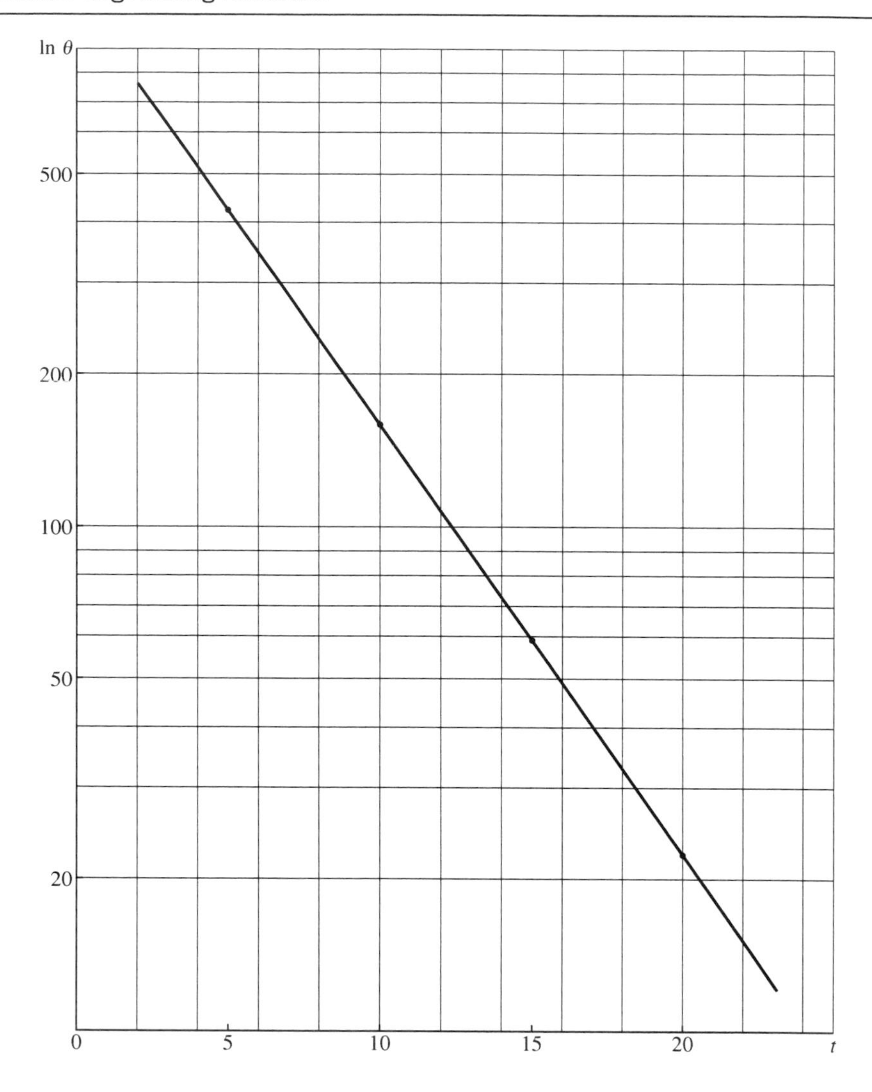

Figure 4.8.
Example of a log–linear plot

Test and assignment exercises 4

1. Express in nested form the following polynomials:

   (a) $P(x) = 2x^3 + 3x^2 - 4x - 7$
   (b) $P(x) = 3x^4 + 2x^3 - x^2 - x + 4$

2. Factorize the following as far as possible:

   (a) $x^2 - 2x - 3$ (b) $x^3 - 27$ (c) $x^4 + 16$
   (d) $x^3 + 9x^2 + 27x + 27$ (e) $x^5 - 32$

3. Find the poles, zeros and asymptotes of the following rational functions and sketch their graphs:

   (a) $\dfrac{x-5}{x}$ (b) $\dfrac{x}{x-5}$ (c) $\dfrac{x+1}{x^2-x}$

   (d) $\dfrac{x+3}{(x-2)(x+1)}$ (e) $\dfrac{x^2+4}{x}$

4. Express the following in partial fractions:

(a) $\dfrac{10x+10}{25-x^2}$

(b) $\dfrac{3x^2}{(x+1)^3}$

(c) $\dfrac{10-11x}{(x^2+1)(x-4)}$

(d) $\dfrac{5}{x^2+x-2}$

(e) $\dfrac{4x}{x^4-16}$

(f) $\dfrac{9}{(x-1)(x+2)^2}$

5. Sketch on the same axes the graphs of:

(a) $e^{-x}+2$ (b) $2-e^{-x}$ (c) $2-e^{x}$

6. The charge q in a capacitor decays according to the law $q=q_0 e^{-\alpha t}$ where t is the time, α is a constant and q_0 is the initial charge. When $t=20$, $q=\frac{1}{2}q_0$. Find the time at which (a) $q=\frac{1}{8}q_0$, (b) $q=\frac{1}{10}q_0$.

7. Simplify the following expressions:

(a) $\log_a x^5-\log_a x^2+3\log_a x$

(b) $\log_a x^2-\log_a\left(\dfrac{1}{x^2}\right)$

(c) $\log_3\left(\dfrac{1}{81}\right)$

8. Solve the following equations:

(a) $e^{2x}-e^{-2x}=4$
(b) $e^{2x}+e^{-2x}=2$
(c) $\ln(x^2)+\ln(x^5)=128$
(d) $3^x=10$

5 Trigonometric and hyperbolic functions

Objectives

This chapter

- defines the circular functions and their inverses
- finds the sine, cosine and tangent of any angle
- explains how to transform circular functions
- finds the general solution of simple trigonometric equations
- provides an introduction to hyperbolic functions

5.1 Circular functions

Modelling oscillations

Many phenomena exhibit a periodic behaviour, that is, a pattern which repeats at regular intervals, for example the voltage produced in a circuit by an alternating current or the displacement of a piston attached to a crank and connecting rod. Such phenomena can be described by periodic functions. A function $f(t)$ is said to be **periodic** with period a if

KEY POINT

$$f(t+a)=f(t) \qquad 5.1$$

for all t in the domain of the function. The **period** is the smallest value of a for which (5.1) is valid.

The most common periodic functions are the sine, cosine and tangent functions. They are known as **trigonometric** functions, because of their definition (see Figure 5.1).

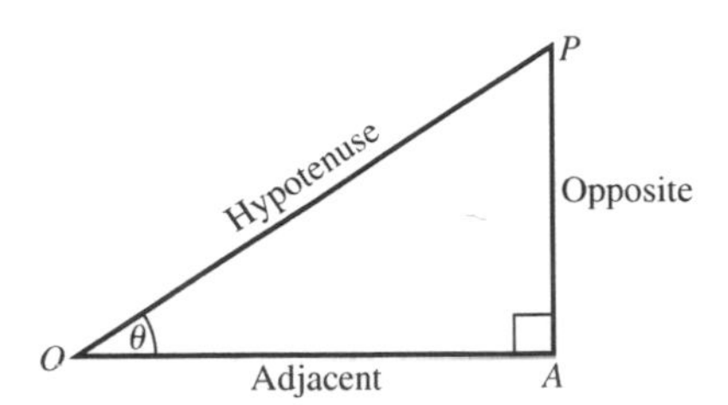

Figure 5.1.
Triangle for defining trigonometric functions

The sides of the right-angled triangle OAP are labelled with reference to $\angle POA$, which has value θ, measured in degrees. We make the definitions in terms of the sides of the triangle.

KEY POINT

$$\sin\theta = \frac{AP}{OP}, \quad \cos\theta = \frac{OA}{OP} \quad \text{and} \quad \tan\theta = \frac{AP}{OA}$$

It follows that

$$\tan\theta = \frac{\sin\theta}{\cos\theta}$$

In modelling oscillations we often measure 'angles' on a new scale: that of radians. A **radian** is defined as the angle subtended at the centre of a circle by an arc of length equal to that of the radius of the circle. In Figure 5.2 $\angle AOB$ is 1 radian. The conversion between degrees and radians is as follows:

KEY POINT

$$180^\circ = \pi \text{ radians} \qquad 5.2$$

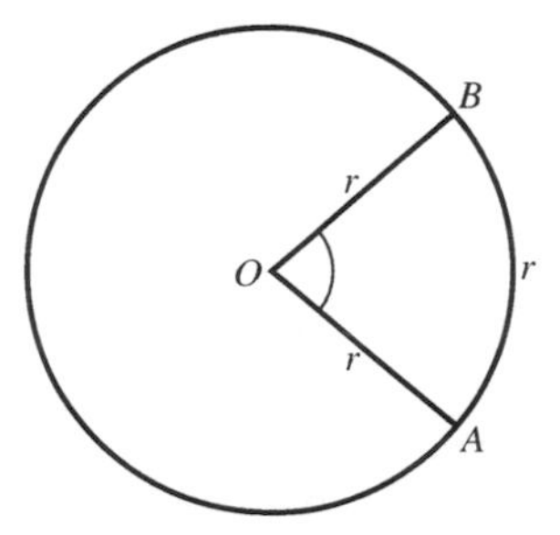

Figure 5.2.
Definition of a radian

Worked examples

5.1 Using an isosceles right-angled triangle find simple expressions for the sine, cosine and tangent of 45°. Then, with reference to an equilateral triangle, find these ratios for angles of 30° and 60°.

Solution In Figure 5.3(a) $OA = AP = 1$ and by Pythagoras' theorem $OP = \sqrt{2}$. Then

$$\sin 45^\circ = \cos 45^\circ = \frac{1}{\sqrt{2}}, \qquad \tan 45^\circ = 1 \tag{5.3}$$

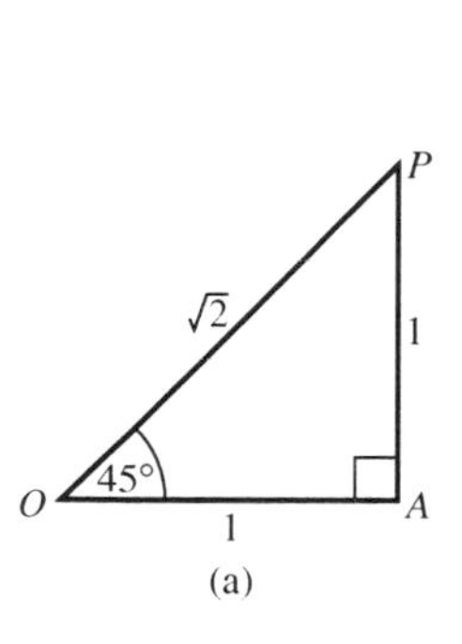

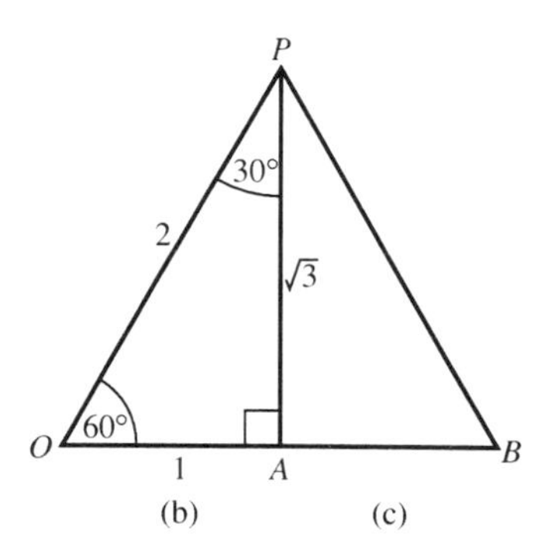

Figure 5.3. Three special angles

In Figure 5.3(b) triangle OPB has sides of length 2 and $OA = AB = 1$. By Pythagoras' theorem, $AP = \sqrt{3}$ and so

$$\sin 30^\circ = \frac{1}{2}, \qquad \cos 30^\circ = \frac{\sqrt{3}}{2}, \qquad \tan 30^\circ = \frac{1}{\sqrt{3}}$$
$$\sin 60^\circ = \frac{\sqrt{3}}{2}, \qquad \cos 60^\circ = \frac{1}{2}, \qquad \tan 60^\circ = \sqrt{3} \tag{5.4}$$

5.2 What is
(a) the value in degrees of 1 radian?
(b) the value in radians of 1°?

Solution (a) 1 radian $= (180/\pi)^\circ \simeq 57.3^\circ$
(b) $1^\circ = (\pi/180)$ radians $\simeq 0.0175$ radians

5.3 Express the following angles in radians:
(a) 90° (b) 270° (c) 360° (d) 30° (e) 45° (f) 60°

Solution (a) $\frac{\pi}{2}$ (b) $\frac{3\pi}{2}$ (c) 2π (d) $\frac{\pi}{6}$ (e) $\frac{\pi}{4}$ (f) $\frac{\pi}{3}$

5.4 Express the following angles in degrees:

(a) $\frac{2\pi}{3}$ radians (b) $\frac{3\pi}{4}$ radians (c) $\frac{5\pi}{4}$ radians (d) 3π radians

Solution (a) 120° (b) 135° (c) 225° (d) 540°

Defining the circular functions

The sine, cosine and tangent functions can be defined on the domain of all real numbers by reference to the rotation of a point around a circle of radius 1. This gives rise to the alternative name of **circular functions**. Figure 5.4 shows the graph of the height AP as P moves anticlockwise around the circle centred at O; P_0 is the initial position of P.

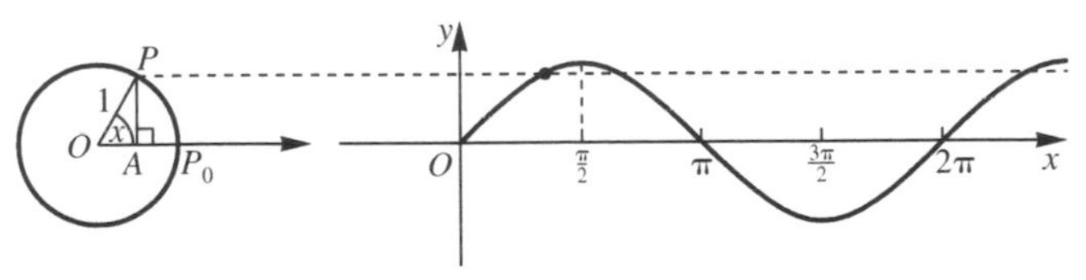

Figure 5.4. Generation of the sine of an angle

When $\angle AOP = x$ then the height $AP = \sin x$, from Figure 5.1 with θ replaced by x.

After one complete rotation P is back at P_0 and the cycle repeats so that we generate the graph of $y = \sin x$ for all positive values of x. If we allow P to rotate clockwise from P_0 then we can generate the graph for negative values of x since angles measured clockwise are designated negative.

The graph of $y = \sin x$ is shown in Figure 5.5. Note that the period of the sine function is 2π and the range is $[-1,1]$.

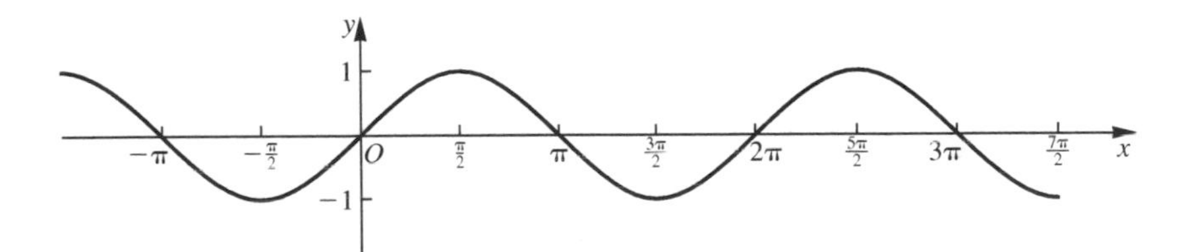

Figure 5.5. The sine of any angle

If P rotates with constant angular frequency ω radians/second then $x = \omega t$. We can describe the position of P relative to OP_0 by the relationship $y = \sin \omega t$. If the circle has radius a then the height AP is described by $y = a \sin \omega t$. The range of this function of t is $[-a,a]$ and we say that the **amplitude** of the oscillation is a and the **period** is $T = 2\pi/\omega$.

The quantity $1/T = \omega/2\pi$ is the **frequency** of the oscillation and gives an indication of how quickly the oscillations occur; the higher the frequency is, the more rapid is the oscillation. The quantity ω is known as the **angular frequency**.

Worked example

5.5 By reference to the rotating point P of Figure 5.4, sketch the graphs of $y = \cos x$ and $y = \tan x$.

Solution Since $OA = \cos x$, the values of OA are plotted against x. We have turned the circle diagram through 90° anticlockwise to allow the plot to be more easily produced: see Figure 5.6.

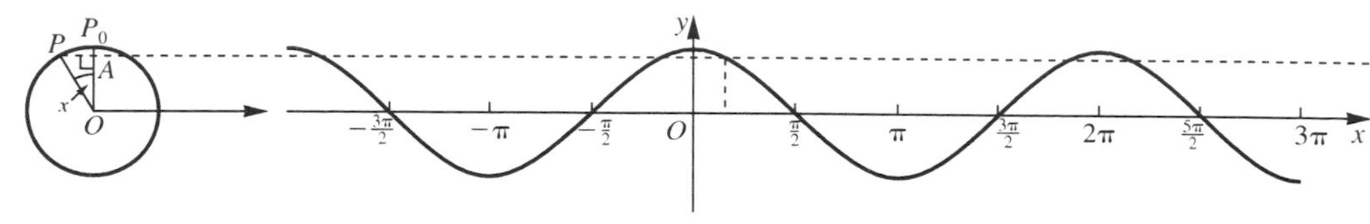

Figure 5.6. The cosine of any angle

Note that this graph is identical to that of $y = \sin x$ moved $\pi/2$ to the left. Hence

KEY POINT

$$\cos x = \sin\left(x + \frac{\pi}{2}\right) \quad \text{or} \quad \sin x = \cos\left(x - \frac{\pi}{2}\right) \qquad 5.5$$

To obtain the graph of the tangent function the tangent to the circle at P_0 is drawn. The radius OP is produced to meet this tangent at M and the distance P_0M is the value of $\tan x$. Note that for some angles it is necessary to produce PO to meet the tangent.

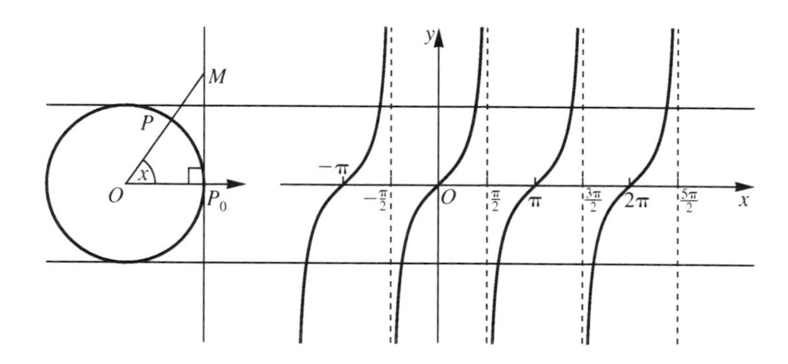

Figure 5.7. The tangent of any angle

Note that the period of $\tan x$ is π. Note also that $f(x) = \cos x$ is an **even function**, that is, $\cos(-x) = \cos x$ and that the sine and tangent functions are **odd functions**, that is, $\sin(-x) = -\sin x$ and $\tan(-x) = -\tan x$.

Finding the trigonometric ratios of any angle

Nowadays, you can find the trigonometric ratios of any angle by pressing appropriate buttons on your calculator. However, the values for certain special angles can be expressed exactly in ways different from the value on the calculator display; for example, a calculator would display $\cos(\pi/4)$ as 0.707 107 (6 d.p.), whereas the exact value is $1/\sqrt{2}$. In some cases it may be easer to work with this latter value.

To find the values of the sine, cosine and tangent functions for any angle we use the CAST rule, the letters representing the ratio which is positive in the appropriate quadrant, so that C refers to cosine, and so on, and A represents All ratios. Refer to Figure 5.8(a).

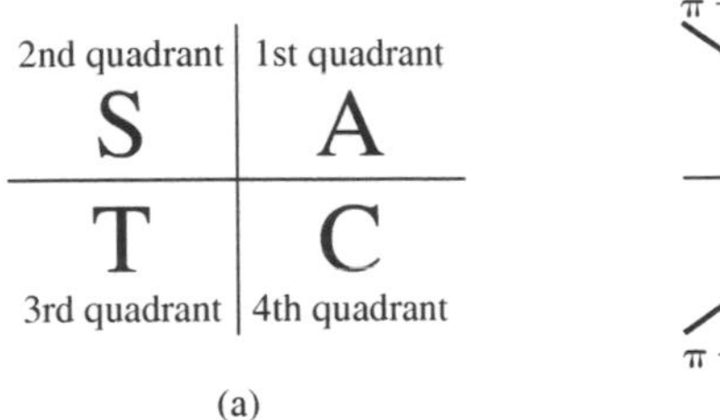

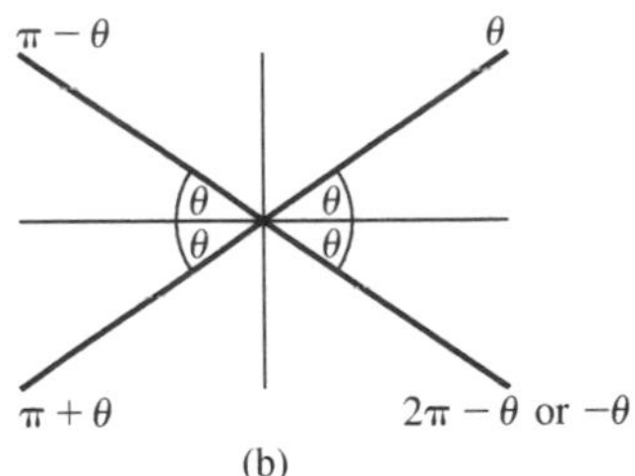

Figure 5.8. Finding trigonometric functions of any angle

For example, to find $\sin(5\pi/4)$ we first note that $5\pi/4$ is in the third quadrant (T) where tangents alone are positive and hence sines are negative: see diagram (a). Next we write $5\pi/4 = \pi + (\pi/4)$. Figure 5.8(b) is an *aide-mémoire* that the magnitudes of the sines of the angles θ, $\pi - \theta$, $\pi + \theta$, $2\pi - \theta$, where θ is an acute angle, are equal; similarly for cosines and tangents. Therefore we can write

$$\sin\frac{5\pi}{4} = \sin\left(\pi + \frac{\pi}{4}\right) = -\sin\frac{\pi}{4} = -\frac{1}{\sqrt{2}}$$

Worked example

5.6 Find the following in terms of the ratios of acute angles:

(a) $\tan\dfrac{7\pi}{4}$ (b) $\sin\dfrac{5\pi}{6}$ (c) $\cos\dfrac{9\pi}{5}$

Solution

(a) $7\pi/4 = 2\pi - (\pi/4)$; this is in the fourth quadrant where tangents are negative. Hence $\tan(7\pi/4) = -\tan(\pi/4)$.

(b) $5\pi/6 = \pi - (\pi/6)$; this is in the second quadrant where cosines are negative. Hence $\cos(5\pi/6) = -\cos(\pi/6)$.

(c) $9\pi/5 = 2\pi - (\pi/5)$, which is in the fourth quadrant where cosines are positive. Hence $\cos(9\pi/5) = \cos(\pi/5)$.

Self-assessment questions 5.1

1. Explain the relationship between degrees and radians.
2. Explain the terms amplitude, period, frequency and angular frequency.

Exercise 5.1

1. Find exact values for the following:

   (a) $\sin\frac{2\pi}{3}$ (b) $\cos\frac{4\pi}{3}$

   (c) $\tan\frac{3\pi}{4}$ (d) $\sin\left(-\frac{3\pi}{4}\right)$

   (e) $\cos\left(-\frac{2\pi}{3}\right)$ (f) $\tan\left(-\frac{4\pi}{3}\right)$

2. The length of arc of a circle of radius r which subtends an angle θ radians at the centre of the circle is $r\theta$. The area of the sector formed by this arc and radii to either end of the arc is $\frac{1}{2}r^2\theta$. Find the angle θ if the magnitude of the arc length is 10π cm and the area is $0.740\pi\ \text{cm}^2$.

5.2 Related circular functions

In this section we examine functions which are related to the simple circular functions of Section 5.1.

Transformations

KEY POINT

The function $y = a\sin(\omega t + \phi)$ represents an oscillation where y is the displacement of some quantity and t is time elapsed. a is the amplitude of the oscillation, ω is the angular frequency of the oscillation, and ϕ is the phase angle.

To obtain the graph of this function from that of $y = \sin t$ we first compress the graphs horizontally by ω to obtain $y = \sin\omega t$, stretch it vertically by a to obtain $y = a\sin\omega t$ and then we move it ϕ/ω to the left if $\phi > 0$ or to the right if $\phi < 0$ to obtain

$$y = a\sin\omega\left(t + \frac{\phi}{\omega}\right) = a\sin(\omega t + \phi)$$

Worked examples

5.7 Sketch on the same axes the graphs of the following:

(a) $y = \sin t,\quad y = 3\sin t,\quad y = \frac{1}{2}\sin t$

(b) $y = \sin t,\quad y = \sin 3t,\quad y = \sin\frac{1}{2}t$

Then sketch the graph of $y = 2\sin(3t + (\pi/2))$.

Solution (a) Figure 5.9 shows the graphs. Note that they have the same period, 2π, but different amplitudes.

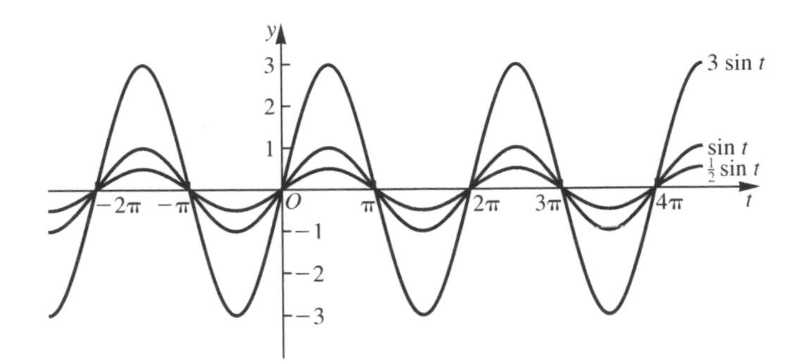

Figure 5.9.
Varying the amplitude

(b) Figure 5.10 shows the graphs; note that the amplitudes are the same, namely 1, but that the frequency and period vary.

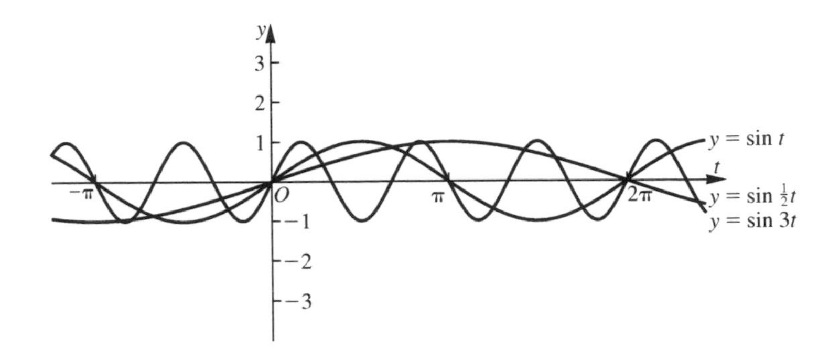

Figure 5.10.
Varying the frequency

Finally, Figure 5.11(a) shows the graph of $y = 2\sin 3t$ and Figure 5.11(b) shows the graph of $y = 2\sin 3(t + (\pi/6)) = 2\sin(3t + (\pi/2))$.

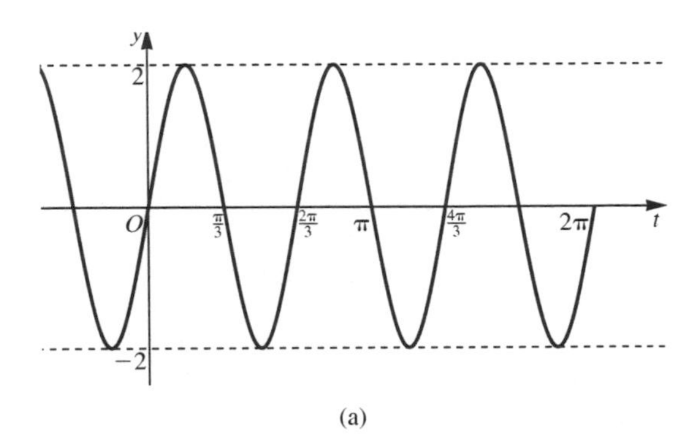

(a)

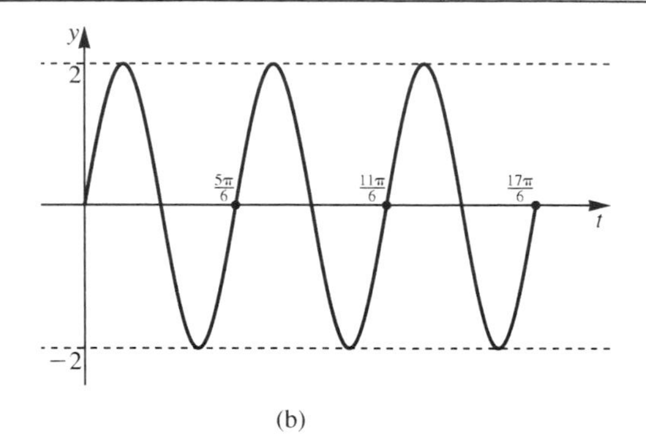

Figure 5.11. Transforming a sine function

5.8 Find the amplitude, angular frequency, period, frequency and phase angle of the following:

(a) $3\sin 2t$ (b) $\frac{1}{2}\cos(3t+\pi)$

Solution (a) Amplitude 3, angular frequency 2, period $(2\pi)/2 = \pi$, frequency $= 1/\pi$, phase angle $= 0$.

(b) Amplitude $= \frac{1}{2}$, angular frequency $= 3$, period $= 2\pi/3$, frequency $= 3/(2\pi)$, phase angle π.

Reciprocal functions

Example 5.9 deals with the functions which are obtained by taking the reciprocals of the sine, cosine and tangent functions.

Worked examples

5.9 Sketch the graphs of the following:

(a) $y = \operatorname{cosec} x \equiv \dfrac{1}{\sin x}$

(b) $y = \sec x \equiv \dfrac{1}{\cos x}$

(c) $y = \cot x \equiv \dfrac{1}{\tan x}$

Solution The graphs are sketched in Figure 5.12. Note that sec like cos is an even function but that cosec and cot are odd. Both cosec and sec have a period of 2π, but cot has a period of π.

The full names for these functions are cosecant, secant and cotangent.

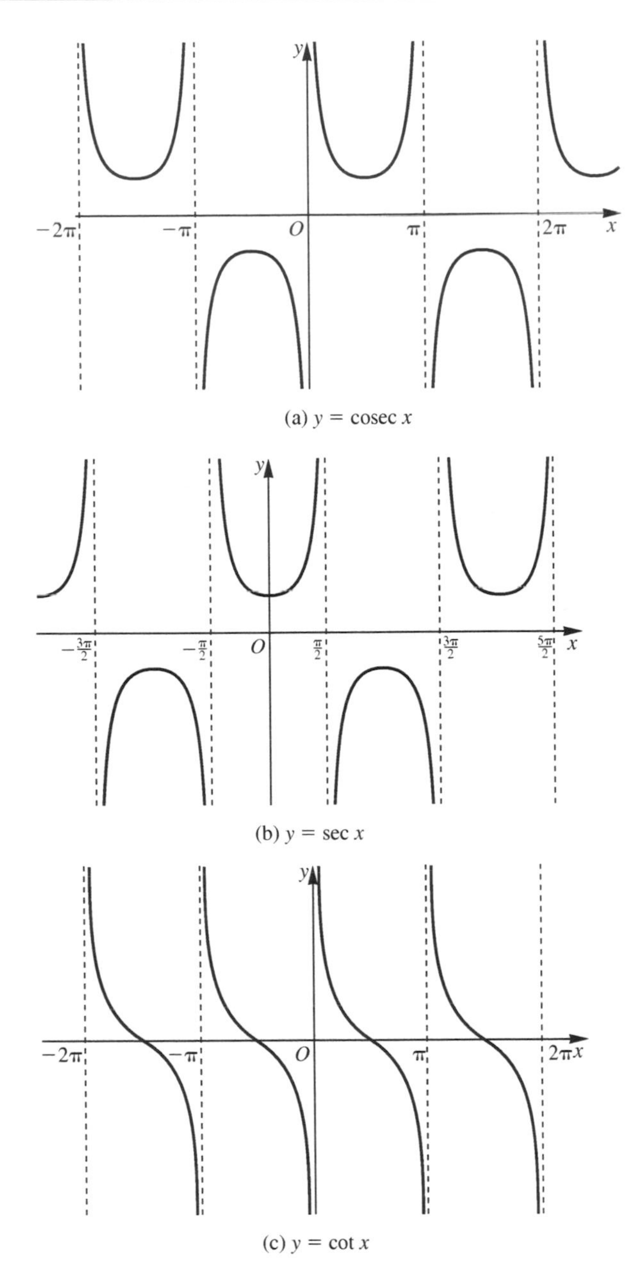

Figure 5.12.
Graphs of cosec, sec and cot

Inverse circular functions

Since the three functions, sine, cosine and tangent, are periodic, inverting them raises difficulties. Figure 5.13 shows the reflection of the graph of $y = \sin x$ in the line $y = x$; it is clearly *not* the graph of a function. Yet if we use the INV and SIN keys on a calculator we get a consistent answer. Since sine takes values between 1 and -1 we take that portion of its graph which goes through the origin and takes all these values once only, that is, the portion for $x = -\pi/2$ to $x = \pi/2$ and reflect *that* in the line $y = x$; see Figure 5.13(b).

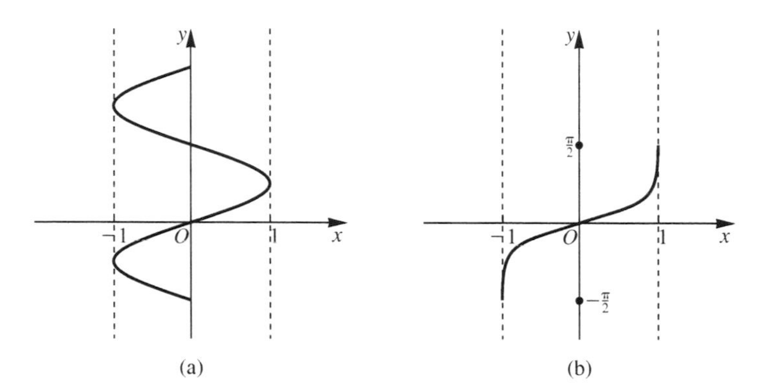

Figure 5.13.
Inverse of the sine function

We write the inverse function as $y = \sin^{-1} x$, or sometimes $y = \arcsin x$. Note that its domain is $-1 \leqslant x \leqslant 1$ and its range is $-\pi/2 \leqslant y \leqslant \pi/2$. Note also that if $y = \sin^{-1} x$ then $x = \sin y$.

Worked example

5.10 Sketch the graphs of (a) $y = \cos^{-1} x$, (b) $\tan^{-1} x$. In each case state the domain and the range.

Solution

(a) The cosine function takes the full range of values $[-1,1]$ once only in the interval $0 \leqslant x \leqslant \pi$. Hence the domain of the inverse function is $-1 \leqslant x \leqslant 1$ and the range is $0 \leqslant y \leqslant \pi$. Its graph is shown in Figure 5.14(a).

(b) The tangent function achieves all the values in its range once only for $-\pi/2 < x < \pi/2$. Hence the inverse function has domain all real values and range $-\pi/2 < y < \pi/2$. The graph is shown in Figure 5.14(b).

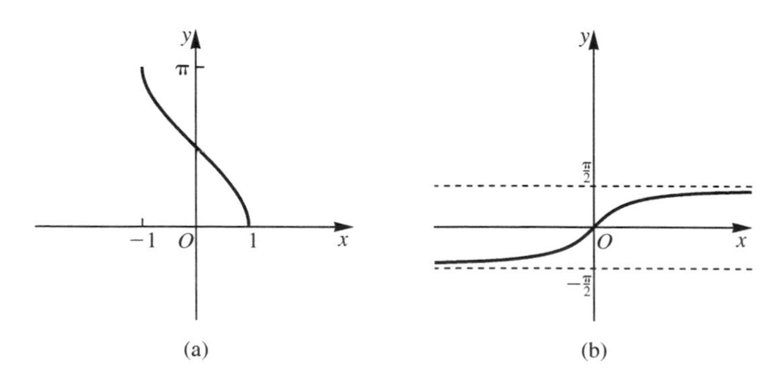

Figure 5.14.
Inverse of the cosine and tangent functions

Self-assessment questions 5.2

1. Explain the term phase angle in relation to oscillations.
2. State the domains of the inverse circular functions.

Exercise 5.2

1. Find the amplitude, angular frequency, period, frequency and phase angle of the following:

   (a) $5\sin\left(3t+\frac{\pi}{2}\right)$

   (b) $\sin\left(\frac{1}{2}t+\pi\right)$

   (c) $2\cos\left(5t-\frac{\pi}{6}\right)$

   (d) $3\cos\left(\frac{1}{3}t+\frac{\pi}{4}\right)$

2. Sketch the graphs of the following:

   (a) $\cos\left(\frac{\pi}{2}-\theta\right)$ (b) $\cos\left(\frac{\pi}{2}+\theta\right)$

   (c) $\cos(\pi+\theta)$

3. Calculate the following values:

   (a) $\sin^{-1}(\frac{1}{2})$ (b) $\cos^{-1}(\frac{1}{2})$

   (c) $\tan^{-1}2$ (d) $\sin^{-1}2$

   (e) $\sin^{-1}(-\frac{1}{2})$ (f) $\cos^{-1}(-\frac{1}{2})$

   (g) $\tan^{-1}(-2)$

4. An oscillation has an amplitude of 4, a period of 4 and a phase angle of $\pi/6$. Write down a possible formula for the oscillation.

5. Find exact values for the following:

   (a) $\sec\frac{\pi}{6}$

   (b) $\sec\frac{7\pi}{6}$

   (c) $\operatorname{cosec}\frac{2\pi}{3}$

   (d) $\operatorname{cosec}\frac{15\pi}{4}$

   (e) $\cot\frac{15\pi}{4}$

   (f) $\cot\frac{13\pi}{6}$

The fully rectified sine wave is effectively the modulus of the sine wave. Plot the graph of $y=|\sin x|$ to see the effect of rectifying the wave.

5.3 Solving trigonometric equations

If we were to look for solutions of the equation $\sin\theta=0.5$ then we could use the INV SIN buttons on a calculator in radian mode. This would give $\theta=0.5236$ (4 d.p.); the exact answer is $\theta=\pi/6$ and this is called the **principal solution**. As Figure 5.15 shows there is a second solution in the interval $[0,2\pi]$; this is $\pi-(\pi/6)=5\pi/6$. Other solutions can be found by adding or subtracting multiples of 2π and hence we generate two sets of solutions: $\pi/6+2n\pi$ and $5\pi/6+2n\pi$ where $n=0, \pm1, \pm2, \ldots$.

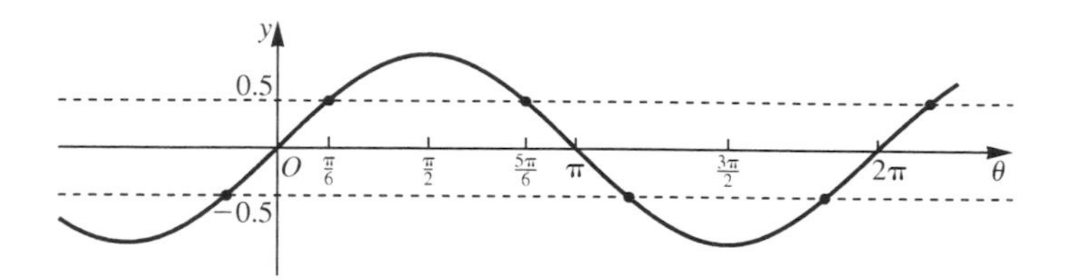

Figure 5.15. Solving the equations $\sin\theta = 0.5$ and $\sin\theta = -0.5$

Worked examples

5.11 Find the solution of the equation $\sin\theta = -0.5$.

Solution A calculator gives the solution $\theta = -0.5236$ (4 d.p.): the exact answer is $\theta = -\pi/6$. The two solutions in the interval $[0,2\pi]$ are found as follows. First, add 2π to get $11\pi/6$; this is the first solution. To find the second solution note that $11\pi/6 = 3\pi/2 + 2\pi/6$; referring to Figure 5.15 we see that the graph is symmetrical about $\theta = 3\pi/2$ and therefore the second solution is $3\pi/2 - 2\pi/6 = 7\pi/6$. We add or subtract multiples of 2π to generate the other solutions. Alternatively, we *could* solve the equation $\sin\theta = 0.5$ to obtain $\theta = \pi/6$ and $5\pi/6$; then using the fact that $\sin\theta$ is an odd function we produce two solutions $\theta = -\pi/6$ and $-5\pi/6$. Adding 2π to each gives $\theta = 11\pi/6$ and $7\pi/6$ and we proceed as before.

5.12 Solve the following equations: (a) $\cos\theta = 0.5$ (b) $\cos\theta = -0.5$

Solution Refer to Figure 5.16.

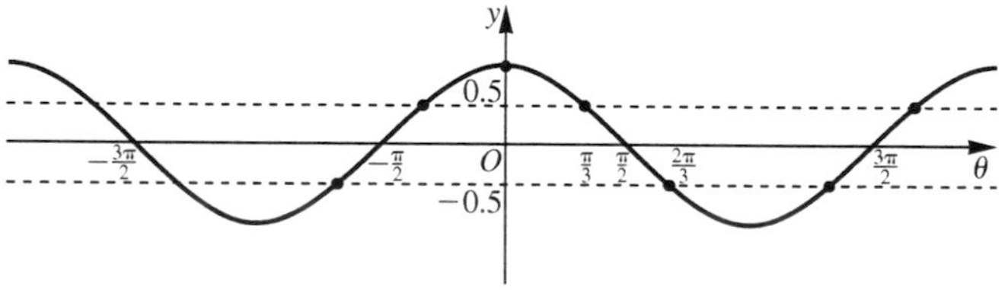

Figure 5.16. Solving the equations $\cos\theta = 0.5$ and $\cos\theta = -0.5$

(a) The 'calculator' solution is its approximation to the exact solution $\theta = \pi/3$. A second solution in $[0,2\pi]$ is

$$\theta = 2\pi - \frac{\pi}{3} = \frac{5\pi}{3}.$$

All solutions are found from $\pi/3 + 2n\pi$ and $5\pi/3 + 2n\pi$, $n = 0, \pm 1, \pm 2, \ldots$.

(b) The 'calculator' solution is its approximation to the exact solution $\theta = 2\pi/3$. A second solution is

$$\theta = 2\pi - \frac{2\pi}{3} = \frac{4\pi}{3}$$

Other solutions are found by adding or subtracting multiples of 2π.

5.13 Solve the following equations: (a) $\tan\theta = 1$ (b) $\tan\theta = -1$

Solution Refer to Figure 5.17.

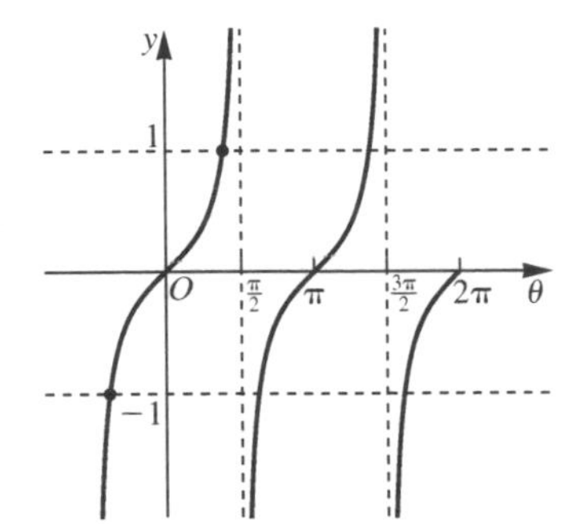

Figure 5.17.
Solving the equations $\tan\theta = 1$ and $\tan\theta = -1$

(a) The 'calculator' solution is its approximation to the exact solution $\theta = \pi/4$. A second solution is found by adding π to give $\theta = 5\pi/4$. All solutions are found from the formula $\pi/4 + n\pi$, $n = 0, \pm 1, \pm 2, \ldots$.

(b) The 'calculator' solution is its approximation to the exact solution $\theta = -\pi/4$. Adding π twice in succession gives the two solutions in $[0,2\pi]$, that is, $\theta = 3\pi/4$ and $7\pi/4$. All solutions can be found from the formula $-\pi/4 + n\pi$, $n = 0, \pm 1, \pm 2, \ldots$.

Self-assessment questions 5.3

1. State how to find the general solution of the equations $\sin\theta = c$, $\cos\theta = c$ and $\tan\theta = c$ where c is a positive constant.
2. Repeat Question 1 when c is a negative constant.

Exercise 5.3

1. Find the solutions in $0 \leqslant \theta \leqslant 2\pi$ of the following equations:

   (a) $\sin\theta = 0.6$ (b) $\sin\theta = -0.6$
   (c) $\cos\theta = 0.8$ (d) $\cos\theta = -0.8$
   (e) $\tan\theta = 3$ (f) $\tan\theta = -3$

2. Find the general solutions of the equations in Question 1.

3. Find, where possible, the general solutions of the following equations:

   (a) $\sec\theta = 2$ (b) $\sec\theta = 0.5$
   (c) $\operatorname{cosec}\theta = 0.6$ (d) $\cot\theta = 0.4$
   (e) $\cot\theta = -0.4$ (f) $\operatorname{cosec}\theta = -0.6$

4. Solve the following equations for $0 \leqslant \theta \leqslant 2\pi$:

   (a) $3\cos^2\theta + 2\cos\theta - 1 = 0$
   (b) $4\cos^2\theta - 5\cos\theta + 1 = 0$

5. Solve the following equations in $0 \leqslant \theta \leqslant 2\pi$:

   (a) $2\cos 3\theta = 1$ (b) $\tan 2\theta = 1$

   (c) $\cos\left(\theta + \dfrac{\pi}{2}\right) = \frac{1}{2}$

   (d) $\cos\left(\theta - \dfrac{\pi}{2}\right) = \frac{1}{2}$

   (*Hint*: in (a) solve for 3θ and then obtain values for θ. The other examples are dealt with in a similar way.)

6. The displacement of an object from equilibrium is given by $x = 3\cos\left(4t + \dfrac{\pi}{2}\right)$. Find the first three times at which the object passes through the equilibrium position $x = 0$.

5.4 Hyperbolic functions

The link to the hyperbola will be made in Section 6.4.

A family of functions related to the exponential function are the **hyperbolic** functions. It is sometimes more convenient to work with them than with exponential functions. The definitions are as follows:

KEY POINT

Hyperbolic cosine $\cosh x = \frac{1}{2}(e^x + e^{-x})$

Hyperbolic sine $\sinh x = \frac{1}{2}(e^x - e^{-x})$ 5.6

Hyperbolic tangent $\tanh x = \frac{\sinh x}{\cosh x}$

The functions are pronounced 'cosh', 'shine' and 'than'.

The properties of these functions can be deduced from their definitions. The graphs are shown in Figure 5.18. Note how different they are from those of the circular functions of similar name.

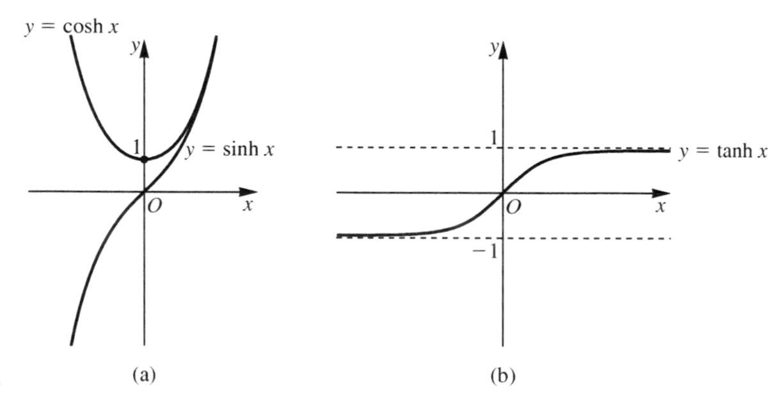

Figure 5.18.
The hyperbolic functions

Worked examples

5.14 Prove the following results:
(a) cosh is an even function but sinh and tanh are odd
(b) $\cosh(0) = 1, \quad \sinh(0) = \tanh(0) = 0$
(c) $\cosh x > \sinh x$ for all x
(d) the domain of the three functions is all real numbers
What is the range of each of the three functions?

Solution (a) $\cosh(-x) = \frac{1}{2}(e^{-x} + e^{+x}) = \cosh x$; cosh is an even function

$\sinh(-x) = \frac{1}{2}(e^{-x} - e^{+x}) = -\sinh x$; sinh is an odd function

$$\tanh(-x) = \frac{\sinh(-x)}{\cosh(-x)} = \frac{-\sinh x}{\cosh x} = -\tanh x \text{ so } \tanh x \text{ is odd}$$

(b) $\cosh(0) = \frac{1}{2}(e^0 + e^0) = \frac{1}{2}(1 + 1) = 1$

$\sinh(0) = \frac{1}{2}(e^0 - e^0) = 0$

$$\tanh(0) = \frac{\sinh(0)}{\cosh(0)} = \frac{0}{1} = 0$$

(c) When $x < 0$, $\sinh x < 0$ and $\cosh x > 0$. As $x \to \infty$ both cosh x and sinh x approach $\frac{1}{2}e^x$. However, $\cosh x > \frac{1}{2}e^x$ and $\sinh x < \frac{1}{2}e^x$ so that $\cosh x > \sinh x$ for all x.

(d) Both e^x and e^{-x} have domain all real numbers so that the combinations $\cosh x$ and $\sinh x$ have the same domain. Also, since cosh x is never zero then tanh x has the domain all real numbers.

Note that the range of sinh is all real numbers, of cosh is all real numbers $\geqslant 1$, and that of tanh is the interval $[-1,1]$.

5.15 Show that

$$\tanh x \equiv \frac{1 - e^{-2x}}{1 + e^{-2x}}$$

and deduce the behaviour of tanh x as $x \to \pm\infty$.

Solution $$\tanh x = \frac{\sinh x}{\cosh x} = \frac{e^x - e^{-x}}{e^x + e^{-x}}$$

Multiplying top and bottom by e^{-x}

$$\tanh x = \frac{1 - e^{-2x}}{1 + e^{-2x}}$$

As $x \to \infty$, $e^{-2x} \to 0$ and tanh $x \to 1$. Since tanh is an odd function, $\tanh x \to -1$ as $x \to -\infty$.

5.16 Solve the equation $5\cosh x + 4\sinh x = 3$.

Solution Replacing cosh x and sinh x by their definitions:

$$\frac{5}{2}(e^x + e^{-x}) + \frac{4}{2}(e^x - e^{-x}) = 3$$

so that

$$\frac{9}{2}e^x + \frac{1}{2}e^{-x} = 3$$

that is

$$9e^{x} - 6 + e^{-x} = 0$$

Multiplying by e^{x} we obtain

$$9(e^{x})^2 - 6e^{x} + 1 = 0$$

that is

$$(3e^{x} - 1)^2 = 0$$

Hence $e^{x} = \frac{1}{3}$ and $x = \ln\frac{1}{3} = -\ln 3$.

(Note that since cosh and sinh are not periodic functions there is not an infinite number of solutions to the equation.)

5.17 (a) By writing the inverse function $y = \sinh^{-1}x$ as $\sinh y = x$ find an expression for $\sinh^{-1}x$ in terms of a natural logarithm.
(b) Similarly, find expressions for $\cosh^{-1}x$ and $\tanh^{-1}x$.
(c) Sketch the graphs of these three inverse functions.

Solution (a) $\sinh y = \dfrac{1}{2}(e^{y} - e^{-y}) = x$

Multiplying by $2e^{y}$ we obtain

$$(e^{y})^2 - 1 = 2xe^{y}$$

so that $(e^{y})^2 - 2xe^{y} - 1 = 0$.
Solving this equation for e^{y} gives

$$e^{y} = \frac{2x \pm \sqrt{4x^2 + 4}}{2}$$

Rejecting the $-\sqrt{\ }$ option since this would give a negative value for e^{y} we have

$$e^{y} = x + \sqrt{x^2 + 1}$$

Taking natural logarithms:

$$y = \sinh^{-1}x = \ln\{x + \sqrt{x^2 + 1}\} \qquad 5.7$$

(b) $\cosh y = \dfrac{1}{2}(e^{y} + e^{-y}) = x$

For cosh to possess an inverse function we need to restrict its domain to $x \geqslant 1$.

Hence $(e^{y})^2 - 2xe^{y} + 1 = 0$. Then

$$e^{y} = x \pm \sqrt{x^2 - 1}$$

giving

$$y = \ln\left\{x \pm \sqrt{x^2 - 1}\right\}$$

From the graph of cosh and its inverse it follows that $\cosh^{-1} x \to \infty$ as $x \to \infty$, so we take the $+\sqrt{\ }$ option. Given that we cannot take the square root of a negative number, the domain of $\cosh^{-1} x$ is restricted to ... $\geqslant 1$.

$$\cosh^{-1} x = \ln\left\{x + \sqrt{x^2 - 1}\right\} \qquad (x \geqslant 1) \tag{5.8}$$

Since

$$\tanh y = \frac{e^y - e^{-y}}{e^y + e^{-y}} = x$$

then

$$e^y(1 - x) = e^{-y}(1 + x)$$

that is

$$e^{2y} = \frac{1 + x}{1 - x}$$

and

$$2y = \ln\left\{\frac{1 + x}{1 - x}\right\}$$

Finally,

KEY POINT

$$\tanh^{-1} x = \frac{1}{2} \ln\left\{\frac{1 + x}{1 - x}\right\}, \qquad -1 < x < 1 \tag{5.9}$$

(This restriction is necessary to ensure that $1 + x$ and $1 - x$ are positive and hence their logarithms can be calculated.)

(c) Figure 5.19 shows the graphs.

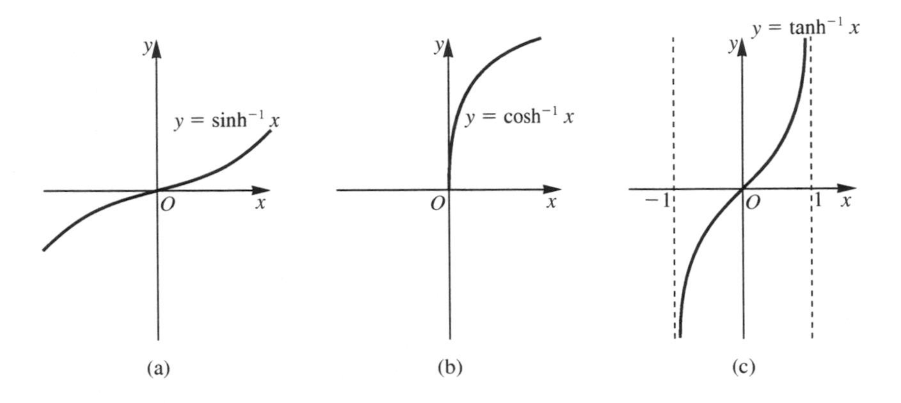

Figure 5.19. Graphs of the inverse hyperbolic functions

Self-assessment questions 5.4

1. Define the functions $f(x) = \cosh x$ and $f(x) = \sinh x$.
2. What restrictions are there on the domains of the inverse hyperbolic functions?

Exercise 5.4

1. Calculate to 3 d.p.

   (a) $\cosh 4$ (b) $\sinh(-0.6)$
   (c) $\tanh 2$

2. Sketch the graphs of the following functions:

   $\operatorname{sech} x \equiv \frac{1}{\cosh x}$,

   $\operatorname{cosech} x \equiv \frac{1}{\sinh x}$,

   $\coth x \equiv \frac{1}{\tanh x}$

3. Calculate

   (a) $\cosh^{-1}(2)$ (b) $\sinh^{-1}(0.8)$
   (c) $\tanh^{-1}(2)$

4. A cable hangs from two points on the same horizontal level a distance 50 m apart. Its profile is described by

   $y = c\cosh\left(\frac{x}{c}\right)$ where c is a constant.

   Sketch the curve representing the profile, showing the position of the axes. Find the sag, that is, the depth of the mid-point of the cable below the level of the points of suspension.

5. Solve the following equations:

   (a) $3\sinh x + 5\cosh x = 4$
   (b) $4\cosh x - 3\sinh x = 4a$

Test and assignment exercises 5

1. Find exact values for the following:

   (a) $\tan\left(\frac{5\pi}{6}\right)$ (b) $\sin\left(-\frac{2\pi}{3}\right)$

   (c) $\cos\left(\frac{3\pi}{2}\right)$ (d) $\tan\left(-\frac{9\pi}{4}\right)$

   (e) $\sin\left(\frac{5\pi}{4}\right)$ (f) $\cos\left(-\frac{2\pi}{3}\right)$

2. Find exact values for the following:

   (a) $\sec\left(\frac{5\pi}{4}\right)$ (b) $\sec\left(-\frac{5\pi}{4}\right)$

   (c) $\operatorname{cosec}\left(\frac{5\pi}{4}\right)$ (d) $\operatorname{cosec}\left(-\frac{5\pi}{4}\right)$

   (e) $\cot\left(\frac{7\pi}{3}\right)$ (f) $\cot\left(-\frac{7\pi}{3}\right)$

3. Find the amplitude, angular frequency, period, frequency and phase angle of the following:

   (a) $4\tan\left(t - \frac{\pi}{6}\right)$

   (b) $\frac{1}{2}\sin\left(4t + \frac{\pi}{4}\right)$

4. Calculate the following:

   (a) $\sin^{-1}\left(\frac{\sqrt{3}}{2}\right)$ (b) $\sin^{-1}\left(-\frac{\sqrt{3}}{2}\right)$

   (c) $\cos^{-1}\left(\frac{\sqrt{3}}{2}\right)$ (d) $\cos^{-1}\left(-\frac{\sqrt{3}}{2}\right)$

   (e) $\tan^{-1}\left(\sqrt{3}\right)$ (f) $\tan^{-1}\left(-\sqrt{3}\right)$

5. Find the general solution of the following equations:

   (a) $\sin\theta = 0.4$
   (b) $\sin\theta = -0.4$
   (c) $\cos\theta = 0.4$
   (d) $\cos\theta = -0.4$
   (e) $\tan\theta = 4$
   (f) $\tan\theta = -4$
   (g) $\sec\theta = \sqrt{3}$
   (h) $\operatorname{cosec}\theta = \sqrt{3}$
   (i) $\cot\theta = \sqrt{3}$

6. Solve the following equations in $0 \leqslant \theta \leqslant 2\pi$:

   (a) $\cos 2\theta = \dfrac{1}{\sqrt{2}}$
   (b) $\cos\left(\theta + \dfrac{\pi}{3}\right) = \dfrac{1}{\sqrt{2}}$
   (c) $\tan 3\theta = 1$

7. Solve the following equations for $0 \leqslant \theta \leqslant 2\pi$:

   (a) $2\cos^2\theta - \cos\theta - 1 = 0$
   (b) $4\sin^2\theta - 7\cos\theta - 3 = 0$

8. Calculate

   (a) $\cosh^{-1}(1.2)$
   (b) $\sinh^{-1}(0.6)$
   (c) $\tanh^{-1}(3)$

9. Solve the following equations:

   (a) $5\sinh x + 3\cosh x = 6$
   (b) $3\cosh x - 4\sinh x = 8$

6 Trigonometric and hyperbolic identities

Objectives

This chapter

- distinguishes between equations and identities
- states and uses identities involving multiple angles
- derives and uses identities based on Pythagoras' theorem
- derives and uses alternative expressions for $a \sin \theta + b \cos \theta$
- solves equations of the form $a \sin \theta + b \cos \theta = c$
- derives and uses hyperbolic identities

6.1 Equations and identities

An equation is a relationship involving one or more variables which is true for certain values of the variables. For example, $x^2 - 5x + 6 = 0$ is an equation with solutions (or roots) $x = 2$ and $x = 3$. On the other hand, $x^2 - 1 = (x - 1)(x + 1)$ is a relationship which is true for all values of x; it is called an **identity**.

To distinguish an identity from an equation we use the **identity sign**, $\equiv$, and write it as $x^2 - 1 \equiv (x - 1)(x + 1)$.

The variable x is a vehicle for expressing the identity. We could equally well write $u^2 - 1 \equiv (u - 1)(u + 1)$ or $v^2 - 1 \equiv (v - 1)(v + 1)$.

The fundamental trigonometric identity is a restatement of Pythagoras' theorem. In triangle ABC of Figure 6.1,

$$(AC)^2 + (CB)^2 = (AB)^2$$

Dividing by $(AB)^2$ we obtain

$$\left(\frac{AC}{AB}\right)^2 + \left(\frac{CB}{AB}\right)^2 = 1$$

that is

$\cos^2 \theta$ is the same as $(\cos \theta)^2$, and so on.

$$\cos^2 \theta + \sin^2 \theta = 1$$

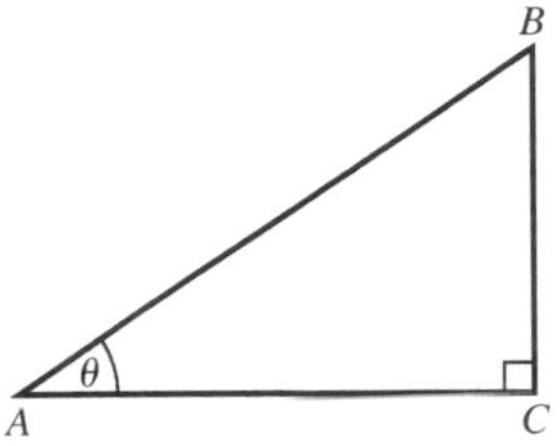

B

A θ C

Figure 6.1.
Deriving the fundamental identity

It can be shown that this relationship is true for *any* angle θ and so we can write the following:

KEY POINT

$$\cos^2\theta + \sin^2\theta \equiv 1 \qquad 6.1$$

Dividing (6.1) by $\cos^2\theta$ we obtain

$$1 + \tan^2\theta \equiv \sec^2\theta \qquad 6.2$$

and dividing (6.1) by $\sin^2\theta$ we obtain

$$\cot^2\theta + 1 \equiv \operatorname{cosec}^2\theta \qquad 6.3$$

We can sometimes use these identities to help solve practical problems, as illustrated in Example 6.1.

Worked examples

6.1 A projectile is launched from level ground at (0,0) with initial velocity u at an angle θ to the horizontal. Subsequently, its position is described by $x = (u\cos\theta)t$, $y = (u\sin\theta)t - \frac{1}{2}gt^2$.

Find the equation of the trajectory of the projectile.

If $\theta = 45°$ find where the projectile hits the ground on its return.

Solution Since $t = x/(u\cos\theta)$ then the equation of the trajectory is

$$y = u\sin\theta\frac{x}{u\cos\theta} - \frac{1}{2}g\frac{x^2}{u^2\cos^2\theta} \equiv x\tan\theta - \frac{1}{2}g\frac{x^2}{u^2}\sec^2\theta$$

If $\theta = 45°$, $\tan\theta = 1$ and $\sec^2\theta \equiv 1 + \tan^2\theta = 2$.
When $y = 0$, $x(1 - (gx/u^2)) = 0$. The required distance is found from $1 - (gx/u^2) = 0$, that is, $x = u^2/g$.

6.2 Simplify the following expressions:

(a) $\sec\theta - \sin\theta\tan\theta$ (b) $\tan\theta + \cot\theta$ (c) $\dfrac{1-\tan^2\theta}{1+\tan^2\theta}$

Solution (a) $$\sec\theta - \sin\theta\tan\theta \equiv \frac{1}{\cos\theta} - \frac{\sin^2\theta}{\cos\theta} \equiv \frac{1-\sin^2\theta}{\cos\theta} \equiv \frac{\cos^2\theta}{\cos\theta} \equiv \cos\theta$$

(b) $$\tan\theta + \cot\theta \equiv \frac{\sin\theta}{\cos\theta} + \frac{\cos\theta}{\sin\theta} \equiv \frac{\sin^2\theta + \cos^2\theta}{\cos\theta\sin\theta} \equiv \frac{1}{\cos\theta\sin\theta}$$

(c) $$\frac{1-\tan^2\theta}{1+\tan^2\theta} \equiv \frac{\cos^2\theta(1-\tan^2\theta)}{\cos^2\theta(1+\tan^2\theta)} \equiv \frac{\cos^2\theta - \sin^2\theta}{\cos^2\theta + \sin^2\theta} \equiv \cos^2\theta - \sin^2\theta$$

6.3 Prove the following identities:

(a) $\dfrac{\sin\theta}{1-\cos\theta} \equiv \dfrac{1+\cos\theta}{\sin\theta}$ (b) $\dfrac{\sin\theta}{1+\cos\theta} + \dfrac{1+\cos\theta}{\sin\theta} \equiv \dfrac{2}{\sin\theta}$

Solution (a) $$\text{LHS} \equiv \frac{\sin\theta}{1-\cos\theta}\left(\frac{1+\cos\theta}{1+\cos\theta}\right) \equiv \frac{\sin\theta(1+\cos\theta)}{1-\cos^2\theta} \equiv \frac{\sin\theta(1+\cos\theta)}{\sin^2\theta} \equiv \frac{1+\cos\theta}{\sin\theta} \equiv \text{RHS}$$

(b) $$\text{LHS} \equiv \frac{\sin^2\theta + (1+\cos\theta)^2}{(1+\cos\theta)\sin\theta} \equiv \frac{\sin^2\theta + 1 + 2\cos\theta + \cos^2\theta}{(1+\cos\theta)\sin\theta} \equiv \frac{2+2\cos\theta}{(1+\cos\theta)\sin\theta} \equiv \frac{2}{\sin\theta} \equiv \text{RHS}$$

6.4 Find all the angles θ in $0 \leqslant \theta \leqslant 2\pi$ which satisfy the following equations:
(a) $2\cos^2\theta = 1 + \sin\theta$ (b) $4\sec^2\theta = 3\tan\theta + 5$

Solution (a) The equation can be written

$$2 - 2\sin^2\theta = 1 + \sin\theta$$

or

$$2\sin^2\theta + \sin\theta - 1 = 0$$

that is

$$(2\sin\theta - 1)(\sin\theta + 1) = 0$$

Hence $\sin\theta = \frac{1}{2}$ or -1.

The first of these equations has solutions $\theta = \pi/6$ and $5\pi/6$ in the interval stated and the second equation has solution $\theta = 3\pi/2$.

(b) The equation can be written:

$$4 + 4\tan^2\theta = 3\tan\theta + 5$$

that is:

$$4\tan^2\theta - 3\tan\theta - 1 = 0$$

or:

$$(4\tan\theta + 1)(\tan\theta - 1) = 0$$

The equation $\tan\theta = 1$ has solutions $\theta = \pi/4$ and $\theta = 5\pi/4$. The equation $\tan\theta = -1/4$ has a 'calculator' solution -0.245 radians. The solutions in the given interval are 2.897 radians and 6.038 radians, found by adding π and 2π respectively.

Self-assessment questions 6.1

1. Distinguish between an equation and an identity.
2. State the three trigonometric identities based on Pythagoras' theorem.

Exercise 6.1

1. Simplify the following expressions:

(a) $\sec\theta - \cos\theta$ (b) $\dfrac{1 - \operatorname{cosec}^2\theta}{1 - \sec^2\theta}$

(c) $\dfrac{1 + \cot^2\theta}{\sin\theta}$

2. Eliminate θ from the following pairs of equations:

(a) $x = 3\tan\theta, y = 3\sec\theta$
(b) $x = 1 + \cos\theta, y = 1 - \sin\theta$
(c) $x = 2\sin\theta, y = 3\sec\theta$

3. Prove the following identities:

(a) $\dfrac{\sin\theta}{1 - \cot\theta} + \dfrac{\cos\theta}{1 - \tan\theta} \equiv \cos\theta + \sin\theta$

(b) $(1 + \sec\theta)(1 - \cos\theta) \equiv \sin\theta\tan\theta$

(c) $\dfrac{\operatorname{cosec}\theta - \sin\theta}{\operatorname{cosec}\theta} \equiv \cos^2\theta$

4. Solve for $0 \leqslant \theta \leqslant 2\pi$ the following equations:

(a) $\sin\theta = \cot\theta$
(b) $4\sin^2\theta + 2.5 = 5\cos\theta$
(c) $3\cot^2\theta + \operatorname{cosec}\theta + 1 = 0$

6.2 Identities involving multiple angles

The following identities, known as the compound angle formulae, are commonly used, for example to find the resultant when two waveforms meet. They are valid for any angles A and B.

KEY POINT

(a) $\sin(A+B) \equiv \sin A \cos B + \cos A \sin B$ 6.3a

(b) $\sin(A-B) \equiv \sin A \cos B - \cos A \sin B$ 6.3b

(c) $\cos(A+B) \equiv \cos A \cos B - \sin A \sin B$ 6.3c

(d) $\cos(A-B) \equiv \cos A \cos B + \sin A \sin B$ 6.3d

By adding formulae (6.3a) and (6.3b) we obtain the identity

$$\sin(A+B) + \sin(A-B) \equiv 2 \sin A \cos B \qquad 6.4a$$

Similar operations give the results below:

$$\sin(A+B) - \sin(A-B) \equiv 2 \cos A \sin B \qquad 6.4b$$

$$\cos(A+B) + \cos(A-B) \equiv 2 \cos A \cos B \qquad 6.4c$$

$$\cos(A+B) - \cos(A-B) \equiv -2 \sin A \sin B \qquad 6.4d$$

The last four identities can be turned round and are most useful if we replace $A+B$ by C and $A-B$ by D; then $A = \frac{1}{2}(C+D)$ and $B = \frac{1}{2}(C-D)$. Hence we obtain the following factor formulae:

KEY POINT

$$\sin C + \sin D \equiv 2 \sin\left(\frac{C+D}{2}\right)\cos\left(\frac{C-D}{2}\right) \qquad 6.5a$$

$$\sin C - \sin D \equiv 2 \cos\left(\frac{C+D}{2}\right)\sin\left(\frac{C-D}{2}\right) \qquad 6.5b$$

$$\cos C + \cos D \equiv 2 \cos\left(\frac{C+D}{2}\right)\cos\left(\frac{C-D}{2}\right) \qquad 6.5c$$

$$\cos C - \cos D \equiv -2 \sin\left(\frac{C+D}{2}\right)\sin\left(\frac{C-D}{2}\right) \qquad 6.5d$$

A very important set of results, known as the double angle formulae, are obtained from (6.3a) and (6.3c) by putting $B = A$.

KEY POINT

$\sin 2A \equiv 2 \sin A \cos A$	6.6a
$\cos 2A \equiv \cos^2 A - \sin^2 A$	6.6b
$\cos 2A \equiv 2\cos^2 A - 1$	6.6c
$\cos 2A \equiv 1 - 2\sin^2 A$	6.6d

Results (6.6c) and (6.6d) were obtained from (6.6b) by using the identity $\cos^2 A + \sin^2 A \equiv 1$.

Worked examples

6.5 Simplify the following formulae:

(a) $\sin\left(\theta + \frac{\pi}{2}\right)$ (b) $\sin\left(\frac{\pi}{2} - \theta\right)$ (c) $\cos(\theta + \pi)$

(d) $\cos\left(\frac{\pi}{2} - \theta\right)$ (e) $1 + \cos 2A$ (f) $1 - \cos 2A$

Solution (a) From (6.3a) with A replaced by θ and B by $\pi/2$,

$$\sin\left(\theta + \frac{\pi}{2}\right) \equiv \sin\theta \cos\frac{\pi}{2} + \cos\theta \sin\frac{\pi}{2}$$

$$\equiv \sin\theta.0 + \cos\theta.1 \equiv \cos\theta$$

(b) $\sin\left(\frac{\pi}{2} - \theta\right) \equiv \sin\frac{\pi}{2}\cos\theta - \cos\frac{\pi}{2}\sin\theta \equiv \cos\theta$

(c) $\cos(\theta + \pi) \equiv \cos(\pi + \theta) \equiv \cos\pi\cos\theta - \sin\pi\sin\theta \equiv -\cos\theta$

(d) $\cos\left(\frac{\pi}{2} - \theta\right) \equiv \cos\frac{\pi}{2}\cos\theta + \sin\frac{\pi}{2}\sin\theta \equiv \sin\theta$

(e) $1 + \cos 2A \equiv 2\cos^2 A$

(f) $1 - \cos 2A \equiv 2\sin^2 A$

6.6 Show that

$$\tan(A \pm B) = \frac{\tan A \pm \tan B}{1 \mp \tan A \tan B}$$

Deduce a formula for $\tan 2A$ in terms of $\tan A$.

Solution $$\tan(A+B) \equiv \frac{\sin(A+B)}{\cos(A+B)} \equiv \frac{\sin A\cos B + \cos A\sin B}{\cos A\cos B - \sin A\sin B}$$

Dividing top and bottom by $\cos A \cos B$ we obtain

$$\tan(A+B) \equiv \frac{\tan A + \tan B}{1 - \tan A\tan B}$$

Replacing B by $(-B)$ and remembering that $\tan(-B) = -\tan B$ gives

$$\tan(A-B) \equiv \frac{\tan A - \tan B}{1 + \tan A\tan B}$$

If $B = A$ then the first identity becomes

$$\tan 2A \equiv \frac{2\tan A}{1 - \tan^2 A}$$

6.7 Two waves, $x_1 = a\sin\omega_1 t$ and $x_2 = a\sin\omega_2 t$, meet. Find the equation for the resultant wave, showing a time-varying amplitude.

If $\omega_1 = 490\,\text{Hz}$ and $\omega_2 = 510\,\text{Hz}$, find the period of the slower resultant oscillation and sketch the resultant waveform.

Solution The waves have equal amplitude. Then

$$x_1 + x_2 = a(\sin\omega_1 t + \sin\omega_2 t)$$

$$\equiv 2a\sin\left[\left(\frac{\omega_1+\omega_2}{2}\right)t\right]\cos\left[\left(\frac{\omega_1-\omega_2}{2}\right)t\right] \qquad \text{(By (6.5a))}$$

With ω_1 and ω_2 as stated,

$$x_1 + x_2 = 2a\sin(500t)\cos 10t$$

The period of the slower oscillation is $2\pi/10$.

Figure 6.2 shows the waveform schematically. Note that the amplitude grows to double the original value and drops to zero and then the cycle repeats. This phenomenon is known as 'beats'.

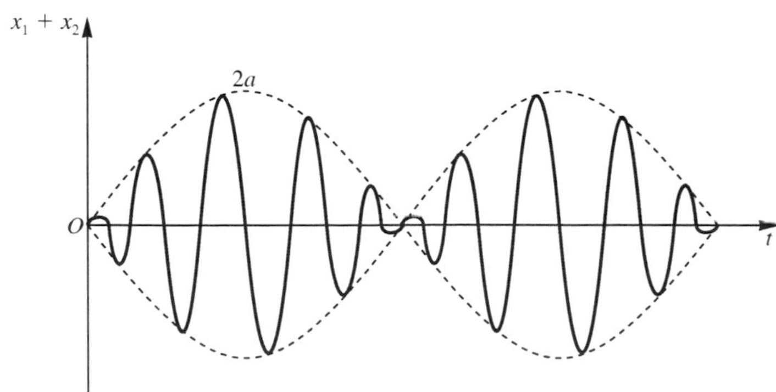

Figure 6.2.
The phenomenon of beats

6.8 Two water waves, $x_1 = a \sin \omega t$ and $x_2 = a \sin(\omega t + \phi)$, meet. Find the equation for the resultant wave and compare its characteristics with those of the original waves. Discuss the following special cases:

(a) $\phi = 0$ (b) $\phi = \frac{\pi}{2}$ (c) $\phi = \pi$

Solution

$$x_1 + x_2 = a[\sin \omega t + \sin(\omega t + \phi)]$$

$$= 2a \sin\left(\omega t + \frac{\phi}{2}\right) \cos \frac{\phi}{2} \qquad \text{(By (6.5a))}$$

Strictly, $x_1 + x_2 = 2a \sin (\omega t + \phi\phi 2) \cos(-\phi/2)$ but $\cos(-\phi/2) \equiv \cos(\phi/2)$.

The period is the same as that of the original waves. The amplitude is $2a \cos(\phi/2)$ and the phase is $\phi/2$.

(a) $x_1 + x_2 = 2a \sin \omega t$. The amplitude is double. The two waves are the same and have reinforced each other.

(b) $x_1 + x_2 = 2a \frac{1}{\sqrt{2}} \sin\left(\omega t + \frac{\pi}{4}\right)$. The amplitude is multiplied by $\sqrt{2}$ and the resultant wave is $\pi/4$ out of phase with the first wave.

(c) $x_1 + x_2 = 0$ $\left(\text{since } \cos \frac{\pi}{2} = 0\right)$. The waves, π out of phase with each other, have completely cancelled each other out.

6.9 Find the solutions in $0 \leqslant \theta \leqslant 2\pi$ of the equation $\sin \theta + \cos 2\theta = 0$.

Solution Using the identity $\cos 2\theta \equiv 1 - 2 \sin^2 \theta$ we can obtain an equation with $\sin \theta$ as the variable. This is

$$\sin \theta + 1 - 2 \sin^2 \theta = 0$$

or

$$1 + \sin \theta - 2 \sin^2 \theta = 0$$

that is

$$(1 + 2 \sin \theta)(1 - \sin \theta) = 0$$

Therefore $\sin \theta = -\frac{1}{2}$ or 1. If $\sin \theta = -\frac{1}{2}$ then $\theta = 7\pi/6$ or $11\pi/6$ and if $\sin \theta = 1$ then $\theta = \pi/2$. The relevant solutions are therefore $\theta = 7\pi/6$, $11\pi/6$ and $\pi/2$.

Self-assessment questions 6.2

1. Write down formulae for $\sin(A + B)$ and $\cos(A + B)$.
2. Find expressions for $\sin 2\theta$ and $\cos 2\theta$

Exercise 6.2

1. If $\cos A = \frac{24}{25}$ and $\cos B = \frac{3}{5}$, find $\cos(A+B)$ and $\cos(A-B)$ given that both A and B are acute angles.

2. Simplify the following expressions:

   (a) $\sin 2\theta \cos\theta + \sin\theta \cos 2\theta$

   (b) $\dfrac{\sin 2\theta}{1+\cos 2\theta}$

   (c) $\dfrac{2\tan\theta}{1+\tan^2\theta}$

3. Solve the following equations for $0 \leqslant \theta \leqslant 2\pi$:

   (a) $\cos 2\theta + \cos 4\theta = 0$
   (b) $\sin 3\theta - \sin\theta = 0$
   (c) $\sin 3\theta - \sin\theta = \cos 2\theta$
   (d) $\cos\theta + \cos 5\theta = \sin\theta + \sin 5\theta$

4. Prove the following identities:

   (a) $\tan\theta + \cot\theta \equiv 2\,\mathrm{cosec}\, 2\theta$
   (b) $\cos 4\theta \equiv 1 - 8\cos^2\theta + 8\cos^4\theta$

5. If $t = \tan(\theta/2)$ find formulae in terms of t for the following:

   (a) $\tan\theta$
   (b) $\sin\theta$
   (c) $\cos\theta$

6. Find the solutions in $0 \leqslant \theta \leqslant 2\pi$ of the following equations:

   (a) $\cos 2\theta = \sin\theta$

   (b) $\sin\left(\theta + \dfrac{\pi}{3}\right) = \sin\theta$

   (c) $\sin 2\theta = 1 + \cos 2\theta$

6.3 The combination $a\sin\theta + b\cos\theta$

An important identity is used to combine the functions $a\sin\omega t$ and $b\cos\omega t$ into a simple term of the form $R\sin(\omega t + \phi)$, which is useful when two waveforms of different amplitude meet. For the moment we replace ωt by θ. Now

$$R\sin(\theta+\phi) \equiv R\sin\theta\cos\phi + R\cos\theta\sin\phi$$

If this is compared with $a\sin\theta + b\cos\theta$ we see that $R\cos\phi = a$ and $R\sin\phi = b$. Squaring these equations and adding we obtain

Remember that $\cos^2\phi + \sin^2\phi = 1$.

$$R^2\cos^2\phi + R^2\sin^2\phi = a^2 + b^2$$

that is

$$R^2 = a^2 + b^2$$

Dividing $R\sin\phi = b$ by $R\cos\phi = a$ we obtain $\tan\phi = b/a$.

Refer to the CAST rule of Section 5.1.

This can cause ambiguity in determining ϕ since, for example, if $a = 2$, $b = -2$ or if $a = -2$, $b = 2$ then $\tan\phi = -1$. We need to be sure to obtain the correct value of ϕ.

We state therefore that:

KEY POINT

To solve equations of the type

$$a \sin \theta + b \cos \theta = c$$

write

$$a \sin \theta + b \cos \theta \equiv R \sin(\theta + \phi)$$

where

$$R = \sqrt{a^2 + b^2}, \quad \cos \phi = \frac{a}{R} \quad \text{and} \quad \sin \phi = \frac{b}{R} \qquad 6.7$$

Worked examples

6.10 Express the following in the form $R \sin(\theta + \phi)$:

(a) $\sin \theta + \cos \theta$ (b) $\sin \theta - \cos \theta$

(c) $\sqrt{3} \cos \theta - \sin \theta$ (d) $2 \sin \theta + 3 \cos \theta$

Solution

(a) Here $a = 1$, $b = 1$ so that $R = \sqrt{2}$ and $\cos \phi = 1/\sqrt{2} = \sin \phi$. Hence $\phi = \pi/4$ and the resultant expression is $\sqrt{2} \sin\left(\theta + \frac{\pi}{4}\right)$.

(b) Here $a = 1$, $b = -1$ so that $R = \sqrt{2}$ and

$$\cos \phi = \frac{1}{\sqrt{2}}, \quad \sin \phi = -\frac{1}{\sqrt{2}}$$

Therefore $\phi = -\pi/4$ and the resultant expression is $\sqrt{2} \sin\left(\phi - \frac{\pi}{4}\right)$.

(c) $a = -1$, $b = \sqrt{3}$, $R = 2$, $\cos \phi = -1/2$ and $\sin \phi = \sqrt{3}/2$. Then $\phi = 2\pi/3$ and the expression is $2 \sin\left(\theta + \frac{2\pi}{3}\right)$.

(d) $R = \sqrt{13}$, $\cos \phi = 2/\sqrt{13}$, $\sin \phi = 3/\sqrt{13}$ so that $\phi = 0.983$ (radians). The expression is $\sqrt{13} \sin(\theta + 0.983)$.

6.11 Express $2 \sin \theta + 3 \cos \theta$ in the form $R \cos(\theta - \phi)$.

Solution Consider the expression $a \sin \theta + b \cos \theta$ and note that

$$R \cos(\theta - \phi) \equiv R \cos \theta \cos \phi + R \sin \theta \sin \phi$$

Hence $R^2 = a^2 + b^2$ and ϕ is given by $\cos \phi = b/R$, $\sin \phi = a/R$. Here $a = 2$, $b = 3$ so that $R = \sqrt{13}$, $\cos \phi = 3/\sqrt{13}$, $\sin \phi = 2/\sqrt{13}$. Therefore $\phi = 0.588$ (radians). The required expression is $\sqrt{13} \cos(\theta - 0.588)$.

6.12 A waveform has the equation

$$x = 3 \sin 2t + 4 \cos 2t$$

where x is the displacement at time t. Find the first two occasions when the displacement is equal to 2.

Solution Let $3\sin 2t + 4\cos 2t \equiv R\sin(2t + \phi)$.
Then $R = \sqrt{3^2 + 4^2} = 5$, $\cos\phi = 3/5$ and $\sin\phi = 4/5$.
Therefore $\phi = 0.9273$ (radians) and

$$3\sin 2t + 4\cos 2t \equiv 5\sin(2t + 0.9273)$$

When $x = 2$

$$\sin(2t + 0.9273) = \frac{2}{5} = 0.4$$

The first two positive solutions of $\sin\theta = 0.4$ are

$$\theta = 0.4115 \text{ and } \theta = 2.7301$$

However, the equation $2t + 0.9273 = 0.4115$ has a negative solution for t and is inadmissible.

We take the third positive solution of $\sin\theta = 0.4$, that is,

$$\theta = 2\pi + 0.4115 = 6.6947$$

Then $2t + 0.9273 = 2.7301$ and 6.6947 so that $t = 0.9014$ and 2.8837.

6.13 The sum of two alternating currents

$$i_1 = 30\sin 4\pi t \text{ and } i_2 = 40\cos 4\pi t$$

has the form $R\sin(4\pi t + \phi)$. Find the amptitude, period and phase angle of the resultant current and the first time that its value is 50 amps.

Solution The amplitude $R = \sqrt{(30)^2 + (40)^2} = 50$; $\cos\phi = 3/5$ and $\sin\phi = 4/5$.

Hence the phase angle $\phi = 0.9273$. The period $T = \dfrac{2\pi}{4\pi} = 0.5$. When

$$50\sin(4\pi t + 0.9273) = 50$$

$$\sin(4\pi t + 0.9273) = 1$$

The first positive solution of $\sin\theta = 1$ is $\theta = \pi/2$.
Hence $4\pi t + 0.9273 = \pi/2$ giving $t = 0.0512$.

Self-assessment questions 6.3

1. Explain how to express $a\sin\theta + b\cos\theta$ in the form $R\sin(\theta + \phi)$.
2. Explain how to solve equations of the form $a\sin\theta + b\cos\theta = c$.

Exercise 6.3

1. Find the greatest value and the least value of the expression $4\sin\theta + 3\cos\theta$. By sketching the curve $y = 4\sin\theta + 3\cos\theta$ find the first positive value of θ at which the greatest value is achieved.

2. Solve the following equations:

   (a) $\cos x - \sin x = 0.5$
   (b) $\cos x - \sin x = 1$
   (c) $\cos x - \sin x = 1.5$

3. Find the general solution of the following equations:

   (a) $5\sin 3\theta + 2\cos 3\theta = 2$
   (b) $\sin x + \cos x = \sqrt{2}$
   (c) $3\sin x + 1 = \cos x$
   (d) $\sin x + \sqrt{3}\cos x = 1$
   (e) $\cos x - 0.5\sin x = 1$

6.4 Hyperbolic identities

For every identity involving cos, sin and tan there is a similar identity involving cosh, sinh and tanh. For example

$$\cosh^2 x \equiv \frac{1}{2}(e^x + e^{-x}) \times \frac{1}{2}(e^x + e^{-x}) \equiv \frac{1}{4}(e^{2x} + 2 + e^{-2x})$$

and

$$\sinh^2 x \equiv \frac{1}{2}(e^x - e^{-x}) \times \frac{1}{2}(e^x - e^{-x}) \equiv \frac{1}{4}(e^{2x} - 2 - e^{-2x})$$

so that:

KEY POINT

$$\cosh^2 x - \sinh^2 x \equiv 1 \qquad 6.8$$

This compares with the trigonometric identity $\cos^2 x + \sin^2 x \equiv 1$.

A quick way to obtain a hyperbolic identity from a trigonometric one is by means of **Osborn's rule**. This states that to obtain the equivalent formula sin can be replaced by sinh, cos by cosh and tan by tanh, *except that* $\sin \times \sin$ must be replaced by $-\sinh \times \sinh$ (or $\tan \times \tan$ by $-\tanh \times \tanh$). Since it is often not easy to see that we have $\sin \times \sin$ implicitly in a formula it is safest to convert the trigonometric formula to one involving sin and cos only.

Worked examples

6.14 Show that $\cosh 2x \equiv 2\cosh^2 x - 1 \equiv 1 + 2\sinh^2 x$.

Solution We could use the definition of $\cosh x$ and $\sinh x$:

$$2\cosh^2 x - 1 \equiv 2 \times \frac{1}{4}(e^{2x} + 2 + e^{-2x}) - 1$$

$$\equiv \frac{1}{2}(e^{2x} + e^{-2x}) \equiv \cosh 2x$$

$$1 + 2\sinh^2 x \equiv 1 + 2 \times \frac{1}{4}(e^{2x} - 2 + e^{-2x})$$

$$\equiv \frac{1}{2}(e^{2x} + e^{-2x}) \equiv \cosh 2x$$

Alternatively, we can apply Osborn's rule to the identities:

$$\cos 2x \equiv 2\cos^2 x - 1 \equiv 1 - 2\sin^2 x$$

noting the presence of the product of two sines.

6.15 Show that the point $(a\cosh\theta, b\sinh\theta)$ lies on the hyperbola

$$\frac{x^2}{a^2} - \frac{y^2}{b^2} = 1$$

Solution Since $x = a\cosh\theta$, $y = a\sinh\theta$,

$$\frac{x^2}{a^2} - \frac{y^2}{b^2} \equiv \cosh^2\theta - \sinh^2\theta \equiv 1$$

This justifies the name *hyperbolic* functions.

Self-assessment questions 6.4

1. Write down the hyperbolic identity corresponding to $\cos^2\theta + \sin^2\theta = 1$.
2. Explain the use of Osborn's rule.

Exercise 6.4

1. Show that:

$$\tanh 2x \equiv \frac{2\tanh x}{1 + \tanh^2 x}$$

2. Using Osborn's rule show that:

(a) $\sinh(x + y) \equiv \sinh x \cosh y + \cosh x \sinh y$

(b) $\cosh(x + y) \equiv \cosh x \cosh y + \sinh x \sinh y$

and verify the results directly.

Test and assignment exercises 6

1. Simplify the following expressions:

(a) $\dfrac{\sec^2\theta - 1}{\operatorname{cosec}^2\theta - 1}$

(b) $\dfrac{\cos\theta}{1 + \tan^2\theta}$

(c) $\dfrac{1}{\cos\theta\sqrt{(1 + \cot^2\theta)}}$

2. Eliminate θ from the following pairs of equations:

(a) $\begin{cases} x = 4\cot\theta \\ y = 4\operatorname{cosec}\theta \end{cases}$

(b) $\begin{cases} x = 3\tan\theta \\ y = 2\cos\theta \end{cases}$

(c) $\begin{cases} x = 1 - \cos\theta \\ y = 1 + \sin\theta \end{cases}$

3. Prove the following identities:

(a) $\sec^2\theta \equiv \dfrac{\operatorname{cosec}\theta}{\operatorname{cosec}\theta - \sin\theta}$

(b) $(\cos\theta - \sin\theta)(\sec\theta - \cos\theta)$
$\equiv \dfrac{1}{\tan\theta + \cot\theta}$

4. Solve for $0 \leqslant \theta \leqslant 2\pi$ the following equations:

(a) $4\cos^2\theta + 5\sin\theta = 3$
(b) $\sec\theta = 1 - 2\tan^2\theta$
(c) $2\cos\theta - 4\sin^2\theta + 2 = 0$

5. Prove the following:

(a) $\dfrac{1 - \cos 2\theta + \sin 2\theta}{1 + \cos 2\theta + \sin 2\theta} \equiv \tan\theta$

(b) $\sec 2\theta + \tan 2\theta \equiv \dfrac{\cos\theta + \sin\theta}{\cos\theta - \sin\theta}$

6. Solve the following equations for $0 \leqslant \theta \leqslant 2\pi$:

(a) $4 - 5\cos\theta = 2\sin^2\theta$
(b) $\sin 2\theta + \cos\theta = 0$
(c) $\cos 2\theta + 3\sin\theta = 2$

7. Solve the following equations:

(a) $2\sin\theta = 3(1 - \cos\theta)$
(b) $\sin\theta = 4 + 5\cos\theta$
(c) $1 + \sin\theta = 2\cos\theta$

8. Using Osborn's rule show that:

(a) $\sinh(2x) \equiv 2\sinh x\cosh x$
(b) $\cosh(2x) \equiv \cosh^2 x + \sinh^2 x$

and verify your results directly.

7 Differentiation I

Objectives

This chapter

- defines a derivative
- provides a table of standard derivatives
- applies the rules for differentiating combinations of standard functions
- defines and determines higher derivatives
- states and uses the chain rule for differentiating composite functions

7.1 Derivative and rate of change

Engineers have to deal with quantities which vary, for example the temperature of a gas, the current in an electrical circuit or the stress in a loaded beam. It is often the rate of change of the quantity which is of interest; for example, in kinematics velocity is the rate of change of displacement and acceleration is the rate of change of velocity. In this chapter we look at methods of calculating that rate of change when the quantity is a known function of one independent variable.

The **average rate of change** of a function $f(x)$ in the interval $a \leqslant x \leqslant a+h$ is given by

$$E = \frac{f(a+h) - f(a)}{h}$$

as illustrated in Figure 7.1(a).

The expression E measures the gradient of the chord joining the points P and Q. As $h \to 0$ Q moves closer to P and in the limit E measures the gradient of the tangent to the curve at P: see Figure 7.1(b). This is the **instantaneous rate of change** of the function at $x = a$.

If E converges to a (finite) limit as $h \to 0$ then the function is said to be **differentiable** at $x = a$ and the value of the limit is the **derivative** of $f(x)$ at $x = a$, denoted by $f'(a)$. If the function is differentiable at all points in its domain then its **derived function** is $f'(x)$. The process of obtaining derivatives is called **differentiation**.

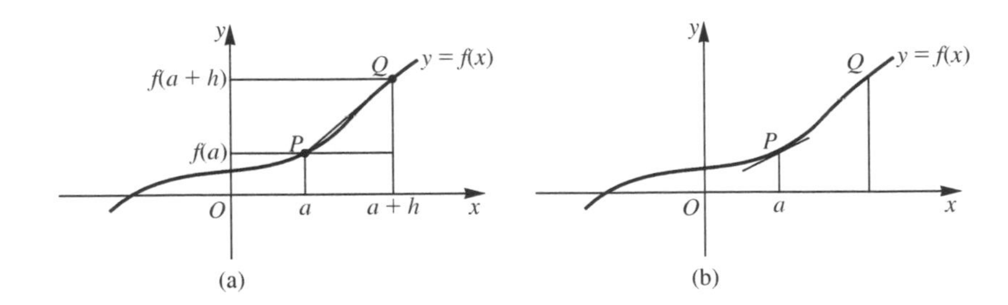

Figure 7.1. Average and instantaneous rates of change

Worked examples

7.1 Find the derivative of $f(x) = x^2$ at $x = 2, 0, -3$. Determine the derived function $f'(x)$.

Solution

$$E = \frac{f(a+h) - f(a)}{h} = \frac{(a+h)^2 - a^2}{h} = \frac{2ah + h^2}{h} = 2a + h$$

As $h \to 0$, $E \to 2a = f'(a)$
At $x = 2, f'(2) = 4$
At $x = 0, f'(0) = 0$
At $x = -3, f'(-3) = -6$
In general, $f'(x) = 2x$.

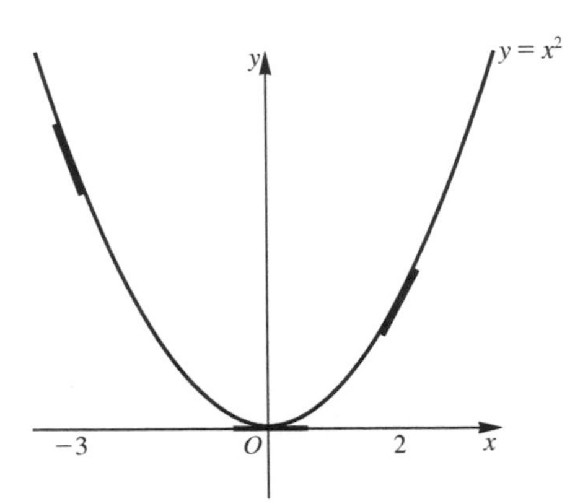

Figure 7.2. Derivatives of $f(x) = x^2$

Figure 7.2 shows the tangents to the curve of $y = x^2$ at $x = -3$, 0 and 2. Compare the gradients of the tangents with the values of $f'(x)$ at each point; they are equal.

KEY POINT

> The gradient of the tangent at a point on the curve $y = f(x)$ is the value of $f'(x)$ at that point.

7.2 By attempting to find $f'(0)$ show that $f(x) = |x|$ is not differentiable at $x = 0$.

Solution In the definition of derivative h can be positive or negative: Figure 7.1(a) showed only the case where h was positive.

If $h > 0$ then

$$E = \frac{f(0+h) - f(0)}{h} = \frac{h-0}{h} = 1$$

If $h < 0$ then

$$E = \frac{-h-0}{h} = -1$$

(These values correspond to the gradients of the straight lines which form the graph of $f(x) = |x|$, which is shown in Figure 7.3.)

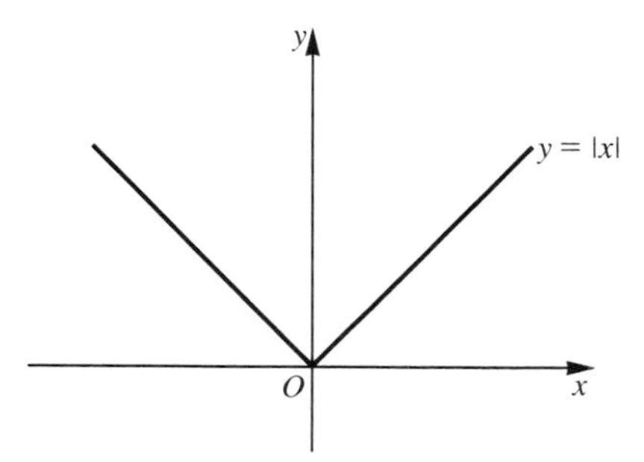

Figure 7.3. $f(x) = |x|$

Since we do not get a consistent value for E as $h \to 0$ the function cannot be differentiated at $x = 0$. (This is the only point at which $f(x)$ is not differentiable.)

7.3 Obtain the derived function of $f(x) = 1/x$.

Solution

$$E = \frac{f(a+h) - f(a)}{h} = \frac{\dfrac{1}{a+h} - \dfrac{1}{a}}{h}$$

$$= \frac{1}{h}\left(\frac{a-(a+h)}{(a+h)a}\right) = \frac{-h}{h(a+h)a}$$

$$= -\frac{1}{(a+h)a}$$

As $h \to 0$, $E \to -1/a^2$.

Note that $x = 0$ is excluded from the domain of $f(x)$, therefore we write $f'(x) = -1/x^2$, $x \neq 0$.

KEY POINT

Where $f'(x) > 0$, $f(x)$ is increasing. Where $f'(x) < 0$, $f(x)$ is decreasing.

Reference to Figure 7.2 shows that $f(x) = x^2$ is increasing for $x > 0$ and decreasing for $x < 0$.

Figure 7.4 shows the graph of $f(x) = 1/x$; it is decreasing everywhere.

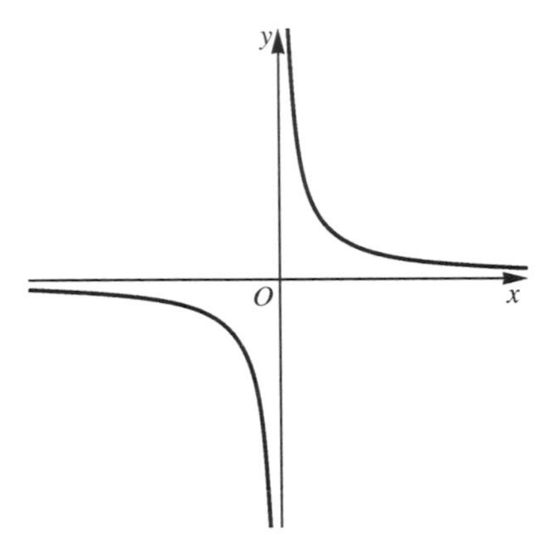

Figure 7.4. $f(x) = 1/x$

KEY POINT

Where $f'(x) = 0$ the function has a **stationary point**.

For example, $f(x) = x^2$ has a stationary point at $x = 0$.

It would be tedious to obtain derived functions from the definition so Table 7.1 gives the derived functions of some of the more common functions $f(x)$.

Worked example

7.4 In the interval $0 \leqslant x \leqslant 2\pi$ find where $f(x) = \sin x$ is increasing and/or decreasing; locate the stationary points. Illustrate your answer by sketching the graph of the function and marking the stationary points.

Solution From Table 7.1, $f'(x) = \cos x$. Now $\cos x > 0$ in $0 < x < \pi/2$ and in $3\pi/2 < x < 2\pi$, hence $f(x)$ is increasing there; see Figure 7.5(a). Also, $\cos x < 0$ in $\pi/2 < x < 3\pi/2$ hence $f(x)$ is decreasing there. $f'(x) = 0$ at $x = \pi/2$ and $x = 3\pi/2$ which are the stationary points.

Table 7.1.
Derived functions

$f(x)$	$f'(x)$	$f(x)$	$f'(x)$
C, constant	0	$\sin x$	$\cos x$
x	1	$\cos x$	$-\sin x$
x^2	$2x$	$\tan x$	$\sec^2 x$
x^n	nx^{n-1}	$\operatorname{cosec} x$	$-\cot x \operatorname{cosec} x$
e^x	e^x	$\sec x$	$\sec x \tan x$
$\ln x, x > 0$	$\frac{1}{x}$	$\cot x$	$-\operatorname{cosec}^2 x$

Figure 7.5(b) illustrates these results. Note that the gradient of the tangents at A and B is zero.

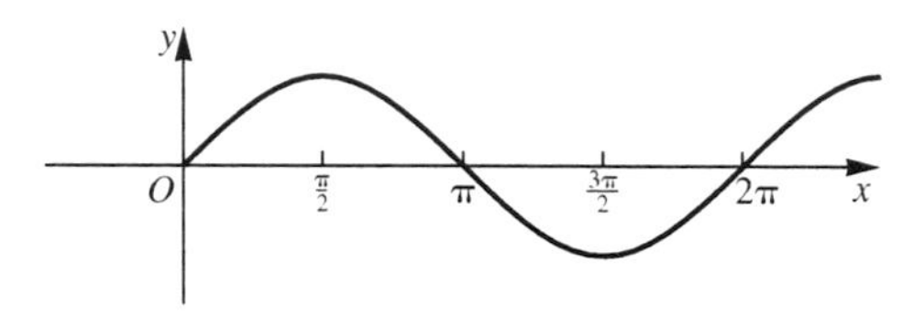

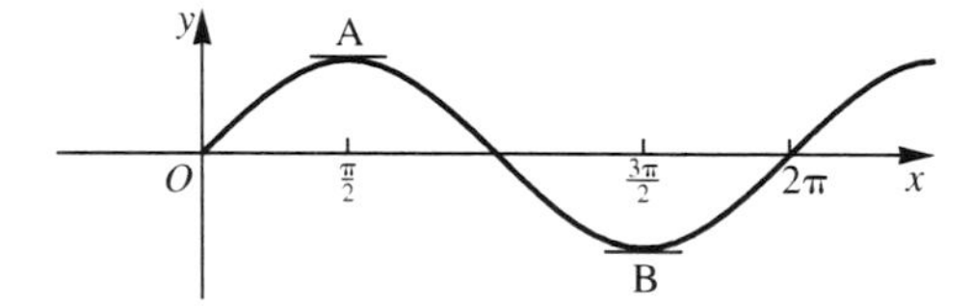

Figure 7.5.
(a) $y = \cos x$;
(b) gradients of $f(x) = \sin x$

Self-assessment questions 7.1

1. Define the derivative of a function.
2. Under what circumstances is a function differentiable?

Exercise 7.1

1. From first principles find the derivative of the function $f(x) = x^3$ for the values $x = -1.0$ and 1.
2. Find the stationary points, if any, of the following functions:

   (a) $\cos x$
   (b) $1/x^2$
   (c) $\ln x$
   (d) $\sec x$
3. Obtain from first principles the derived function of $f(x) = 1/x^2$.
4. Is the unit step function differentiable?
5. An object is projected vertically upwards. Its height above ground level at time t is given by $h = 40t - 9.8t^2$. From first principles find $h'(t)$ and determine the time at which the speed of the object is instantaneously zero.

7.2 Rules for differentiation

First, we introduce an alternative notation which is sometimes more convenient. If $y = f(x)$ then the derived function is denoted

$\frac{dy}{dx}$ is pronounced 'dee y by dee x'.

$$\frac{dy}{dx}$$

For example

$$\text{if} \quad y = x^2 \quad \text{then} \quad \frac{dy}{dx} = 2x$$

$$\text{if} \quad y = \sin x \quad \text{then} \quad \frac{dy}{dx} = \cos x$$

The following rules are valid if all relevant derivatives exist:

KEY POINT

Multiple

If the derived function of $f(x)$ is $f'(x)$ and a is a constant then the derived function of $af(x)$ is the function $af'(x)$ 7.1

For example

$$\text{if} \quad y = 3\sin x \quad \text{then} \quad \frac{dy}{dx} = 3\cos x$$

KEY POINT

Sum

The derivative of the sum of two functions is the sum of their derivatives. Hence

$$\text{if} \quad y = f(x) + g(x) \quad \text{then} \quad \frac{dy}{dx} = f'(x) + g'(x) \qquad 7.2$$

For example

$$\text{if} \quad y = x^2 + \sin x \quad \text{then} \quad \frac{dy}{dx} = 2x + \cos x$$

KEY POINT

Linear combination

The derivative of the linear combination of two functions is the same linear combination of their derivatives. Hence

if $y = af(x) + bg(x)$

where a and b are constants then

$$\frac{dy}{dx} = af'(x) + bg'(x) \qquad 7.3$$

For example

The multiple and sum rules are special cases of this rule.

$$\text{if} \quad y = 3x^2 - 5\sin x \quad \text{then} \quad \frac{dy}{dx} = 3 \times 2x - 5\cos x$$

$$= 6x - 5\cos x$$

KEY POINT

Product

$$\text{If} \quad y = f(x).g(x) \quad \text{then} \quad \frac{dy}{dx} = f(x).g'(x) + f'(x).g(x) \qquad 7.4$$

In words, the derivative of a product is calculated as 'the first term multiplied by the derivative of the second plus the derivative of the first term multiplied by the second'.

A popular alternative notation is

$$\text{if} \quad y = u(x).v(x) \quad \text{then} \quad \frac{dy}{dx} = u.\frac{dv}{dx} + \frac{du}{dx}.v \qquad 7.5$$

For example

$$\text{if} \quad y = x^2 \sin x \quad \text{then} \quad \frac{dy}{dx} = x^2 \cos x + 2x\sin x$$

KEY POINT

Quotient

$$\text{If} \quad y = \frac{u(x)}{v(x)} \quad \text{then} \quad \frac{dy}{dx} = \frac{v\frac{du}{dx} - u\frac{dv}{dx}}{v^2} \qquad 7.6$$

For example

Remember that $(\sin x)^2$ is written $\sin^2 x$.

$$\text{if} \quad y = \frac{x^2}{\sin x} \quad \text{then} \quad \frac{dy}{dx} = \frac{\sin x.2x - x^2\cos x}{\sin^2 x} = \frac{2x\sin x - x^2\cos x}{\sin^2 x}$$

Worked examples

7.5 Differentiate the following functions, simplifying the results where feasible. (Refer to Table 7.1.)

(a) $x^2 + x^3$ (b) $x^5 - \dfrac{1}{x^2}$ (c) $9\sin x + 2e^x$ (d) $x^3 \ln x$

(e) $x^3 . e^x . \sin x$ (f) $\dfrac{x+2}{e^x - 1}$ (g) $\dfrac{x^2 \cos x + \sin x}{4x + e^{-x}}$

Solution (a) The derived function of x^2 is $2x$ and of x^3 is $3x^2$ (x^n with $n = 2$ and 3). The sum rule gives the result as

$$2x + 3x^2 \qquad \text{or} \qquad x(2 + 3x)$$

(b) The derived functions of x^5 and x^{-2} are $5x^4$ and $-2x^{-3}$ (or $-2/x^3$), respectively. Hence the result is $5x^4 + (2/x^3)$.

(c) The derived functions of $\sin x$ and e^x are $\cos x$ and e^x, respectively. For the given linear combination the result is $9\cos x + 2e^x$.

(d) The derived function of x^3 is $3x^2$ and of $\ln x$ is $1/x$. Applying the product rule, we obtain

$$x^3\left(\frac{1}{x}\right) + 3x^2(\ln x)$$

This simplifies to $x^2 + 3x^2 \ln x = x^2(1 + 3\ln x)$.

(e) The derived functions of x^3, e^x and $\sin x$ are $3x^2$, e^x and $\cos x$, respectively. The derived function of $e^{-x} \equiv \dfrac{1}{e^x}$ can be obtained for the quotient rule as $-e^{-x}$.

First we apply the product rule to x^3e^x to obtain the result

$$x^3(e^x) + (3x^2)e^x$$

which simplifies to $x^2e^x(x + 3)$. However, for subsequent work this is not really helpful.

Applying the product to the given function we obtain

$$(x^3e^x).(\cos x) + (x^3e^x + 3x^2e^x).\sin x$$

Since x^2e^x is common to all terms the expression simplifies to $x^2e^x(x\cos x + (x + 3)\sin x)$.

(It would be instructive for you to write the original function as $x^3(e^x \sin x)$ and work from there to obtain this result again.)

(f) The derived function of $x + 2$ is $1 + 0 = 1$ (sum rule), and of $e^x - 1$ is $e^x - 0 = e^x$ (sum rule).

The quotient rule gives

$$\frac{(e^x - 1).(1) - (x + 2)e^x}{(e^x - 1)^2}$$

This simplifies to

$$\frac{e^x - 1 - xe^x - 2e^x}{(e^x - 1)^2} = \frac{-1 - e^x(x + 1)}{(e^x - 1)^2}$$

although it is not much of a simplification.

(g) The expression looks fearsome, but it can be dealt with if we are patient. First, we notice that it is a quotient, and to apply the quotient rule we need to differentiate both the numerator, u, and the denominator, v. The denominator is a sum so that

$$\frac{dv}{dx} = 4 + (-e^{-x}) = 4 - e^{-x}$$

The numerator is also a sum but it involves a product. Let $w = x^2 \cos x$ then, by the product rule,

$$\frac{dw}{dx} = x^2(-\sin x) + 2x\cos x = -x^2 \sin x + 2x\cos x$$

Let $u = x^2 \cos x + \sin x$ then, by the sum rule,

$$\frac{du}{dx} = -x^2 \sin x + 2x\cos x + \cos x$$

Finally, applying the quotient rule, the result is

$$\frac{(4x + e^{-x})(-x^2 \sin x + 2x\cos x + \cos x) - (x^2 \cos x + \sin x)(4 - e^{-x})}{(4x + e^{-x})^2}$$

This is an awful expression and we will not attempt to simplify it. At least we have carried out the differentiation!

7.6 Find the equation of the tangent and normal to the curve $y = x + \frac{1}{x}$ at the points $(\frac{1}{2}, 2\frac{1}{2})$, $(2, 2\frac{1}{2})$, $(1, 2)$, $(-1, -2)$; the graph is shown in Figure 7.6.

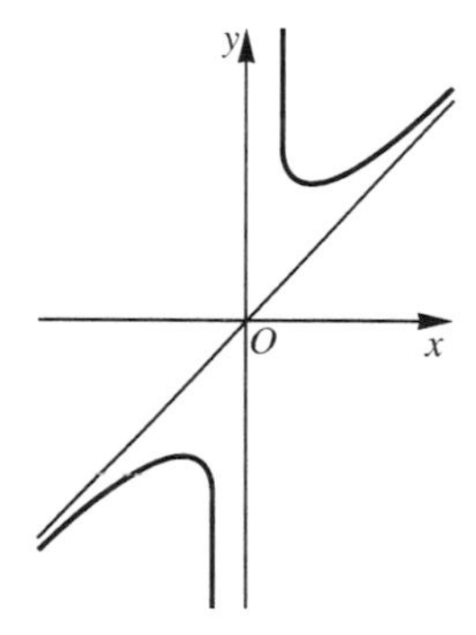

Figure 7.6. The function $f(x) = x + 1/x$

Solution As was said earlier, the gradient of the tangent at a point on the curve is given by the value of dy/dx at that point. The normal to the curve at a point is perpendicular to the tangent. Now

$$\frac{dy}{dx} = 1 - \frac{1}{x^2}$$

When $x = \frac{1}{2}$, $\frac{dy}{dx} = -3$ and the equation of the tangent is

$$y - 2\tfrac{1}{2} = -3\left(x - \tfrac{1}{2}\right) \qquad \text{that is,} \qquad y = -3x + 4$$

The normal, being perpendicular to the tangent, has a gradient $\frac{-1}{(-3)} = \frac{1}{3}$. Its equation is

$$y - 2\tfrac{1}{2} = \tfrac{1}{3}\left(x - \tfrac{1}{2}\right) \qquad \text{that is,} \qquad y = \tfrac{1}{3}x + \tfrac{7}{3}$$

When $x = 2$, $\frac{dy}{dx} = \frac{3}{4}$ and the equation of the tangent is

$$y - 2\tfrac{1}{2} = \tfrac{3}{4}(x - 2) \qquad \text{that is,} \qquad y = \tfrac{3}{4}x + 1$$

The gradient of the normal is $-1/\left(\frac{3}{4}\right) = -\frac{4}{3}$ and its equation is

$$y - 2\tfrac{1}{2} = -\tfrac{4}{3}(x - 2) \qquad \text{that is,} \qquad y = -\tfrac{4}{3}x + \tfrac{23}{6}$$

At $x = \pm 1$, $\frac{dy}{dx} = 0$.

Hence the tangent at $(1, 2)$ is $y = 2$ and at $(-1, -2)$ is $y = -2$. The normal at $(1, 2)$ is $x = 1$ and at $(-1, -2)$ is $x = -1$.

7.7 A projectile is thrown vertically upwards. Its distance above the point of projection at time t is given by $s = 20t - 4.9t^2$.

When does the projectile reach its greatest height and what is its speed when it reaches the point of projection on the way down?

Solution The projectile reaches its greatest height when it comes temporarily to rest, in other words its speed $\frac{ds}{dt}$ is zero. Now its speed at time t is

$$\frac{ds}{dt} = 20 - 9.8t$$

and

$$\frac{ds}{dt} = 0 \qquad \text{when} \qquad t = \frac{20}{9.8} = 2.04$$

Also $s = 0$ when $20t - 4.9t^2 = 0$, that is, $t(20 - 4.9t) = 0$.

The time taken to reach the point of projection again is $t = 20/4.9$ and its speed is then

$$\frac{ds}{dt} = 20 - 9.8 \times \frac{20}{4.9} = 20 - 40 = -20$$

Self-assessment questions 7.2

1. State the product rule and the quotient rule.
2. Explain how to use the linear combination rule.

Exercise 7.2

1. The diode current I in a semiconductor at room temperature is given by $I = I_s(e^{40V} - 1)$ where V is the applied voltage and I_s is a constant. Find an equation for I which involves its derivative with respect to V.

2. Find the equations of the tangent and the normal to the curve $y = \sin x$ at $x = 0$, $\frac{\pi}{4}$ and $\frac{\pi}{2}$.

3. Differentiate the following functions with respect to x:

(a) $7x^6$ (b) $3\sin x + x^4$

(c) $\sin x \cos x$ (d) $\dfrac{x^2}{1+x^2}$

(e) $e^x \cos x$ (f) $x \ln x$

(g) $\dfrac{\ln x}{x^3}$ (h) $\dfrac{x^2 \sin x}{1+\cos x}$

7.3 Higher derivatives

Acceleration is the rate of change of velocity and velocity is the rate of change of displacement, that is

$$a = \frac{dv}{dt} \quad \text{where} \quad v = \frac{ds}{dt}$$

Since v is the derivative of s and a is the derivative of v we say that a is the **second derivative** of s and write

$$a = \frac{d^2s}{dt^2}$$

In general, if y is a function of x given by $y = f(x)$ then the (first) derivative of y is

$$\frac{dy}{dx} = f'(x)$$

and the second derivative of y is

$$\frac{d^2y}{dx^2} = f''(x)$$

$\frac{d^2y}{dx^2}$ is pronounced 'dee two y by dee x squared'.

(Strictly, we should always refer to $\frac{dy}{dx}$ as the first derivative, but we often omit the word 'first' if it is the only derivative involved in a problem.)

Since

$$\frac{d^2y}{dx^2} \equiv \frac{d}{dx}\left(\frac{dy}{dx}\right)$$

it can be found by repeated differentiation of y with respect to x.

The third derivative of y with respect to x is denoted

$$\frac{d^3y}{dx^3} \quad \text{or} \quad f'''(x)$$

and in general the nth derivative is denoted

$$\frac{d^ny}{dx^n} \quad \text{or} \quad f^{(n)}(x)$$

Worked examples

7.8 Find the second derivative of each of the following functions $y = f(x)$:
(a) x^3 (b) x^n (c) $\sin x$ (d) $\ln x$ (e) e^x (f) e^{-x}

Solution

(a) $\frac{dy}{dx} = 3x^2$ and $\frac{d^2y}{dx^2} = 3 \times 2x = 6x$

(b) $\frac{dy}{dx} = nx^{n-1}$ and $\frac{d^2y}{dx^2} = n(n-1)x^{n-2}$

(c) $\frac{dy}{dx} = \cos x$ and $\frac{d^2y}{dx^2} = -\sin x$

(d) $\frac{dy}{dx} = \frac{1}{x}$ and $\frac{d^2y}{dx^2} = -\frac{1}{x^2}$

(e) $\frac{dy}{dx} = e^x$ and $\frac{d^2y}{dx^2} = e^x$

(f) $\frac{dy}{dx} = -e^{-x}$ and $\frac{d^2y}{dx^2} = e^{-x}$

7.9 The displacement of a particle along a straight line at time t is given by $s = a\cos t + b\sin t$ where a and b are constants. Find the velocity v and acceleration a at time t and relate the acceleration directly to the displacement.

Solution

$$v = \frac{ds}{dt} = -a\sin t + b\cos t \qquad \text{(linear combination rule)}$$

$$a = \frac{dv}{dt} = \frac{d^2s}{dt^2} = -a\cos t - b\sin t = -s$$

Motion in which a is proportional to $-s$ is called simple harmonic motion (SHM).

7.10 For the following functions find the derivative shown:
(a) x^6 (3rd) (b) $\sin x$ (4th) (c) e^x (nth)

Solution (a) $\dfrac{dy}{dx} = 6x^5, \quad \dfrac{d^2y}{dx^2} = 30x^4, \quad \dfrac{d^3y}{dx^3} = 120x^3$

(b) $\dfrac{dy}{dx} = \cos x, \quad \dfrac{d^2y}{dx^2} = -\sin x, \quad \dfrac{d^3y}{dx^3} = -\cos x, \quad \dfrac{d^4y}{dx^4} = \sin x$

(c) Since $\dfrac{dy}{dx} = e^x$ all subsequent derivatives are also equal to e^x.

7.11 The **radius of curvature** ρ at any point on a curve $y = f(x)$ is an indication of the rate at which it is bending: the smaller ρ the more rapid the bending. It is given by

$$\rho = \frac{\left[1 + \left(\dfrac{dy}{dx}\right)^2\right]^{3/2}}{\dfrac{d^2y}{dx^2}}$$

Find the radius of curvature at the point (1,1) on the curve $y = 1/x$.

Solution If $y = 1/x$,

$$\frac{dy}{dx} = -\frac{1}{x^2} \quad \text{and} \quad \frac{d^2y}{dx^2} = \frac{2}{x^3}$$

Then

$$\rho = \frac{\left(1 + \dfrac{1}{x^4}\right)^{3/2}}{\dfrac{2}{x^3}} = \frac{x^3}{2}\left(1 + \frac{1}{x^4}\right)^{3/2} = \frac{1}{2}\left(x^2 + \frac{1}{x^2}\right)^{3/2}$$

At (1,1), $\rho = \frac{1}{2}(2)^{3/2} = 2^{\frac{1}{2}} = \sqrt{2}$.

Self-assessment questions 7.3

1. Explain the meaning of the second derivative.
2. Explain the meaning of an infinite radius of curvature.

Exercise 7.3

1. Find the second derivatives of the functions of Question 2 of Exercise 7.2.

2. Find the radius of curvature at any point on the curve

$$y = c\cosh\frac{x}{c}$$

What is its least value and where does this occur?

3. Show that $y = A\cos\lambda x + B\sin\lambda x + C\cosh\lambda x + D\sinh\lambda x$, where A, B, C and D are constants, satisfies the equation

$$\frac{d^4y}{dx^4} = \lambda^4 y$$

4. The angular displacement of a rotating cylinder is given by $\theta = 2\cos t + 3\sin t$. Find the angular speed and the magnitude of the angular acceleration of the cylinder. When is the angular speed first zero?

7.4 Chain rule

A **composite function** arises when one function is applied to another, for example $\sin(2x)$ and $(2x + 4)^3$. We can write

$$y = f(u) \qquad \text{where} \qquad u = g(x)$$

Hence the second composite function above can be written as

$$y = u^3 \qquad \text{where} \qquad u = 2x + 4$$

The **chain rule** provides a means of finding the derivative of a composite function.

KEY POINT

The chain rule states that if $y = f(u)$ where $u = g(x)$ then

$$\frac{dy}{dx} = \frac{dy}{du} \times \frac{du}{dx} \tag{7.7}$$

Hence if $y = (2x + 4)^3$, as above, we write $y = u^3$ where $u = 2x + 4$. Then

$$\frac{dy}{du} = 3u^2 \qquad \text{and} \qquad \frac{du}{dx} = 2$$

so that

$$\frac{dy}{dx} = \frac{dy}{du} \times \frac{du}{dx} = 3u^2 \times 2 = 6u^2 = 6(2x + 4)^2$$

Worked examples

7.12 Differentiate the following functions:

(a) $y = (3x - 2)^6$ (b) $y = \dfrac{1}{x+3}$

(c) $y = (3x + 1)^{\frac{1}{2}}$ (d) $y = \ln(2x + 3)$

(e) $y = e^{x^2}$ (f) $y = \left(\dfrac{1-x}{1+x}\right)^{\frac{1}{2}}$

Solution (a) Put $u = 3x - 2$, then $y = u^6$.

$$\frac{du}{dx} = 3, \qquad \frac{dy}{du} = 6u^5$$

Hence

$$\frac{dy}{dx} = \frac{dy}{du} \times \frac{du}{dx} = 6u^5 \times 3 = 18u^5 = 18(3x-2)^5$$

(b)
$$y = \frac{1}{u} \quad \text{where} \quad u = x + 3$$

$$\frac{dy}{du} = -\frac{1}{u^2} \quad \text{and} \quad \frac{du}{dx} = 1$$

Hence

$$\frac{dy}{dx} = \frac{dy}{du} \times \frac{du}{dx} = -\frac{1}{u^2} \times 1 = -\frac{1}{(x+3)^2}$$

(c)
$$y = u^{\frac{1}{2}} \quad \text{where} \quad u = 3x + 1$$

$$\frac{dy}{du} = \frac{1}{2}u^{-\frac{1}{2}} = \frac{1}{2u^{1/2}} \quad \text{and} \quad \frac{du}{dx} = 3$$

Hence

$$\frac{dy}{dx} = \frac{1}{2u^{1/2}} \times 3 = \frac{3}{2u^{1/2}} = \frac{3}{2(3x+1)^{1/2}}$$

(d)
$$y = \ln u \quad \text{where} \quad u = 2x + 3$$

$$\frac{dy}{du} = \frac{1}{u} \quad \text{and} \quad \frac{du}{dx} = 2$$

Hence

$$\frac{dy}{dx} = \frac{1}{u} \times 2 = \frac{2}{u} = \frac{2}{2x+3}$$

(e) $y = e^u$ where $u = x^2$

$$\frac{dy}{du} = e^u \quad \text{and} \quad \frac{du}{dx} = 2x$$

Hence

$$\frac{dy}{dx} = e^u \times 2x = 2x\,e^u = 2x\,e^{x^2}$$

(f) $y = u^{1/2}$ where $u = \left(\dfrac{1-x}{1+x}\right)$

$$\frac{dy}{du} = \frac{1}{2u^{\frac{1}{2}}}$$

To find $\frac{du}{dx}$ we use the quotient rule:

$$\frac{du}{dx} = \frac{(1+x).(-1) - (1-x)(1)}{(1+x)^2}$$

$$= \frac{-2}{(1+x)^2}$$

Then

$$\frac{dy}{dx} = \frac{1}{2u^{\frac{1}{2}}}\left(-\frac{2}{(1+x)^2}\right) = -\frac{1}{u^{1/2}(1+x)^2}$$

$$= -\left(\frac{1+x}{1-x}\right)^{\frac{1}{2}} \frac{1}{(1+x)^2} = -\frac{1}{(1-x)^{1/2}(1+x)^{3/2}}$$

KEY POINT

Note the general result that

if $y = \ln\{f(x)\}$ then $\dfrac{dy}{dx} = \dfrac{f'(x)}{f(x)}$

7.13 Find the acceleration of an object oscillating in a straight line whose displacement is given by $s = a\cos\omega t + b\sin\omega t$, where ω, a and b are constants.

In the case where $a = 0$ find when (i) the velocity is zero, (ii) the acceleration is zero.

Solution Let $y = a\cos\omega t$ be written as $y = a\cos u$, $u = \omega t$.
Then

$$\frac{dy}{du} = a(-\sin u) \quad \text{and} \quad \frac{du}{dt} = \omega$$

hence

$$\frac{dy}{dt} = \frac{dy}{du} \times \frac{du}{dt} = -a \sin u \times \omega = -\omega a \sin u = -\omega a \sin \omega t$$

Similarly, if

$$\frac{dy}{dt} = z = -\omega a \sin u \qquad \text{and} \qquad u = \omega t$$

then

$$\frac{dz}{du} = -\omega a \cos u \qquad \text{and} \qquad \frac{du}{dt} = \omega$$

so that

$$\frac{dz}{dt} = \frac{dz}{du} \times \frac{du}{dt} = -\omega a \cos u \times \omega = -\omega^2 a \cos u = -\omega^2 a \cos \omega t$$

Also, if $y = b \sin \omega t$,

$$\frac{dy}{dt} = \omega b \cos \omega t \qquad \text{and} \qquad \frac{d^2 y}{dt^2} = -\omega^2 b \sin \omega t$$

Hence the acceleration is given by

$$\frac{d^2 s}{dt^2} = -\omega^2 a \cos \omega t - \omega^2 b \sin \omega t = -\omega^2 s$$

This is simple harmonic motion again.

In the case where $a = 0$,

$$v = \frac{ds}{dt} = \omega b \cos \omega t$$

Hence $v = 0$ when $\cos \omega t = 0$, that is at

$$\omega t = \frac{\pi}{2}, \quad \frac{3\pi}{2}, \quad \frac{5\pi}{2}, \ \ldots$$

and therefore at

$$t = \frac{\pi}{2\omega}, \quad \frac{3\pi}{2\omega}, \quad \frac{5\pi}{2\omega}, \ \ldots$$

The acceleration is

$$\frac{d^2 s}{dt^2} = -\omega^2 b \sin \omega t$$

and is zero where $\sin \omega t = 0$, that is at

$$\omega t = 0, \quad \pi, \quad 2\pi, \ \ldots$$

Therefore

$$t = 0, \quad \frac{\pi}{\omega}, \quad \frac{2\pi}{\omega}, \ \ldots$$

Self-assessment questions 7.4

1. State the chain rule.
2. Explain how the chain rule can be used to find the derivative of a composite function.

Exercise 7.4

1. Differentiate the following functions:

   (a) $(2-x)^4$ (b) $(5x-4)^4$
   (c) $(x^2+1)^4$ (d) $(x^2+4)^{-1}$
   (e) $\sin(4x+3)$ (f) $\ln(1+x^4)$
   (g) $\sin \ln x$ (h) $\cos(x^2+1)$
   (i) $(1+e^{3x})^2$

2. Find the first derivatives of $\sinh^{-1} x$, $\cosh^{-1} x$ and $\tanh^{-1} x$.
   (*Hint*: use the logarithmic equivalent given in Section 5.4.)

3. Find the equations of the tangent and the normal to the curve $y = 4\sin(\omega t + \frac{\pi}{6})$ at $t = \frac{\pi}{4\omega}$.

4. Differentiate the function $\sin(1 + (\ln x)^2)$.

Test and assignment exercises 7

1. From first principles obtain the derivatives of the function

   $f(x) = x^2 + 2x + 1$ at $x = -1$, 0 and 1

2. Find where the following functions are increasing and/or decreasing and identify any stationary points:

   (a) x^4 (b) x^5
   (c) $\tan x$, $-\frac{\pi}{2} \leqslant x \leqslant \frac{\pi}{2}$

3. Differentiate the following functions with respect to x:

   (a) $4x^5$ (b) x^{-4}
   (c) $x^{\frac{1}{2}}$ (d) $4x^3 - 3x^{-\frac{2}{3}}$

4. Differentiate the following functions with respect to x:

   (a) $2\cos x + 4\sin x$ (b) $e^x \sin x$
   (c) $x \ln x$ (d) $\dfrac{x}{1+x^4}$
   (e) $\dfrac{1+\ln x}{x^2+1}$ (f) $\dfrac{x\cos x}{1+\sin x}$

5. Find the equations of the tangent and the normal to the following curves at the point indicated:

   (a) $y = \cos x$ at $x = 0$
   (b) $y = \ln x$ at $x = 1$
   (c) $y = x \sin x$ at $x = \frac{\pi}{2}$

6. Find the second and third derivatives of the functions of Question 3.

7. Find the second derivatives of functions (a), (b), (c) and (d) of Question 4.

8. Find the radius of curvature at $x = 0$ of the following curves:

   (a) $y = x^2$ (b) $y = x^4$ (c) $y = \dfrac{1}{x^2}$

9. Differentiate the following functions:

   (a) $(4-x)^3$ (b) $(2x^2-3)^{\frac{1}{2}}$
   (c) $(x^2+2)^3$ (d) $\dfrac{4}{(3-x^2)^{\frac{1}{2}}}$
   (e) $\sin(3x-4)$ (f) $\ln(1+2x^2)$
   (g) $(1+e^{2x})^3$ (h) $\ln(1+e^{-x})$

8 Maximum and minimum values

Objectives

This chapter

- distinguishes between absolute and local extrema
- defines a stationary point of a function
- defines concavity of a function and a point of inflection on a curve
- uses the first derivative to classify stationary points
- uses the second derivative to classify stationary points

8.1 Greatest and least values

In many problems in engineering we are concerned with the optimization of quantities: the speed at which a belt drive transmits maximum power, the dimensions of a rectangular container of fixed volume which minimize the heat loss through the faces and so on.

The greatest value that a function takes in a given interval of its domain is known as the **absolute maximum** on that interval. The least value which it takes in the interval is the **absolute minimum**. (The terms global maximum and minimum are also used.)

If a function is increasing throughout an interval then the absolute minimum occurs at the left-hand end of the interval and the absolute maximum occurs at the right-hand end. If the function is decreasing throughout the interval then the reverse occurs.

Consider the function $f(x) = x^2$. Reference to Figure 8.1(a) shows that for $x < 0$ it is decreasing and for $x > 0$ it is increasing. This is confirmed by looking at the derivative $f'(x) = 2x$ which is positive for $x > 0$ and negative for $x < 0$. Figure 8.1(b) depicts the function $f(x) = x - x^2$; $f'(x) = 1 - 2x$ and therefore $f'(x) > 0$ if $1 - 2x > 0$, that is if $x < \frac{1}{2}$ so that $f(x)$ is increasing in this interval, whereas $f'(x) < 0$ if $x > \frac{1}{2}$ and $f(x)$ is decreasing there.

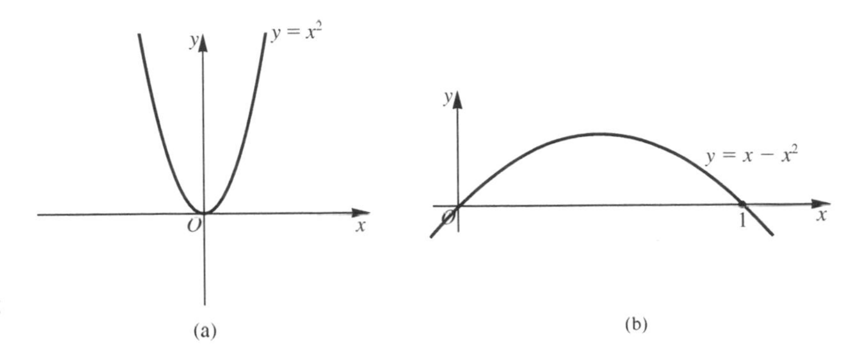

Figure 8.1.
Increasing and decreasing regions

For the first function $f'(x) = 0$ when $x = 0$ and the curve has a horizontal tangent there; for the second function $f'(x) = 0$ when $x = \frac{1}{2}$ and the curve has a horizontal tangent there.

KEY POINT

Points where the tangent is horizontal are called **stationary points**.

In the case of $f(x) = x^2$ the stationary point is the absolute minimum of the function; in the case of $f(x) = x - x^2$ the stationary point is the absolute maximum of the function. In general, neither situation need arise with a stationary point. Figure 8.2 depicts the graph of $y = x^3 - 3x$.

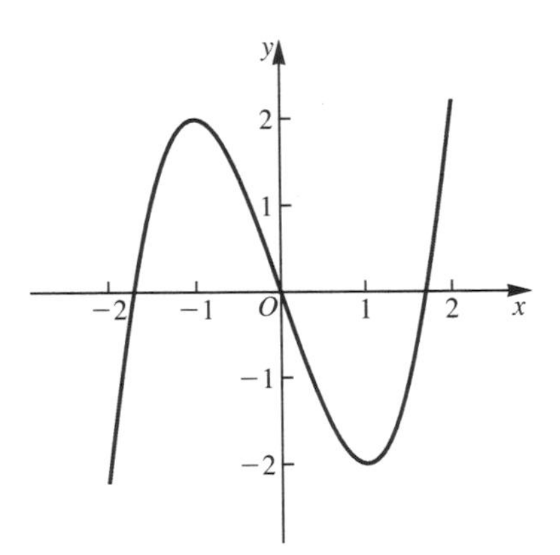

Figure 8.2.
$y = x^3 - 3x$

dy/dx is being used in this example to represent the derivative

The derivative is $\mathrm{d}y/\mathrm{d}x = 3x^2 - 3$; this is zero when $3x^2 = 3$, that is, $x = \pm 1$, and these are the stationary points. At $x = -1$, $y = 2$; but at $x = 2$, $y = 2$ and for $x > 2$, $y > 2$; hence $x = -1$ is not the absolute maximum. At $x = 1$, $y = -2$; but for $x < -2$, $y < -2$; hence $x = 1$ is not the absolute minimum. Whether the stationary point at $x = -1$ is an absolute maximum depends on which interval of x is being considered. Similar remarks apply to the stationary point at $x = 1$.

Worked examples

8.1 Where do the following functions take their absolute maximum and minimum values on the intervals given?

(a) $x^3, 1 \leqslant x \leqslant 3$ (b) $e^{-x}, 0 \leqslant x \leqslant 2$

(c) $e^{-x}, 0 \leqslant x$ (d) $\tan x, 0 \leqslant x \leqslant \pi$

Solution (a) If $y = x^3$ then $dy/dx = 3x^2$, which is positive for all x and therefore the function is increasing over any interval. Hence its least value in the interval quoted is at $x = 1$, and this value is 1; the greatest value, 27, occurs at $x = 3$. See Figure 8.3(a).

(b) If $y = e^{-x}$ then $dy/dx = -e^{-x}$, hence $dy/dx < 0$ for all x. The function is decreasing over any interval and takes its greatest value on the given interval when $x = 0$ (the value is 1) and its least value at $x = 2$ (the value is $e^{-2} = 1/e^2$). See Figure 8.3(b).

(c) The greatest value is again 1 when $x = 0$. As $x \to \infty$, e^{-x} decreases towards zero, but never reaches that value; it has *no* least value.

(d) If $y = \tan x$ then $dy/dx = \sec^2 x$; hence $dy/dx \geqslant 1$ for $x \neq \pi/2$. Because the function is discontinuous at $x = \pi/2$ (see Figure 8.3c) it is increasing on each of the intervals $0 \leqslant x < \pi/2$ and $\pi/2 < x \leqslant \pi$. It does not, however, possess either a greatest or a least value on the given interval.

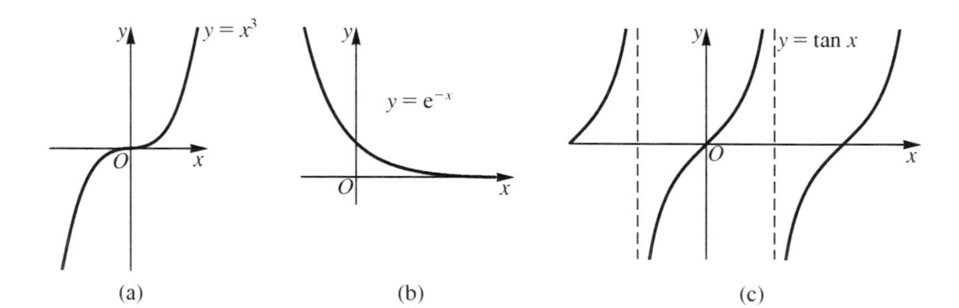

Figure 8.3.
(a) $y = x^3$
(b) $y = e^{-x}$
(c) $y = \tan x$

8.2 Find the absolute maximum and minimum values of the function

$$f(x) = \frac{1}{(x^2 + 2)^2}$$

Solution $f(x) > 0$ for all values of x. For large values of $|x|$, $x^2 \gg 2$ so that

$$f(x) \simeq \frac{1}{(x^2)^2} = \frac{1}{x^4}$$

As $x \to \pm\infty$, $f(x) \to 0$ but for no value of x does $f(x) = 0$. The function has no absolute minimum. Now $f(x)$ is greatest when $x^2 + 2$ is least, that is, when $x = 0$. Therefore the absolute maximum value is $f(0) = 1/2^2 = 1/4$.

8.3 Determine the intervals on which each of the following functions is increasing and on which it is decreasing. Where are the stationary points?

(a) $f(x) = 2x - x^2$ (b) $f(x) = x^2 + 4x + 3$

(c) $f(x) = 4 - x + 2x^2$ (d) $f(x) = x^3 - 4x^2 + 4x + 2$

(e) $f(x) = 1 + 9x - \dfrac{x^3}{3}$ (f) $f(x) = -x^3$

Solution

(a) $f'(x) = 2 - 2x$. Hence, when $x < 1$, $2 - 2x > 0$ and the function is increasing; when $x > 1$ the function is decreasing; $x = 1$ is the stationary point.

(b) $f'(x) = 2x + 4$. Hence, when $x < -2$, $2x + 4 < 0$ and the function is decreasing; when $x > -2$ the function is increasing; $x = -2$ is the stationary point.

(c) $f'(x) = -1 + 4x$. When $x < \frac{1}{4}$, $-1 + 4x < 0$ and the function is decreasing; when $x > \frac{1}{4}$ the function is increasing; $x = \frac{1}{4}$ is the stationary point.

(d) $f'(x) = 3x^2 - 8x + 4 = (3x - 2)(x - 2)$. When $x = \frac{2}{3}$ or $x = 2$ the function has a stationary point. When $x < \frac{2}{3}$ or $x > 2$, $f'(x) > 0$ and the function is increasing; when $\frac{2}{3} < x < 2$ then $f'(x) < 0$ and the function is decreasing.

(e) $f'(x) = 9 - x^2 = (3 - x)(3 + x)$. Stationary points occur where $x = -3$ and $x = 3$. The function is decreasing for $x < -3$ or $x > 3$ and increasing when $-3 < x < 3$.

(f) $f'(x) = -3x^2$. So $x = 0$ is a stationary point. For all other values of x, $f'(x) < 0$ and the function is decreasing.

Self-assessment questions 8.1

1. Explain the terms absolute maximum and absolute minimum of a function.
2. Explain how to locate the stationary points of a function.

Exercise 8.1

1. Find the absolute minimum and maximum values of the following functions on the intervals given:

   (a) x^2 $\quad -2 \leqslant x \leqslant 1$
   (b) e^x $\quad -1 \leqslant x \leqslant 2$
   (c) $\cos x$ $\quad -2\pi \leqslant x \leqslant \pi$
   (d) $\tan x$ $\quad -\pi \leqslant x \leqslant \frac{\pi}{3}$

2. Find the absolute maximum and minimum values of the following functions:

   (a) $f(x) = \dfrac{1}{(x-2)^2 + 4}$

   (b) $f(x) = \dfrac{1}{5 - (x+4)^2}$

3. Determine the intervals on which each of the following functions is (i) increasing and (ii) decreasing. Locate any stationary points:

   (a) $f(x) = 4 - (x + 2)^2$
   (b) $f(x) = (x - 1)(x + 5)$
   (c) $f(x) = (x + 1)^3 - 4$
   (d) $f(x) = x^4 - 4x^2 + 3$
   (e) $f(x) = 1 - e^{-x}$
   (f) $f(x) = \ln x$

8.2 Local minima and maxima and points of inflection

KEY POINT

The function $f(x)$ has a **local minimum** at $x = a$ if $f(x) \geqslant f(a)$ for all x near a.

The function $f(x)$ has a **local maximum** at $x = a$ if $f(x) \leqslant f(a)$ for all x near a.

A local maximum may also be the absolute maximum of a function in an interval. To locate the absolute maximum we should consider the end points of the interval as well as try local maxima.

(Local maxima and minima are known collectively as local extrema; *extrema* is the plural of *extremum*.)

With the exception of functions whose graphs have sharp bends (for example $f(x) = |x|$) or have discontinuous jumps (for example the unit step function) local maxima and minima are stationary points, at which $f'(x) = 0$. The converse is not always true. Look again at the graph of $f(x) = x^3$ in Figure 8.3(a). Although $f'(x) = 0$ at $x = 0$, the feature there is neither a local minimum nor a local maximum; to understand what is happening at $x = 0$ we need to use the second derivative of $f(x)$.

Consider the graphs of Figure 8.4. Part (a) shows a **concave upwards** function while part (b) shows a **concave downwards** function.

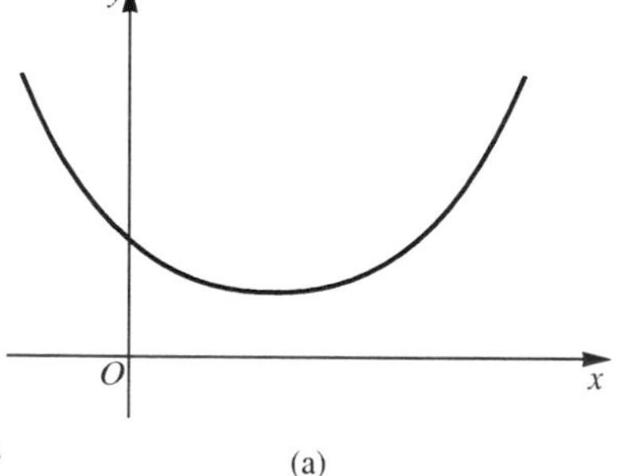

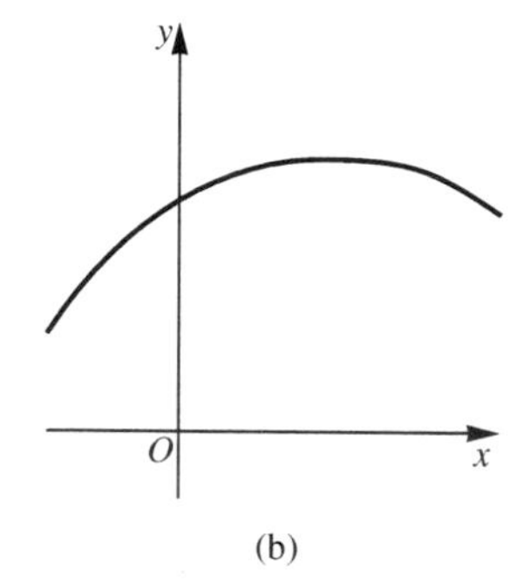

Figure 8.4.
(a) Concave upwards function
(b) Concave downwards function

Notice that in part (a) as we move from left to right along the graph the gradient goes from steep and negative, to shallow and negative, through zero, to shallow and positive, then steep and positive; hence $f'(x)$ is increasing everywhere. Since the second derivative measures the rate of change of the first derivative then the second derivative is positive everywhere. A similar argument applied to part (b) shows that $f'(x)$ is decreasing everywhere and therefore the second derivative is negative everywhere. If we drive along a road in the shape of the graph of part (a) in the direction of increasing x we would turn the steering wheel to the left; travelling along the curve in part (b) the steering wheel would be turned to the right. Many functions have a graph which is concave upwards on part of its domain and concave downwards elsewhere; see for example Figure 8.2.

The second derivative allows us to determine concavity.

KEY POINT

Where $f''(x) > 0$ the graph of $f(x)$ is concave upwards.
Where $f''(x) < 0$ the graph of $f(x)$ is concave downwards.

Looking again at the graph of $f(x) = x^3$ in Figure 8.3(a) we see that the concavity of the function changes at $x = 0$, from concave downwards to concave upwards.

KEY POINT

A point where the concavity changes is known as a **point of inflection**.

Worked examples

8.4 Locate the stationary points of each of the following functions. By evaluating the function at nearby points on either side of each stationary point determine the nature of that stationary point.
(a) $f(x) = 3 + x^2$ (b) $f(x) = \sin x, \quad 0 \leqslant x \leqslant \pi$

Solution (a) $f'(x) = 2x$; $f'(x) = 0$ where $x = 0$, which is the stationary point and $f(0) = 3$. Let h be a small positive number. To the right of the stationary point $f(h) = 3 + h^2 > f(0)$. To the left, $f(-h) = 3 + h^2 > f(0)$. Hence the stationary point is a local minimum.

(b) $f'(x) = \cos x$; $f'(x) = 0$ where $\cos x = 0$, that is, at $x = \pi/2$; note that $\sin(\pi/2) = 1$.

Remember that $\sin(\pi/2) = 1$, $\cos(\pi/2) = 0$.

To the right,

$$\sin\left(\frac{\pi}{2} + h\right) = \sin\frac{\pi}{2}\cos h + \cos\frac{\pi}{2}\sin h$$
$$= \cos h < 1 = f\left(\frac{\pi}{2}\right)$$

To the left,

$$\sin\left(\frac{\pi}{2} - h\right) = \sin\frac{\pi}{2}\cos h - \cos\frac{\pi}{2}\sin h$$
$$= \cos h < 1 = f\left(\frac{\pi}{2}\right)$$

Hence the stationary point is a local maximum.

8.5 For each of the following functions use the second derivative to determine the intervals on which the function is concave upwards and on which it is concave downwards; identify any points of inflection.
(a) $f(x) = 3x - 2x^2$ (b) $f(x) = -x^3 + 12x^2 + 2x - 3$
(c) $f(x) = (x + 2)^3$ (d) $f(x) = x(x - 4)^3$
(e) $f(x) = \tan x, -\pi/2 < x < \pi/2$ (f) $f(x) = x\,e^{2x}$

Solution

(a) $f'(x) = 3 - 4x$; $f''(x) = -4$. Since $f''(x) < 0$ everywhere then the function is concave downwards everywhere.

(b) $f'(x) = -3x^2 + 24x + 2$; $f''(x) = -6x + 24$. When $x < 4$, $f''(x) > 0$ and the function is concave upwards; when $x > 4$ $f''(x) < 0$ and the function is concave downwards. At $x = 4$ there is a point of inflection.

(c) $f'(x) = 3(x + 2)^2$; $f''(x) = 6(x + 2)$. When $x < -2$ $f''(x) < 0$ and the function is concave downwards; when $x > -2$ the function is concave upwards. There is a point of inflection when $x = -2$.

(d)
$$\begin{aligned} f'(x) &= x \times 3(x - 4)^2 + (x - 4)^3 = (3x + x - 4)(x - 4)^2 \\ &= (4x - 4)(x - 4)^2 \end{aligned}$$
$f''(x) = 12(x - 2)(x - 4)$ so that if $x < 2$ or $x > 4$, $f''(x) > 0$ and the function is concave upwards. If $2 < x < 4$, $f''(x) < 0$ and the function is concave downwards. Points of inflection are found when $x = 2$ and $x = 4$.

(e) $f'(x) = \sec^2 x$, $f''(x) = 2\sec^2 x \tan x$. Note that $\sec^2 x \geqslant 1$. If $-\pi/2 < x < 0$, $\tan x < 0$ so that $f''(x) < 0$ and the function is concave downwards. If $0 < x < \pi/2$, $\tan x > 0$ so that $f''(x) > 0$ and the function is concave upwards. There is a point of inflection at $x = 0$.

(f) $f'(x) = (1 + 2x)e^{2x}$, $f''(x) = (4 + 4x)e^{2x}$. Note that $e^{2x} > 0$ for all x. If $x < -1$, $f''(x) < 0$ and the function is concave downwards. If $x > -1$, $f''(x) > 0$ and the function is concave upwards. There is a point of inflection where $x = -1$.

Self-assessment questions 8.2

1. Explain the terms concave upwards and concave downwards with regard to a function.
2. Explain the terms local maximum and local minimum in relation to functions.
3. Give the definition of a point of inflection.

Exercise 8.2

1. Locate the stationary points of each of the following functions. Evaluate the function at nearby points on either side of the stationary point and hence determine the nature of the stationary point:

   (a) $f(x) = 9 - x^2$
   (b) $f(x) = \cos x \quad 0 \leqslant x \leqslant \pi$

2. Determine the intervals on which each of the following functions is (i) concave upwards, (ii) concave downwards. Identify any points of inflection.

   (a) $f(x) - 4x - 3x^2$
   (b) $f(x) = x^3 - 6x^2 - x + 3$
   (c) $f(x) = x^3(x + 2)$
   (d) $f(x) = x^2(x + 2)^2$
   (e) $f(x) = \cos x \quad 0 \leqslant x \leqslant \pi$
   (f) $f(x) = xe^{-x}$

8.3 First derivative test and applications

The **first derivative test** determines the nature of a stationary point by examining the first derivative of the function in the neighbourhood of the stationary point. At a local maximum or a local minimum the first derivative of the function changes sign, hence the name **turning point**.

Table 8.1 summarizes the situation for a stationary point where $x = a$.

The sketches give the approximate behaviour of the function very near the stationary point. We may summarize the test for a local maximum and a local minimum as follows:

KEY POINT

If $f'(a) = 0$ and $f'(x)$ changes sign from negative to positive then the function has a local minimum at $x = a$.

If $f'(a) = 0$ and $f'(x)$ changes sign from positive to negative then the function has a local maximum at $x = a$.

Worked examples

8.6 Find and classify the stationary points of the following functions:
(a) $f(x) = x^3 + x^2$ (b) $f(x) = (x - 2)(x + 1)^2$ (c) $f(x) = 2x^2 - x^4$
(d) $f(x) = x^4 - 6x^2 + 8x + 10$ (e) $f(x) = x^5 - 5x^4 + 5x^3 - 1$

Solution (a) $f'(x) = 3x^2 + 2x = x(3x + 2)$. $f'(x) = 0$ when $x = 0$ or $x = -\frac{2}{3}$. These are the only places where $f'(x)$ *might* change sign. We try some simple values of x on either side of each stationary point.

At $x = -1$, $f'(x) > 0$; at $x = -\frac{2}{3}$, $f'(x) = 0$ and at $x = -\frac{1}{3}$, $f'(x) = \frac{3}{9} - \frac{2}{3} < 0$. Hence from Table 8.1 there is a local maximum at $x = -\frac{2}{3}$.

At $x = -\frac{1}{3}$, $f'(x) < 0$; at $x = 0$, $f'(x) = 0$ and at $x = 1$, $f'(x) > 0$. Hence there is a local minimum at $x = 0$.

(b)
$$f'(x) = (x - 2) . 2(x + 1) + 1 . (x + 1)^2 \quad \text{(by the product rule)}$$
$$= (x + 1)(3x - 3) = 3(x + 1)(x - 1)$$

Hence $f'(x) = 0$ at $x = \pm 1$.

At $x = -2$, $f'(x) = 9 > 0$ and at $x = 0$, $f'(x) = -3 < 0$ so that there is a local maximum at $x = -1$.

At $x = 2$, $f'(x) = 9$ so that there is a local minimum at $x = 1$.

(c) $f'(x) = 4x - 4x^3 = 4x(1 - x^2)$; $f'(x) = 0$ at $x = 0, \pm 1$. Now $f'(-2) > 0, f'\left(-\frac{1}{2}\right) < 0, f\left(\frac{1}{2}\right) > 0$ and $f(2) < 0$. Hence there is a local maximum at $x = -1$ and at $x = 1$ and a local minimum at $x = 0$.

(d) $f'(x) = 4x^3 - 12x + 8 = 4(x - 1)^2(x + 2)$. Hence $f'(x) = 0$ when $x = 1$ or $x = -2$.

Now $f'(-3) < 0, f'(0) > 0$ and $f'(2) > 0$. When $x = -2$ there is a local minimum and when $x = 1$ there is a horizontal point of inflection.

Table 8.1.

Sign of $f'(x)$ to the left of $x = a$	Sign of $f'(x)$ to the right of $x = a$	Sketch	Deduction
$-$	$+$	$\cup$	Local minimum
$+$	$-$	$\frown$	Local maximum
$-$	$-$		Horizontal point of inflection
$+$	$+$		

(e)
$$f'(x) = 5x^4 - 20x^3 + 15x^2 = 5x^2(x^2 - 4x + 3)$$
$$= 5x^2(x - 1)(x - 3)$$

There are stationary points at $x = 0$, $x = 1$ and $x = 3$. Now $f'(-1) > 0$, $f'(\frac{1}{2}) > 0$, $f'(2) < 0$ and $f'(4) > 0$. Therefore there is a horizontal point of inflection at $x = 0$, a local maximum at $x = 1$ and a local minimum at $x = 3$.

8.7 Find the stationary values of the functions in Example 8.6(c) and (d) and sketch the graphs of these functions.

Solution For $f(x) = 2x^2 - x^4$ the stationary values are $f(-1) = 1$, $f(0) = 0$ and $f(1) = 1$.

The graph is symmetrical about the y-axis since $f(-x) = f(x)$. Now $f(x) = x^2(2 - x^2)$ so that $f(x) = 0$ when $x = 0$, $\pm\sqrt{2}$ and as $x \to \pm\infty$, $f(x) \to -\infty$. The sketch is shown in Figure 8.5.

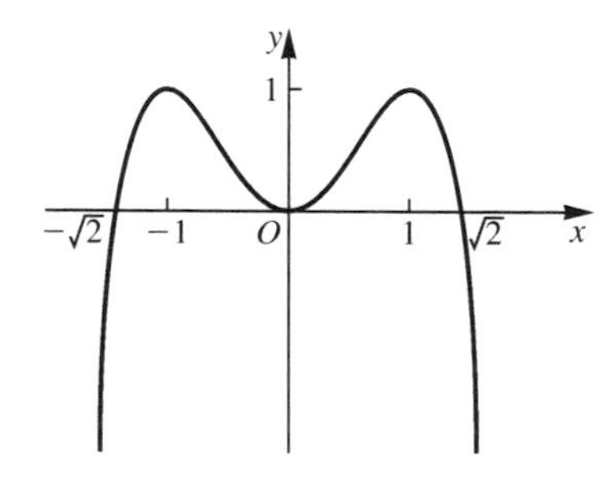

Figure 8.5. $y = 2x^2 - x^4$

For $f(x) = x^4 - 6x^2 + 8x + 10$ the stationary values are $f(-2) = -14$ and $f(1) = 13$.

When $x = 0$, $f(x) = 10$. The sketch is shown in Figure 8.6.

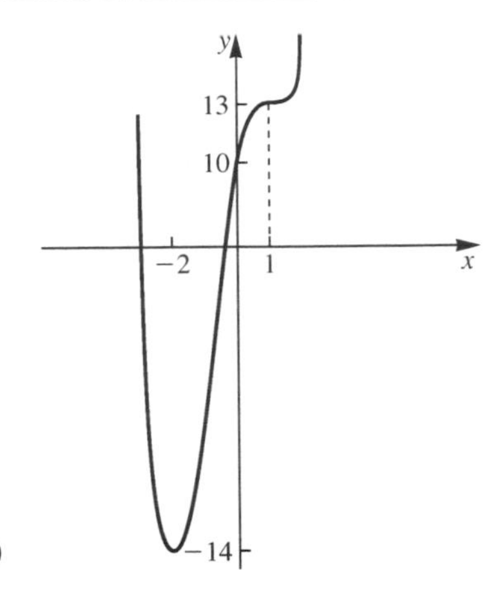

Figure 8.6.
$f(x) = x^4 - 6x^2 + 8x + 10$

8.8 The amplitude of the alternating current in a series *LCR* circuit is given by

$$I = \frac{V}{\left[R^2 + \left(\omega L - \dfrac{1}{\omega C}\right)^2\right]^{\frac{1}{2}}}$$

where V is the amplitude of the applied voltage and $\omega/(2\pi)$ is its frequency. Find the value of ω for which I is at a maximum, given fixed values for L, C and R.

Solution

$$\frac{\mathrm{d}I}{\mathrm{d}\omega} = \frac{-V\left(L + \dfrac{1}{\omega^2 C}\right)\left(\omega L - \dfrac{1}{\omega C}\right)}{\left[R^2 + \left(\omega L - \dfrac{1}{\omega C}\right)^2\right]^{\frac{3}{2}}}$$

found by using the chain rule.

Now $\mathrm{d}I/\mathrm{d}\omega = 0$ when $L = 1/(\omega^2 C)$, that is

$$\omega^2 = \frac{1}{LC}, \qquad \text{or} \qquad \omega = \frac{1}{\sqrt{LC}}$$

For

$$\omega < \frac{1}{\sqrt{LC}}, \qquad L - \frac{1}{\omega^2 C} < 0 \qquad \text{and} \qquad \frac{\mathrm{d}I}{\mathrm{d}\omega} > 0$$

and for

$$\omega > \frac{1}{\sqrt{LC}}, \qquad \frac{\mathrm{d}I}{\mathrm{d}\omega} < 0$$

Hence I has a maximum value when $\omega = 1/\sqrt{LC}$. This maximum value is V/R.

Self-assessment questions 8.3

1. Explain how to use the first derivative test to find the nature of a stationary point of a function.
2. Explain the difference between a stationary point and a turning point.

Exercise 8.3.

1. Find and classify the stationary points of the following functions:

   (a) $f(x) = x^3 - x^2$
   (b) $f(x) = (x-1)^2(x+2)$
   (c) $f(x) = 4x^2 - x^4$
   (d) $f(x) = x^4 - 6x^2 - 8x - 3$

   Hence sketch the graphs of these functions.

2. The horizontal range of a projectile launched with velocity u at an angle θ to the horizontal is

   $$\frac{2u^2 \cos\theta \sin\theta}{g}$$

   At what angle should it be launched to achieve maximum horizontal range?

3. The power P transmitted by a belt drive is given by $P = Tv - cv^3$ where T is the tension in the belt, v is the speed of the belt and c is a constant. Find the speed at which the belt transmits maximum power. Hence sketch the graph of P against v.

4. An open box is made from a rectangular sheet of metal 80 cm by 100 cm. From each corner a square of metal is cut off, then the remainder is folded to form the box. What are the dimensions of the squares cut off if the volume contained by the box is to be maximized?

8.4 Second derivative test and applications

The **second derivative test** is an alternative strategy for classifying stationary points and identifying points of inflection. It is often simpler to use the second derivative if its calculation does not require too much effort.

KEY POINT

If $f'(a) = 0$ and $f''(a) > 0$ then the function has a local minimum at $x = a$.

If $f'(a) = 0$ and $f''(a) < 0$ then the function has a local maximum at $x = a$.

If $f''(a) = 0$ and $f''(x)$ changes sign at $x = a$ then the function has a point of inflection at $x = a$.

The case where *both* $f'(a) = 0$ *and* $f''(a) = 0$ needs further consideration.

Although it is true that if $x = a$ is a point of inflection then $f''(a) = 0$, the converse is not necessarily true. We say that the condition $f''(a) = 0$ is *necessary* but not *sufficient* for the function to have a point of inflection at $x = a$.

(The function $f(x) = x^4$ has $f'(0) = 0$ and $f''(0) = 0$ but it has a local minimum at $x = 0$.)

This case is best handled by checking the sign of the first derivative on either side of the stationary point.

Worked examples

8.9 Find the points of inflection of the following functions:
(a) $f(x) = x^2 - x^3$ (b) $3x^5 - 5x^3$

Solution (a) $f'(x) = 2x - 3x^2, f''(x) = 2 - 6x$
$f'(x) = 0$ when $x(2 - 3x) = 0$, that is, $x = 0$ or $x = \frac{2}{3}$
$f''(x) = 0$ when $x = \frac{1}{3}$ and this is where the function has a point of inflection.

(b) $f'(x) = 15x^4 - 15x^2 = 15x^2(x^2 - 1)$
$f''(x) = 60x^3 - 30x = 30x(2x^2 - 1)$
When $x = \pm(1/\sqrt{2})$, $2x^2 = 1$ and $f''(x) = 0$ but $f'(x) \neq 0$. Hence the function has points of inflection where $x = \pm(1/\sqrt{2})$. At $x = 0$ both $f'(x)$ and $f''(x)$ are zero and we need further investigation.

The gradient of the function does not change sign between $x = -(1/\sqrt{2})$ and $x = 0$; we take $x = -1/2$ for simplicity. Similarly, $f'(x)$ does not change sign between $x = 0$ and $x = 1/\sqrt{2}$ so we take $x = 1/2$. Now $f'(-\frac{1}{2}) = \frac{15}{4}(-\frac{3}{4}) < 0$ and $f'(\frac{1}{2}) = \frac{15}{4}(-\frac{3}{4}) < 0$. Hence the function has a point of inflection where $x = 0$.

8.10 The deflection downwards y at a distance x from the left-hand end of a beam of length L built in at both ends and subject to a uniformly distributed load w/unit length is given by

$$y = \frac{w}{24EI}(x^4 - 2Lx^3 + L^2x^2)$$

where E and I are constants.

Find the location of the greatest deflection and of the points of inflection.

Solution

$$\frac{dy}{dx} = \frac{w}{24EI}(4x^3 - 6Lx^2 + 2L^2x) = \frac{wx}{12EI}(2x^2 - 3Lx + L^2)$$

$$= \frac{wx}{12EI}(2x - L)(x - L)$$

$$\frac{d^2y}{dx^2} = \frac{w}{24EI}(12x^2 - 12Lx + 2L^2) = \frac{w}{12EI}(6x^2 - 6xL + L^2)$$

Now $dy/dx = 0$ when $x = 0$, $x = L/2$ or $x = L$.

The ends $x = 0$ and $x = L$ are built in and there is no deflection. At $x = L/2$,

$$\frac{d^2y}{dx^2} = -\frac{wL^2}{24EI} < 0$$

and this a local maximum on the curve. Therefore this represents the maximum deflection of the beam. Further,

$$\frac{d^2y}{dx^2} = 0 \qquad \text{where} \qquad 6x^2 - 6Lx + L^2 = 0$$

that is

$$x = \frac{6L \pm \sqrt{(36L^2 - 24L^2)}}{12} = \frac{6L \pm 2\sqrt{3}L}{12}$$

To 2 d.p. these values are $0.21L$ and $0.79L$.

8.11 An open tank is in the form of a cylinder with circular cross-section. If the tank is formed from sheet metal and the area of metal used is S, show that the volume of the tank is given by

$$V = \frac{1}{2}(Sr - \pi r^3)$$

Hence, for a given S, show that the volume of the tank is greatest when the ratio of height h to diameter d is $1:2$.

Solution The volume of a tank of height h and radius r is $V = \pi r^2 h$. The surface area is $S = 2\pi rh + \pi r^2$. The expression $\frac{1}{2}(Sr - \pi r^3)$ is equal to

$$\tfrac{1}{2}(2\pi r^2 h + \pi r^3 - \pi r^3) = \pi r^2 h = V$$

To find the maximum volume we first find $dV/dr = \frac{1}{2}(S - 3\pi r^2)$. Now

$$\frac{dV}{dr} = 0 \qquad \text{when} \qquad r^2 = \frac{S}{3\pi} \qquad \text{that is,} \qquad r = \left(\frac{S}{3\pi}\right)^{\frac{1}{2}}$$

Further,

$$\frac{d^2V}{dr^2} = -3\pi r < 0$$

so that the stationary point is a local maximum (the negative square root is ignored because the radius must be a positive number). When

$$r = \left(\frac{S}{3\pi}\right)^{\frac{1}{2}}, \qquad S = 2\pi\left(\frac{S}{3\pi}\right)^{\frac{1}{2}} h + \pi \,.\, \frac{S}{3\pi}$$

that is,

$$2\pi\left(\frac{S}{3\pi}\right)^{\frac{1}{2}} h = \frac{2}{3}S$$

Hence

$$h = \frac{2S}{3 \times 2\pi}\left(\frac{3\pi}{S}\right)^{\frac{1}{2}} = \frac{S^{\frac{1}{2}}}{3^{\frac{1}{2}}\pi^{\frac{1}{2}}} = \left(\frac{S}{3\pi}\right)^{\frac{1}{2}} = r$$

Therefore

$$\frac{h}{r} = 1 \qquad \text{and} \qquad \frac{h}{d} = \frac{1}{2}$$

Self-assessment questions 8.4

1. Explain how to use the second derivative test to find the nature of a stationary point of a function.
2. Explain how to determine the nature of a stationary point if both the first and second derivatives of the function are zero at that point.

Exercise 8.4

1. Find the points of inflection of the following functions:

   (a) $f(x) = x^2 - x^4$
   (b) $f(x) = xe^{-x}$
   (c) $f(x) = x^2e^{-x}$
   (d) $f(x) = x^ne^{-x} \qquad n > 2$

2. The displacement at time t of an object from equilibrium is given by $x = 4e^{-\alpha t} \cos \omega t$, where α and ω are positive constants. Find the first two times at which the object is temporarily at rest. How do these values relate to the local maxima and minima on the graph of $x(t)$?

3. A stable point of equilibrium occurs when the potential energy V is at a minimum. The potential energy of one atom due to the presence of another is given by

$$V = \frac{A}{r^{12}} - \frac{B}{r^6}$$

   where r is the distance between the atoms and A and B are positive constants. Find the distance between the atoms at the point of stable equilibrium.

Test and assignment exercises 8

1. Find the absolute maximum and minimum values of the following functions on the intervals given:

   (a) $\frac{1}{x}, \quad 1 \leqslant x \leqslant 2$

   (b) $\frac{1}{x}, \quad -2 \leqslant x \leqslant 1$

   (c) $\frac{1}{x}, \quad -2 \leqslant x \leqslant 2$ (d) $\sin x, \quad$ all x

2. Find the absolute maximum and minimum values of the function

$$f(x) = \frac{1}{x^2 + 4x + 8}$$

3. An open tank which has vertical sides and a square base is to be constructed from metal so as to use the minimum amount of material. If the capacity of the tank is to be 8 m^3 find its dimensions.

4. Determine the intervals on which each of the following functions is increasing and on which it is decreasing. Where are the stationary points?

   (a) $f(x) = x - x^2$
   (b) $f(x) = x^2 - 4x + 3$
   (c) $f(x) = x + \frac{1}{x}$
   (d) $f(x) = (x - 1)^2(x + 2)$

5. Determine the intervals on which each of the following functions is (i) concave upwards, (ii) concave downwards. Identify any points of inflection.

   (a) $f(x) = 2x - x^2$
   (b) $f(x) = x^3 - 12x^2 + x - 8$
   (c) $f(x) = x^2(x + 2)$
   (d) $f(x) = x^{-1}$
   (e) $f(x) = x + x^{-1}$
   (f) $f(x) = \ln x, \qquad x > 0$

6. Use the first derivative test to locate and classify the stationary points of the following functions:

   (a) $x^4 - 4x$

   (b) $x^2(x^2 + 8)$

   (c) $x^2 + \frac{1}{x^2}$

   (d) $3x^4 - 2x^3 - 6x^2 + 2$

7. Use the second derivative test to locate and classify the stationary points of the following functions. Identify any points of inflection.

   (a) $x^4 - 6x^2$

   (b) $x^2(x - 4)$

   (c) $x - \frac{1}{x}$

   (d) $x^4 - 4x^3$

9 Integration

Objectives

This chapter

- defines definite and indefinite integration
- states the fundamental theorem of calculus
- states the basic properties of definite integrals
- integrates some simple functions
- carries out integration by substitution
- carries out integration by parts

9.1 Definite and indefinite integrals

The process of integration is used in two contexts: where the process of differentiation needs to be reversed, and where an infinite summation is carried out. We look at an example of the latter.

The work done against a constant force is the product of the magnitude of the force and the distance moved in a direction which is opposite to the direction of the force. According to Coulomb's law the force F between two charges q_1 and q_2 which are separated by a distance r is given by

$$F(r) = \frac{q_1 q_2}{4\pi\varepsilon_0 r^2}$$

where ε_0 is the permittivity of free space. We want to calculate the work done in moving a unit charge against the field of a charge q_2 from a distance a to a distance b from that charge. In this example the force is not constant. If we divide the journey into n subintervals we can approximate the force by a constant value, say the value at the end of the subinterval. See Figure 9.1.

Figure 9.1. Subintervals used to approximate work done

Let the force in the subinterval r_{k-1} to r_k be $F(r_k)$. The work done is approximately $F(r_k).\delta r$ and the total work done going from a distance a to a distance b is given by

$$W \cong \sum_{k=1}^{n} F(r_k)\delta r$$

If we let the number of subintervals increase progressively then $\delta r \to 0$ and

$$W = \lim_{\delta r \to 0} \left[\sum_{k=1}^{n} F(r_k)\delta r \right] \qquad 9.1$$

Such a limit as that on the right-hand side of (9.1) is called a **definite integral** and is written

$$W = \int_a^b F(r)\mathrm{d}r \qquad 9.2$$

In this expression $F(r)$ is known as the **integrand**, a and b are the **lower** and **upper limits of integration**, respectively, and r is the **variable of integration**. The symbol $\int$ is known as the **integral sign** and is a variation on an upper-case S (for sum) and the symbol dr is sometimes called the element of integration. Note that the definite integral is a *value* (in this example, measured in kN). Graphically, the work done is the area under the force–distance graph between $r = a$ and $r = b$.

To evaluate (9.2) we need a second concept. If $f(x)$ is a specified function and $F(x)$ is *any* function for which $F'(x) = f(x)$ then $F(x)$ is a **primitive** of $f(x)$ or an **indefinite integral** of $f(x)$. We write $F(x) = \int f(x)\mathrm{d}x$.

For example, if $f(x) = 3x^2$ then we could take $F(x) = x^3$ since $F'(x) = 3x^2$. However the functions $x^3 + 2, x^3 + \frac{1}{2}, x^3 - 1996$ are all possible candidates for $F(x)$ since all have derivatives of $3x^2$. To cover all cases we write $F(x) = x^3 + C$ where C is an unspecified constant, known as the **constant of integration**. In future, in all indefinite integrals in this chapter we shall use C for the constant of integration.

Worked examples

9.1 Find the indefinite integrals

(a) $\int 4x^3\,\mathrm{d}x$ (b) $\int 4u^3\,\mathrm{d}u$ (c) $\int x^3\,\mathrm{d}x$

Solution (a) $\dfrac{\mathrm{d}}{\mathrm{d}x}(x^4) = 4x^3$ so that $\int 4x^3\,\mathrm{d}x = x^4 + C$

The last statement is read 'the indefinite integral of $4x^3$ with respect to x is $x^4 + C$'.

(b) $\frac{d}{du}(u^4) = 4u^3$ so that $\int 4u^3\,du = u^4 + C$

(c) $\frac{d}{dx}\left(\frac{1}{4}x^4\right) = x^3$ so that $\int x^3\,dx = \frac{x^4}{4} + C$

In a sense, indefinite integration *reverses* the process of differentiation; not completely, however, because of the need to add '$+C$'.

Note that if $F(x)$ and $G(x)$ are indefinite integrals of a function $f(x)$ then $F(x) = G(x) + C$ for some constant C.

9.2 Verify the following results by differentiating the right-hand side:

(a) $\int \frac{1}{\sqrt{x+1}}\,dx = 2\sqrt{x+1} + C$

(b) $\int \frac{\cos x}{(1+\sin x)^2}\,dx = -\frac{1}{1+\sin x} + C$

Solution (a) If $y = 2(x+1)^{\frac{1}{2}}$ then

$$\frac{dy}{dx} = 2.\frac{1}{2}(x+1)^{-\frac{1}{2}} = \frac{1}{(x+1)^{\frac{1}{2}}}$$

Hence

$$\int \frac{1}{\sqrt{x+1}}\,dx = 2\sqrt{x+1} + C$$

(b) The derivative of the right-hand side is found by the chain rule:

$$\frac{d}{dx}\left((1+\sin x)^{-1}\right) = -1(1+\sin x)^{-2}.\cos x = \frac{-\cos x}{(1+\sin x)^2}$$

Hence the result.

Evaluation of definite integrals

You will have noticed the link between the notation for indefinite and definite integrals. What then is the link between $\int 3x^2 dx$ and $\int_a^b 3x^2\,dx$? The first is another function (of x) and the second is a value. The link is provided by the **fundamental theorem of calculus:**

KEY POINT

If $F(x)$ is any indefinite integral of $f(x)$ so that $F'(x) = f(x)$ then

$$\int_a^b f(x)\,dx = F(b) - F(a)$$

We write $F(b) - F(a)$ as $[F(x)]_a^b$ when we evaluate the integral.

Worked example

9.3 Use the fundamental theorem of calculus to evaluate the following:

(a) $\int_1^2 5x^4\,\mathrm{d}x$ (b) $\int_1^2 5u^4\,\mathrm{d}u$

Solution (a) If $f(x) = 5x^4$ then $F(x) = x^5 + C$ since

$$\frac{\mathrm{d}}{\mathrm{d}x}(x^5 + C) = 5x^4$$

Hence

$$\int_1^2 5x^4\,\mathrm{d}x = \left[x^5\right]_1^2 = 2^5 - 1^5 = 31$$

(Note that we omitted the term '$+C$' since it would cancel: $2^5 + C - (1^5 + C) = 2^5 - 1^5$, as before.)

(b) The variable of integration may have changed but the value is the same as in (a). Since

$$\frac{\mathrm{d}}{\mathrm{d}u}(u^5) = 5u^4$$

then

$$\int_1^2 5u^4\,\mathrm{d}u = \left[u^5\right]_1^2 = 31$$

Properties of definite integrals

The following properties of definite integrals are stated without proof. (Properties 4 and 5 apply to indefinite integrals also.)

1. $\int_a^a f(x)\,\mathrm{d}x = 0$
2. $\int_b^a f(x)\,\mathrm{d}x = -\int_a^b f(x)\,\mathrm{d}x$
3. $\int_a^c f(x)\,\mathrm{d}x + \int_c^b f(x)\,\mathrm{d}x = \int_a^b f(x)\,\mathrm{d}x$ where $a \leqslant c \leqslant b$
4. $\int_a^b \{f(x) + g(x)\}\,\mathrm{d}x = \int_a^b f(x)\,\mathrm{d}x + \int_a^b g(x)\,\mathrm{d}x$
   (The integral of a sum is the sum of the separate integrals.)
5. $\int_a^b k\,f(x)\,\mathrm{d}x = k\int_a^b f(x)\,\mathrm{d}x$ for constant k
6. If $f(x) \geqslant 0$ for all x in $a \leqslant x \leqslant b$ then $\int_a^b f(x)\,\mathrm{d}x \geqslant 0$
7. If $m \leqslant f(x) \leqslant M$ for all x in $a \leqslant x \leqslant b$ then

$$m(b - a) \leqslant \int_a^b f(x)\,\mathrm{d}x \leqslant M(b - a)$$

Properties 4 and 5 can be combined into the rule that the integral of a linear combination is the same linear combination of the integrals.

Worked example

9.4 By finding the minimum and maximum of $(x^2+4)^{\frac{1}{2}}$ on the interval $1 \leqslant x \leqslant 2$ show that

$$\sqrt{5} \leqslant \int_1^2 (x^2+4)^{\frac{1}{2}}\,dx \leqslant 2\sqrt{2}$$

Solution Note that x^2+4 and hence $(x^2+4)^{\frac{1}{2}}$ increase as x increases from 1 to 2. Hence the least value of $(x^2+4)^{\frac{1}{2}}$ in the given interval is $(1^2+4)^{\frac{1}{2}} = \sqrt{5}$ and the greatest value is $(2^2+4)^{\frac{1}{2}} = \sqrt{8} = 2\sqrt{2}$. Also, $(b-a) = 2-1 = 1$. Using property 7,

$$\sqrt{5} \leqslant \int_1^2 (x^2+4)^{\frac{1}{2}}\,dx \leqslant 2\sqrt{2}$$

Self-assessment questions 9.1

1. Explain the terms indefinite integral and definite integral.
2. Explain how the fundamental theorem of calculus can be used to evaluate definite integrals.
3. State the properties of integrals which apply to indefinite integrals

Exercise 9.1

1. Find the following indefinite integrals:

   (a) $\int 6x^5\,dx$ (b) $\int u^5\,du$

   (c) $\displaystyle\int \frac{1}{x^2}\,dx$

2. Use the fundamental theorem of calculus to evaluate the following:

   (a) $\int_1^3 4x^3\,dx$

   (b) $\int_1^3 x^3\,dx$

3. Verify, by differentiation, the following results:

   (a) $\displaystyle\int \frac{1}{(x-2)^{3/2}}\,dx = -\frac{2}{\sqrt{x-2}} + C$

   (b) $\displaystyle\int \frac{\sin x}{(1+\cos x)^3}\,dx = \frac{1}{2(1+\cos x)^2} + C$

4. Without evaluating the integrals, show that

$$\int_a^b (x^2+9)\,dx \geqslant 6\int_a^b x\,dx$$

9.2 Integrating simple functions

Table 9.1 gives some standard indefinite integrals. Note that for the trigonometric functions angles are measured in radians.

Table 9.1. Standard indefinite integrals

$f(x)$	$\int f(x)\,dx$	$f(x)$	$\int f(x)\,dx$
1	$x + C$	$\operatorname{cosec} x$	$\ln\left\|\tan\left(\frac{x}{2}\right)\right\| + C$
$x^n,\ n \neq -1$	$\frac{x^{n+1}}{n+1} + C$	$\sec x$	$\ln\|\sec x + \tan x\| + C$
$\frac{1}{x},\ x \neq 0$	$\ln\|x\| + C$	$\cot x$	$\ln\|\sin x\| + C$
$\cos x$	$\sin x + C$	$\sec^2 x$	$\tan x + C$
$\sin x$	$-\cos x + C$	e^x	$e^x + C$
$\tan x$	$\ln\|\sec x\| + C$	$\ln x$	$x\ln x - x + C$

Worked examples

9.5 Using Table 9.1, find the following integrals:

(a) $\int 1\,du$ (b) $\int_a^b 1\,dx$ (c) $\int 4\,dx$

Solution (a) From the table, $\int 1\,dx = x + C$ so that $\int 1\,du = u + C$.

(b) $\int_a^b 1\,dx = [x]_a^b = b - a$. (Remember that $[x + C]_a^b = (b + C) - (a + C)$ $= b - a$ because the $+C$ cancels.)

(c) From property 5 of Section 9.1:

$$\int 4\,dx = 4\int 1\,dx = 4x + C$$

(You might have thought that the answer was $4x + 4C$, but, remember, C is an arbitrary constant whose value is not specified. Why then write an unknown number as $4C$ when C will do?)

9.6 Find the following integrals:

(a) $\int_{-1}^1 x^2\,dx$ (b) $\int_{-1}^0 x^2\,dx$ (c) $\int_0^1 x^2\,dx$

(d) $\int_{-1}^1 x^3\,dx$ (e) $\int_{-1}^0 x^3\,dx$ (f) $\int_0^1 x^3\,dx$

Solution From Table 9.1,

$$\int x^2\,dx = \frac{x^3}{3} + C \quad \text{and} \quad \int x^3\,dx = \frac{x^4}{4} + C$$

(a) $$\int_{-1}^1 x^2\,dx = \left[\frac{x^3}{3}\right]_{-1}^1 = \frac{1}{3} - \left(-\frac{1}{3}\right) = \frac{2}{3}$$

(b) $$\left[\frac{x^3}{3}\right]_{-1}^0 = 0 - \left(-\frac{1}{3}\right) = \frac{1}{3}$$

(c) $\left[\frac{x^3}{3}\right]_0^1 = \frac{1}{3} - 0 = \frac{1}{3}$

(Note that property 3 is demonstrated.)

(d) $\left[\frac{x^4}{4}\right]_{-1}^1 = \frac{1}{4} - \left(-\frac{1}{4}\right) = 0$

(e) $\left[\frac{x^4}{4}\right]_{-1}^0 = 0 - \left(\frac{1}{4}\right) = -\frac{1}{4}$

(f) $\left[\frac{x^4}{4}\right]_0^{-1} = \frac{1}{4} - 0 = \frac{1}{4}$

(Again, property 3 is demonstrated.)

9.7 Evaluate the following integrals:

(a) $\int x^{\frac{1}{2}}(x^2 - 4)\,dx$ (b) $\int \frac{u^2 + 1}{u^4}\,du$ (c) $\int_0^{\frac{\pi}{2}} (3\cos x - 2\sin x)\,dx$

(d) $\int_1^2 (e^x + \ln x)\,dx$ (e) $\int_0^{\frac{\pi}{3}} \tan^2\theta\,d\theta$

Solution (a) $\int x^{\frac{1}{2}}(x^2 - 4)\,dx = \int (x^{\frac{5}{2}} - 4x^{\frac{1}{2}})\,dx$

$$= \frac{2}{7}x^{\frac{7}{2}} - 4 \times \frac{2x^{\frac{3}{2}}}{3} + C \qquad \text{(linear combination)}$$

$$= \frac{2}{7}x^{\frac{7}{2}} - \frac{8}{3}x^{\frac{3}{2}} + C$$

(b) $\int \frac{u^2 + 1}{u^4}\,du = \int \left(\frac{1}{u^2} + \frac{1}{u^4}\right)du$

$$= \int (u^{-2} + u^{-4})\,du = -\frac{1}{u} - \frac{1}{3u^3} + C$$

(c) $\int_0^{\frac{\pi}{2}} (3\cos x - 2\sin x)\,dx = \Big[3\sin x + 2\cos x\Big]_0^{\frac{\pi}{2}}$

$$= (3 \times 1 + 2 \times 0) - (3 \times 0 + 2 \times 1) = 3 - 2 = 1$$

(d) $\int_1^2 (e^x + \ln x)\,dx = \Big[e^x + x\ln x - x\Big]_1^2$

$$= (e^2 + 2\ln 2 - 2) - (e + \ln 1 - 1)$$

$$= e^2 - e + 2\ln 2 - 1$$

(e) We resort to a trick; using $\tan^2\theta \equiv \sec^2\theta - 1$

$$\int_0^{\frac{\pi}{3}} \tan^2\theta\,d\theta = \int_0^{\frac{\pi}{3}} (\sec^2\theta - 1)d\theta = \Big[\tan\theta - \theta\Big]_0^{\frac{\pi}{3}}$$

$$= \left(\sqrt{3} - \frac{\pi}{3}\right) - (0 - 0) = \sqrt{3} - \frac{\pi}{3}$$

Self-assessment questions 9.2

1. Explain how you would use the properties of definite integrals in combination with Table 9.1 to evaluate integrals.

Exercise 9.2

1. Use Table 9.1 to find the following integrals:

   (a) $\int \sec^2 x\,dx$ (b) $\int_0^{\frac{\pi}{4}} \sec^2 x\,dx$

   (c) $\int_1^3 \ln x\,dx$

2. Evaluate the following integrals:

   (a) $\int_{-\frac{\pi}{2}}^{\frac{\pi}{2}} \cos x\,dx$ (b) $\int_{-\frac{\pi}{2}}^{0} \cos x\,dx$

   (c) $\int_0^{\frac{\pi}{2}} \cos x\,dx$ (d) $\int_{-\frac{\pi}{2}}^{\frac{\pi}{2}} \sin x\,dx$

   (e) $\int_{-\frac{\pi}{2}}^{0} \sin x\,dx$ (f) $\int_0^{\frac{\pi}{2}} \sin x\,dx$

3. By differentiation, verify the following integrals:

   (a) $\int \cos(ax+b)\,dx = \frac{1}{a}\sin(ax+b) + C$

   (b) $\int \sin(ax+b)\,dx = -\frac{1}{a}\cos(ax+b) + C$

   (c) $\int \sec^2(ax+b)\,dx = \frac{1}{a}\tan(ax+b) + C$

   (d) $\int e^{ax+b}\,dx = \frac{1}{a}e^{ax+b} + C$

   Assume that a and b are constants and that $a \neq 0$.

4. Find the following integrals:

   (a) $\int x^{\frac{1}{3}}(x^2+2)\,dx$

   (b) $\int \frac{v+1}{v^2}\,dv$

   (c) $\int_0^{\frac{\pi}{4}} (3\sec^2 x - 2e^x)\,dx$

   (d) $\int_0^{\frac{\pi}{4}} \tan^2 x\,dx$

5. Evaluate the following integrals:

   (a) $\int_{\frac{\pi}{6}}^{\frac{\pi}{3}} (2\sec x + 3\,\mathrm{cosec}\,x)\,dx$

   (b) $\int_0^{\frac{\pi}{4}} \tan x\,dx$ (c) $\int_{\frac{\pi}{4}}^{\frac{\pi}{2}} 2\cot x\,dx$

   (d) $\int_1^2 \ln x\,dx$

6. (a) Use the result of Question 4(d) to show that $\int e^{-x}\,dx = -e^{-x} + C$.
   (b) Use the definitions of $\cosh x$ and $\sinh x$ in terms of exponentials, Table 9.1 and the result of part (a) to show that

   $$\int \cosh x\,dx = \sinh x + C$$

   and

   $$\int \sinh x\,dx = \cosh x + C$$

9.3 Integration by substitution

The aim of the method of **integration by substitution** is to transform the given integral into a related but simpler integral which we can evaluate by means of Table 9.1 and the properties of Section 9.1.

In the case of indefinite integrals we need to find a function which differentiates to the integrand. The guiding principle is to simplify the most complicated term in the integrand. The method is best illustrated by examples.

Worked examples

9.8 Find the following indefinite integrals:

(a) $\int x\sqrt{x^2+4}\,\mathrm{d}x$ (b) $\int \frac{x-2}{x^2-4x+17}\,\mathrm{d}x$

Solution (a) The awkward customer is $\sqrt{x^2+4}$. To simplify the integral we put $u = x^2 + 4$. Differentiating this with respect to x we obtain

$$\frac{\mathrm{d}u}{\mathrm{d}x} = 2x$$

Although the next step is not rigorous it is what many people do in practice. Remember though it is only a means to an end.

Now $\mathrm{d}u/\mathrm{d}x$ is *not* a fraction, but if it were $\mathrm{d}u/\mathrm{d}x = 2x$ *could* be rewritten $\mathrm{d}u = 2x\,\mathrm{d}x$ and then $\frac{1}{2}\,\mathrm{d}u = x\,\mathrm{d}x$. What we really mean is that we replace the combination $x\,\mathrm{d}x$ in the given integral by $\frac{1}{2}\,\mathrm{d}u$. To be somewhat safer we write '$\frac{1}{2}\,\mathrm{d}u = x\,\mathrm{d}x$' to show that it is not really a valid equation.

For ease of checking back we introduce the notation I for the integral with which we are working. Then

$$I = \int x\sqrt{x^2+4}\,\mathrm{d}x = \int \sqrt{x^2+4}\,x\,\mathrm{d}x$$

becomes

$$I = \int \sqrt{u}\,\frac{1}{2}\,\mathrm{d}u = \int u^{\frac{1}{2}}\frac{1}{2}\,\mathrm{d}u = \frac{1}{3}u^{\frac{3}{2}} + C$$

However 'u' is our invention and we express the result back in terms of x so that

$$I = \frac{1}{3}(x^2+4)^{\frac{3}{2}} + C$$

(Note that we used a *simple* substitution $u = x^2 + 4$, which made the differentiation relatively straightforward.)

(b) Put $u = x^2 - 4x + 17$ then $du/dx = 2x - 4 = 2(x - 2)$.
Hence '$\frac{1}{2}\,du = (x - 2)\,dx$' and

$$I = \int \frac{x - 2}{x^2 - 4x + 17}\,dx$$

becomes

$$I = \frac{1}{2}\int \frac{1}{u}\,du = \frac{1}{2}\ln|u| + C$$

Reverting to the original variable,

$$I = \frac{1}{2}\ln|x^2 - 4x + 17| + C$$

KEY POINT

Note the general result that the integral of $\dfrac{f'(x)}{f(x)}$ is $\ln|f(x)|$.

9.9 Find the following indefinite integrals:

(a) $\displaystyle\int \frac{x^2}{(1 + x^3)^2}\,dx$ (b) $\int \cos x\,(1 + \sin x)^{\frac{1}{2}}\,dx$

Solution (a) The substitution we make is $u = 1 + x^3$. Then $du/dx = 3x^2$ so that '$\frac{1}{3}\,du = x^2\,dx$'. Then

$$I = \int \frac{x^2}{(1 + x^2)^3}\,dx = \int \frac{\frac{1}{3}}{u^2}\,du = \int \frac{1}{3u^2}\,du$$

Hence

$$I = -\frac{1}{3u} + C$$

and, reverting to x via the substitution, we obtain

$$I = -\frac{1}{3(1 + x^3)} + C$$

(We *could* verify this result by differentiating to recover the integrand but it is not usual to do so.)

(b) Put $u = 1 + \sin x$ so that

$$\frac{du}{dx} = \cos x \text{ and hence '}du = \cos x\,dx\text{'}$$

Then

$$I = \int \cos x(1 + \sin x)^{\frac{1}{2}}\,dx = \int (1 + \sin x)^{\frac{1}{2}} \cos x\,dx$$

becomes

$$I = \int u^{\frac{1}{2}}\,du = \frac{2}{3}u^{\frac{3}{2}} + C$$

Finally,

$$I = \frac{2}{3}(1 + \sin x)^{\frac{3}{2}} + C$$

The decision on the substitution to be made is largely a matter of judgement based on experience. Similar-looking integrals can require quite different substitutions. Sometimes, one substitution may not be enough; often other results have to be employed. Here is a selection of further examples.

Worked example

9.10 By using the substitution suggested find the following indefinite integrals:

(a) $\int \sin^2\theta \cos^3\theta \, d\theta \qquad s = \sin\theta$

(b) $\int \sin^2\theta \cos^2\theta \, d\theta \qquad s = \sin\theta$

(c) $\int x^2\sqrt{4 - x^2}\, dx \qquad x = 2\sin\theta$ (Very hard!)

Solution (a) $\dfrac{ds}{d\theta} = \cos\theta$ so '$\cos\theta\, d\theta = ds$'

Also, $\cos^2\theta \equiv 1 - \sin^2\theta = 1 - s^2$. Then

$$I = \int \sin^2\theta \cos^3\theta\, d\theta = \int \sin^2\theta \cos^2\theta . \cos\theta\, d\theta$$

becomes

$$I = \int s^2(1 - s^2)\, ds = \int (s^2 - s^4)\, ds$$

$$= \frac{s^3}{3} - \frac{s^5}{5} + C$$

$$= \frac{1}{3}\sin^3\theta - \frac{1}{5}\sin^5\theta + C$$

(b) Using the results of part (a) we obtain

$$I = \int \sin^2\theta \cos^2\theta\, d\theta = \int s^2(1 - s^2)^{\frac{1}{2}}\, ds$$

and this is very awkward.

Recalling the identity $\sin 2\theta \equiv 2\sin\theta\cos\theta$ it follows that

$$\sin\theta\cos\theta = \frac{1}{2}\sin 2\theta$$

and therefore

$$\sin^2\theta\cos^2\theta = \frac{1}{4}\sin^2 2\theta$$

But $\cos 2\theta \equiv 1 - 2\sin^2\theta$ and so

$$\cos 4\theta \equiv 1 - 2\sin^2 2\theta$$

and

$$2\sin^2 2\theta \equiv 1 - \cos 4\theta$$

Then

$$\frac{1}{4}\sin^2 2\theta \equiv \frac{1}{8}(1 - \cos 4\theta)$$

Since

$$\frac{\mathrm{d}}{\mathrm{d}\theta}\sin 4\theta = 4\cos 4\theta \qquad \text{then} \qquad \frac{\mathrm{d}}{\mathrm{d}\theta}\left(\frac{1}{4}\sin 4\theta\right) = \cos 4\theta$$

and therefore

$$\int \cos 4\theta \,\mathrm{d}\theta = \frac{1}{4}\sin 4\theta + C$$

Then

$$I = \int \frac{1}{8}(1 - \cos 4\theta)\,\mathrm{d}\theta = \frac{1}{8}\left(\theta - \frac{1}{4}\sin 4\theta\right) + C$$

(It is worth pointing out that: (i) we could never have obtained this result by trial and error and (ii) this was a complicated process. The whole art of the method of substitution relies on acquiring such 'tricks of the trade'. In this book we merely introduce you to some of the more common kinds of substitution and show you how the method works, in the hope that you can gather further substitutions later.)

(c) If $x = 2\sin\theta$ then $4 - x^2 = 4 - 4\sin^2\theta \equiv 4\cos^2\theta$ and

$$\frac{\mathrm{d}x}{\mathrm{d}\theta} = 2\cos\theta \text{ so that '}\mathrm{d}x = 2\cos\theta\,\mathrm{d}\theta\text{'}$$

Then

$$\begin{aligned} I &= \int x^2\sqrt{4 - x^2}\,\mathrm{d}x \\ &= \int 4\sin^2\theta . 2\cos\theta . 2\cos\theta\,\mathrm{d}\theta \\ &= 16\int \sin^2\theta\cos^2\theta\,\mathrm{d}\theta \end{aligned}$$

This is similar to part (b) and further work would be required to evaluate the integral. We can use the result of (b) directly to obtain

$$I = 2\theta - \frac{1}{2}\sin 4\theta + C$$

We now need to express the result in terms of x. If $x = 2\sin\theta$ then

$$\sin\theta = \frac{x}{2} \qquad \text{and} \qquad \theta = \sin^{-1}\left(\frac{x}{2}\right)$$

Also

$$\begin{aligned}\sin 4\theta &\equiv 2\sin 2\theta \cos 2\theta \\ &\equiv 2(2\sin\theta\cos\theta)(1 - 2\sin^2\theta)\end{aligned}$$

We need therefore to find an expression in x for $\cos\theta$. Since

$$\cos^2\theta \equiv 1 - \sin^2\theta = 1 - \frac{x^2}{4} = \frac{4 - x^2}{4}$$

we take $\cos\theta = \sqrt{4 - x^2}/2$. Then

$$\begin{aligned}\sin 4\theta &= 2\left(2 \times \frac{x}{2} \times \frac{\sqrt{4 - x^2}}{2}\right)\left(1 - 2 \times \frac{x^2}{4}\right) \\ &= \frac{x\sqrt{4 - x^2}}{2}(2 - x^2)\end{aligned}$$

Finally,

$$I = 2\sin^{-1}\left(\frac{x}{2}\right) - \frac{1}{4}x\sqrt{4 - x^2}(2 - x^2) + C$$

This was an awful example and you should not expect to have to handle one like this at this stage. You should, however, be aware that some relatively harmless-looking integrals do require a great deal of patience and ingenuity to handle.

Definite integrals

There are two main methods of attack: (i) to obtain the indefinite integral and then substitute the limits of integration; (ii) to find the limits on the substitution variable and obtain a new definite integral which has the same value as the original one and there is no need to substitute back. Example 9.11 illustrates the two methods. Remember that at the heart of definite integration is the reversal of differentiation.

Worked example

9.11 Evaluate $I = \int_0^2 x^2(1 + x^3)^{\frac{1}{2}}\,dx$.

Solution Put $u = 1 + x^3$. Then $du/dx = 3x^2$ so that '$\frac{1}{3}\, du = x^2\, dx$'.

Method (i) We use substitution to find the corresponding indefinite integral J, then apply the limits to evaluate the given definite integral, I. First,

$$J = \int x^2(1 + x^3)^{\frac{1}{2}}\, dx = \int \frac{1}{3} u^{\frac{1}{2}}\, du$$

$$= \frac{1}{3} \cdot \frac{2}{3} u^{\frac{3}{2}}\, du + C$$

$$= \frac{2}{9}(1 + x^3)^{\frac{3}{2}} + C$$

Then

$$I = \left[\frac{2}{9}(1 + x^3)^{\frac{3}{2}}\right]_0^2 = \left(\frac{2}{9}\, 9^{\frac{3}{2}} - \frac{2}{9}\, 1^{\frac{3}{2}}\right)$$

$$= \frac{2}{9} \times 27 - \frac{2}{9} \times 1 = \frac{52}{9}$$

Method (ii) When $x = 0$, $u = 1$ and when $x = 2$, $u = 9$. Hence

$$I = \int_1^9 \frac{1}{3} u^{\frac{1}{2}}\, du = \left[\frac{2}{9} u^{\frac{3}{2}}\right]_1^9 = \frac{2}{9} \cdot 27 - \frac{2}{9} \cdot 1 = \frac{52}{9}$$

Self-assessment questions 9.3

1. Explain the principle behind the method of substitution for finding indefinite integrals.
2. How would you apply the method of substitution to evaluate a definite integral?

Exercise 9.3

1. Find the following indefinite integrals:

(a) $\int x^2(x^3 - 8)^{1/2}\, dx$ (b) $\int \frac{e^x}{1 + 2e^x}\, dx$

(c) $\int \frac{u}{(4 + u^2)^3}\, du$

(d) $\int \sec^2 x(1 + \tan x)^{3/2}\, dx$

2. Evaluate the following integrals:

(a) $\int_3^4 x^2(x^3 - 8)^{1/2}\, dx$

(b) $\int_0^{\frac{\pi}{4}} \sec^2 x(1 + \tan x)^{3/2}\, dx$

(c) $\int_0^1 \frac{e^x}{1 + 2e^x}\, dx$

3. By using the substitution suggested, find the following indefinite integrals:

(a) $\int xe^{x^2}\, dx, \quad u = x^2$

(b) $\int \sin^3 \theta \cos^2 \theta\, d\theta, \quad c = \cos \theta$

(c) $\int \frac{x^{1/2}}{2 + x^{3/2}}\, dx, \quad u = 2 + x^{3/2}$

(d) $\int \frac{1}{(1 + x^2)^2}\, dx, \quad x = \tan u$

(e) $\int \frac{1}{x\sqrt{x^2 + 4}}\, dx, \quad u = \frac{1}{x}$

(f) $\int (x - 2)\sqrt{x + 2}\, dx, \quad u = x + 2$

9.4 Integration by parts

The method of **integration by parts** is useful when integrating a product where one of the terms simplifies on differentiation. It can also be used for integrating some standard functions: see Exercise 9.4, Question 3 at the end of this section.

The method may be symbolized by writing the product as

$$u\frac{dv}{dx}$$

KEY POINT

$$\int u\left(\frac{dv}{dx}\right)dx = uv - \int \frac{du}{dx}v\,dx \qquad 9.4$$

In words, 'the integral of a product is the first term times the integral of the second minus the integral of (the derivative of the first term times the integral of the second)'.

The recommended approach is to apply the rule, tidy up the result and review the situation.

Example 9.12 shows a simple application of the method. Example 9.13 provides instances where a second application of the rule is required.

Worked examples

9.12 Find (a) $\int x\,e^x\,dx$ (b) $\int x\cos x\,dx$

Solution (a) Applying the rule with $u = x$ and $dv/dx = e^x$ (so that $v = e^x$ and $du/dx = 1$) we obtain

$$I = \int xe^x\,dx = x(e^x) - \int 1(e^x)\,dx$$
$$= x\,e^x - \int e^x\,dx$$

We can carry out the remaining integral easily (we have already performed it) so that

$$I = x\,e^x - e^x + C$$

(Note that in the first line we had to integrate e^x once only: we copied the result for the second occasion. Also note that had we been given $I = \int e^x . x\,dx$ it would have been necessary to write the product the other way around.)

(b) Here $u = x$, $dv/dx = \cos x$ and hence $du/dx = 1$, $v = \sin x$. Then

$$\begin{aligned} I &= \int x \cos x \, dx = x(\sin x) - \int 1(\sin x) \, dx \\ &= x \sin x - \int \sin x \, dx \\ &= x \sin x - (-\cos x) + C \\ &= x \sin x + \cos x + C \end{aligned}$$

9.13 Find (a) $\int x^2 \cos x \, dx$ (b) $\int e^x \cos x \, dx$

Solution (a) First, note that $u = x^2$, $dv/dx = \cos x$. Then

$$\begin{aligned} I &= \int x^2 \cos x \, dx = x^2 \sin x - \int 2x \sin x \, dx \\ &= x^2 \sin x - 2 \int x \sin x \, dx \end{aligned}$$

Let

$$\begin{aligned} J &= \int x \sin x \, dx \\ &= x(-\cos x) - \int 1(-\cos x) \, dx \\ &= -x \cos x + \int \cos x \, dx \\ &= -x \cos x + \sin x \end{aligned}$$

Then

$$\begin{aligned} I &= x^2 \sin x - 2(-x \cos x + \sin x) + C \\ &= x^2 \sin x + 2x \cos x - 2 \sin x + C \end{aligned}$$

(Note that we add '$+C$' only at the end of our calculations and re-adjustments.)

(b) First, note that $u = e^x$, $dv/dx = \cos x$. Then

$$I = \int e^x \cos x \, dx = e^x \sin x - \int e^x \sin x \, dx$$

Let

$$\begin{aligned} J &= \int e^x \sin x \, dx = e^x(-\cos x) - \int e^x(-\cos x) \, dx \\ &= -e^x \cos x + \int e^x \cos x \, dx \\ &= e^x \cos x + I \end{aligned}$$

Therefore

$$\begin{aligned} I &= e^x \sin x - (-e^x \cos x + I) \\ &= e^x \sin x + e^x \cos x - I \end{aligned}$$

Hence

$$2I = e^x(\sin x + \cos x)$$

so that

$$I = \frac{1}{2} e^x(\sin x + \cos x) + C$$

(Again, note when we add '$+C$'.)

Definite integrals

We could find the corresponding indefinite integral by the method of parts and then insert the limits of integration. Sometimes applying the limits 'as we go' can lead to a simplification.

Worked example

9.14 Evaluate $\int_0^{\frac{\pi}{2}} x \sin x \, \mathrm{d}x$.

Solution We *could* find

$$J = \int x \sin x \, \mathrm{d}x = -x \cos x - \int \cos x \, \mathrm{d}x \quad \text{and therefore}$$
$$= -x \cos x + \sin x + C$$

Then

$$I = \int_0^{\frac{\pi}{2}} x \sin x \, \mathrm{d}x = \Big[-x \cos x + \sin x\Big]_0^{\frac{\pi}{2}}$$
$$= (0 + 1) - (0 + 0) = 1$$

Alternatively,

$$I = \int_0^{\frac{\pi}{2}} x \sin x \, \mathrm{d}x = \Big[-x \cos x\Big]_0^{\frac{\pi}{2}} - \int_0^{\frac{\pi}{2}} 1(-\cos x) \, \mathrm{d}x$$
$$= 0 + \int_0^{\frac{\pi}{2}} \cos x \, \mathrm{d}x$$
$$= \Big[\sin x\Big]_0^{\frac{\pi}{2}} = 1$$

Self-assessment questions 9.4

1. Explain the method of integration by parts.
2. How would you adapt the method of integration by parts to deal with a definite integral?

Exercise 9.4

1. Find the following integrals:

   (a) $\int xe^{2x} \, \mathrm{d}x$ (b) $\int x \cos 2x \, \mathrm{d}x$
   (c) $\int x^2 e^{-x} \, \mathrm{d}x$ (d) $\int x(x+1)^{1/2} \, \mathrm{d}x$
   (e) $\int x \ln x \, \mathrm{d}x$ (f) $\int e^x \cos x \, \mathrm{d}x$

2. Evaluate the following integrals:

   (a) $\int_0^{\frac{\pi}{2}} x \cos 2x \, \mathrm{d}x$ (b) $\int_0^1 x(x+2)^{1/2} \, \mathrm{d}x$

3. (a) Find $\int \ln x \, \mathrm{d}x$ by writing $\ln x$ as $\ln x \times 1$ and using 'parts'.
   (b) Find $\int \sin^{-1} x \, \mathrm{d}x$ by writing $\sin^{-1} x$ as $\sin^{-1} x \times 1$ and using 'parts'.

9.5 Improper integrals

We consider two kinds of **improper integrals**

1. The **integration** is **over an infinite range**. For example, the energy required to project a rocket of mass m to escape from the Earth's gravitational pull is given by

$$\int_R^\infty \frac{GmM}{r^2}\,\mathrm{d}r$$

where G is the gravitational constant, M is the mass of the Earth and R is the radius of the Earth.

To evaluate such an integral we evaluate

$$\int_R^X \frac{GmM}{r^2}\,\mathrm{d}r$$

for a value $X > R$ and consider the limit of the result as $X \to \infty$. If the limit does not exist then we cannot evaluate the integral.

2. The **integrand is not defined at all points in the range of integration.** For example, in $\int_{-1}^{1} 1/x^2\,\mathrm{d}x$ the integrand $1/x^2$ is infinitely large when $x = 0$. To evaluate it we consider

$$E = \int_{-1}^{-\varepsilon} \frac{1}{x^2}\,\mathrm{d}x + \int_{\varepsilon}^{1} \frac{1}{x^2}\,\mathrm{d}x$$

and consider the limit of E as $\varepsilon \to 0$.

Again, if the limits do not exist then the integral cannot be evaluated.

Worked example

9.15 Evaluate the following, if possible:

(a) $\displaystyle\int_R^\infty \frac{GmM}{r^2}\,\mathrm{d}r$ (b) $\displaystyle\int_{-1}^{2} \frac{1}{x^{\frac{1}{3}}}\,\mathrm{d}x$ (c) $\displaystyle\int_0^1 \frac{1}{x}\,\mathrm{d}x$

Solution (a) $$\int_R^X \frac{GmM}{r^2}\,\mathrm{d}r = \left[-\frac{GmM}{r}\right]_R^X = \left[\frac{GmM}{r}\right]_X^R = \frac{GmM}{R} - \frac{GmM}{X}$$

As $X \to \infty$, $GmM/X \to 0$ and hence

$$\int_R^\infty \frac{GmM}{r^2}\,\mathrm{d}r = \frac{GmM}{R}$$

(b) Consider

$$\int_{-1}^{-\varepsilon} \frac{1}{x^{\frac{1}{3}}}\,\mathrm{d}x + \int_{\varepsilon}^{2} \frac{1}{x^{\frac{1}{3}}}\,\mathrm{d}x = \left[\frac{3}{2}x^{\frac{2}{3}}\right]_{-1}^{-\varepsilon} + \left[\frac{3}{2}x^{\frac{2}{3}}\right]_{\varepsilon}^{2}$$

$$= \frac{3}{2}(-\varepsilon)^{\frac{2}{3}} - \frac{3}{2} + \frac{3}{2}\cdot 2^{\frac{2}{3}} - \frac{3}{2}\varepsilon^{\frac{2}{3}}$$

$$= \frac{3}{2}\left(2^{\frac{2}{3}} - 1\right)$$

As $\varepsilon \to 0$ this value remains constant and the integral

$$\int_{-1}^{2} \frac{1}{x^{\frac{1}{3}}}\,\mathrm{d}x$$

exists with this value.

(c) $\int_{\varepsilon}^{1} \frac{1}{x} = \left[\ln|x|\right]_{\varepsilon}^{1} = \ln 1 - \ln \varepsilon = -\ln \varepsilon$

As $\varepsilon \to 0$ $\ln \varepsilon$ decreases without limit ($\to -\infty$). Hence the integral does not exist.

For parts (b) and (c) evaluate the integrals taking in turn $\varepsilon = 0.1, 0.01, 0.001, 0.0001$.

Self-assessment questions 9.5

1. State two types of improper integral.

Exercise 9.5

1. Evaluate, if possible:

(a) $\int_{0}^{\infty} \mathrm{e}^{-x}\,\mathrm{d}x$ (b) $\int_{0}^{1} \frac{1}{x^2}\,\mathrm{d}x$

Test and assignment exercises 9

1. By differentiation, verify the following results:

(a) $\int \frac{1}{1+x^2}\,\mathrm{d}x = \tan^{-1}x + C$

(b) $\int \frac{\mathrm{e}^x}{\mathrm{e}^x + 1}\,\mathrm{d}x = \ln(\mathrm{e}^x + 1) + C$

2. Use the fundamental theorem of calculus to evaluate the following:

(a) $\int_{1}^{2} 5x^4\,\mathrm{d}x$

(b) $\int_{0}^{\frac{\pi}{2}} \cos u\,\mathrm{d}u$

3. Find the following integrals:

(a) $\int_2^4 \ln x \, dx$

(b) $\int_0^{\frac{\pi}{4}} \sec^2 x \, dx$

(c) $\int_0^{\frac{\pi}{4}} \sec x \, dx$

(d) $\int x^{\frac{3}{2}}(x^2 - 6) \, dx$

(e) $\int \frac{u^2 - 1}{u} \, du$

(f) $\int_0^2 \cosh x \, dx$

4. Evaluate the following integrals:

(a) $\int_0^1 x\sqrt{x^2 + 1} \, dx$

(b) $\int_0^1 \frac{2x + 3}{x^2 + 3x + 10} \, dx$

(c) $\int_0^{\frac{\pi}{4}} \cos x(1 - \sin x)^2 \, dx$

5. Use the substitution suggested to find the following indefinite integrals:

(a) $\int \sin^2 \theta \cos^5 \theta \, d\theta, \quad s = \sin \theta$

(b) $\int \frac{x^2}{(4 + x^3)^2} \, dx, \quad u = 4 + x^3$

(c) $\int (x + 3)(x - 1)^{1.5} \, dx, \quad u = x - 1$

6. Find the following integrals:

(a) $\int xe^{-x} \, dx$ (b) $\int (x + 1) \sin x \, dx$

(c) $\int x(x + 1)^{1.5} \, dx$

(d) $\int x^2 \ln x \, dx$ (e) $\int \tan^{-1} x \, dx$

7. Evaluate the following integrals:

(a) $\int_0^1 x\sqrt{(x + 2)} \, dx$

(b) $\int_0^{\frac{\pi}{4}} x \cos x \, dx$

8. Evaluate, if possible:

(a) $\int_0^2 \frac{1}{x^4} \, dx$ (b) $\int_1^{\infty} e^{-x} \, dx$

10 Applications of integration

Objectives

This chapter

- expresses plane areas as definite integrals
- expresses the volume of a solid of revolution as a definite integral
- expresses arc length and curved surface area as definite integrals
- expresses mean and RMS values of a function as a definite integral
- evaluates examples of the above integrals
- finds first and second moments of area and moments of inertia
- finds centroids of both plane areas and solids of revolution

10.1 Plane areas

The area under the curve $y = f(x)$ between $x = a$ and $x = b$ is given by $\int_a^b f(x)\,\mathrm{d}x$. (We divide the area into thin vertical strips; a typical strip has thickness δx, height y and approximate area $y\,\delta x$. We take the sum of such areas and let $\delta x \to 0$.)

Consider

$$I_1 = \int_0^2 x^3\,\mathrm{d}x, \quad I_2 = \int_{-1}^2 x^3\,\mathrm{d}x, \quad I_3 = \int_{-2}^2 x^3\,\mathrm{d}x$$

It is readily seen that

$$I_1 = \left[\frac{x^4}{4}\right]_0^2 = \frac{16}{4} = 4, \quad I_2 = \left[\frac{x^4}{4}\right]_{-1}^2 = \frac{16}{4} - \frac{1}{4} = \frac{15}{4}$$

This is perhaps surprising since we might expect the area to **increase** when we widen the range of the integration. Worse is to follow since

$$I_3 = \left[\frac{x^4}{4}\right]_{-2}^2 = \frac{16}{4} - \frac{16}{4} = 0$$

A graph of $y = x^3$, Figure 10.1, explains what has happened.

Now consider

$$I_4 = \int_{-1}^{0} x^3 \, dx = \left[\frac{x^4}{4}\right]_{-1}^{0} = 0 - \frac{1}{4} = -\frac{1}{4}$$

This implies that *areas below the* x*-axis are assigned a negative value.*

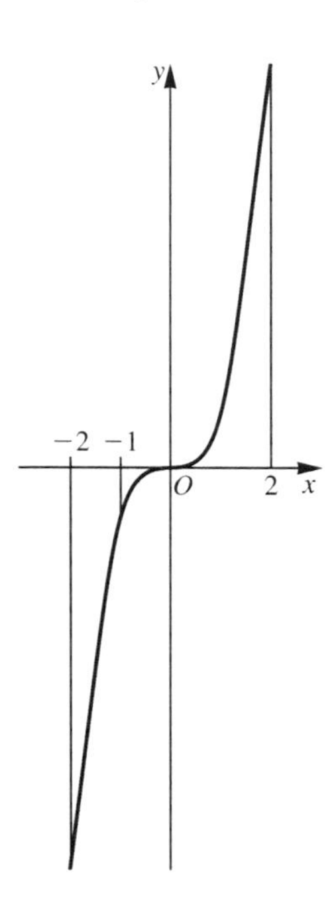

Figure 10.1. $y = x^3$

If we write

$$I_2 = \int_{-1}^{0} x^3 \, dx + \int_{0}^{2} x^3 \, dx = -\frac{1}{4} + 4$$

then the magnitude of the area between the x-axis, the curve $y = x^3$ and the ordinates $x = -1$ and $x = 2$ is

$$\frac{1}{4} + 4 = \frac{17}{4}$$

Worked examples

10.1 The velocity of an object is given by $v = 6\cos 2t$. Find the displacement x from its starting position and the distance travelled by the object at

(a) $t = \frac{\pi}{4}$ (b) $t = \frac{\pi}{2}$ (c) $t = \frac{3\pi}{4}$

Solution Since $v = dx/dt$ then $x = \int v dt = \int 6\cos 2t \, dt$ so that $x = 3\sin 2t + C$.

Let the object be at $x = 0$ initially, that is, at $t = 0$; then $C = 0$, and $x = 3\sin 2t$. The displacement of the object is the area under its velocity–time graph.

Now the displacement of an object is the distance between its current position and its starting position. This may not be the same as the distance it has travelled. For example, if you walk in a straight line from one point to another 100 metres away, both your displacement and the distance you have travelled are 100 metres, but if you now walk back to the starting point your displacement is zero but the distance you have travelled is 200 metres.

(a) When $t = \pi/4$, $x = 3\sin\pi/2 = 3$.

Is this the distance travelled? Figure 10.2 provides the answer. The distance travelled is indeed 3, since in $0 \leqslant t \leqslant \pi/4$ the velocity of the object is positive so that it is always moving in the same direction.

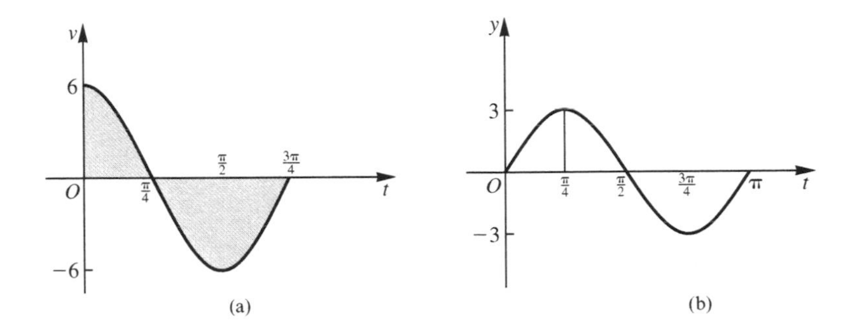

Figure 10.2.
(a) Velocity
(b) displacement

(b) When $t = \pi/2$ $x = 3\sin\pi = 0$. Although the displacement is zero, being the sum of 3 and -3, the distance travelled is 3 on the way out plus 3 on the way back, giving a total distance of 6.

(c) When $t = 3\pi/4$ $x = 3\sin(3\pi/2) = -3$. The distance travelled between $t = \pi/2$ and $t = 3\pi/4$ is 3. The total distance travelled is therefore 9.

10.2 A solid has a cross-section which is in the shape of the region between the curves $y = x^2$ and $y^2 = 8x$ as in Figure 10.3. Find the area of cross-section.

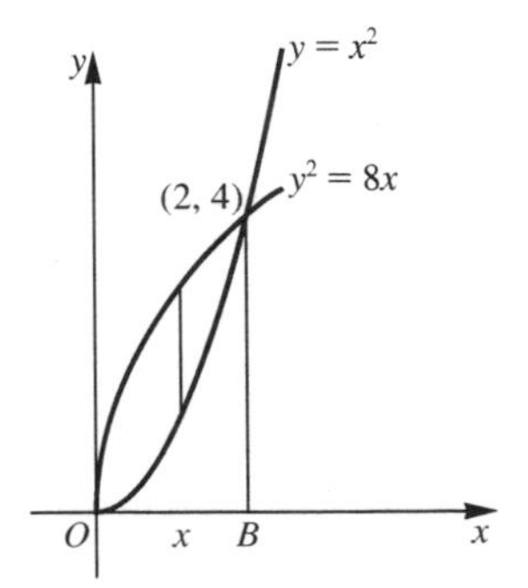

Figure 10.3.
Area between two curves

Solution The curves intersect at (0,0) and (2,4). Note that the part of the curve $y^2 = 8x$ above the x-axis can be written as $y = 2\sqrt{2x}$.

The elementary strip shown has thickness δx and length equal to the difference in heights of the two curves, that is

$$2\sqrt{2x} - x^2 = 2\sqrt{2}x^{\frac{1}{2}} - x^2$$

Hence the area between the curves is

$$\int_0^2 \left(2\sqrt{2}x^{\frac{1}{2}} - x^2\right) \mathrm{d}x = \left[\frac{4\sqrt{2}}{3}x^{\frac{3}{2}} - \frac{x^3}{3}\right]_0^2$$

$$= \frac{4\sqrt{2}}{3} \times 2\sqrt{2} - \frac{8}{3} = \frac{16}{3} - \frac{8}{3} = 2.667 \quad \text{(3 d.p.)}$$

(Note that we could find the area as the difference between the area under the curve $y = 2\sqrt{2x}$ and that under $y = x^2$, that is, as $\int_0^2 2\sqrt{2x}\,\mathrm{d}x - \int_0^2 x^2\,\mathrm{d}x$.)

Self-assessment questions 10.1

1. Explain how to interpret a definite integral as a plane area and vice versa.
2. Explain how to deal with areas below the x-axis.

Exercise 10.1

1. The velocity of an object is $v = 2\cos t + 3\sin t$. Find the displacement from its starting point at:

   (a) $t = \frac{\pi}{2}$

   (b) $t = \pi$

   (c) $t = 2\pi$

2. Find the area between the x-axis and the curve $y = x(x - 4)$.

3. Find the area between the following pairs of curves:

   (a) $y = \cos x$ and $y = 1 - \frac{2}{\pi}$

   (b) $y = 1 - x^2$ and $y = (1 - x^2)^{\frac{1}{2}}$

10.2 Volume of a solid of revolution

When that area between the curve $y = f(x)$, the x-axis and the lines $x = a$ and $x = b$ is rotated about the x-axis through a complete revolution it generates a **solid of revolution** having the x-axis as an axis of symmetry: see Figure 10.4(a).

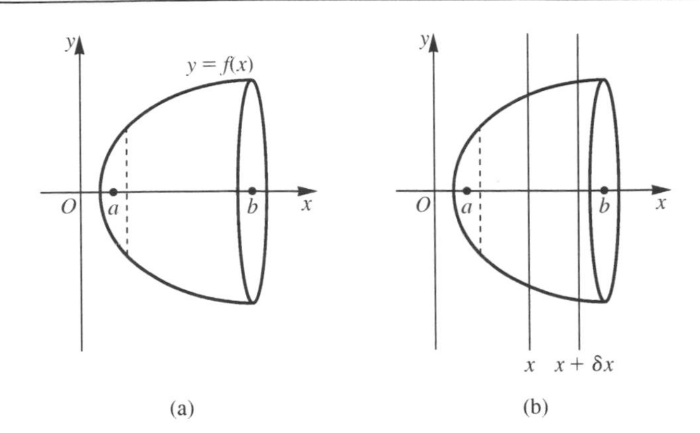

Figure 10.4.
Volume of a solid of revolution

If this solid is cut by two planes perpendicular to the x–y plane a distance δx apart then the solid cut off by them is a thin disc. Each cross-section produced by such a plane cut is a circle and we assume that the cross-sectional area of the disc is approximately constant throughout and equal to that at the left-hand edge, namely πy^2.

Hence the volume of the thin disc is approximately $\pi y^2\,\delta x$. The total volume is therefore approximately $\sum \pi y^2\,\delta x$ where the summation is taken over the collection of the discs into which the solid can be cut. As the discs are cut ever thinner, the approximation to the total volume improves. Then, as $\delta x \to 0$ the volume is given *exactly* by:

KEY POINT

$$V = \int_a^b \pi y^2\,dx \qquad 10.1$$

Worked examples

10.3 A paraboloid is formed by rotating that part of the parabola $y^2 = 4ax$ between $x = 0$ and $x = a$ about the x-axis. Find the volume of the solid generated.

Solution The volume is given by

$$V = \int_0^a \pi\, y^2\,dx = \int_0^a \pi \times 4\,ax\,dx$$

$$= \left[2\pi\, ax^2\right]_0^a = 2\pi a^3$$

10.4 Water is poured into a hemispherical bowl whose axis is vertical; its radius is a. If the depth of water measured from the water surface to the lowest point of the bowl is x, show that the area of the water surface is $\pi x(2a - x)$. Hence find the volume of the water in the bowl at this instant.

Solution

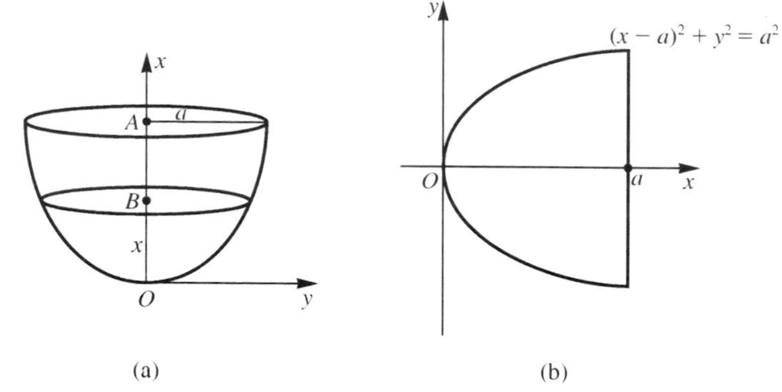

Figure 10.5.
(a) Hemispherical bowl
(b) cross-section

Refer to Figure 10.5(a). The water surface is a circle, centre B. Figure 10.5(b) shows a cross-section in the x–y plane. The equation of the circle which generates the hemisphere is $(x-a)^2+y^2=a^2$ or $y^2=a^2-(x-a)^2$, that is $y^2=2ax-x^2$.

The area of the water surface is $\pi(2ax-x^2)=\pi x(2a-x)$.

The volume of water then in the bowl is

$$\int_0^x \pi x(2a-x)\,\mathrm{d}x=\pi\int_0^x (2ax-x^2)\,\mathrm{d}x$$

$$=\pi\left[ax^2-\frac{x^3}{3}\right]_0^x=\frac{\pi x^2}{3}(3a-x)$$

10.5 The area between the curves $x^2+y^2=1$ and $4x^2+y^2=4$ which lies above the x-axis is rotated about the x-axis. Find the volume of the solid formed.

Solution Figure 10.6 shows the area concerned. Note the symmetry about the y-axis.

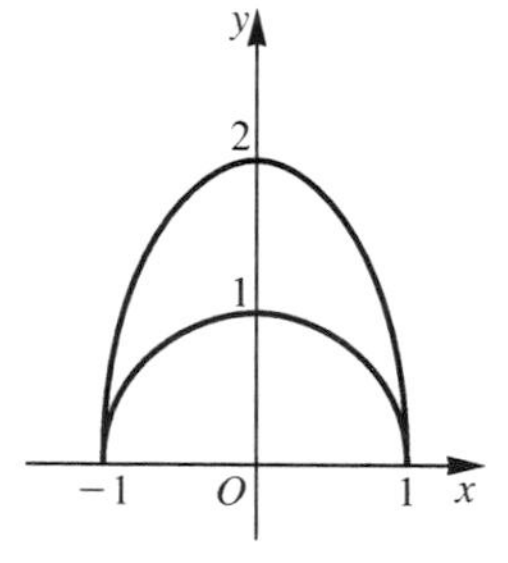

Figure 10.6.
Area between two curves

The volume required is the difference between the volume formed by the curve $4x^2+y^2=4$ and that formed by the curve $x^2+y^2=1$, that is, between $y^2=4-4x^2$ and $y^2=1-x^2$ so that

$$\begin{aligned}V &= \pi\int_{-1}^{1}(4-4x^2)\,dx - \pi\int_{-1}^{1}(1-x^2)\,dx\\ &= 2\pi\left\{\int_0^1(4-4x^2)\,dx - \int_0^1(1-x^2)\,dx\right\} \qquad \text{(by symmetry)}\\ &= 2\pi\left\{\left[\left(4x-\frac{4}{3}x^3\right)\right]_0^1 - \left[\left(x-\frac{x^3}{3}\right)\right]_0^1\right\}\\ &= 2\pi\left\{\left(4-\frac{4}{3}\right)-\left(1-\frac{1}{3}\right)\right\}\\ &= 2\pi\times 2 = 4\pi\end{aligned}$$

Self-assessment questions 10.2

1. Explain how to formulate the volume of a solid of revolution as a definite integral.

Exercise 10.2

1. Find the volume generated when the area enclosed between that portion of the curve $xy = 8$ from $x = 2$ to $x = 4$ and the x-axis is rotated about the x-axis.
2. The area between the curve $y = x(a - x)$ and the x-axis is rotated about the x-axis. Find the volume of the solid which is formed.
3. The area between the curves $y = 2 + x^2$ and $y = 3 + x^2$, the ordinates $x = 0$ and $x = 1$ is rotated about the x-axis. Find the volume of the resulting solid.
4. Find the volume generated when that portion of the curve $y = \cos x$ between $x = -\pi/2$ and $x = \pi/2$ is rotated about the x-axis.

10.3 Arc length and surface area

In Figure 10.7(a) consider that portion of the straight line $y = mx + c$ between $x = a$ and $x = b$. The length of the segment RS is $l = RM\sec\theta$. Now $\tan\theta = m$ and hence

$$\sec\theta = (1+\tan^2\theta)^{\frac{1}{2}} = (1+m^2)^{\frac{1}{2}}$$

Then the length of RS is $l = (b-a).(1+m^2)^{\frac{1}{2}}$.

Figure 10.7(b) shows a general curve $y = f(x)$. PQ is a small part of the arc RS and is approximately a straight line with gradient equal to $dy/dx = f'(x)$ evaluated at P. The length PQ is given approximately by

$$\delta x\left(1+\left(\frac{dy}{dx}\right)^2\right)^{\frac{1}{2}}$$

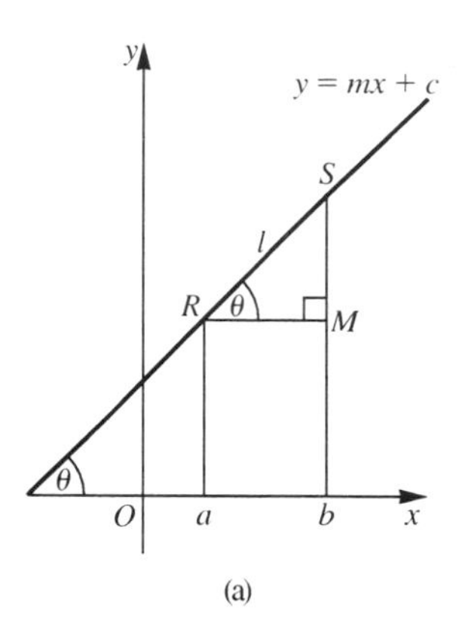

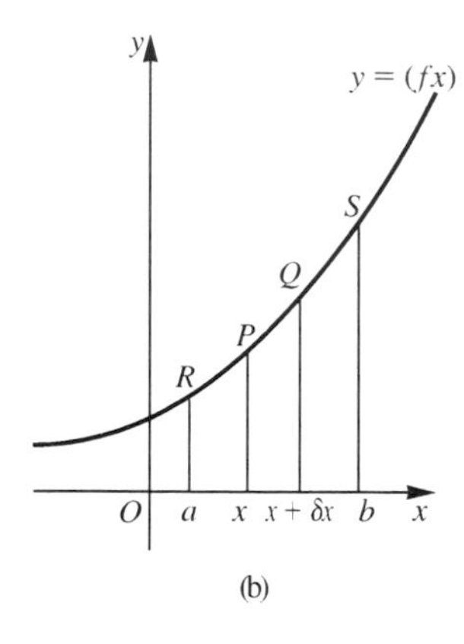

Figure 10.7. Arc length

The arc length RS is approximately

$$\sum \delta x\left(1+\left(\frac{\mathrm{d}y}{\mathrm{d}x}\right)^2\right)^{\frac{1}{2}}$$

If we divide RS into ever shorter sections then the arc length RS is given *exactly* by a definite integral.

KEY POINT

The length along the curve $y = f(x)$ between $x = a$ and $x = b$ is

$$\int_a^b \left(1+\left(\frac{\mathrm{d}y}{\mathrm{d}x}\right)^2\right)^{\frac{1}{2}} \mathrm{d}x \qquad 10.2$$

The curved surface area of the solid of revolution formed by rotating the curve $y = f(x)$ between $x = a$ and $x = b$ about the x-axis is

$$\int_a^b 2\pi y\left(1+\left(\frac{\mathrm{d}y}{\mathrm{d}x}\right)^2\right)^{\frac{1}{2}} \mathrm{d}x \qquad 10.3$$

Worked examples

10.6 The equation for the profile of a heavy chain hanging between two supports on the same horizontal level a distance $2a$ apart is given by

$$y = c\cosh\frac{x}{c}$$

(see Figure 10.8). Find the length of the chain.

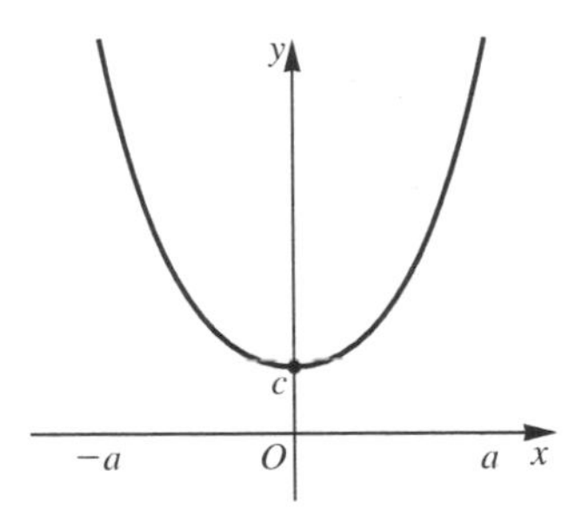

Figure 10.8.
A hanging chain

Solution

$$\frac{dy}{dx} = c\frac{1}{c}\sinh\frac{x}{c} = \sinh\frac{x}{c}$$

$$1 + \left(\frac{dy}{dx}\right)^2 = 1 + \sinh^2\frac{x}{c} = \cosh^2\frac{x}{c}$$

The length of the chain is

$$\int_{-a}^{a}\cosh\frac{x}{c}\,dx = 2\int_0^a \cosh\frac{x}{c}\,dx \qquad \text{(by symmetry)}$$

$$= 2\left[c\sinh\frac{x}{c}\right]_0^a$$

$$= 2c\sinh\frac{a}{c}$$

10.7 A reflecting telescope is in the form of a paraboloid of revolution obtained by rotating the parabola $y^2 = 4ax$ between $x = 0$ and $x = a$ about the axis. Find the area of coating required to cover the interior of the paraboloid.

Solution Differentiating the equation of the parabola in the form $y = 2\sqrt{ax}$ we obtain

$$\frac{dy}{dx} = 2.\frac{1}{2}\sqrt{\frac{a}{x}} = \frac{2a}{2\sqrt{ax}} \qquad \text{that is} \qquad \frac{dy}{dx} = \frac{2a}{y}$$

Thus

$$1 + \left(\frac{dy}{dx}\right)^2 = 1 + \frac{4a^2}{y^2} = \frac{y^2 + 4a^2}{y^2} = \frac{4ax + 4a^2}{y^2}$$

The area required is given by

$$\int_0^a 2\pi y\,.\,\frac{(4ax + 4a^2)^{\frac{1}{2}}}{y}\,dx = \int_0^a 2\pi\, 2a^{\frac{1}{2}}(x + a)^{\frac{1}{2}}\,dx$$

$$= 4\pi a^{\frac{1}{2}}\int_0^a (x + a)^{\frac{1}{2}}\,dx$$

$$= 4\pi a^{\frac{1}{2}}\frac{2}{3}\left[(x + a)^{\frac{3}{2}}\right]_0^a$$

$$= \frac{8\pi a^{\frac{1}{2}}}{3}\left[(2a)^{\frac{3}{2}} - a^{\frac{3}{2}}\right] = \frac{8\pi a^{\frac{1}{2}}}{3}a^{\frac{3}{2}}(2\sqrt{2} - 1)$$

$$= \frac{8}{3}\pi\left(2\sqrt{2} - 1\right)a^2$$

Self-assessment questions 10.3

1. Explain how to formulate the length along a plane curve as a definite integral.
2. Explain how to formulate the curved surface area of a solid of revolution as a definite integral.

Exercise 10.3

1. Sketch the cycloid $x = a(\theta - \sin\theta)$, $y = a(1 - \cos\theta)$. Show that the arch between $x = 0$ and $x = \pi$ has length $8a$.
2. A segment of thickness h is taken from a sphere of radius a. Find the curved surface area of the segment. (The sphere can be obtained by rotating the curve $x^2 + y^2 = a^2$ about the x-axis.)
3. Find the length along the curve $\ln \sec x$ between $x = 0$ and $x = \pi/4$.
4. Find the curved surface area of the solid produced when that part of the curve
$$y = \frac{x^3}{6} + \frac{1}{2x}$$
between $x = 1$ and $x = 2$ is rotated about the x-axis.

10.4 Further applications

Let $f(t)$ be a function defined on $a \leqslant t \leqslant b$.

KEY POINT

The **average value** of $f(t)$ is

$$\bar{f} = \frac{1}{(b-a)} \int_a^b f(t)\,dt \qquad 10.4$$

Figure 10.9 shows that the rectangle with base $[a,b]$ and height given by (10.4) has the same area as that under the curve in the interval $a \leqslant t \leqslant b$.

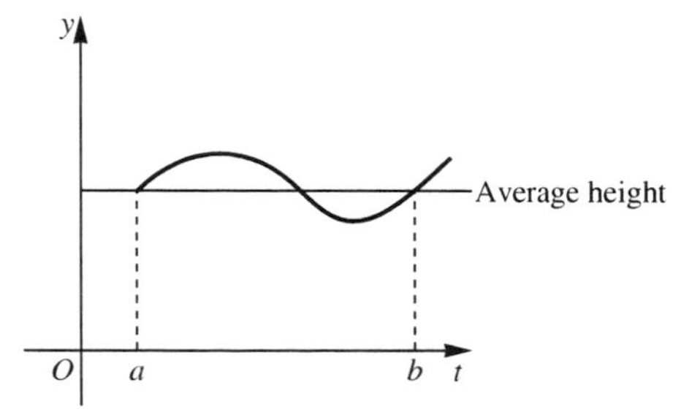

Figure 10.9. Average height of a function

KEY POINT

The root mean square (RMS) value of $f(t)$ is

$$\left\{\frac{1}{(b-a)} \int_a^b [f(t)]^2\,dt\right\}^{\frac{1}{2}} \qquad 10.5$$

Worked example

10.8 Find the mean value and RMS value of the waveform $x = A\cos\omega t$

(a) over $\left[0, \dfrac{\pi}{\omega}\right]$ (b) over $\left[0, \dfrac{2\pi}{\omega}\right]$

Interpret your results graphically.

Solution (a)

$$\bar{f} = \frac{1}{\left(\frac{\pi}{\omega}\right)} \int_0^{\pi/\omega} A\cos\omega t\,\mathrm{d}t$$

$$= \frac{\omega A}{\pi}\left[\frac{\sin\omega t}{\omega}\right]_0^{\pi/\omega}$$

$$= \frac{A}{\pi}\left[\sin\omega t\right]_0^{\pi/\omega} = \frac{A}{\pi}(\sin\pi - 0) = 0$$

$$(\mathrm{RMS})^2 = \frac{1}{\left(\frac{\pi}{\omega}\right)} \int_0^{\pi/\omega} A^2\cos^2\omega t\,\mathrm{d}t$$

$$= \frac{\omega A^2}{\pi}\int_0^{\pi/\omega} \cos^2\omega t\,\mathrm{d}t$$

$$= \frac{\omega A^2}{\pi}\int_0^{\pi/\omega} \left\{\frac{1+\cos 2\omega t}{2}\right\}\mathrm{d}t$$

$$= \frac{\omega A^2}{2\pi}\left[t + \frac{\sin 2\omega t}{2\omega}\right]_0^{\pi/\omega}$$

$$= \frac{\omega A^2}{2\pi}\left[\left(\frac{\pi}{\omega} + 0\right) - 0\right] = \frac{A^2}{2}$$

Therefore RMS $= A/\sqrt{2} \approx 0.707A$.

(b)

$$\bar{f} = \frac{1}{\left(2\frac{\pi}{\omega}\right)} \int_0^{2\pi/\omega} A\cos\omega t\,\mathrm{d}t$$

$$= \frac{\omega A}{2\pi}\left[\frac{\sin\omega t}{\omega}\right]_0^{2\pi} = 0 \quad \text{(again)}$$

$$\begin{aligned}(\text{RMS})^2 &= \frac{\omega A^2}{2\pi}\int_0^{2\pi/\omega}\cos^2 \omega t\,\mathrm{d}t\\ &= \frac{\omega A^2}{4\pi}\left[t + \frac{\sin 2\omega t}{2\omega}\right]_0^{2\pi/\omega} \quad \text{(by analogy with part (a))}\\ &= \frac{\omega A^2}{4\pi}\left[\left(\frac{2\pi}{\omega}+0\right)-0\right]\\ &= \frac{A^2}{2}\end{aligned}$$

Hence RMS $= A/\sqrt{2}$, again.

Figure 10.10 shows the graphs of $A \cos \omega t$ and $A^2 \cos^2 \omega t$ respectively.

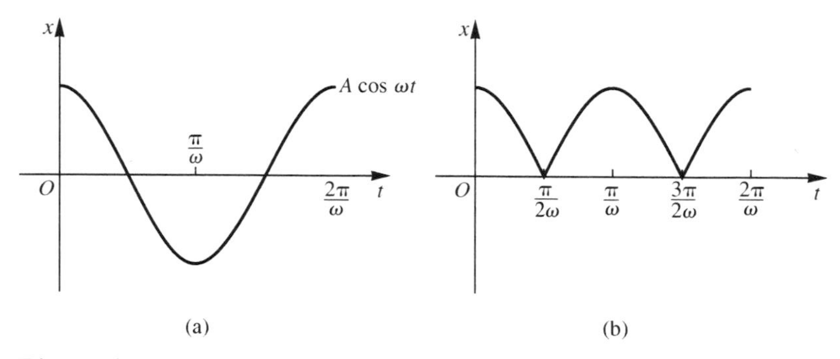

Figure 10.10.
(a) $A \cos \omega t$
(b) $A^2 \cos^2 \omega t$

The period of the waveform is $T = 2\pi/\omega$ and over an interval of this length the mean value is zero. Since the waveform is symmetrical about $t = \pi/\omega$ it follows that over $[0, \pi/\omega]$ *and* over $[0, 2\pi/\omega]$ and over $[\pi/\omega, 2\pi/\omega]$ the mean value is zero also.

Further, the RMS values for $[0, \pi/\omega]$, $[\pi/\omega, 2\pi/\omega]$ and $[0, 2\pi/\omega]$ are all equal.

Note the general result that the RMS value of a sinusoidal waveform taken over a period is $1/\sqrt{2} \times$ the amplitude.

Centroids of plane areas and solids of revolution

Consider the plane area shown in Figure 10.11.

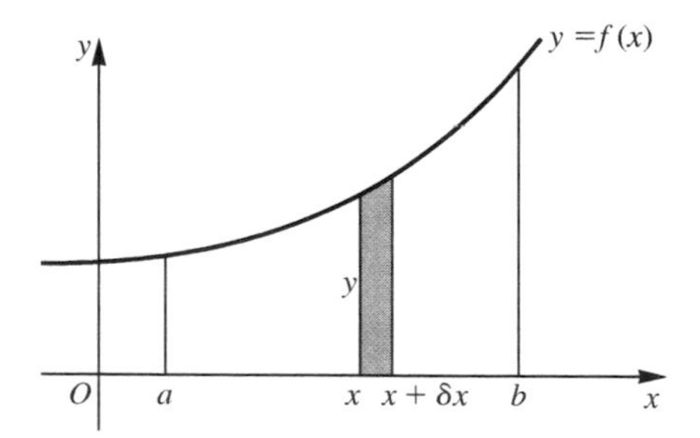

Figure 10.11.
Moment of area about the y-axis

The thin shaded strip consists of points which are approximately the same distance x from the y-axis. Its **moment** about the y-axis is $xy\,\delta x$. As usual, we divide the interval $[a,b]$ into several subintervals of width δx and the total moment about the y-axis of the area under $y = f(x)$ between $x = a$ and $x = b$ is approximately

$$\sum xy\,\delta x$$

As $\delta x \to 0$ the moment of the area about the y-axis is *exactly*

$$\text{Moment of area about } y\text{-axis} = \int_a^b xy\,\mathrm{d}x \qquad 10.6$$

(Note that the moment of area is strictly the *first* moment of area, but we usually drop the word 'first'.) To find the moment of the area about the x-axis we return to the elementary strip of Figure 10.11. Its centroid is approximately half-way up, that is, at a height $y/2$ above the x-axis. The moment of this strip about the x-axis is approximately $\frac{1}{2}y \,.\, y\,\delta x$. Hence the total moment of the area about the x-axis is approximately $\sum \frac{1}{2}y \,.\, y\,\delta x$.

As $\delta x \to 0$ the total moment is *exactly*

$$\int_a^b \frac{1}{2}y^2\,\mathrm{d}x \qquad 10.7$$

If the area concerned is A then we can find the coordinates $(\bar{x},\bar{y})$ of the **centroid** of the area. We use the following results:

KEY POINT

The position $(\bar{x},\bar{y})$ of the centroid of a plane region between $x = a$ and $x = b$ is given by

$$A\bar{x} = \int_a^b xy\,\mathrm{d}x$$

$$A\bar{y} = \int_a^b \frac{1}{2}y^2\,\mathrm{d}x \qquad 10.8$$

(A word of warning. The results above apply only when the area rests on the x-axis. In other cases it is necessary to get back to basics.)

It *may* be more convenient to use an alternative approach. In the case shown in Figure 10.12 the horizontal strip consists of points which are approximately the same distance, y, from the x-axis. Its moment about the x-axis is approximately $xy\,\delta y$. The total moment of the area trapped between $y = c$ and $y = d$ is exactly

$$\int_c^d xy\,\mathrm{d}y$$

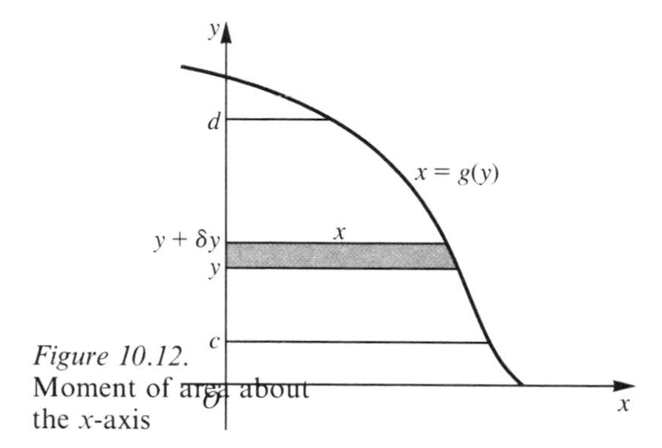

Figure 10.12. Moment of area about the x-axis

Centroid of a volume

To find the position of the centroid of a solid of revolution which has the x-axis as its axis of symmetry we use the following results, noting that because of symmetry the centroid lies on the x-axis. The volume of the solid is V.

Two other concepts which are important are the second moment of area and the moment of inertia. To find the second moment of area we proceed in a similar way to that for the (first) moment of area except that before integrating we multiply the area of the strip by the *square* of the distance from the relevant axis. To find the moment of inertia we multiply also by the density of the material before integrating; see Example 10.11.

KEY POINT

The position $(\bar{x},\bar{y})$ of the centroid of a solid of revolution of volume V between $x = a$ and $x = b$ is given by

$$V\bar{x} = \int_a^b x.\pi y^2\,\mathrm{d}x \qquad 10.9$$

$$\bar{y} = 0$$

Worked examples

10.9 Find the position of the centroid of the triangular area shown in Figure 10.13.

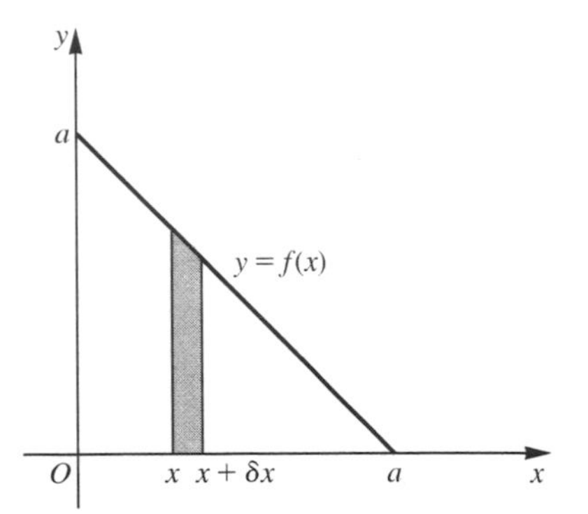

Figure 10.13. Triangular area for Example 10.9

Solution Before we embark on any calculation we note that $\bar{x} < \frac{1}{2}a$, $\bar{y} < \frac{1}{2}a$.

First, we find the area of the triangle. We do not need to use integration in this case; using geometry the answer is $\frac{1}{2}a^2$.

Using the strip shown, the moment of area about the y-axis is

$$\int_0^a xy\,dx = \int_0^a x(a-x)\,dx$$

$$= \int_0^a (ax - x^2)\,dx$$

$$= \left[\frac{ax^2}{2} - \frac{x^3}{3}\right]_0^a$$

$$= \frac{1}{6}a^3$$

Hence from (10.8)

$$\bar{x} = \frac{1}{6}a^3 \div \frac{1}{2}a^2 = \frac{1}{3}a$$

Taking moments about the x-axis, we obtain

$$\int_0^a \frac{1}{2}y^2\,dx = \frac{1}{2}\int_0^a (a-x)^2\,dx$$

$$= \frac{1}{2}\left[-\frac{(a-x)^3}{3}\right]_0^a$$

$$= \frac{1}{6}\Big[-(a-x)^3\Big]_0^a = \frac{1}{6}a^3$$

Hence $\bar{y} = \frac{1}{3}a$. (We could have obtained the result for $\bar{y}$ by symmetry.)

10.10 Find the position of the centroid of a hemisphere of radius a.

Solution By symmetry, the centroid lies on the axis of the hemisphere. The thin disc shown in Figure 10.14 has an approximate moment about the y-axis of $x \,.\, \pi y^2 \,\delta x$.

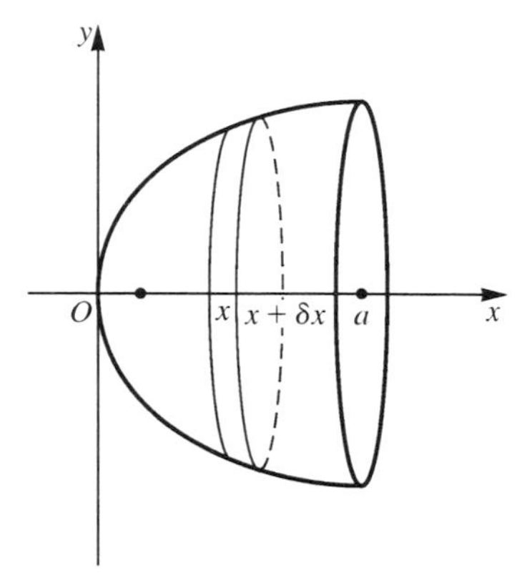

Figure 10.14.
Centroid of a hemisphere

Hence the moment of the hemisphere about the y-axis is

$$M_y = \int_0^a \pi x y^2 \,dx$$

From Example 10.4 the semicircle which generates the hemisphere has equation $y^2 = 2ax - x^2$. Hence

$$\begin{aligned} M_y &= \int_0^a \pi x(2ax - x^2)\,dx \\ &= \pi \int_0^a (2ax^2 - x^3)\,dx \\ &= \pi \left[\frac{2ax^3}{3} - \frac{x^4}{4}\right]_0^a \\ &= \pi\left(\frac{2}{3}a^4 - \frac{a^4}{4}\right) = \frac{5\pi a^4}{12} \end{aligned}$$

Now $M_y = V\bar{x}$ so that

$$\bar{x} = \frac{5\pi a^4}{12} \div \frac{2}{3}\pi a^3 = \frac{5}{8}a$$

10.11 Find the moment of inertia of a uniform square lamina of side a and density ρ about (a) one of its sides, (b) a line through its centre parallel to two of its edges.

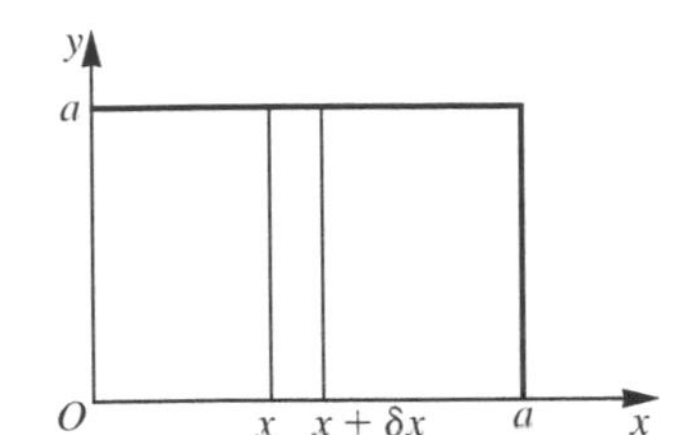

Figure 10.15.
Second moment of a rectangular area

Solution (a) Refer to Figure 10.15. The second moment of the strip about the y-axis is $x^2 a\,\delta x$. Hence the total second moment of the square about the y-axis is

$$\int_0^a x^2 a\,\mathrm{d}x = a\frac{a^3}{3} = \frac{a^4}{3}$$

and the moment of inertia is $\rho(a^4/3)$. The mass of the area $M = \rho a^2$. Hence the moment of inertia is $(Ma^2)/3$.

(b) The second moment of the strip about the line $x = a/2$ is

$$\left(\frac{a}{2} - x\right)^2 a\,\delta x$$

The moment of inertia of the square about this line is

$$\rho\int_0^a \left(\frac{a}{2} - x\right)^2 a\,\mathrm{d}x = \rho a\int_0^a \left(\frac{a^2}{4} - ax + x^2\right)\mathrm{d}x$$

$$= \rho a\left[\frac{a^2}{4}x - a\frac{x^2}{2} + \frac{x^3}{3}\right]_0^a$$

$$= \rho a\left(\frac{a^3}{4} - \frac{a^3}{2} + \frac{a^3}{3}\right)$$

$$= \rho a\frac{a^3}{12} = M\frac{a^2}{12}$$

Self-assessment questions 10.4

1. Explain how to formulate the mean value of a function over a given interval as a definite integral.
2. Explain how to formulate the root mean square value of a function over a given interval as a definite integral.
3. Explain how to formulate the moment of a plane area about an axis as a definite integral.
4. Explain how to locate the centroid of a plane area using definite integrals.
5. Explain how to locate the centroid of a solid of revolution using definite integrals.

Exercise 10.4

1. Find the mean value of the following:

   (a) $\sin\theta$ over $0 \leqslant \theta \leqslant \pi$
   (b) $\sin\omega t$ over $0 \leqslant t \leqslant \pi/\omega$
   (c) $1 + t^2$ over $0 \leqslant t \leqslant 2$

2. The temperature of a body at time t is given by $\theta(t) = 40 + 60e^{-0.1t}$. Find its average temperature from $t = 0$ to $t = 5$.

3. Find the position of the centroid of the plane area enclosed by the following:

   (a) the curve $y = x^2$, the x-axis and the ordinates $x = 0$ and $x = 2$
   (b) the curve $y = x^3$, the y-axis and the lines $y = 0$ and $y = 8$
   (c) the curve $y = \sin x$, the x-axis and the ordinates $x = 0$ and $x = \pi$

4. Find the root mean square values of the functions of Question 1 over the intervals given. Find also the root mean square value of $3\cos\omega t + 4\sin\omega t$ over $0 \leqslant t \leqslant \pi/\omega$.

5. Find the position of the centroid of the solid formed by rotating about the x-axis that part of the given curve between the given ordinates:

   (a) $y^2 = 4ax$ between $x = 0$ and $x = 2a$
   (b) $y = \sin x$ between $x = 0$ and $x = \pi$

6. Find the position of the centroid of a solid cone of uniform density.

Test and assignment exercises 10

1. Find the area above the x-axis and below the following curves:

   (a) $y = x(5 - x)$
   (b) $y = \sin x, \quad 0 \leqslant x \leqslant \pi$

2. Find the area between the curves $y = x^2$ and $y = 1 - x^2$.

3. The area above the x-axis and below the curve $y = \sqrt{(a^2 - x^2)}$ is rotated about the x-axis through an angle 2π. Find the volume of the resulting solid.

4. Find the volume of the solid formed when the area between the curves $y = 4 - x^2$ and $y = x^2$ is rotated about the x-axis through an angle 2π.

5. Find the length of arc of the curve

   $$y = \frac{1}{2x} + \frac{x^3}{6}$$

   between $x = 1$ and $x = 4$.

6. The curve

   $$x^{\frac{1}{2}} - \frac{1}{3}x^{\frac{3}{2}}$$

   between $x = 0$ and $x = 3$ is rotated about the x-axis through an angle 2π. Find the surface area of the solid formed.

7. Find the mean value of the function $f(x) = 1 - e^{-x}$ in the interval $0 \leqslant x \leqslant 2$. Find its root mean square value in the same interval.

8. Find the area between the x-axis and the curve $y = x(4 - x)$. Determine the first moment of the area about the x-axis and hence find the coordinates of the centroid of the area.

9. Find the position of the centroid of the solid which is formed when the area bounded by the x-axis, the ordinate $x = 2$ and the curve $y^2 = 4x$ is rotated about the x-axis through an angle 2π.

11 Vectors I – algebra and 2D geometry

Objectives

This chapter

- distinguishes between vectors and scalars
- defines the resultant of two vectors
- states the rules of vector algebra
- uses vectors to prove results in geometry
- derives the vector equation of a straight line
- obtains results in the Cartesian coordinate systems in two dimensions and three dimensions

11.1 Vectors and scalars

A **scalar** quantity can be completely described by a magnitude, which is a real number in suitable units. A **vector** quantity requires both a magnitude and a direction for a complete description. Vectors have special rules for their combination.

We can sometimes *represent* a vector geometrically by a line segment. The length of the segment represents the magnitude of the vector in suitable units and the direction of the segment represents the direction of the vector.

An arrowhead is placed on the segment to indicate the *sense* of the vector. The two line segments in Figure 11.1(a) represent the same vector since they have the same length and are parallel and the arrowheads point in the same direction.

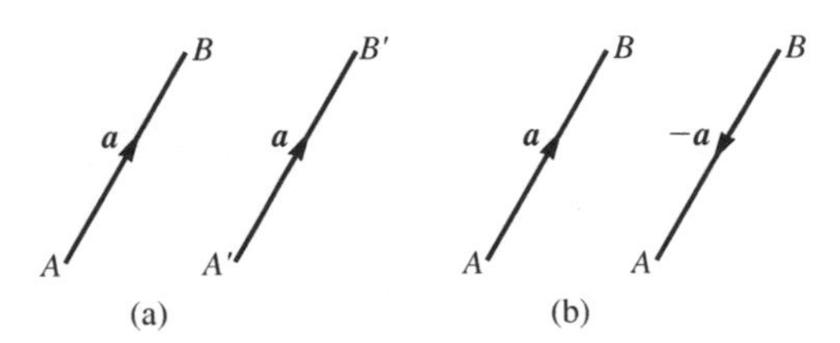

Figure 11.1.
Representation of vectors

Two common notations for a vector are a lower-case bold letter, for example $\boldsymbol{a}$ (the handwritten version is $\underline{a}$), or as $\overline{AB}$ or $\overrightarrow{AB}$ which indicates the direction. Upper-case bold italic letters are usually used when representing forces, such as $\boldsymbol{F}$.

The **magnitude** of a vector is written as $|\boldsymbol{a}|$ or $|\underline{a}|$ or AB. A vector whose magnitude is 1 is a **unit vector** and is written $\hat{\boldsymbol{a}}$ (pronounced 'a hat').

Two vectors $\boldsymbol{a}$ and $\boldsymbol{b}$ are equal if (and only if) they have the same magnitude, direction *and* sense.

The vector which has the same magnitude and direction as $\boldsymbol{a}$ but has the opposite sense is denoted $-\boldsymbol{a}$; see Figure 11.1(b).

The **zero vector** is a vector with zero magnitude.

Worked examples

11.1 Which of the following are scalars and which are vectors?

(a) area (b) temperature (c) force
(d) pressure (e) displacement (f) distance
(g) work (h) velocity (i) mass (j) acceleration

Solution (a), (b), (d), (f), (g), (i) are scalars.
(c), (e), (h), (j) are vectors.

11.2 In Figure 11.2 the line segments are scaled so that 1 cm represents a magnitude of 1 kN or $1\,\mathrm{m\,s^{-1}}$, as appropriate. Specify the vectors completely.

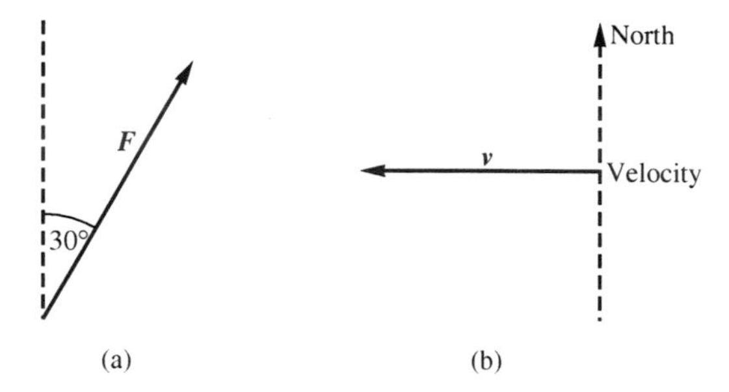

Figure 11.2. Line segments for Example 11.2

Solution (a) $\boldsymbol{F}$ is a force of 3 kN acting at 30° to the vertical.
(b) $\boldsymbol{v}$ is a velocity of $2.5\,\mathrm{m\,s^{-1}}$ due west.

Self-assessment questions 11.1

1. Distinguish between a vector quantity and a scalar quantity.
2. What does it mean when we say that two vectors are equal?

Exercise 11.1

1. Classify the following as vectors or scalars:

   (a) volume
   (b) angular velocity
   (c) momentum
   (d) energy

2. Draw line segments to represent the following:

   (a) a force of magnitude 1 kN acting vertically upwards
   (b) a horizontal force of magnitude 2 kN acting to the right
   (c) a velocity of 3 $m\,s^{-1}$ north-east
   (d) a velocity of 2 $m\,s^{-1}$ due south

11.2 Vector algebra - sums and scalar multiples

The system shown in Figure 11.3(a) represents a pulley of weight $\boldsymbol{W}$ supported by two ropes in which the magnitudes of the tensions are $\boldsymbol{T}_1$ and $\boldsymbol{T}_2$.

If the pulley is in equilibrium then the joint effect of the two tensions is to support the weight of the pulley. We say that the **resultant** or sum of the two tension forces is equal and opposite to that of the weight (force). Clearly the resultant acts vertically upwards and has magnitude $\boldsymbol{W}$.

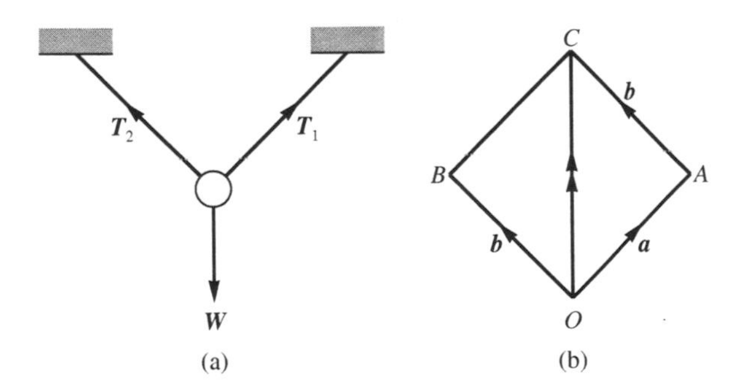

Figure 11.3. Addition of two vectors

In general, to find the resultant or sum of two vectors $\boldsymbol{a}$ and $\boldsymbol{b}$ we proceed as in Figure 11.3(b): we draw a line segment OA in the direction of $\boldsymbol{a}$ and of length scaled to the magnitude of $\boldsymbol{a}$ and a line segment OB in the direction of $\boldsymbol{b}$ drawn to the same scale as for $\boldsymbol{a}$. The parallelogram $OBCA$ is completed. The diagonal OC represents the resultant $\boldsymbol{a} + \boldsymbol{b}$. This is known as the **parallelogram law of addition.**

Alternatively, we draw OA as before, then draw AC in the direction of $\boldsymbol{b}$ with the same scaling as OA. The third side, OC, represents the resultant. This process is known as the **triangle law of addition.**

KEY POINT

To add two vectors we use the parallelogram law of addition.

From the parallelogram law it follows that vector addition is **commutative**, that is $\boldsymbol{a} + \boldsymbol{b} = \boldsymbol{b} + \boldsymbol{a}$.

Vector addition is also **associative**, that is $(\boldsymbol{a} + \boldsymbol{b}) + \boldsymbol{c} = \boldsymbol{a} + (\boldsymbol{b} + \boldsymbol{c})$. In effect, the parentheses are unnecessary and we could write simply $\boldsymbol{a} + \boldsymbol{b} + \boldsymbol{c}$.

The vector $\lambda\boldsymbol{b}$ where λ is a scalar is called a **scalar multiple** of $\boldsymbol{b}$.

KEY POINT

The vector $\lambda\boldsymbol{b}$ is in a direction parallel to that of $\boldsymbol{b}$. If $\lambda > 0$ then $\lambda\boldsymbol{b}$ is in the same direction as $\boldsymbol{b}$ and its magnitude is λ times that of $\boldsymbol{b}$, that is $|\lambda\boldsymbol{b}| = \lambda\,|\boldsymbol{b}|$. If $\lambda < 0$ then $\lambda\boldsymbol{b}$ is in the opposite direction to $\boldsymbol{b}$ and $|\lambda\boldsymbol{b}| = |\lambda||\boldsymbol{b}|$.

An example of a scalar multiple occurs in the vector equation $\boldsymbol{B} = \mu\boldsymbol{H}$ where $\boldsymbol{B}$ is the magnetic flux density in webers per square metre, $\boldsymbol{H}$ is the magnetic field intensity in amperes per meter and μ is the (scalar) permeability in webers per ampere per metre.

The **distributive** law also holds, that is $\lambda(\boldsymbol{a} + \boldsymbol{b}) = \lambda\boldsymbol{a} + \lambda\boldsymbol{b}$.

Worked examples

11.3 Use a diagram to illustrate vector subtraction.

Solution We can write $\boldsymbol{a} - \boldsymbol{b}$ as $\boldsymbol{a} + (-\boldsymbol{b})$. Referring to Figure 11.4, we first draw OA and OB to represent $\boldsymbol{a}$ and $\boldsymbol{b}$, respectively. BO is produced to OC so that $BO = OC$; OC then represents $-\boldsymbol{b}$. The parallelogram law is then applied to $\boldsymbol{a}$ and $-\boldsymbol{b}$.

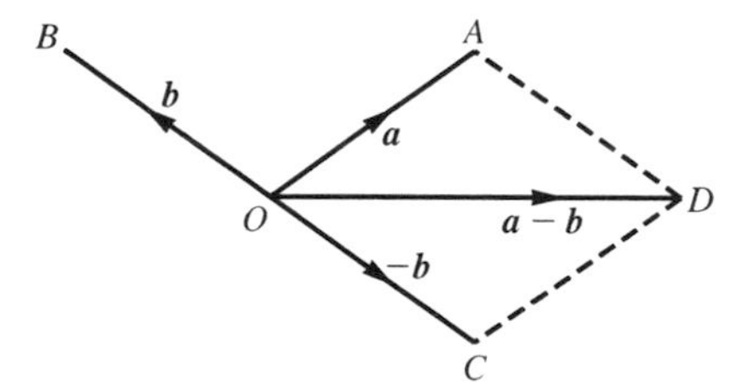

Figure 11.4. Vector subtraction

11.4 A plane flies at a speed of 300 mph in a wind coming from the north with a speed of 40 mph. If the pilot wishes to fly due east, in which direction should he head and what will be his actual speed relative to the ground?

Solution In Figure 11.5 $\overrightarrow{OA}$ represents the velocity of the wind and OE is drawn in the required direction of flight. AB is a line segment drawn from A to the same scale as OA to meet OE at B.

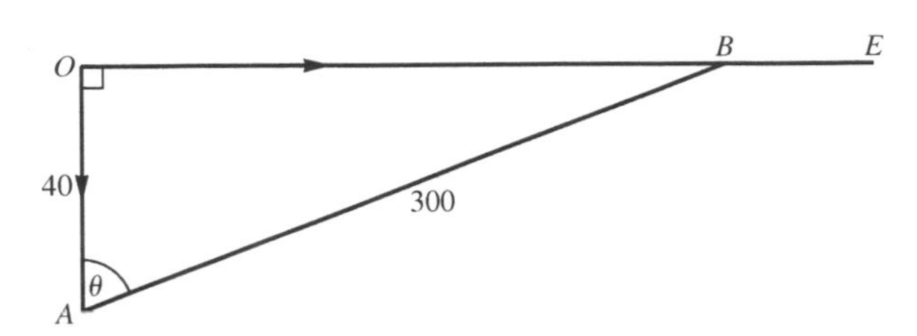

Figure 11.5. Vector diagram for Example 11.4

From triangle OAB, $\cos\theta = 40/300 = 2/15$ so that $\theta = 82.3°$. The direction in which the plane should head is N82.3°E. The magnitude of OB is $\{(300)^2 - (40)^2\}^{\frac{1}{2}} = (88\,400)^{\frac{1}{2}} = 297$ mph.

11.5 In Figure 11.6 the side OA of the rectangle $OACB$ represents the vector $\boldsymbol{a}$ and the side OB represents the vector $\boldsymbol{b}$. E is the mid-point of OC (and AB). What vectors are represented by OC, AB, OE, AE and EB?

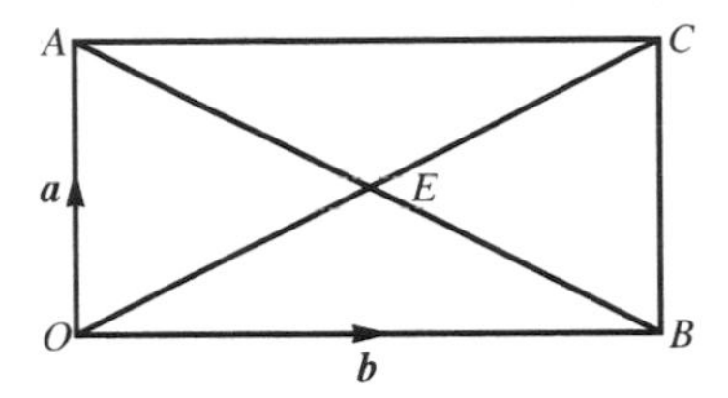

Figure 11.6. Diagram for Example 11.5

Solution $\overrightarrow{BC} = \boldsymbol{a}$ since BC is parallel to OA and has the same length as OA

$\overrightarrow{OC} = \overrightarrow{OB} + \overrightarrow{BC} = \boldsymbol{b} + \boldsymbol{a}$ (or $\boldsymbol{a} + \boldsymbol{b}$)

$\overrightarrow{OA} + \overrightarrow{AB} = \overrightarrow{OB}$ so that $\overrightarrow{AB} = \overrightarrow{OB} - \overrightarrow{OA} = \boldsymbol{b} - \boldsymbol{a}$

$\overrightarrow{OE} = \frac{1}{2}\overrightarrow{OC} = \frac{1}{2}(\boldsymbol{b} + \boldsymbol{a})$

$\overrightarrow{AE} = \overrightarrow{EB} = \frac{1}{2}\overrightarrow{AB} = \frac{1}{2}(\boldsymbol{b} - \boldsymbol{a})$

Self-assessment questions 11.2

1. Explain the relationship between the vectors $\boldsymbol{a}$ and $\lambda\boldsymbol{a}$ where λ is a scalar.
2. Explain the geometrical interpretation of the addition of two vectors.

Exercise 11.2

1. A boat has a speed of 7 km per hour and is set to head north-east but a tide of 2 km per hour flows from a north-west direction. What is the actual speed and direction of the boat?
2. A rocket falls with acceleration 9.81 m s^{-2} then experiences a side thrust which causes a horizontal acceleration of 1 m s^{-2}. What is the resultant acceleration?
3. A boat has to make a journey from a point on one bank of a river to a point on the other bank, directly opposite. If the current in the river has a speed of 2 m s^{-1} and the maximum speed of the boat is 5 m s^{-1}, in which direction should the boat be headed if the river is 30 m wide? How long does the journey take if the boat travels at its maximum speed?

4. What is the relative velocity of a motorist travelling south along a motorway lane at 45 km per hour relative to a second motorist travelling south at 60 km per hour?
5. Draw a diagram to illustrate the resultant of three vectors ***a***, ***b*** and ***c***.
6. Motorist A approaches a crossroads from the west at 50 km per hour and a second motorist B approaches the same crossroads from the south at 30 km per hour. What is the velocity of A relative to B?

11.3 Applications in geometry

The essence of the application of vectors in geometry is that two vectors are regarded as equal if they have the same magnitude, direction and sense. These are **free** vectors; in mechanics we may also need to specify the precise line of action of the vector.

Worked examples

11.6 Figure 11.7 shows a triangle ABC with D on AB and E on AC such that $AD = \frac{3}{5}AB$ and $AE = \frac{3}{5}AC$. Show that DE is parallel to BC and that the length of DE is $\frac{3}{5}$ times the length of BC.

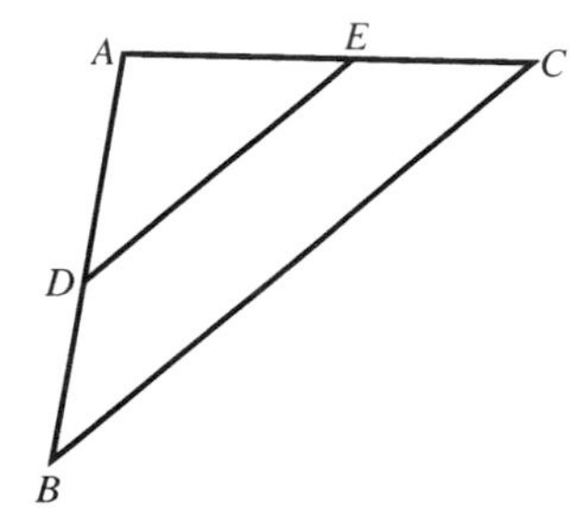

Figure 11.7. Diagram for Example 11.6

Solution The statement $AD = \frac{3}{5}AB$ can be expressed as a vector equation; since AD is part of AB it is parallel to it. Then

$$\overrightarrow{AD} = \tfrac{3}{5}\overrightarrow{AB}$$

Similarly,

$$\overrightarrow{AE} = \tfrac{3}{5}\overrightarrow{AC}$$

From triangle ABC,

$$\overrightarrow{AB} + \overrightarrow{BC} = \overrightarrow{AC} \qquad \text{i}$$

(Note that this states that the *displacement* from A to B followed by the *displacement* from B to C is equivalent to the *displacement* from A to C; it is *not* a statement about distances.)

From triangle ADE,

$$\overrightarrow{AD} + \overrightarrow{DE} = \overrightarrow{AE}$$

that is

$$\tfrac{3}{5}\overrightarrow{AB} + \overrightarrow{DE} = \tfrac{3}{5}\overrightarrow{AC} \qquad \text{ii}$$

Comparing (i) and (ii) we find that $\overrightarrow{DE} = \frac{3}{5}\overrightarrow{BC}$. This single vector equation means both that the side DE is parallel to the side BC and that the length of DE is $\frac{3}{5}$ the length of BC.

11.7 Show that the diagonals of a parallelogram bisect each other.

Solution Figure 11.8 shows the parallelogram $OACB$ with the diagonals AB and OC meeting at D.

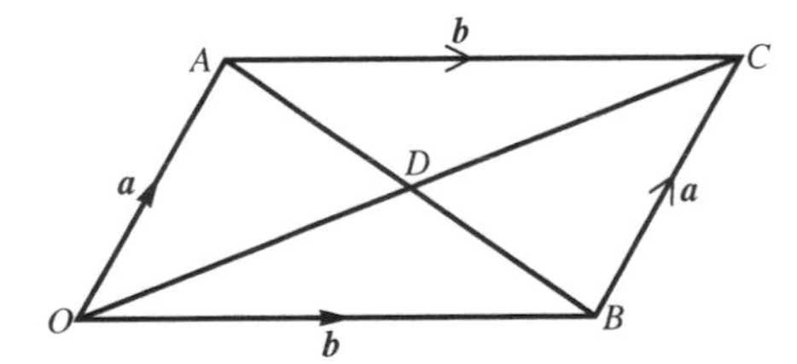

Figure 11.8. Diagram for Example 11.7

We denote the side $\overrightarrow{OA}$ by $\boldsymbol{a}$ and the side $\overrightarrow{OB}$ by $\boldsymbol{b}$. Then we can represent AC by $\boldsymbol{b}$ since it has the same length as $\overrightarrow{OB}$ and is parallel to it. Similarly, we represent $\overrightarrow{BC}$ by $\boldsymbol{a}$.

Now $\overrightarrow{OC} = \overrightarrow{OB} + \overrightarrow{BC} = \boldsymbol{b} + \boldsymbol{a}$. Also, $\overrightarrow{OD}$ is in the same direction as $\overrightarrow{OC}$; hence $\overrightarrow{OD} = \lambda(\boldsymbol{b} + \boldsymbol{a})$ for some value λ such that $0 < \lambda < 1$. Further, $\overrightarrow{OA} + \overrightarrow{AB} = \overrightarrow{OB}$ so that $\overrightarrow{AB} = \overrightarrow{OB} - \overrightarrow{OA} = \boldsymbol{b} - \boldsymbol{a}$. Since D is a point on the line AB, $\overrightarrow{AD} = \mu\overrightarrow{AB} = \mu(\boldsymbol{b} - \boldsymbol{a})$ for $0 < \mu < 1$.

Finally $\overrightarrow{OA} + \overrightarrow{AD} = \overrightarrow{OD}$, that is

$$\boldsymbol{a} + \mu(\boldsymbol{b} - \boldsymbol{a}) = \lambda(\boldsymbol{a} + \boldsymbol{b})$$

so that

$$(1 - \mu)\boldsymbol{a} + \mu\boldsymbol{b} = \lambda\boldsymbol{a} + \lambda\boldsymbol{b}$$

Since $\boldsymbol{a}$ and $\boldsymbol{b}$ can be varied independently it is valid to compare the coefficients of $\boldsymbol{a}$ and $\boldsymbol{b}$ separately to obtain, respectively

Note that, in general, a vector equation cannot be solved in a similar way to an algebraic equation

$$1 - \mu = \lambda \qquad \text{i}$$

and

$$\mu = \lambda \qquad \text{ii}$$

Solving (i) and (ii) we obtain $\lambda = \mu = \frac{1}{2}$. Hence D is the mid-point of OC and of AB and therefore the diagonals bisect each other.

11.8 With reference to a fixed origin O the vector $\overrightarrow{OP} = \boldsymbol{r}$ is called the **position vector** of the point P. Find the position vector of a point on the line

(a) passing through a point A and having the direction of the vector $\boldsymbol{u}$

(b) passing through two given points A and B

Solution

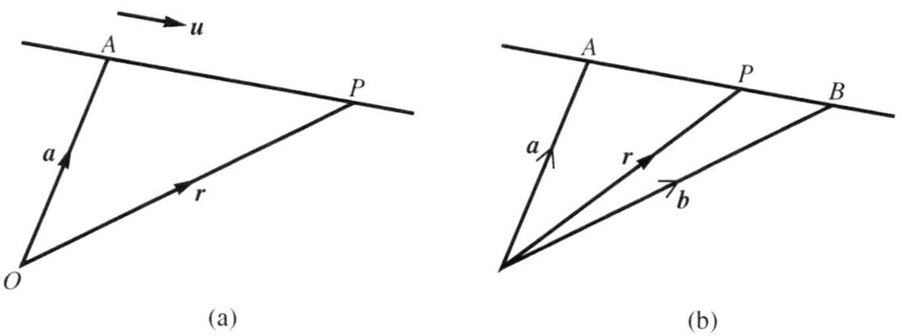

Figure 11.9. Vector equations of a straight line

(a) In Figure 11.9(a) $\overrightarrow{OA} = \boldsymbol{a}$ and P is some point on the line whose direction is that of the vector $\boldsymbol{u}$. Since the direction of $\overrightarrow{AP}$ is parallel to that of the vector $\boldsymbol{u}$, $\overrightarrow{AP} = \lambda\boldsymbol{u}$ for some scalar λ. Now $\overrightarrow{OP} = \overrightarrow{OA} + \overrightarrow{AP}$, that is

$$\boldsymbol{r} = \boldsymbol{a} + \lambda\boldsymbol{u} \qquad 11.1$$

This is the vector equation of the line.

(b) In Figure 11.9(b) A and B are the two given points and P is any point on the line. Then

$$\overrightarrow{AB} = \boldsymbol{b} - \boldsymbol{a} \quad \text{(from triangle } OAB\text{)}$$

AP is on AB and hence

$$\overrightarrow{AP} = \lambda(\boldsymbol{b} - \boldsymbol{a})$$

for some λ. Also,

$$\overrightarrow{OP} = \overrightarrow{OA} + \overrightarrow{AP}$$

so that

$$\boldsymbol{r} = \boldsymbol{a} + \lambda(\boldsymbol{b} - \boldsymbol{a})$$

or

$$\boldsymbol{r} = (1 - \lambda)\boldsymbol{a} + \lambda\boldsymbol{b} \qquad 11.2$$

Each value of λ corresponds to a unique point on the line; $\lambda = 0$ is the point A, $\lambda = 1$ is B. Hence $\lambda < 0$ represents points 'to the left' of A, $0 < \lambda < 1$ represents points between A and B and $\lambda > 0$ represents points 'to the right' of B.

Alternative forms of (11.2) are

$$\boldsymbol{r} = \mu\boldsymbol{a} + (1 - \mu)\boldsymbol{b} \qquad 11.3$$

and

$$\boldsymbol{r} = \mu\boldsymbol{a} + \lambda\boldsymbol{b}, \qquad \lambda + \mu = 1 \qquad 11.4$$

KEY POINT

The vector equation of the line through the point A in the direction of $\boldsymbol{u}$ is $\boldsymbol{r} = \boldsymbol{a} + \lambda\boldsymbol{u}$.
The vector equation of the line through the points A and B is $\boldsymbol{r} = \boldsymbol{a} + \lambda(\boldsymbol{b} - \boldsymbol{a})$.

Self-assessment questions 11.3

1. If $\boldsymbol{a}$ and $\boldsymbol{b}$ represent the sides AB and BC, respectively, of a triangle ABC explain which vector represents the sides AC and CA and why.
2. What is the vector equation of the line joining the two points with position vectors $\boldsymbol{a}$ and $\boldsymbol{b}$?

Exercise 11.3

1. In triangle ABC $AB = BC$; M is a point on AC such that BM is perpendicular to AC. If $\overrightarrow{AB} = \boldsymbol{a}$ and $\overrightarrow{BC} = \boldsymbol{b}$ find expressions in terms of $\boldsymbol{a}$ and $\boldsymbol{b}$ for

   (a) $\overrightarrow{AC}$ (b) $\overrightarrow{CM}$ (c) $\overrightarrow{BM}$

2. $ABCDEF$ is a regular hexagon. If $\overrightarrow{AB} = \boldsymbol{a}$, $\overrightarrow{BC} = \boldsymbol{b}$ and $\overrightarrow{CD} = \boldsymbol{c}$, express the following in terms of $\boldsymbol{a}$, $\boldsymbol{b}$ and $\boldsymbol{c}$:

   (a) $\overrightarrow{DC}$ (b) $\overrightarrow{DE}$
   (c) $\overrightarrow{EF}$ (d) $\overrightarrow{AD}$
   (e) $\overrightarrow{BE}$ (f) $\overrightarrow{CF}$
   (g) $\overrightarrow{AC}$ (h) $\overrightarrow{AE}$

3. $ABCD$ is a quadrilateral and E, F, G and H are the mid-points of the sides AB, BC, CD and DA, respectively. The sides AB, BC, CD and DA are represented by $\boldsymbol{a}$, $\boldsymbol{b}$, $\boldsymbol{c}$ and $\boldsymbol{a} - \boldsymbol{b}$, respectively.

   (a) What is the resultant $\boldsymbol{a} + \boldsymbol{b} + \boldsymbol{c} + \boldsymbol{d}$?
   (b) Consider triangle FBE. Show that $\overrightarrow{EF} = \frac{1}{2}\boldsymbol{a} + \frac{1}{2}\boldsymbol{b}$.
   Similarly, using triangle GHD, find an expression in terms of $\boldsymbol{c}$ and $\boldsymbol{d}$ for $\overrightarrow{GH}$. Using part (a) compare $\overrightarrow{EF}$ and $\overrightarrow{GH}$. What can you deduce about the line segments EF and HG?
   (c) What do you deduce about the quadrilateral $EFGH$?

4. $ABCD$ is a quadrilateral and P is the mid-point of CD. If $\overrightarrow{AB} = \boldsymbol{a}$, $\overrightarrow{BC} = \boldsymbol{b}$ and $\overrightarrow{CD} = \boldsymbol{c}$ express the following vectors in terms of $\boldsymbol{a}$, $\boldsymbol{b}$ and $\boldsymbol{c}$:

   (a) $\overrightarrow{AD}$ (b) $\overrightarrow{BA}$
   (c) $\overrightarrow{CP}$ (d) $\overrightarrow{PC}$
   (e) $\overrightarrow{BP}$ (f) $\overrightarrow{AP}$

5. ABC is a triangle and the mid-points of sides AB, BC and AC are F, E and D, respectively. The medians AD and BE meet at G. Let $\overrightarrow{AB} = \boldsymbol{a}$, $\overrightarrow{BC} = \boldsymbol{b}$ and $\overrightarrow{CA} = \boldsymbol{c}$.

   (a) Express $\boldsymbol{c}$ in terms of $\boldsymbol{a}$ and $\boldsymbol{b}$.
   (b) Find a vector to represent $\overrightarrow{AD}$, noting that D is on BC.
   (c) Find a vector to represent $\overrightarrow{BE}$.
   (d) Noting that G is on AD and on BE, find vector expressions for $\overrightarrow{AG}$ and $\overrightarrow{BG}$.
   (e) Using the fact that $\overrightarrow{AG} = \overrightarrow{AB} + \overrightarrow{BG}$ find a second vector expression for $\overrightarrow{AG}$. Deduce the position of G on AD and on BF.
   (f) Show that $\overrightarrow{CG}$ is a scalar multiple of $\overrightarrow{CF}$ and hence that G lies on the median CF. Where?
   (g) What conclusions can you draw about the medians of a triangle?

11.4 Components

For many applications it is useful to provide a frame of reference in which vectors are described. In two dimensions we define a pair of perpendicular axes meeting at a fixed point O (the origin of coordinates).

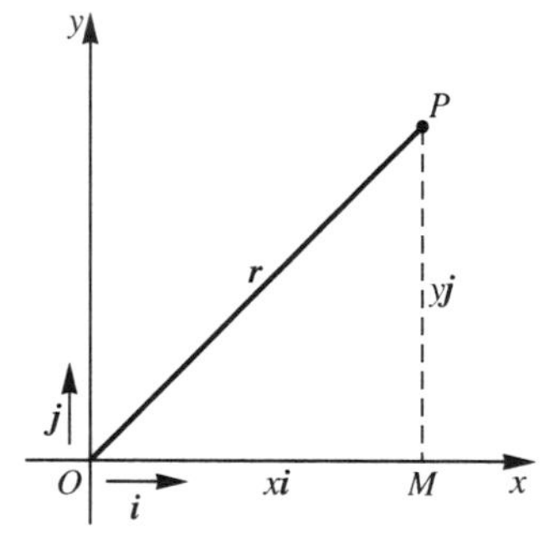

Figure 11.10. Cartesian components of a vector

Unit vectors in the direction of the x- and y-axes are denoted $\boldsymbol{i}$ and $\boldsymbol{j}$ as in Figure 11.10. Consider the point P: if the distances OM and MP are x and y, respectively, then $\overrightarrow{OM} = x\boldsymbol{i}$ and $\overrightarrow{MP} = y\boldsymbol{j}$.

By the triangle law of addition,

$$\overrightarrow{OP} = \overrightarrow{OM} + \overrightarrow{MP}$$

that is

$$r = xi + yj$$

We say that the (Cartesian) **coordinates** of P are (x, y).

Sometimes we write r as a **column vector**

$$\begin{pmatrix} x \\ y \end{pmatrix}$$

We can combine vectors by using the component form:

If $\boldsymbol{a} = a_1\boldsymbol{i} + a_2\boldsymbol{j}$ and $\boldsymbol{b} = b_1\boldsymbol{i} + b_2\boldsymbol{j}$ then

$$\boldsymbol{a} + \boldsymbol{b} = (a_1 + b_1)\boldsymbol{i} + (a_2 + b_2)\boldsymbol{j} \qquad 11.5$$

$$\lambda\boldsymbol{a} = \lambda a_1\boldsymbol{i} + \lambda a_2\boldsymbol{j}$$

KEY POINT

The magnitude of $\boldsymbol{a}$ is $(a_1^2 + a_2^2)^{\frac{1}{2}}$. 11.6

The **direction cosines** of a vector are the values of $\cos\alpha$ and $\cos\beta$ where α and β are as shown in Figure 11.11.

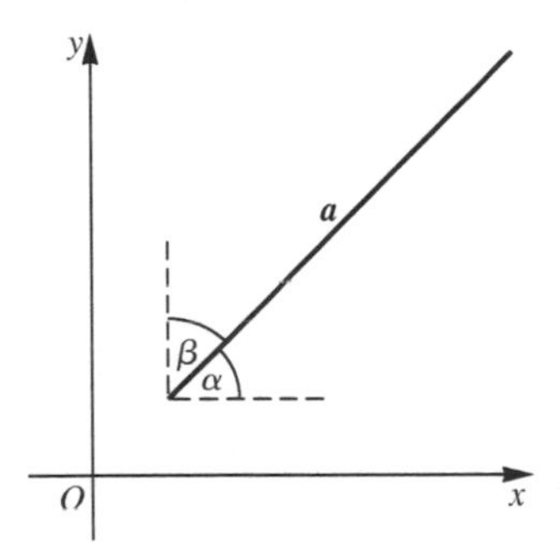

Figure 11.11.
Direction cosines

The angle α is measured anticlockwise from the positive x-direction and the angle $\beta(= 90° - \alpha)$ is measured clockwise from the positive y-direction. In Figure 11.11, both α and β are acute angles.

Worked examples

11.9 Given the vectors

$$\boldsymbol{a} = \begin{pmatrix} 1 \\ 1 \end{pmatrix}, \boldsymbol{b} = \begin{pmatrix} -1 \\ 2 \end{pmatrix}, \boldsymbol{c} = \begin{pmatrix} 3 \\ 0 \end{pmatrix}$$

find:

(a) $\boldsymbol{a} + \boldsymbol{b}$ (b) $3\boldsymbol{a} - 2\boldsymbol{b}$ (c) $\boldsymbol{a} + \boldsymbol{b} - 2\boldsymbol{c}$

(d) the magnitudes of the vectors $\boldsymbol{a}$, $\boldsymbol{b}$ and $\boldsymbol{c}$

(e) a unit vector in the direction of $\boldsymbol{a} + \boldsymbol{b} - 2\boldsymbol{c}$

Write the results to parts (a), (b) and (c) in $\boldsymbol{i}, \boldsymbol{j}$ notation.

Solution (a) $a+b=\begin{pmatrix}1+(-1)\\1+2\end{pmatrix}=\begin{pmatrix}0\\3\end{pmatrix}$

(b) $3a-2b=\begin{pmatrix}3\times 1-2\times(-1)\\3\times 1-2\times 2\end{pmatrix}=\begin{pmatrix}5\\-1\end{pmatrix}$

(c) $a+b-2c=\begin{pmatrix}1+(-1)-2\times 3\\1+2-2\times 0\end{pmatrix}=\begin{pmatrix}-6\\3\end{pmatrix}$

(d) $|a|=\sqrt{(1)^2+(1)^2}=\sqrt{2}; |b|=\sqrt{(-1)^2+(2)^2}=\sqrt{5}$

$|c|=\sqrt{(3)^2+(0)^2}=3$

(e) $|a+b-2c|=\sqrt{(-6)^2+(3)^2}=\sqrt{36+9}=3\sqrt{5}$

The unit vector in this direction is

$$\begin{pmatrix}\dfrac{-6}{3\sqrt{5}}\\\dfrac{3}{3\sqrt{5}}\end{pmatrix}=\begin{pmatrix}-\dfrac{2}{\sqrt{5}}\\\dfrac{1}{\sqrt{5}}\end{pmatrix}$$

In the i, j notation $a=i+j$, $b=-i+2j$, $c=3i$, $a+b=3j$, $3a-2b=5i-j$, $a+b-2c=-6i+3j$.

11.10 Three forces act at a point on an object. Their magnitudes and lines of action are shown in Figure 11.12(a). Find the resultant force.

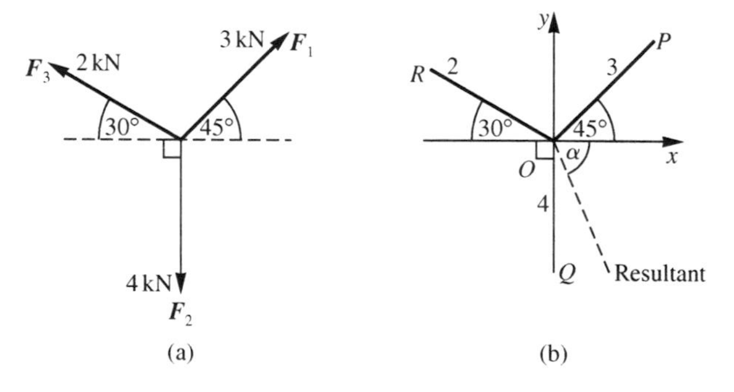

Figure 11.12. Resultant of forces

Solution We represent the forces F_1, F_2 and F_3 by vectors $\overrightarrow{OP}$, $\overrightarrow{OQ}$ and $\overrightarrow{OR}$, respectively, as in Figure 11.12(b). Then

$$\overrightarrow{OP}=3\cos 45^\circ i+3\sin 45^\circ j$$
$$=\frac{3}{\sqrt{2}}i+\frac{3}{\sqrt{2}}j$$

Similarly, $\overrightarrow{OQ}=-4j$ and

$$\overrightarrow{OR}=2\cos 150^\circ i+2\sin 150^\circ j=-2.\frac{\sqrt{3}}{2}i+2.\frac{1}{2}j$$
$$=-\sqrt{3}i+j$$

(This process of replacing a vector by two components at right angles to each other is called **resolving**.)

Then

$$\overrightarrow{OP} + \overrightarrow{OQ} + \overrightarrow{OR} = \left(\frac{3}{\sqrt{2}} + 0 - \sqrt{3}\right)\boldsymbol{i} + \left(\frac{3}{\sqrt{2}} - 4 + 1\right)\boldsymbol{j}$$

$$\simeq 0.39\boldsymbol{i} - 0.88\boldsymbol{j}$$

The resultant force $\boldsymbol{F} \simeq 0.39\boldsymbol{i} - 0.88\boldsymbol{j}$. Its magnitude is $\sqrt{(0.39)^2 + (-0.88)^2} \simeq 0.96\,\text{kN}$.

The direction is indicated by the dashed line in Figure 11.11(b). Now

$$\tan\alpha = \frac{0.88}{0.39} \quad \text{so that } \alpha \simeq 66.1^\circ$$

Much of engineering is concerned with three dimensions. Many of the results we have collected so far extend easily, often by the addition of an extra component.

Figure 11.13 shows the Cartesian axes and the unit vectors in the direction of the axes, namely $\boldsymbol{i}$, $\boldsymbol{j}$, $\boldsymbol{k}$. In the x–y plane $\overrightarrow{OM} = \overrightarrow{ON} + \overrightarrow{NM} = x\boldsymbol{i} + y\boldsymbol{j}$. In the vertical plane containing the triangle OMP,

$$\overrightarrow{OP} = \overrightarrow{OM} + \overrightarrow{MP}$$

that is

$$\boldsymbol{r} = x\boldsymbol{i} + y\boldsymbol{j} + z\boldsymbol{k} \tag{11.7}$$

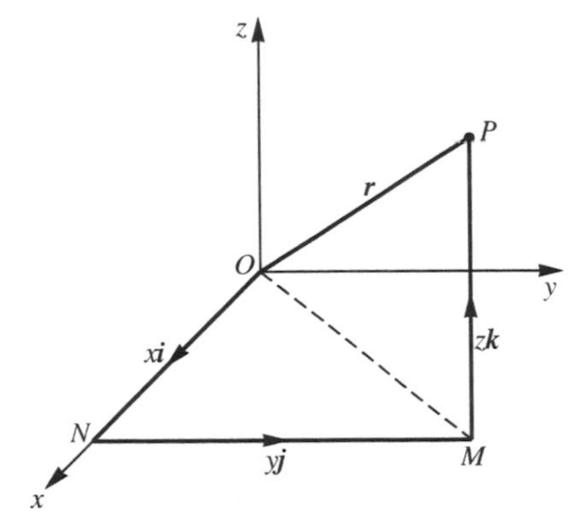

Figure 11.13. Vectors in three-dimensional Cartesian coordinates

We say that the **coordinates** of P in this system are (x, y, z). Applying Pythagoras' theorem to triangle ONM and then to triangle OMP we obtain

$$(OM)^2 = (ON)^2 + (NM)^2$$

and

$$(OP)^2 = (OM)^2 + (MP)^2 = (ON)^2 + (NM)^2 + (MP)^2 = x^2 + y^2 + z^2$$

KEY POINT

The magnitude of OP is $|\boldsymbol{r}| = (x^2 + y^2 + z^2)^{\frac{1}{2}}$. 11.7

To add two vectors we add them by components as in the two-dimensional case.

Worked examples

11.11 Given the vectors $\boldsymbol{a} = 3\boldsymbol{i} + 2\boldsymbol{j} - \boldsymbol{k}$, $\boldsymbol{b} = -3\boldsymbol{i} + \boldsymbol{j} + 2\boldsymbol{k}$, $\boldsymbol{c} = 2\boldsymbol{i} + 3\boldsymbol{j} - 2\boldsymbol{k}$ find:
(a) $\boldsymbol{a} + \boldsymbol{b}$ (b) $\boldsymbol{a} - \boldsymbol{b}$ (c) $\boldsymbol{a} + \boldsymbol{b} - 2\boldsymbol{c}$
(d) the magnitudes of the vectors $\boldsymbol{a}$, $\boldsymbol{b}$, $\boldsymbol{c}$, $\boldsymbol{a} + \boldsymbol{b}$.
Write the results of (a), (b) and (c) as column vectors.

Solution (a) $\boldsymbol{a} + \boldsymbol{b} = 3\boldsymbol{j} + \boldsymbol{k}$ (b) $\boldsymbol{a} - \boldsymbol{b} = 6\boldsymbol{i} + \boldsymbol{j} - 3\boldsymbol{k}$
(c) $\boldsymbol{a} + \boldsymbol{b} - 2\boldsymbol{c} = -4\boldsymbol{i} - 3\boldsymbol{j} + 5\boldsymbol{k}$
(d) $|\boldsymbol{a}| = \left((3)^2 + (2)^2 + (-1)^2\right)^{\frac{1}{2}} = \sqrt{14}$

$|\boldsymbol{b}| = \left((-3)^2 + (1)^2 + (2)^2\right)^{\frac{1}{2}} = \sqrt{14}$

$|\boldsymbol{c}| = \left((2)^2 + (3)^2 + (-2)^2\right)^{\frac{1}{2}} = \sqrt{17}$

$|\boldsymbol{a} + \boldsymbol{b}| = \left(0^2 + (3)^2 + (1)^2\right)^{\frac{1}{2}} = \sqrt{10}$

$$\boldsymbol{a} + \boldsymbol{b} = \begin{pmatrix} 0 \\ 3 \\ 1 \end{pmatrix}, \quad \boldsymbol{a} - \boldsymbol{b} = \begin{pmatrix} 6 \\ 1 \\ -3 \end{pmatrix}, \quad \boldsymbol{a} + \boldsymbol{b} - 2\boldsymbol{c} = \begin{pmatrix} -4 \\ -3 \\ 5 \end{pmatrix}$$

The **direction cosines** of a vector in 3D are $\cos\alpha$, $\cos\beta$ and $\cos\gamma$ where α, β and γ are shown in Figure 11.14. (Note that if one of the angles is obtuse then its direction cosine is negative.)

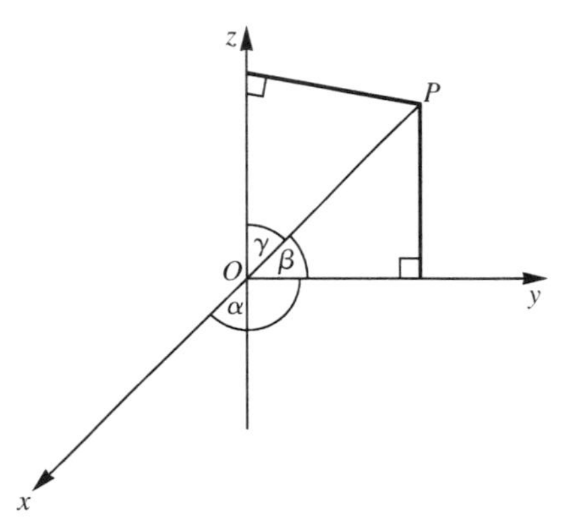

Figure 11.14. Direction cosines in three dimensions

KEY POINT

For the vector $\boldsymbol{a} = a_1\boldsymbol{i} + a_2\boldsymbol{j} + a_3\boldsymbol{k}$ the direction cosines are given by

$$\cos\alpha = \frac{a_1}{|\boldsymbol{a}|}, \cos\beta = \frac{a_2}{|\boldsymbol{a}|}, \cos\gamma = \frac{a_3}{|\boldsymbol{a}|}$$

Worked examples

11.12 Given the vectors $\boldsymbol{a} = 2\boldsymbol{i} + \boldsymbol{j} + \boldsymbol{k}$ and $\boldsymbol{b} = -\boldsymbol{i} + 2\boldsymbol{j} - \boldsymbol{k}$ find the vector $\boldsymbol{a} - \boldsymbol{b}$, its direction cosines and the angles made by the vector with the axes.

Solution

$$\boldsymbol{a} - \boldsymbol{b} = (2 - (-1))\boldsymbol{i} + (1 - 2)\boldsymbol{j} + (1 - (-1))\boldsymbol{k} = 3\boldsymbol{i} - \boldsymbol{j} + 2\boldsymbol{k}$$
$$|\boldsymbol{a} - \boldsymbol{b}| = \left((3)^2 + (-1)^2 + (2)^2\right)^{\frac{1}{2}} = \sqrt{14}$$

Then

$$\cos\alpha = \frac{3}{\sqrt{14}}, \quad \cos\beta = -\frac{1}{\sqrt{14}}, \quad \cos\gamma = \frac{2}{\sqrt{14}}$$

Hence $\alpha \simeq 36.7°$, $\beta \simeq 105.5°$, $\gamma \simeq 57.7°$.

11.13 The robot arm shown schematically in Figure 11.15 has three links OA, AB and BP which can be represented by vectors $\overrightarrow{OA} = \boldsymbol{a}$, $\overrightarrow{AB} = \boldsymbol{b}$ and $\overrightarrow{BP} = \boldsymbol{c}$, respectively.

If $\boldsymbol{a} = 10\boldsymbol{i} + 3\boldsymbol{j} + 5\boldsymbol{k}$, $\boldsymbol{b} = 8\boldsymbol{i} - 6\boldsymbol{j} + 12\boldsymbol{k}$, $\boldsymbol{c} = 4\boldsymbol{i} + 3\boldsymbol{j} - 2\boldsymbol{k}$, find the length of each link and the position of P relative to O.

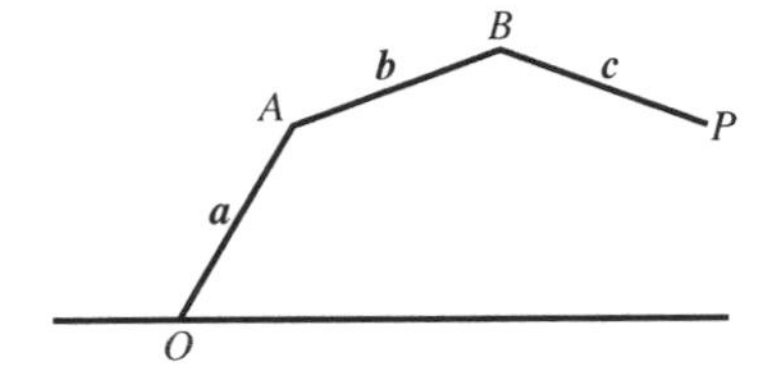

Figure 11.15. Representation of a robot arm

Solution

$$|\boldsymbol{a}| = \left((10)^2 + (3)^2 + (5)^2\right)^{\frac{1}{2}} = \sqrt{134} \simeq 11.6$$
$$|\boldsymbol{b}| = \left((8)^2 + (-6)^2 + (12)^2\right)^{\frac{1}{2}} = \sqrt{244} \simeq 15.6$$
$$|\boldsymbol{c}| = \left((4)^2 + (3)^2 + (-2)^2\right)^{\frac{1}{2}} = \sqrt{29} \simeq 5.4$$
$$\overrightarrow{OP} = \boldsymbol{a} + \boldsymbol{b} + \boldsymbol{c} = 22\boldsymbol{i} + 15\boldsymbol{k}$$

(OP lies in the plane containing the x- and z-axes, since it has a zero y-component.)

Self-assessment questions 11.4

1. Explain how to add two vectors in (3D) Cartesian component form.
2. What are the direction cosines of a vector in 3D Cartesian coordinates?

Exercise 11.4

1. Given the following vectors:

   (a) $\boldsymbol{a} = \begin{pmatrix} 1 \\ 2 \end{pmatrix}$ (b) $\boldsymbol{b} = \begin{pmatrix} 2 \\ -1 \end{pmatrix}$

   (c) $\boldsymbol{c} = \begin{pmatrix} 0 \\ 5 \end{pmatrix}$

   Find

   (a) $\boldsymbol{a} - \boldsymbol{b}$
   (b) $\boldsymbol{a} + \boldsymbol{b} + \boldsymbol{c}$
   (c) $\boldsymbol{b} + 2\boldsymbol{c}$
   (d) the magnitudes of the vectors $\boldsymbol{a}$, $\boldsymbol{b}$ and $\boldsymbol{c}$
   (e) a unit vector in the direction of $\boldsymbol{a} - \boldsymbol{b} + 2\boldsymbol{c}$

2. Repeat Question 1 for the following vectors:

   (a) $\boldsymbol{a} = \begin{pmatrix} 1 \\ 1 \\ 1 \end{pmatrix}$ (b) $\boldsymbol{b} = \begin{pmatrix} 2 \\ 0 \\ -2 \end{pmatrix}$

   (c) $\boldsymbol{c} = \begin{pmatrix} 3 \\ -1 \\ 2 \end{pmatrix}$

3. Find the direction cosines of the vectors $\boldsymbol{a}$, $\boldsymbol{b}$ and $\boldsymbol{c}$ in Question 1.

4. Find the direction cosines of the vectors $\boldsymbol{a}$, $\boldsymbol{b}$ and $\boldsymbol{c}$ in Question 2.

5. Find the vector representation of a point on the following lines:

   (a) the line joining the points

   $\boldsymbol{a} = \begin{pmatrix} 1 \\ 1 \end{pmatrix}$ and $\boldsymbol{b} = \begin{pmatrix} 2 \\ -3 \end{pmatrix}$

   (b) the line through the point

   $\boldsymbol{a} = \begin{pmatrix} 1 \\ -2 \end{pmatrix}$ in the direction given by

   $\begin{pmatrix} -1 \\ 3 \end{pmatrix}$

   (c) the line joining the points

   $\boldsymbol{a} = \begin{pmatrix} 1 \\ 2 \\ 3 \end{pmatrix}$ and $\boldsymbol{b} = \begin{pmatrix} 2 \\ 0 \\ -1 \end{pmatrix}$

Test and assignment exercises 11

1. Draw line segments to represent the following:

   (a) a force of magnitude 1.5 kN acting vertically downwards
   (b) a force of magnitude 2.2 kN acting at an angle of 30° to the horizontal
   (c) a velocity of $4\,\mathrm{m\,s^{-1}}$ north-west
   (d) a velocity of $2.5\,\mathrm{m\,s^{-1}}$ due east

2. Draw a diagram to represent the combination $\boldsymbol{a} + \boldsymbol{b} + \boldsymbol{c} + \boldsymbol{d}$.

3. An aircraft of top speed 250 km per hour sets a course to the north-east but, owing to the prevailing wind, its actual speed relative to the ground is 260 km per hour in a direction 40° east of north. Find the speed and direction of the wind.

4. A motorist travels at 60 km per hour due north. The wind blows rain from the south-east at 20 km per hour. At what speed and from which direction is the rain coming relative to the motorist?

5. $ABCDHGFE$ is a cube as shown in Figure 11.16. If $\overrightarrow{AB} = \boldsymbol{a}$, $\overrightarrow{AD} = \boldsymbol{b}$, and $\overrightarrow{AE} = \boldsymbol{c}$, find vector expressions for the following:

(a) $\overrightarrow{AC}$ (b) $\overrightarrow{BD}$ (c) $\overrightarrow{AG}$
(d) $\overrightarrow{DG}$ (e) $\overrightarrow{CG}$ (f) $\overrightarrow{AF}$
(g) $\overrightarrow{FG}$ (h) $\overrightarrow{EG}$

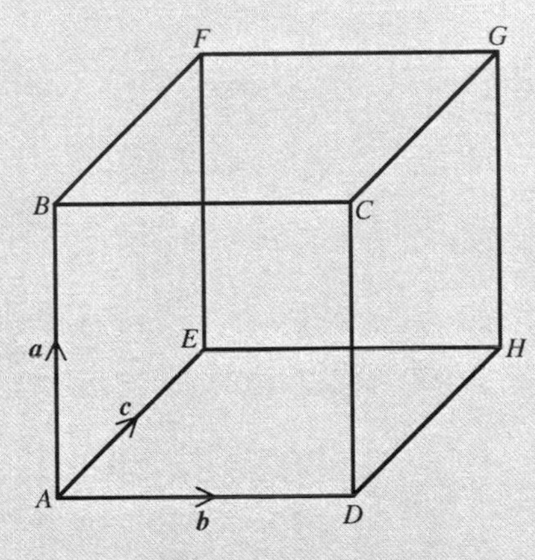

Figure 11.16. Diagram for Question 5

6. $ABCD$ is a parallelogram and E is the mid-point of AD. The line segment BE cuts AC at F. Find the ratio $AF{:}FC$.

7. In triangle ABC the mid-points of AB, BC and CA are D, E and F, respectively. O is any point. Prove that
$\overrightarrow{OA} + \overrightarrow{OB} + \overrightarrow{OC} = \overrightarrow{OD} + \overrightarrow{OE} + \overrightarrow{OF}$

8. If $\boldsymbol{a} = \boldsymbol{i} + \boldsymbol{j} + 2\boldsymbol{k}$, $\boldsymbol{b} = -\boldsymbol{i} + 4\boldsymbol{j} + \boldsymbol{k}$ and $\boldsymbol{c} = 3\boldsymbol{i} + 5\boldsymbol{j} + 2\boldsymbol{k}$ write down expressions for the following:

(a) $\boldsymbol{a} + \boldsymbol{b}$
(b) $\boldsymbol{b} - \boldsymbol{c}$
(c) $\boldsymbol{a} + \boldsymbol{b} + \boldsymbol{c}$
(d) $\boldsymbol{a} - \boldsymbol{b} + \boldsymbol{c}$
(e) $2\boldsymbol{a} + 3\boldsymbol{b} - 4\boldsymbol{c}$

Find the magnitudes of each of these vectors and the direction cosines of $\boldsymbol{a}$, $\boldsymbol{b}$ and $\boldsymbol{c}$.

9. A triangle ABC has vertices at the points

$$A = \begin{pmatrix} -2 \\ 8 \\ 4 \end{pmatrix},$$

$$B = \begin{pmatrix} -4 \\ 10 \\ 6 \end{pmatrix} \quad \text{and}$$

$$C = \begin{pmatrix} 12 \\ 4 \\ -2 \end{pmatrix}$$

Find in terms of $\boldsymbol{i}$, $\boldsymbol{j}$ and $\boldsymbol{k}$ the vectors $\overrightarrow{AB}$, $\overrightarrow{BC}$ and $\overrightarrow{CA}$ and hence find the lengths of the sides of the triangle.

12 Vectors II – products and 3D geometry

Objectives

This chapter

- defines the scalar product of two vectors
- gives a test to determine whether two vectors are perpendicular
- defines the vector product of two vectors
- uses the products in mechanical applications
- develops the geometry of lines and planes in three dimensions
- generalizes the ideas on vectors to n dimensions

12.1 The scalar product

Figure 12.1 shows the field of a constant force $\boldsymbol{F}$ in which a small object is moved from A to B.

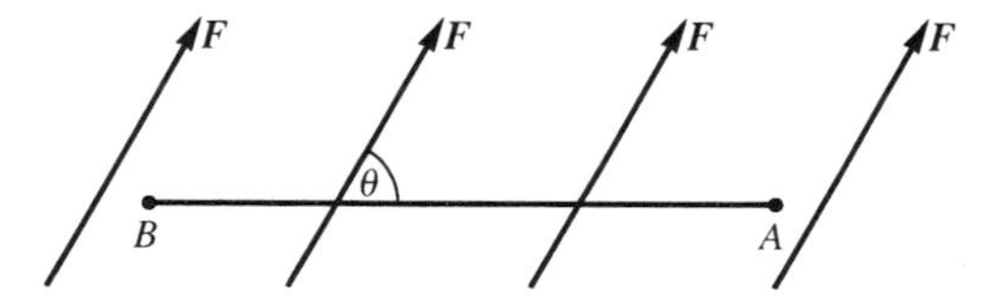

Figure 12.1. Work done in moving from A to B

The fact that $\boldsymbol{F}$ is constant means that it has the same magnitude and direction at all points in its field of influence. If the line segment BA makes an angle θ with the direction of $\boldsymbol{F}$ then the component of $\boldsymbol{F}$ in the direction of $\overrightarrow{BA}$ is $F\cos\theta$. The work done in moving from A to B against the force is $Fd\cos\theta$, where d is the length of $\overrightarrow{AB}$. This 'product' of the magnitudes of $\boldsymbol{F}$ and $\overrightarrow{AB}$ is now generalized.

KEY POINT

> The **scalar product**, or dot product, of two vectors $\boldsymbol{a}$ and $\boldsymbol{b}$ is written $\boldsymbol{a}\cdot\boldsymbol{b}$ and defined by
>
> $$|\boldsymbol{d}| = \boldsymbol{a}\cdot\boldsymbol{b} = |\boldsymbol{a}|\,|\boldsymbol{b}|\cos\theta \qquad 12.1$$
>
> where θ is the angle between the vectors.

The four results below follow from the definition:

$$\boldsymbol{a}\cdot\boldsymbol{b} = \boldsymbol{b}\cdot\boldsymbol{a} \qquad 12.2a$$

$$\lambda(\boldsymbol{a}\cdot\boldsymbol{b}) = (\lambda\boldsymbol{a}\cdot\boldsymbol{b}) = (\boldsymbol{a}\cdot\lambda\boldsymbol{b}) \qquad 12.2b$$

$$(\boldsymbol{a}+\boldsymbol{b})\cdot\boldsymbol{c} = \boldsymbol{a}\cdot\boldsymbol{c}+\boldsymbol{b}\cdot\boldsymbol{c} \qquad 12.2c$$

$$\boldsymbol{a}\cdot(\boldsymbol{b}+\boldsymbol{c}) = \boldsymbol{a}\cdot\boldsymbol{b}+\boldsymbol{a}\cdot\boldsymbol{c} \qquad 12.2d$$

Note the two following special cases: if $\boldsymbol{a}$ and $\boldsymbol{b}$ are parallel then $\boldsymbol{a}\cdot\boldsymbol{b} = |\boldsymbol{a}|\,|\boldsymbol{b}|$;

KEY POINT

> if $\boldsymbol{a}$ and $\boldsymbol{b}$ are perpendicular then $\boldsymbol{a}\cdot\boldsymbol{b} = 0$. 12.3

This latter result can be used to test whether two vectors are perpendicular. Note that the equation $\boldsymbol{a}\cdot\boldsymbol{b} = 0$ does not necessarily imply that either $\boldsymbol{a}$ or $\boldsymbol{b}$ is the zero vector.

Worked examples

12.1 If $\boldsymbol{a} = a_1\boldsymbol{i}+a_2\boldsymbol{j}+a_3\boldsymbol{k}$ and $\boldsymbol{b} = b_1\boldsymbol{i}+b_2\boldsymbol{j}+b_3\boldsymbol{k}$ find an expression for $\boldsymbol{a}\cdot\boldsymbol{b}$ and hence show that the vectors $\boldsymbol{a} = 2\boldsymbol{i}-3\boldsymbol{j}+\boldsymbol{k}$ and $\boldsymbol{b} = 2\boldsymbol{i}+\boldsymbol{j}-\boldsymbol{k}$ are perpendicular.

Solution $\boldsymbol{a}\cdot\boldsymbol{b} = (a_1\boldsymbol{i}+a_2\boldsymbol{j}+a_3\boldsymbol{k})\cdot(b_1\boldsymbol{i}+b_2\boldsymbol{j}+b_3\boldsymbol{k})$

A term such as $a_1\boldsymbol{i}\cdot b_2\boldsymbol{j}$ can be re-expressed as $a_1b_2\boldsymbol{i}\cdot\boldsymbol{j}$.

Now $\boldsymbol{i}, \boldsymbol{j}$ and $\boldsymbol{k}$ are mutually perpendicular so that

$$\boldsymbol{i}\cdot\boldsymbol{j} = \boldsymbol{j}\cdot\boldsymbol{i} = 0, \quad \boldsymbol{i}\cdot\boldsymbol{k} = \boldsymbol{k}\cdot\boldsymbol{i} = 0, \quad \boldsymbol{j}\cdot\boldsymbol{k} = \boldsymbol{k}\cdot\boldsymbol{j} = 0$$

Further, $\boldsymbol{i}\cdot\boldsymbol{i} = 1\times 1 = 1$; similarly $\boldsymbol{j}\cdot\boldsymbol{j} = \boldsymbol{k}\cdot\boldsymbol{k} = 1$. Then

$$\begin{aligned}\boldsymbol{a}\cdot\boldsymbol{b} = {} & a_1b_1\boldsymbol{i}\cdot\boldsymbol{i}+a_1b_2\boldsymbol{i}\cdot\boldsymbol{j}+a_1b_3\boldsymbol{i}\cdot\boldsymbol{k} \\ & +a_2b_1\boldsymbol{j}\cdot\boldsymbol{i}+a_2b_2\boldsymbol{j}\cdot\boldsymbol{j}+a_2b_3\boldsymbol{j}\cdot\boldsymbol{k} \\ & +a_3b_1\boldsymbol{k}\cdot\boldsymbol{i}+a_3b_2\boldsymbol{k}\cdot\boldsymbol{j}+a_3b_3\boldsymbol{k}\cdot\boldsymbol{k}\end{aligned}$$

that is

$$\boldsymbol{a}\cdot\boldsymbol{b} = a_1b_1+a_2b_2+a_3b_3$$

For the given vector

$$\boldsymbol{a} \cdot \boldsymbol{b} = 2 \times 2 + (-3) \times 1 + 1 \times (-1) = 4 - 3 - 1 = 0$$

Hence $\boldsymbol{a}$ and $\boldsymbol{b}$ are perpendicular.

12.2 Find expressions for $\boldsymbol{a} \cdot \boldsymbol{a}$ and $|\boldsymbol{a}|^2$.

Solution $\boldsymbol{a} \cdot \boldsymbol{a} = a_1 \cdot a_1 + a_2 \cdot a_2 + a_3 \cdot a_3$

$|\boldsymbol{a}| = (a_1^2 + a_2^2 + a_3^2)^{1/2}$ so that

$|\boldsymbol{a}|^2 = a_1^2 + a_2^2 + a_3^2 = \boldsymbol{a} \cdot \boldsymbol{a}$

12.3 Find the angle between the vectors $\boldsymbol{a} = \boldsymbol{i} - \boldsymbol{j} + 3\boldsymbol{k}$ and $\boldsymbol{b} = \boldsymbol{i} + 2\boldsymbol{j} + 2\boldsymbol{k}$.

Solution $\boldsymbol{a} \cdot \boldsymbol{b} = 1 \times 1 + (-1) \times 2 + 3 \times 2 = 1 - 2 + 6 = 5$

$|\boldsymbol{a}| = \left((1)^2 + (-1)^2 + (3)^2\right)^{\frac{1}{2}} = \sqrt{11}$

$|\boldsymbol{b}| = \left((1)^2 + (2)^2 + (2)^2\right)^{\frac{1}{2}} = \sqrt{9} = 3$

Since $\boldsymbol{a} \cdot \boldsymbol{b} = |\boldsymbol{a}|\,|\boldsymbol{b}| \cos\theta$ then

$$\cos\theta = \frac{\boldsymbol{a} \cdot \boldsymbol{b}}{|\boldsymbol{a}|\,|\boldsymbol{b}|} = \frac{5}{\sqrt{11} \times 3} = \frac{5}{3\sqrt{11}}$$

and therefore $\theta = 59.8^\circ$.

12.4 Show that the diagonals of a rhombus intersect at right angles.

Solution The quadrilateral $ABCD$ shown in Figure 12.2 is a rhombus.

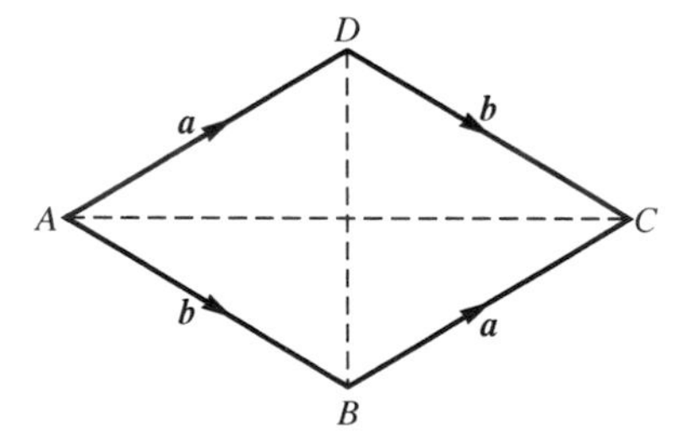

Figure 12.2.
Diagonals of a rhombus

A rhombus is a parallelogram with all sides of equal length. Let $\overrightarrow{AB}$ be denoted by $\boldsymbol{b}$ and $\overrightarrow{AD}$ be denoted by $\boldsymbol{a}$. Then $\overrightarrow{DC}$ is also denoted by $\boldsymbol{b}$ and $\overrightarrow{BC}$ is also denoted by $\boldsymbol{a}$. Further, $\overrightarrow{AC} = \boldsymbol{a} + \boldsymbol{b}$ and $\overrightarrow{BD} = \boldsymbol{a} - \boldsymbol{b}$.

$$\begin{aligned}\overrightarrow{AC} \cdot \overrightarrow{BD} &= (\boldsymbol{a} + \boldsymbol{b}) \cdot (\boldsymbol{a} - \boldsymbol{b}) \\ &= \boldsymbol{a} \cdot \boldsymbol{a} + \boldsymbol{b} \cdot \boldsymbol{a} - \boldsymbol{a} \cdot \boldsymbol{b} - \boldsymbol{b} \cdot \boldsymbol{b} \\ &= \boldsymbol{a} \cdot \boldsymbol{a} + \boldsymbol{a} \cdot \boldsymbol{b} - \boldsymbol{a} \cdot \boldsymbol{b} - \boldsymbol{b} \cdot \boldsymbol{b} \\ &= a^2 \qquad\qquad\qquad\quad - b^2\end{aligned}$$

But $AD = AB$, that is $|\boldsymbol{a}| = |\boldsymbol{b}|$ and hence $a = b$. Then $\overrightarrow{AC} \cdot \overrightarrow{BD} = 0$ so that $\overrightarrow{AC}$ is perpendicular to $\overrightarrow{BD}$.

12.5 Find the component of the force $\boldsymbol{F} = 2\boldsymbol{i} - \boldsymbol{j} + 3\boldsymbol{k}$ in the direction of the vector $3\boldsymbol{i} - 2\boldsymbol{j} - \boldsymbol{k}$. Hence find the work done by the force in moving a particle from the point $A = (1,1,1)$ to $B = (4,-1,0)$.

Solution The unit vector in the direction of $3\boldsymbol{i} - 2\boldsymbol{j} - \boldsymbol{k}$ is

$$\frac{1}{\left((3)^2 + (-2)^2 + (-1)^2\right)^{1/2}}(3\boldsymbol{i} - 2\boldsymbol{j} - \boldsymbol{k})$$

that is

$$\frac{1}{\sqrt{14}}(3\boldsymbol{i} - 2\boldsymbol{j} - \boldsymbol{k}) \qquad \text{or} \qquad \frac{3}{\sqrt{14}}\boldsymbol{i} - \frac{2}{\sqrt{14}}\boldsymbol{j} - \frac{1}{\sqrt{14}}\boldsymbol{k}$$

The required component is

$$\begin{aligned}(2\boldsymbol{i} - \boldsymbol{j} + 3\boldsymbol{k}) \cdot \frac{1}{\sqrt{14}}(3\boldsymbol{i} - 2\boldsymbol{j} - \boldsymbol{k}) &= \frac{1}{\sqrt{14}}(2 \times 3 + (-1) \times (-2) + 3 \times (-1)) \\ &= \frac{1}{\sqrt{14}} \times 5 = \frac{5}{\sqrt{14}}\end{aligned}$$

Now $\overrightarrow{AB} = 4\boldsymbol{i} - \boldsymbol{j} - (\boldsymbol{i} + \boldsymbol{j} + \boldsymbol{k}) = 3\boldsymbol{i} - 2\boldsymbol{j} - \boldsymbol{k} = \boldsymbol{d}$, say, so that the work done is

$$\begin{aligned}\boldsymbol{F} \cdot \boldsymbol{d} &= (2\boldsymbol{i} - \boldsymbol{j} + 3\boldsymbol{k}) \cdot (3\boldsymbol{i} - 2\boldsymbol{j} - \boldsymbol{k}) \\ &= 2 \times 3 + (-1) \times (-2) + 3 \times (-1) \\ &= 6 + 2 - 3 = 5\end{aligned}$$

The following results, deduced from the examples, are most important.

KEY POINT

$\boldsymbol{a} \cdot \boldsymbol{b} = a_1b_1 + a_2b_2 + a_3b_3$	12.4
$\boldsymbol{i} \cdot \boldsymbol{j} = \boldsymbol{j} \cdot \boldsymbol{k} = \boldsymbol{k} \cdot \boldsymbol{i} = 0$	12.5
$\boldsymbol{i} \cdot \boldsymbol{i} = \boldsymbol{j} \cdot \boldsymbol{j} = \boldsymbol{k} \cdot \boldsymbol{k} = 1$	12.6
$\boldsymbol{a} \cdot \boldsymbol{a} = \lvert\boldsymbol{a}\rvert^2$	12.7

Self-assessment questions 12.1

1. How is the scalar product of two vectors $\boldsymbol{a}$ and $\boldsymbol{b}$ defined?
2. How would you calculate the scalar product of two vectors in Cartesian component form?
3. If the scalar product of two vectors is zero, what can be concluded about the vectors?

Exercise 12.1

1. Which pairs of the following vectors are mutually perpendicular? $\boldsymbol{a} = \boldsymbol{i} + \boldsymbol{j} - \boldsymbol{k}$, $\boldsymbol{b} = 2\boldsymbol{i} - \boldsymbol{j} + \boldsymbol{k}$, $\boldsymbol{c} = 2\boldsymbol{i} - \boldsymbol{j} - \boldsymbol{k}$, $\boldsymbol{d} = \boldsymbol{i} + \boldsymbol{j} - 2\boldsymbol{k}$, $\boldsymbol{e} = \boldsymbol{i} + 3\boldsymbol{j} - 2\boldsymbol{k}$

2. If the sides of a triangle are $\boldsymbol{a}$, $\boldsymbol{b}$ and $\boldsymbol{c}$, use the concept of a scalar product to prove the cosine rule.

3. Find the angles between the vectors $\boldsymbol{b}$, $\boldsymbol{c}$ and $\boldsymbol{d}$ in Question 1.

4. Prove Pythagoras' theorem for the right-angled triangle ABC where $\angle B = 90°$. (*Hint*: let $\boldsymbol{AB} = \boldsymbol{a}$ and $\boldsymbol{BC} = \boldsymbol{b}$.)

5. Find the work done in moving an object from the point $A = (-1,0,1)$ to $B = (3,1,4)$ against a force $\boldsymbol{F} = \boldsymbol{i} - \boldsymbol{j} + 2\boldsymbol{k}$.

12.2 The vector product

The **torque** or moment of force about an axis is illustrated in Figure 12.3. The force is $\boldsymbol{F}$ and the axis is OQ. ON is the shortest distance from the axis to the line of action of the force. If $OP = r$ then $ON = r \sin\theta$. The magnitude of the torque is $|\boldsymbol{F}|\,|\boldsymbol{r}| \sin\theta$. In three dimensions torques are added by treating them as vectors and using the parallelogram law. The direction associated with the torque in Figure 12.3 is that of $\overrightarrow{OQ}$. If $\hat{\boldsymbol{n}}$ is a unit vector in the direction of $\overrightarrow{OQ}$ then the vector representing the torque can be written

$$|\boldsymbol{F}| \cdot |\boldsymbol{r}| \cdot \sin\theta\, \hat{\boldsymbol{n}}$$

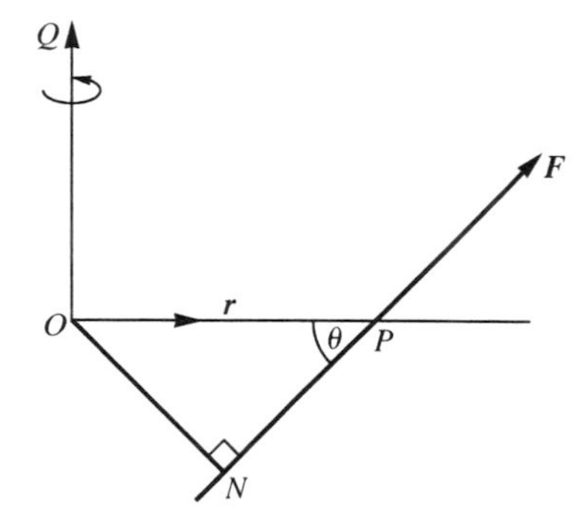

Figure 12.3.
Moment of a force

This idea is generalized as follows.

The **vector product** of two vectors $\boldsymbol{a}$ and $\boldsymbol{b}$ is written $\boldsymbol{a} \times \boldsymbol{b}$. The definition of the vector product is:

KEY POINT

$$\boldsymbol{a} \times \boldsymbol{b} = |\boldsymbol{a}|\,|\boldsymbol{b}| \sin\theta\, \hat{\boldsymbol{n}} \qquad 12.8$$

where θ is the angle between the vectors $\boldsymbol{a}$ and $\boldsymbol{b}$, and the unit vector $\hat{\boldsymbol{n}}$ is in the direction indicated in Figure 12.4.

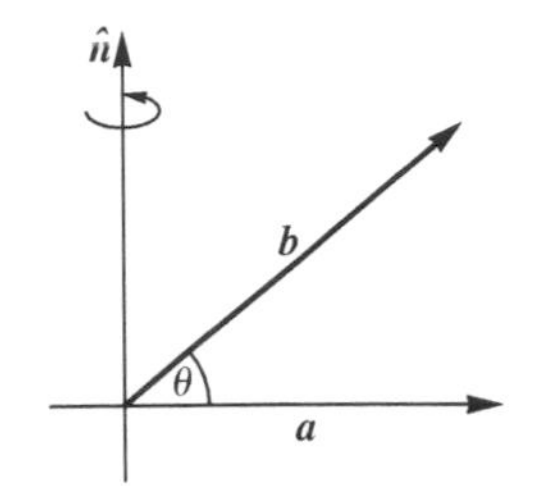

Figure 12.4.
Vector product

The convention for $\hat{\boldsymbol{n}}$ is known as the right-hand screw rule since in turning a screwdriver from the direction of $\boldsymbol{a}$ to that of $\boldsymbol{b}$ with the right hand the screw advances in the direction of $\hat{\boldsymbol{n}}$. Note that $\hat{\boldsymbol{n}}$ and therefore $\boldsymbol{a} \times \boldsymbol{b}$ are perpendicular to both $\boldsymbol{a}$ and $\boldsymbol{b}$.

The following results are important:

$$\boldsymbol{b} \times \boldsymbol{a} = -\boldsymbol{a} \times \boldsymbol{b} \qquad 12.9$$

$$\boldsymbol{a} \times (\boldsymbol{b} + \boldsymbol{c}) = \boldsymbol{a} \times \boldsymbol{b} + \boldsymbol{a} \times \boldsymbol{c} \qquad 12.10$$

$$\lambda(\boldsymbol{a} \times \boldsymbol{b}) = (\lambda \boldsymbol{a}) \times \boldsymbol{b} = \boldsymbol{a} \times (\lambda \boldsymbol{b}) \qquad 12.11$$

Worked examples

12.6 (a) If $\boldsymbol{a}$ and $\boldsymbol{b}$ are parallel find $\boldsymbol{a} \times \boldsymbol{b}$.
(b) Find $\boldsymbol{i} \times \boldsymbol{j}$, $\boldsymbol{j} \times \boldsymbol{k}$ and $\boldsymbol{k} \times \boldsymbol{i}$.
(c) Find $\boldsymbol{i} \times \boldsymbol{i}$, $\boldsymbol{j} \times \boldsymbol{j}$ and $\boldsymbol{k} \times \boldsymbol{k}$.

Solution (a) If $\boldsymbol{a}$ and $\boldsymbol{b}$ are parallel, $\theta = 0$ so that $\sin\theta = 0$ and $\boldsymbol{a} \times \boldsymbol{b} = \boldsymbol{0}$.

(b)

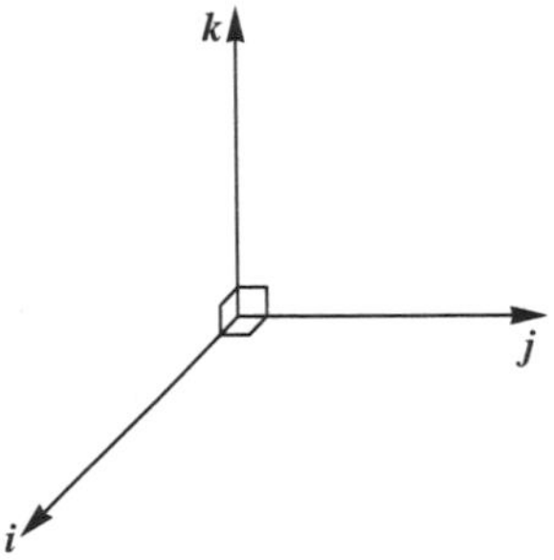

Figure 12.5.
Diagram for Example 12.6

Figure 12.5 can be used by first noting that the angle between any two of $\boldsymbol{i}$, $\boldsymbol{j}$ and $\boldsymbol{k}$ is $\theta = 90°$ so that $\sin\theta = 1$. Also, $|\boldsymbol{i}| = |\boldsymbol{j}| = |\boldsymbol{k}| = 1$. Hence

$$\boldsymbol{i} \times \boldsymbol{j} = \boldsymbol{k}, \qquad \boldsymbol{j} \times \boldsymbol{k} = \boldsymbol{i} \qquad \text{and} \qquad \boldsymbol{k} \times \boldsymbol{i} = \boldsymbol{j}$$

(c) Using the result of part (a) with $\boldsymbol{b} = \boldsymbol{a}$

$$\boldsymbol{a} \times \boldsymbol{a} = \boldsymbol{0}$$

Therefore $\boldsymbol{i} \times \boldsymbol{i} = \boldsymbol{j} \times \boldsymbol{j} = \boldsymbol{k} \times \boldsymbol{k} = \boldsymbol{0}$.

12.7 If $\boldsymbol{a} = a_1\boldsymbol{i} + a_2\boldsymbol{j} + a_3\boldsymbol{k}$ and $\boldsymbol{b} = b_1\boldsymbol{i} + b_2\boldsymbol{j} + b_3\boldsymbol{k}$ find an expression for $\boldsymbol{a} \times \boldsymbol{b}$.

Solution

$$\begin{aligned}\boldsymbol{a} \times \boldsymbol{b} &= (a_1\boldsymbol{i} + a_2\boldsymbol{j} + a_3\boldsymbol{k}) \times (b_1\boldsymbol{i} + b_2\boldsymbol{j} + b_3\boldsymbol{k}) \\ &= a_1b_1\boldsymbol{i} \times \boldsymbol{i} + a_1b_2\boldsymbol{i} \times \boldsymbol{j} + a_1b_3\boldsymbol{i} \times \boldsymbol{k} \\ &\quad + a_2b_1\boldsymbol{j} \times \boldsymbol{i} + a_2b_2\boldsymbol{j} \times \boldsymbol{j} + a_2b_3\boldsymbol{j} \times \boldsymbol{k} \\ &\quad + a_3b_1\boldsymbol{k} \times \boldsymbol{i} + a_3b_2\boldsymbol{k} \times \boldsymbol{j} + a_3b_3\boldsymbol{k} \times \boldsymbol{k} \\ &= a_1b_2\boldsymbol{k} + a_1b_3(-\boldsymbol{j}) + a_2b_1(-\boldsymbol{k}) + a_2b_3\boldsymbol{i} + a_3b_1\boldsymbol{j} + a_3b_2(-\boldsymbol{i})\end{aligned}$$

KEY POINT

$$\boldsymbol{a} \times \boldsymbol{b} = (a_2b_3 - a_3b_2)\boldsymbol{i} + (a_3b_1 - a_1b_3)\boldsymbol{j} + (a_1b_2 - a_2b_1)\boldsymbol{k} \qquad 12.12$$

12.8 Given $\boldsymbol{a} = 3\boldsymbol{i} + 2\boldsymbol{j} + 5\boldsymbol{k}$ and $\boldsymbol{b} = 7\boldsymbol{i} - \boldsymbol{j} + \boldsymbol{k}$, find $\boldsymbol{a} \times \boldsymbol{b}$ and $\boldsymbol{b} \times \boldsymbol{a}$.

Solution Using (12.12)

$$\begin{aligned}\boldsymbol{a} \times \boldsymbol{b} &= (2 \times 1 - 5 \times (-1))\boldsymbol{i} + (5 \times 7 - 3 \times 1)\boldsymbol{j} + (3 \times (-1) - 2 \times 7)\boldsymbol{k} \\ &= 7\boldsymbol{i} + 32\boldsymbol{j} - 17\boldsymbol{k} \\ \boldsymbol{b} \times \boldsymbol{a} &= -\boldsymbol{a} \times \boldsymbol{b} = -7\boldsymbol{i} - 32\boldsymbol{j} + 17\boldsymbol{k}\end{aligned}$$

Other situations in which the vector product arises include:

1. The force on a charged particle moving in a magnetic field:

   $$\boldsymbol{F} = q\boldsymbol{v} \times \boldsymbol{B}$$

   where $\boldsymbol{F}$ is the force, $\boldsymbol{v}$ is the particle's velocity, $\boldsymbol{B}$ is the magnetic field and q is the charge.

2. The velocity of a point on a body rotating with angular speed about a fixed axis is given by

   $$\boldsymbol{v} = \omega \times \boldsymbol{r}$$

   where $\boldsymbol{r}$ is the position vector of P relative to a fixed point O on the axis.

Worked example

12.9 A force of magnitude 6 acts through the point $Q(3,2,1)$ in the direction of the vector (1,2,2). Find the moment of the force about the point $P(2,-3,1)$ and deduce the moment of the force about axes through P parallel to the coordinate axes.

Solution A unit vector in the direction of the force is $\left(\frac{1}{3}, \frac{2}{3}, \frac{2}{3}\right)$. Hence the force $\boldsymbol{F}$ is given by

$$6\left(\tfrac{1}{3}, \tfrac{2}{3}, \tfrac{2}{3}\right) = (2,4,4) \qquad \text{that is} \quad 2\boldsymbol{i} + 4\boldsymbol{j} + 4\boldsymbol{k}$$

The vector $\overrightarrow{PQ}$ is $(3,2,1) - (2,-3,1) = \boldsymbol{i} + 5\boldsymbol{j}$. The moment of the force is

$$\overrightarrow{PQ} \times \boldsymbol{F} = (5 \times 4 - 0)\boldsymbol{i} + (0 - 1 \times 4)\boldsymbol{j} + (1 \times 4 - 5 \times 2)\boldsymbol{k}$$
$$= 20\boldsymbol{i} - 4\boldsymbol{j} - 6\boldsymbol{k}$$

The moments about the axes through 0 are, respectively, 20, 4, 6.

Self-assessment questions 12.2

1. How is the vector product of the vectors $\boldsymbol{a}$ and $\boldsymbol{b}$ defined?
2. If the vector product of two vectors is $\boldsymbol{O}$ what can be concluded about the vectors?
3. What is the component form of $\boldsymbol{a} \times \boldsymbol{b}$ when $\boldsymbol{a} = a_1\boldsymbol{i} + a_2\boldsymbol{j}$ and $\boldsymbol{b} = b_1\boldsymbol{i} + b_2\boldsymbol{j}$?

Exercise 12.2

1. For the vectors $\boldsymbol{a} = (2,1,-1)$ and $\boldsymbol{b} = (1,-1,3)$ find $\boldsymbol{a} \times \boldsymbol{b}$ and hence calculate the angle between them.
2. A triangle ABC has vertices $A = (2,0,6)$, $B = (-2,4,-2)$ and $C = (2,4,2)$. Find the area of the triangle, which is given by $\frac{1}{2}\left|\overrightarrow{AB} \times \overrightarrow{AC}\right|$.
3. For the vectors $\boldsymbol{a} = (2,1,-1)$, $\boldsymbol{b} = (0,3,-2)$, $\boldsymbol{c} = (1,-1,1)$, find $\boldsymbol{a} \,.\, (\boldsymbol{b} \times \boldsymbol{c})$ and $(\boldsymbol{a} \times \boldsymbol{b}) \,.\, \boldsymbol{c}$.
   Further, find $(\boldsymbol{a} \times \boldsymbol{b}) \times \boldsymbol{c}$ and $(\boldsymbol{a} \,.\, \boldsymbol{c})\boldsymbol{b} - (\boldsymbol{a} \,.\, \boldsymbol{b})\boldsymbol{c}$. Comment on your results.
4. Calculate the magnitude of the following vector products for the vectors
   $\boldsymbol{a} = \boldsymbol{i} + \boldsymbol{j} - \boldsymbol{k}$,
   $\boldsymbol{b} = \boldsymbol{i} - \boldsymbol{j} - \boldsymbol{k}$,
   $\boldsymbol{c} = 3\boldsymbol{i} + \boldsymbol{j} - 2\boldsymbol{k}$
   (a) $\boldsymbol{a} \times \boldsymbol{b}$ (b) $\boldsymbol{a} \times \boldsymbol{c}$
   (c) $\boldsymbol{c} \times \boldsymbol{a}$ (d) $\boldsymbol{b} \times \boldsymbol{c}$
   (e) $\boldsymbol{a} \times \boldsymbol{a}$ (f) $\boldsymbol{c} \times \boldsymbol{b}$
5. A force of magnitude 5 N acts in the direction from $A = (1,1,1)$ to $B = (3,2,4)$ the units being metres. Find a vector representation of the force and find its moment about the origin. What is the magnitude of this moment?

12.3 Lines and planes

In Section 11.3 we gave the equation for a straight line through a given point A in a given direction. It is

$$\boldsymbol{r} = \boldsymbol{a} + \lambda\boldsymbol{u} \qquad 11.1$$

In three dimensions the general position vector is

$$\boldsymbol{r} = \begin{pmatrix} x \\ y \\ z \end{pmatrix}$$

the fixed point is

$$A = \begin{pmatrix} a_1 \\ a_2 \\ a_3 \end{pmatrix}$$

and the vector is

$$\boldsymbol{u} = \begin{pmatrix} u_1 \\ u_2 \\ u_3 \end{pmatrix}$$

Then the equation of the line becomes

$$\begin{pmatrix} x \\ y \\ z \end{pmatrix} = \begin{pmatrix} a_1 \\ a_2 \\ a_3 \end{pmatrix} + \lambda \begin{pmatrix} u_1 \\ u_2 \\ u_3 \end{pmatrix}$$

This single vector equation can be written as three separate component equations:

$$x = a_1 + \lambda u_1, \quad y = a_2 + \lambda u_2, \quad z = a_3 + \lambda u_3$$

Rearranging these equations we obtain

$$\lambda = \frac{x - a_1}{u_1}, \quad \lambda = \frac{y - a_2}{u_2}, \quad \lambda = \frac{z - a_3}{u_3}$$

Equating the three expressions for λ we obtain the general equation of a line in three dimensions

KEY POINT

The general equation of a line is

$$\frac{x - a_1}{u_1} = \frac{y - a_2}{u_2} = \frac{z - a_3}{u_3} \qquad 12.13$$

Worked examples

12.10 Find the equation of the line through the points (1,−1,−1) and (1,1,1). Where does it intersect the line through the points (−1,3,1) and (0,1,0)?

Solution Let the first line be denoted by l_1. The direction of the line is given by the vector

$$\begin{pmatrix} 1 \\ 1 \\ 1 \end{pmatrix} - \begin{pmatrix} 1 \\ -1 \\ -1 \end{pmatrix} = \begin{pmatrix} 0 \\ 2 \\ 2 \end{pmatrix}$$

and its equation by

$$\boldsymbol{r} = \begin{pmatrix} 1 \\ -1 \\ -1 \end{pmatrix} + \lambda \begin{pmatrix} 0 \\ 2 \\ 2 \end{pmatrix} = \begin{pmatrix} 1 \\ -1+2\lambda \\ -1+2\lambda \end{pmatrix}$$

Let the second line be denoted l_2. The direction of the line is given by the vector

$$\begin{pmatrix} 0 \\ 1 \\ 0 \end{pmatrix} - \begin{pmatrix} -1 \\ 3 \\ 1 \end{pmatrix} = \begin{pmatrix} 1 \\ -2 \\ -1 \end{pmatrix}$$

and its equation by

$$\boldsymbol{r} = \begin{pmatrix} -1 \\ 3 \\ 1 \end{pmatrix} + \mu \begin{pmatrix} 1 \\ -2 \\ -1 \end{pmatrix} = \begin{pmatrix} -1+\mu \\ 3-2\mu \\ 1-\mu \end{pmatrix}$$

The lines intersect if there is a value for λ and a value for μ which give the same $\boldsymbol{r}$.

Comparing first components: $1 = -1 + \mu; \quad \mu = 2$

Comparing second components: $-1 + 2\lambda = 3 - 2\mu = -1; \quad \lambda = 0$

The third components are $-1 + 2\lambda = -1$ and $1 - \mu = -1$

Hence when $\lambda = 0$ and $\mu = 2$ the equations of l_1 and l_2 each give

$$\boldsymbol{r} = \begin{pmatrix} 1 \\ -1 \\ -1 \end{pmatrix}$$

which is the point of intersection.

(Had the values of λ and μ which matched both the first and second components not made the third components match then the lines would not intersect. They could have been parallel or they could have been **skew**.)

12.11 Find the shortest distance between the lines

$$l_1: \frac{x-1}{2} = \frac{y-2}{3} = \frac{z-3}{1} \quad \text{and} \quad l_2: \frac{x}{1} = \frac{y+1}{2} = \frac{z+2}{3}$$

Solution For l_1 we write

$$\frac{x-1}{2} = \frac{y-2}{3} = \frac{z-3}{1} = \lambda$$

Then we write $x - 1 = 2\lambda$, $y - 2 = 3\lambda$, $z - 3 = \lambda$ and rearrange these equations to obtain $x = 1 + 2\lambda$, $y = 2 + 3\lambda$, $z = 3 + \lambda$. In vector form, the equation of the lines is

$$l_1: \ \boldsymbol{r} = \begin{pmatrix} 1 \\ 2 \\ 3 \end{pmatrix} + \lambda \begin{pmatrix} 2 \\ 3 \\ 1 \end{pmatrix} = \boldsymbol{a} + \lambda \boldsymbol{u}$$

Similarly,

$$l_2:\ \boldsymbol{r} = \begin{pmatrix} 0 \\ -1 \\ -2 \end{pmatrix} + \mu \begin{pmatrix} 1 \\ 2 \\ 3 \end{pmatrix} = \boldsymbol{b} + \mu\boldsymbol{v}$$

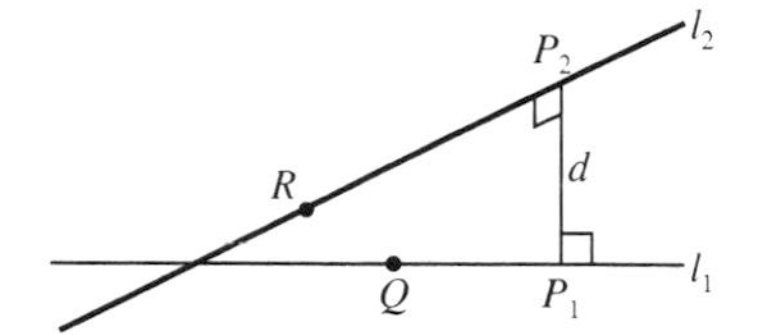

Figure 12.6. The skew lines of Example 12.11

Let the shortest distance between the lines be d, shown as P_1P_2 in Figure 12.6. Then $\overrightarrow{P_1P_2}$ is perpendicular both to $\boldsymbol{u}$ and to $\boldsymbol{v}$. A vector in the direction of $\overrightarrow{P_1P_2}$ is $\boldsymbol{u} \times \boldsymbol{v}$, that is

$$\begin{pmatrix} 7 \\ -5 \\ 1 \end{pmatrix}$$

the unit vector in this direction is

$$\hat{\boldsymbol{n}} = \frac{1}{\sqrt{75}} \begin{pmatrix} 7 \\ -5 \\ 1 \end{pmatrix}$$

Then

$$\overrightarrow{P_1P_2} = d\hat{\boldsymbol{n}} = \frac{d}{\sqrt{75}} \begin{pmatrix} 7 \\ -5 \\ 1 \end{pmatrix}$$

Let Q be a point on l_1 and R be a point on l_2. Then $\overrightarrow{RQ}$ is given by $\boldsymbol{a} - \boldsymbol{b} + \lambda\boldsymbol{u} - \mu\boldsymbol{v}$ for some λ and some μ, that is

$$\overrightarrow{RQ} = \begin{pmatrix} 1 \\ 3 \\ 5 \end{pmatrix} + \lambda \begin{pmatrix} 2 \\ 3 \\ 1 \end{pmatrix} - \mu \begin{pmatrix} 1 \\ 2 \\ 3 \end{pmatrix}$$

Now, from the geometry of Figure 12.6 it follows that $\overrightarrow{RQ}.\hat{\boldsymbol{n}} = d$. Taking the scalar product of $\overrightarrow{RQ}$ with the vector

$$\begin{pmatrix} 7 \\ -5 \\ 1 \end{pmatrix}$$

gives

$$(7 - 15 + 5) + \lambda \cdot 0 - \mu \cdot 0 = -3$$

Hence $\overrightarrow{RQ}.\hat{\boldsymbol{n}} = -3/\sqrt{75}$. Therefore $d = -3/\sqrt{75}$. The minus sign indicates that we travel along one line 'left to right' rather than the other way; it can be ignored. The shortest distance is therefore $3/\sqrt{75}$. (Note that if the lines had intersected then d would have been 0.)

The **vector equation of a plane** is found by noting that the line which joins any two points in the plane is perpendicular to the normal to the plane.

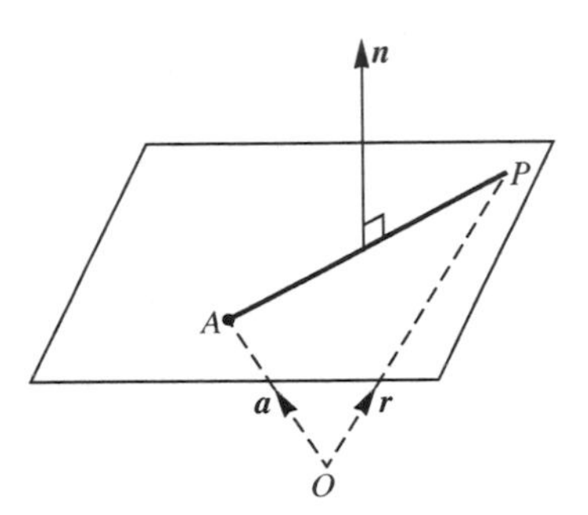

Figure 12.7. Vector equation of a plane

In Figure 12.7 A is a given point in the plane and P is any other point in the plane; $\boldsymbol{n}$ is a vector perpendicular to the plane. Then $\overrightarrow{AP} = \boldsymbol{r} - \boldsymbol{a}$ is perpendicular to $\boldsymbol{n}$. Hence

$$(\boldsymbol{r} - \boldsymbol{a}) \cdot \boldsymbol{n} = 0$$

or

$$\boldsymbol{r} \cdot \boldsymbol{n} = \boldsymbol{a} \cdot \boldsymbol{n} \qquad 12.14$$

If

$$\boldsymbol{n} = \begin{pmatrix} p \\ q \\ r \end{pmatrix}$$

then

$$px + qy + rz = \boldsymbol{a} \cdot \boldsymbol{n}$$

Such an equation is called a *linear* equation in x, y and z.

KEY POINT

The general equation of a plane is $px + qy + rz = d$, where p, q, r and d are constants. The vector

$$\boldsymbol{n} = \begin{pmatrix} p \\ q \\ r \end{pmatrix}$$

is perpendicular to the plane.

Worked examples

12.12 Find a unit vector perpendicular to the plane $3x - 2y + 4z = 1$.

Solution Here $p = 3$, $q = -2$, $r = 4$ and $d = 1$. A vector perpendicular to the plane is

$$\begin{pmatrix} 3 \\ -2 \\ 4 \end{pmatrix}$$

Now $|\boldsymbol{n}| = \sqrt{(3)^2 + (-2)^2 + (4)^2} = \sqrt{29}$. A unit vector perpendicular to the plane is

$$\begin{pmatrix} \frac{3}{\sqrt{29}} \\ \frac{-2}{\sqrt{29}} \\ \frac{4}{\sqrt{29}} \end{pmatrix}$$

12.13 Find the equation of the plane through the points

$$\boldsymbol{a} = \begin{pmatrix} 0 \\ 1 \\ 1 \end{pmatrix}, \quad \boldsymbol{b} = \begin{pmatrix} 1 \\ 1 \\ 1 \end{pmatrix} \quad \text{and} \quad \boldsymbol{c} = \begin{pmatrix} 1 \\ 1 \\ 0 \end{pmatrix}$$

Solution The vectors

$$\boldsymbol{b} - \boldsymbol{a} = \begin{pmatrix} 1 \\ 0 \\ 0 \end{pmatrix} \quad \text{and} \quad \boldsymbol{c} - \boldsymbol{a} = \begin{pmatrix} 1 \\ 0 \\ -1 \end{pmatrix}$$

lie in the plane and the normal to the plane is

$$(\boldsymbol{b} - \boldsymbol{a}) \times (\boldsymbol{c} - \boldsymbol{a}) = \begin{pmatrix} 0 \\ 1 \\ 0 \end{pmatrix}$$

The equation of the plane is

$$\begin{pmatrix} x \\ y \\ z \end{pmatrix} \cdot \begin{pmatrix} 0 \\ 1 \\ 0 \end{pmatrix} = \begin{pmatrix} 0 \\ 1 \\ 1 \end{pmatrix} \cdot \begin{pmatrix} 0 \\ 1 \\ 0 \end{pmatrix} \qquad \text{that is} \quad y = 1$$

(x and z can take any value)

12.14 Find the shortest distance from the point $P(3,2,1)$ to the plane $x + 2y + z = 6$.

Solution A vector perpendicular to the plane is

$$\begin{pmatrix} 1 \\ 2 \\ 1 \end{pmatrix}$$

Hence the equation of the plane can be written as

$$\boldsymbol{r} \cdot \begin{pmatrix} 1 \\ 2 \\ 1 \end{pmatrix} = 6$$

The equation of a line perpendicular to the plane through the point P is

$$\boldsymbol{r} = \begin{pmatrix} 3 \\ 2 \\ 1 \end{pmatrix} + \lambda \begin{pmatrix} 1 \\ 2 \\ 1 \end{pmatrix}$$

This will meet the plane where

$$\boldsymbol{r} \cdot \begin{pmatrix} 1 \\ 2 \\ 1 \end{pmatrix} = \begin{pmatrix} 3 \\ 2 \\ 1 \end{pmatrix} \cdot \begin{pmatrix} 1 \\ 2 \\ 1 \end{pmatrix} + \lambda \begin{pmatrix} 1 \\ 2 \\ 1 \end{pmatrix} \cdot \begin{pmatrix} 1 \\ 2 \\ 1 \end{pmatrix} = 6$$

Therefore $8 + 6\lambda = 6$, that is $\lambda = -\frac{1}{3}$.

The line meets the plane at the point M with position vector

$$\boldsymbol{r} = \begin{pmatrix} 3 \\ 2 \\ 1 \end{pmatrix} - \frac{1}{3} \begin{pmatrix} 1 \\ 2 \\ 1 \end{pmatrix} = \begin{pmatrix} \frac{8}{3} \\ \frac{4}{3} \\ \frac{2}{3} \end{pmatrix}$$

The shortest distance is

$$PM = \sqrt{\left(\tfrac{8}{3} - 3\right)^2 + \left(\tfrac{4}{3} - 2\right)^2 + \left(\tfrac{2}{3} - 1\right)^2}$$
$$= \sqrt{\tfrac{2}{3}}$$

12.15 Find the equation of the line in which the planes $x + y + z = 4$ and $x + 2y + 4z = 8$ meet.

Solution The equations can be solved simultaneously to give $y + 3z = 4$. If we take $z = 0$ then $y = 4$ and hence, from the equation of either plane, $x = 0$. Therefore a point on both planes is

$$\begin{pmatrix} 0 \\ 4 \\ 0 \end{pmatrix}$$

The line of intersection lies in both planes and is therefore perpendicular to each of the normals to the planes, that is to

$$\begin{pmatrix} 1 \\ 1 \\ 1 \end{pmatrix} \text{ and } \begin{pmatrix} 1 \\ 2 \\ 4 \end{pmatrix}$$

Such a vector is

$$\begin{pmatrix}1\\1\\1\end{pmatrix} \times \begin{pmatrix}1\\2\\4\end{pmatrix} = \begin{pmatrix}2\\-3\\1\end{pmatrix}$$

The equation of the line is

$$\boldsymbol{r} = \begin{pmatrix}0\\4\\0\end{pmatrix} + \lambda\begin{pmatrix}2\\-3\\1\end{pmatrix} \quad \text{or} \quad \frac{x}{2} = \frac{y-4}{-3} = \frac{z}{1}$$

Self-assessment questions 12.3

1. What are the possibilities for two lines in space as regards their meeting?
2. What form does the vector equation of a plane take?

Exercise 12.3

1. Find the equation of the line through the points (3,0,2) and (2,−1,1). Where does it intersect the line through the points (2,2,1) and (3,1,2)?
2. Find the shortest distance between the lines

$$l_1: \frac{x+1}{1} = \frac{y+2}{2} = \frac{z+3}{3}$$

and

$$l_2: \frac{x}{2} = \frac{y-3}{3} = \frac{z-4}{1}$$

3. Find a unit vector perpendicular to the plane $2x + 3y - z = 5$.
4. Find the equation of the plane through the points

$$\boldsymbol{a} = \begin{pmatrix}1\\0\\-1\end{pmatrix},$$

$$\boldsymbol{b} = \begin{pmatrix}1\\1\\1\end{pmatrix} \quad \text{and}$$

$$\boldsymbol{c} = \begin{pmatrix}0\\1\\0\end{pmatrix}$$

5. Find the shortest distance from the point $P(1,2,1)$ to the plane $2x + y - z = 4$.
6. Find the equation of the line in which the planes $2x + y - z = 2$ and $x + y + 2z = 4$ meet.

12.4 Vectors in *n* dimensions

We have so far considered vectors in two and three dimensions which can be drawn respectively in a plane and (less easily) in three-dimensional space. If we have more than three quantities it is still possible under certain conditions to treat them as a single entity, namely as a vector; for example, the mesh currents in various parts of a circuit. In Chapter 14 we shall see how this approach allows us to find the mesh currents in a network.

KEY POINT

A set of vectors $\{\boldsymbol{x}_1, \boldsymbol{x}_2, \ldots, \boldsymbol{x}_n\}$ is **linearly independent** if and only if the equation

$$\alpha_1 \boldsymbol{x}_1 + \alpha_2 \boldsymbol{x}_2 + \ldots + \alpha_n \boldsymbol{x}_n = \boldsymbol{0}$$

implies that the scalars $\alpha_1, \alpha_2, \ldots, \alpha_n$ are all zero.

If scalars $\beta_1, \beta_2, \ldots, \beta_m$ can be found such that

$$\boldsymbol{x} = \beta_1 \boldsymbol{x}_1 + \beta_2 \boldsymbol{x}_2 + \ldots + \beta_m \boldsymbol{x}_m$$

then $\boldsymbol{x}$ is **linearly dependent** on $\boldsymbol{x}_1, \boldsymbol{x}_2, \ldots, \boldsymbol{x}_m$ (that is, $\boldsymbol{x}$ is a **linear combination** of $\boldsymbol{x}_1, \boldsymbol{x}_2, \ldots, \boldsymbol{x}_m$).

Worked examples

12.16 Given the vectors

$$\boldsymbol{a} = \begin{pmatrix} 1 \\ 2 \\ 3 \\ 4 \end{pmatrix}, \quad \boldsymbol{b} = \begin{pmatrix} 1 \\ 1 \\ 1 \\ 1 \end{pmatrix}, \quad \boldsymbol{c} = \begin{pmatrix} 1 \\ -1 \\ 1 \\ -1 \end{pmatrix}, \quad \boldsymbol{d} = \begin{pmatrix} 2 \\ 0 \\ 0 \\ 2 \end{pmatrix}$$

(a) find $\boldsymbol{a} + \boldsymbol{b}$, $\boldsymbol{a} - \boldsymbol{b}$, $3\boldsymbol{a} + 2\boldsymbol{b}$

(b) find the magnitude or **norm** of $\boldsymbol{a}$, $\boldsymbol{b}$, $\boldsymbol{c}$ and $\boldsymbol{d}$

(c) find which vectors are **mutually orthogonal** (their scalar product is zero)

Solution (a) We add by components and we scalar multiply by components. Hence

$$\boldsymbol{a} + \boldsymbol{b} = \begin{pmatrix} 1+1 \\ 2+1 \\ 3+1 \\ 4+1 \end{pmatrix} = \begin{pmatrix} 2 \\ 3 \\ 4 \\ 5 \end{pmatrix}, \quad \boldsymbol{a} - \boldsymbol{b} = \begin{pmatrix} 0 \\ 1 \\ 2 \\ 3 \end{pmatrix},$$

$$3\boldsymbol{a} + 2\boldsymbol{b} = \begin{pmatrix} 3+2 \\ 6+2 \\ 9+2 \\ 12+2 \end{pmatrix} = \begin{pmatrix} 5 \\ 8 \\ 11 \\ 14 \end{pmatrix}$$

(b) The magnitude of a vector is the square root of the sum of squares of its components. Hence

$$|\boldsymbol{a}| = \sqrt{(1)^2 + (2)^2 + (3)^2 + (4)^2} = \sqrt{30}$$

$$|\boldsymbol{b}| = \sqrt{(1)^2 + (1)^2 + (1)^2 + (1)^2} = \sqrt{4} = 2$$

$$|\boldsymbol{c}| = \sqrt{(1)^2 + (-1)^2 + (1)^2 + (-1)^2} = 2 = |\boldsymbol{b}|$$

$$|\boldsymbol{d}| = \sqrt{(2)^2 + (0)^2 + (0)^2 + (2)^2} = 2\sqrt{2}$$

(c) Scalar products are found by multiplying corresponding components and adding.

$$\boldsymbol{a}\cdot\boldsymbol{b} = 1\times 1 + 2\times 1 + 3\times 1 + 4\times 1 = 10$$
$$\boldsymbol{a}\cdot\boldsymbol{c} = 1 - 2 + 3 - 4 = -2$$
$$\boldsymbol{a}\cdot\boldsymbol{d} = 2 + 0 + 0 + 8 = 10$$
$$\boldsymbol{b}\cdot\boldsymbol{c} = 1 - 1 + 1 - 1 = 0 \qquad \boldsymbol{b} \text{ and } \boldsymbol{c} \text{ are mutually orthogonal}$$
$$\boldsymbol{b}\cdot\boldsymbol{d} = 2 + 0 + 0 + 2 = 4$$
$$\boldsymbol{c}\cdot\boldsymbol{d} = 2 - 0 + 0 - 2 = 0 \qquad \boldsymbol{c} \text{ and } \boldsymbol{d} \text{ are mutually orthogonal}$$

12.17 Show that the vectors

$$\boldsymbol{e}_1 = \begin{pmatrix}1\\0\\0\end{pmatrix},\quad \boldsymbol{e}_2 = \begin{pmatrix}0\\1\\0\end{pmatrix} \quad\text{and}\quad \boldsymbol{e}_3 = \begin{pmatrix}0\\0\\1\end{pmatrix}$$

are linearly independent. Show further that any vector

$$\begin{pmatrix}a\\b\\c\end{pmatrix}$$

with three components can be expressed as a linear combination of $\boldsymbol{e}_1$, $\boldsymbol{e}_2$ and $\boldsymbol{e}_3$.

Solution

$$\alpha_1\begin{pmatrix}1\\0\\0\end{pmatrix} + \alpha_2\begin{pmatrix}0\\1\\0\end{pmatrix} + \alpha_3\begin{pmatrix}0\\0\\1\end{pmatrix} = \begin{pmatrix}0\\0\\0\end{pmatrix}$$

can be rewritten

$$\begin{pmatrix}\alpha_1\\\alpha_2\\\alpha_3\end{pmatrix} = \begin{pmatrix}0\\0\\0\end{pmatrix}$$

Hence $\alpha_1 = 0$, $\alpha_2 = 0$ and $\alpha_3 = 0$ so that $\boldsymbol{e}_1$, $\boldsymbol{e}_2$ and $\boldsymbol{e}_3$ are linearly independent.

For the given vector we can write

$$\begin{pmatrix}a\\b\\c\end{pmatrix} = a\begin{pmatrix}1\\0\\0\end{pmatrix} + b\begin{pmatrix}0\\1\\0\end{pmatrix} + c\begin{pmatrix}0\\0\\1\end{pmatrix} = a\boldsymbol{e}_1 + b\boldsymbol{e}_2 + c\boldsymbol{e}_3$$

Self-assessment questions 12.4

1. What is the test for a set of vectors $\{\boldsymbol{x}_1, \boldsymbol{x}_2, \ldots, \boldsymbol{x}_n\}$ to be linearly independent?
2. Write down an expression for the norm of the vector $(a_1, a_2, \ldots, a_n)$.

Exercise 12.4

1. Given the vectors

$$a = \begin{pmatrix} 1 \\ 2 \\ 3 \\ 4 \\ 5 \end{pmatrix}, \quad b = \begin{pmatrix} 1 \\ -1 \\ 1 \\ -1 \\ 1 \end{pmatrix},$$

$$c = \begin{pmatrix} 1 \\ 0 \\ 1 \\ 0 \\ 1 \end{pmatrix}, \quad d = \begin{pmatrix} 5 \\ 4 \\ 3 \\ 2 \\ 1 \end{pmatrix}$$

(a) find $\boldsymbol{a} + \boldsymbol{b}$, $\boldsymbol{a} - \boldsymbol{b}$, $2\boldsymbol{a} - 3\boldsymbol{b} + \boldsymbol{c} - \boldsymbol{d}$
(b) find the norms of $\boldsymbol{a}$, $\boldsymbol{b}$, $\boldsymbol{c}$, $\boldsymbol{d}$
(c) which of the vectors $\boldsymbol{a}$, $\boldsymbol{b}$, $\boldsymbol{c}$, $\boldsymbol{d}$ are orthogonal to

$$\text{(i)} \begin{pmatrix} 1 \\ 1 \\ 0 \\ 1 \\ 1 \end{pmatrix} \quad \text{(ii)} \begin{pmatrix} 1 \\ -1 \\ 0 \\ -1 \\ 1 \end{pmatrix}?$$

2. Show that the vectors

$$a = \begin{pmatrix} 1 \\ 1 \\ 0 \end{pmatrix}, \quad b = \begin{pmatrix} 1 \\ 0 \\ 1 \end{pmatrix}$$

and $$c = \begin{pmatrix} 0 \\ 1 \\ 1 \end{pmatrix}$$

are linearly independent, but the set $\{\boldsymbol{a},\boldsymbol{b},\boldsymbol{c},\boldsymbol{d}\}$ is linearly dependent for any vectors

$$d = \begin{pmatrix} d_1 \\ d_2 \\ d_3 \end{pmatrix}$$

Test and assignment exercises 12

1. Which pairs of the following vectors are perpendicular?

(a) $\boldsymbol{a} = 3\boldsymbol{i} - 4\boldsymbol{j}$, $\quad \boldsymbol{b} = 2\boldsymbol{i} - \boldsymbol{j}$,
$\quad \boldsymbol{c} = \boldsymbol{i} - 2\boldsymbol{j}$
(b) $\boldsymbol{a} = \boldsymbol{i} - \boldsymbol{j} + \boldsymbol{k}$, $\quad \boldsymbol{b} = \boldsymbol{i} + 2\boldsymbol{j} + \boldsymbol{k}$,
$\quad \boldsymbol{c} = \boldsymbol{i} + \boldsymbol{j} - \boldsymbol{k}$, $\quad \boldsymbol{d} = \boldsymbol{i} - \boldsymbol{j}$

Find the angles between the pairs of vectors which are not mutually perpendicular.

2. Prove that the sum of the squares of the lengths of the diagonals of a parallelogram is equal to the sum of the squares of the lengths of the four sides.

3. Find the work done in moving a magnetic pole from the point $A = (2,0,3)$ to $B = (3,1,1)$ against a force $\boldsymbol{F} = (2,-2,1)$.

4. Find the vector products $\boldsymbol{a} \times \boldsymbol{b}$ and $\boldsymbol{b} \times \boldsymbol{c}$ for

$\boldsymbol{a} = (1,-1,2)$, $\quad \boldsymbol{b} = (2,2,1)$,
$\boldsymbol{c} = (1,2,-1)$

5. The moment of the force $\boldsymbol{F}$ acting through the point A is given by $\overrightarrow{AP} \times \boldsymbol{F}$. If $\boldsymbol{F} = (4,2,1)$, $A = (1,2,-3)$ and $P = (-1,4,7)$ find the magnitude of the force and its moment about each of the coordinate axes.

6. Find the vector equation of the following lines:

(a) through the points $A = (2,1,-1)$ and $B = (-1,-1,0)$
(b) through A and the point $C = (1,-2,1)$.

Find the equation of the line through A perpendicular to these two lines.

7. Find the equation of the plane through the following points:

$$A = \begin{pmatrix} 1 \\ 1 \\ 1 \end{pmatrix}, \quad B = \begin{pmatrix} 1 \\ -2 \\ 1 \end{pmatrix}, \quad C = \begin{pmatrix} 0 \\ 4 \\ 0 \end{pmatrix}$$

8. Given the vectors

$$\boldsymbol{a} = \begin{pmatrix} 1 \\ 0 \\ 1 \\ 0 \end{pmatrix}, \quad \boldsymbol{b} = \begin{pmatrix} 0 \\ 1 \\ 0 \\ 1 \end{pmatrix}$$

$$\boldsymbol{c} = \begin{pmatrix} 1 \\ 1 \\ 1 \\ 1 \end{pmatrix}, \quad \boldsymbol{d} = \begin{pmatrix} 1 \\ -1 \\ -1 \\ 1 \end{pmatrix}$$

determine whether the set $\{\boldsymbol{a},\boldsymbol{b},\boldsymbol{c},\boldsymbol{d}\}$ is linearly independent or not.

13 Matrices and determinants

Objectives

This chapter

- defines a matrix and basic kinds of matrix
- defines the operations of matrix addition, subtraction and multiplication
- illustrates that the order of multiplication of two matrices matters
- defines and evaluates the determinant of a square matrix
- finds, where it exists, the inverse of a square matrix

13.1 Introduction to matrices

A **matrix** is an ordered array of numbers or **elements**, arranged in rows and columns. For example,

$$\boldsymbol{A} = \begin{bmatrix} 1 & 2 \\ 3 & 4 \end{bmatrix}, \quad \boldsymbol{B} = \begin{bmatrix} 1 & -2 & 3 \\ -4 & 5 & -6 \end{bmatrix}, \quad \boldsymbol{C} = \begin{bmatrix} 1 & 0 \\ -2 & 3 \\ 4 & 5 \end{bmatrix}$$

are matrices; note the use of a capital letter to denote a matrix (sometimes round brackets are used instead of square brackets).

The size of a matrix is specified by the number of its rows and the number of its columns. $\boldsymbol{A}$ is called a 2×2 matrix, $\boldsymbol{B}$ is 2×3 and $\boldsymbol{C}$ is 3×2.

In general, an m by n matrix (one with m rows and n columns) can be written

$$\boldsymbol{A} = \begin{bmatrix} a_{11} & a_{12} & a_{13} & \dots & a_{1n} \\ a_{21} & a_{22} & a_{23} & \dots & a_{2n} \\ \vdots & \vdots & \vdots & & \vdots \\ a_{m1} & a_{m2} & a_{m3} & \dots & a_{mn} \end{bmatrix} \qquad 13.1$$

Hence the element a_{12} is in row 1 and column 2. In general, the element in row i and column j can be written a_{ij}.

A matrix for which $m = n$ is said to be a **square matrix**. In a square matrix, the elements $a_{11}, a_{22}, a_{33}, \dots, a_{nn}$ form the **leading diagonal** or **principal diagonal**. The sum of these elements is the **trace** of the matrix.

Any column of a matrix is called a **column vector**; any row is called a **row vector**.

(For simplicity, in most of the examples the elements are integers.)

Worked example

13.1 Consider the following matrices:

$$A = \begin{bmatrix} 1 & -1 \\ 2 & 3 \end{bmatrix}, \quad B = \begin{bmatrix} 1 & 0 & 0 \\ -2 & 4 & 5 \end{bmatrix}, \quad C = \begin{bmatrix} 1 & 2 & 3 \\ 3 & 2 & 1 \\ 4 & 4 & 4 \end{bmatrix},$$

$$D = \begin{bmatrix} 1 & 2 & 3 & 4 & 5 \\ 6 & 5 & 4 & 3 & 2 \end{bmatrix}$$

(a) What are the sizes of the matrices?
(b) Write down a_{12}, b_{22}, c_{23} and d_{24}.
(c) Which matrices are square? For those which are, find the trace.

Solution (a) $\boldsymbol{A}$ is 2×2, $\boldsymbol{B}$ is 2×3, $\boldsymbol{C}$ is 3×3, $\boldsymbol{D}$ is 2×5.
(b) $a_{12} = -1$, $b_{22} = 4$, $c_{23} = 1$, $d_{24} = 3$.
(c) $\boldsymbol{A}$ and $\boldsymbol{C}$ are square. The trace of $\boldsymbol{A}$ is $1 + 3 = 4$ and the trace of $\boldsymbol{C}$ is $1 + 2 + 4 = 7$.

A **zero matrix** is one with all elements zero; we write it as **0**.

A **diagonal matrix** is a square matrix with all elements not on the leading diagonal equal to zero.

KEY POINT

> A **unit matrix** or **identity matrix** is a diagonal matrix with $a_{11} = a_{22} = \ldots = 1$.

A matrix for which $a_{ij} = 0$ if $i < j$ is called **lower triangular** and one for which $a_{ij} = 0$ if $i > j$ is called **upper triangular**.

Worked example

13.2 Classify the following matrices:

$$A = \begin{bmatrix} 1 & -1 \\ 2 & 3 \end{bmatrix}, \quad B = \begin{bmatrix} 0 & 0 \\ 0 & 0 \end{bmatrix}, \quad C = \begin{bmatrix} 1 & 0 \\ 0 & 1 \end{bmatrix}, \quad D = \begin{bmatrix} 0 & 0 & 0 \\ 0 & 0 & 0 \end{bmatrix}$$

$$E = \begin{bmatrix} 1 & 1 & 1 \\ 2 & 2 & 2 \\ 3 & 3 & 3 \end{bmatrix}, \quad F = \begin{bmatrix} 1 & 0 & 0 \\ 2 & 3 & 0 \\ 3 & 4 & 5 \end{bmatrix}, \quad G = \begin{bmatrix} 1 & 2 & 3 \\ 0 & 3 & 4 \\ 0 & 0 & 5 \end{bmatrix},$$

$$H = \begin{bmatrix} 1 & 0 & 0 \\ 0 & 1 & 0 \\ 0 & 0 & 1 \end{bmatrix}, \quad J = \begin{bmatrix} 1 & 0 & 0 \\ 0 & 0 & 0 \\ 0 & 0 & 0 \end{bmatrix}, \quad K = \begin{bmatrix} 1 & 0 & 0 \\ 0 & 2 & 0 \\ 0 & 0 & 3 \end{bmatrix}$$

Solution $\boldsymbol{D}$ is the only matrix which is not square.
$\boldsymbol{B}$ and $\boldsymbol{D}$ are zero matrices.
$\boldsymbol{C}$, $\boldsymbol{H}$, $\boldsymbol{J}$ and $\boldsymbol{K}$ are diagonal matrices.
$\boldsymbol{C}$ and $\boldsymbol{H}$ are identity matrices.
(An identity matrix is also a diagonal matrix.)
$\boldsymbol{F}$ is a lower triangular matrix and $\boldsymbol{G}$ is an upper triangular matrix.
(A diagonal matrix is strictly both a lower triangular matrix and an upper triangular matrix.)

An identity matrix is usually denoted $\boldsymbol{I}$. (If we want to emphasize its size we can write $\boldsymbol{I}_2$ for the 2×2 identity matrix, $\boldsymbol{I}_3$ for the 3×3 identity matrix and so on.)

KEY POINT

> The **transpose** of a matrix $\boldsymbol{A}$ is obtained by writing the rows of $\boldsymbol{A}$ as columns (and therefore its columns as rows). It is denoted by $\boldsymbol{A}^{\mathrm{T}}$.

Hence, if

$$\boldsymbol{A} = \begin{bmatrix} 1 & 2 \\ 3 & 4 \end{bmatrix}$$

then

$$\boldsymbol{A}^{\mathrm{T}} = \begin{bmatrix} 1 & 3 \\ 2 & 4 \end{bmatrix}$$

Note that $(\boldsymbol{A}^{\mathrm{T}})^{\mathrm{T}} = \boldsymbol{A}$.

Two matrices are **equal** if and only if corresponding elements are equal; hence they must be of the same size.

A **symmetric matrix** is one for which $\boldsymbol{A}^{\mathrm{T}} = \boldsymbol{A}$. Note that $\boldsymbol{A}$ must be square.

A **skew-symmetric matrix** is one for which $\boldsymbol{A}^{\mathrm{T}} = -\boldsymbol{A}$, that is, the signs of all the elements of $\boldsymbol{A}$ are reversed by transposing $\boldsymbol{A}$. Note that $\boldsymbol{A}$ must be square.

Worked examples

13.3 Express the definitions of transpose, equality, symmetry and skew-symmetry in terms of typical elements of an $m \times n$ matrix $\boldsymbol{A}$.

Solution If $\boldsymbol{B} = \boldsymbol{A}^{\mathrm{T}}$ then $b_{ij} = a_{ji}$ for $1 \leqslant i \leqslant n$, $1 \leqslant j \leqslant m$. Note that $\boldsymbol{B}$ is $n \times m$.
If $\boldsymbol{A} = \boldsymbol{B}$ then $b_{ij} = a_{ij}$ for $1 \leqslant i \leqslant m$, $1 \leqslant j \leqslant n$.
If $\boldsymbol{A}$ is symmetric then $a_{ij} = a_{ji}$ for $1 \leqslant i \leqslant n$, $1 \leqslant j \leqslant n$. Note that $\boldsymbol{A}$ must be $n \times n$.
If $\boldsymbol{A}$ is skew-symmetric then $a_{ij} = -a_{ji}$ for $1 \leqslant i \leqslant n$, $1 \leqslant j \leqslant n$.

13.4 By taking transposes, classify the following matrices as symmetric, skew-symmetric or neither:

$$A=\begin{bmatrix}1 & 2\\ 3 & 4\end{bmatrix},\quad B=\begin{bmatrix}0 & 1\\ -1 & 0\end{bmatrix},\quad C=\begin{bmatrix}1 & 2\\ 2 & 1\end{bmatrix},\quad D=\begin{bmatrix}-1 & -2\\ -2 & -3\end{bmatrix},$$

$$E=\begin{bmatrix}1 & 2 & 3\\ 4 & 5 & 6\end{bmatrix},\quad F=\begin{bmatrix}1 & 2 & 3\\ 4 & 5 & 6\\ 3 & 6 & 9\end{bmatrix},\quad G=\begin{bmatrix}3 & 1 & 1\\ -1 & 2 & -3\\ -1 & 3 & 1\end{bmatrix},$$

$$H=\begin{bmatrix}0 & -1 & 1\\ 1 & 0 & -2\\ -1 & 2 & 0\end{bmatrix}$$

Solution

$$A^{\mathrm{T}}=\begin{bmatrix}1 & 3\\ 2 & 4\end{bmatrix},\quad B^{\mathrm{T}}=\begin{bmatrix}0 & -1\\ 1 & 0\end{bmatrix},\quad C^{\mathrm{T}}=\begin{bmatrix}1 & 2\\ 2 & 1\end{bmatrix}$$

$$D^{\mathrm{T}}=\begin{bmatrix}1 & -2\\ -2 & -3\end{bmatrix},\quad E^{\mathrm{T}}=\begin{bmatrix}1 & 4\\ 2 & 5\\ 3 & 6\end{bmatrix},\quad F^{\mathrm{T}}=\begin{bmatrix}1 & 4 & 3\\ 2 & 5 & 6\\ 3 & 6 & 9\end{bmatrix}$$

$$G^{\mathrm{T}}=\begin{bmatrix}3 & -1 & -1\\ 1 & 2 & 3\\ 1 & -3 & 1\end{bmatrix},\quad H^{\mathrm{T}}=\begin{bmatrix}0 & 1 & -1\\ -1 & 0 & 2\\ 1 & -2 & 0\end{bmatrix}$$

Since $C^{\mathrm{T}}=C$ and $D^{\mathrm{T}}=D$ then C and D are symmetric matrices.

Since $B^{\mathrm{T}}=-B$ and $H^{\mathrm{T}}=-H$ then B and H are skew-symmetric matrices.

13.5 Show that if A is an $n\times n$ skew-symmetric matrix then its diagonal elements must be all zero. Is the converse true?

Solution Since $a_{ij}=-a_{ji}$ for a skew-symmetric matrix, when $i=j$ this implies that $a_{ii}=-a_{ii}$ and this can happen only if $a_{ii}=0$, that is if the diagonal elements are zero.

Consider for example

$$A=\begin{bmatrix}0 & 1\\ 1 & 0\end{bmatrix}$$

It is symmetric and therefore having zero diagonal elements does not guarantee skew-symmetry.

Self-assessment questions 13.1

1. Explain the terms matrix, trace, diagonal matrix and triangular matrix.
2. Explain the terms transpose, symmetric matrix and skew-symmetric matrix.

Exercise 13.1

1. Consider the following matrices:

$$A = \begin{bmatrix} 1 & 2 \\ 3 & 4 \\ 4 & 6 \end{bmatrix}, \quad B = \begin{bmatrix} 1 & 2 & 3 & 4 \\ 5 & 6 & 7 & 8 \end{bmatrix},$$

$$C = \begin{bmatrix} 1 & 1 \\ 1 & 1 \end{bmatrix}, \quad D = \begin{bmatrix} 4 & 5 & 6 \\ 1 & 2 & 3 \\ 7 & 8 & 9 \end{bmatrix}$$

(a) What are the sizes of the matrices?
(b) Write down a_{32}, b_{14}, c_{12} and d_{22}.
(c) Which matrices are square?
For those which are, find the trace.

2. Classify the following matrices as square, non-square, zero, identity, diagonal:

$$A = \begin{bmatrix} 0 & 1 \\ 1 & 0 \end{bmatrix}, \quad B = \begin{bmatrix} 1 & 2 \\ 3 & 4 \end{bmatrix},$$

$$C = \begin{bmatrix} 0 & 0 \\ 0 & 0 \\ 0 & 0 \end{bmatrix}, \quad D = [1],$$

$$E = \begin{bmatrix} 1 & 2 & 3 \\ 1 & 2 & 3 \\ 1 & 2 & 3 \end{bmatrix}, \quad F = \begin{bmatrix} 1 & 1 \\ 0 & 1 \end{bmatrix},$$

$$G = \begin{bmatrix} 1 & 0 \\ 1 & 1 \end{bmatrix}, \quad H = \begin{bmatrix} 0 & 0 & 0 \\ 0 & 1 & 0 \\ 0 & 0 & 0 \end{bmatrix}$$

$$J = \begin{bmatrix} 1 & 1 & 0 \\ 0 & 1 & 1 \\ 0 & 0 & 1 \end{bmatrix}, \quad K = \begin{bmatrix} 3 & 0 & 0 \\ 0 & -1 & 0 \\ 0 & 0 & -2 \end{bmatrix}$$

3. Find the transposes of the following matrices and hence classify them as symmetric, skew-symmetric or neither:

$$A = \begin{bmatrix} 0 & -2 \\ 2 & 0 \end{bmatrix}, \quad B = \begin{bmatrix} 4 & 3 \\ 1 & 2 \end{bmatrix},$$

$$C = \begin{bmatrix} -1 & -1 \\ -1 & 1 \end{bmatrix}, \quad D = \begin{bmatrix} 2 & 1 \\ 1 & 2 \end{bmatrix},$$

$$E = \begin{bmatrix} 1 & 4 \\ 2 & 5 \\ 3 & 6 \end{bmatrix}, \quad F = \begin{bmatrix} 1 & 3 & 3 \\ -3 & 2 & -1 \\ -3 & 1 & 3 \end{bmatrix},$$

$$G = \begin{bmatrix} 4 & 3 & 1 \\ 5 & 6 & 2 \\ 6 & 9 & 3 \end{bmatrix}, \quad H = \begin{bmatrix} 0 & 2 & -2 \\ -2 & 0 & 4 \\ 2 & -4 & 0 \end{bmatrix}$$

4. How many elements does an $n \times n$ matrix have? How many of these elements must be zero if:
(a) the matrix is diagonal?
(b) the matrix is triangular?
(c) the matrix is skew-symmetric?
(d) the matrix is symmetric?

13.2 Addition, subtraction and multiplication of matrices

Two matrices A and B can be added only if they have the same size. Corresponding elements are added to get the result. For example if

$$A = \begin{bmatrix} 1 & 2 \\ 3 & 4 \end{bmatrix} \quad \text{and} \quad B = \begin{bmatrix} 5 & -6 \\ -7 & 8 \end{bmatrix}$$

then

$$A + B = \begin{bmatrix} 1+5 & 2+(-6) \\ 3+(-7) & 4+8 \end{bmatrix} = \begin{bmatrix} 6 & -4 \\ -4 & 12 \end{bmatrix} = B + A$$

Subtraction follows a similar approach.

For the matrices on the previous page

$$A - B = \begin{bmatrix} 1-5 & 2-(-6) \\ 3-(-7) & 4-8 \end{bmatrix} = \begin{bmatrix} -4 & 8 \\ 10 & -4 \end{bmatrix}$$

$$B - A = \begin{bmatrix} 5-1 & -6-2 \\ -7-3 & 8-4 \end{bmatrix} = \begin{bmatrix} 4 & -8 \\ -10 & 4 \end{bmatrix} = -(A - B)$$

The matrix λA has all its elements multiplied by the scalar λ. Hence if

$$A = \begin{bmatrix} 2 & 4 \\ 6 & 8 \end{bmatrix}$$

then

$$\tfrac{1}{2}A = \begin{bmatrix} 1 & 2 \\ 3 & 4 \end{bmatrix}$$

The following laws hold where addition is possible:

$$(A + B)^{\mathrm{T}} = A^{\mathrm{T}} + B^{\mathrm{T}} \qquad 13.2$$

$$A + B = B + A \quad \text{(commutative law)} \qquad 13.3$$

$$(A + B) + C = A + (B + C) \quad \text{(associative law)} \qquad 13.4$$

$$\lambda(A + B) = \lambda A + \lambda B \quad \text{(distributive law)} \qquad 13.5$$

Worked examples

13.6 If

$$A = \begin{bmatrix} 1 & 1 & 1 \\ 1 & 2 & 1 \\ 1 & 1 & 2 \end{bmatrix}, \quad B = \begin{bmatrix} 1 & 1 \\ 1 & 0 \\ 2 & 1 \end{bmatrix}, \quad C = \begin{bmatrix} 1 & 0 & 0 \\ 0 & 1 & 1 \\ 0 & 0 & 1 \end{bmatrix}, \quad D = \begin{bmatrix} 1 & 2 & 3 \\ 4 & 5 & 6 \end{bmatrix}$$

find, where possible, $A + B$, $A + C$, $C - A$, $B^{\mathrm{T}} + D$, $B + D^{\mathrm{T}}$, $3A$, $2A - 3C$.

Solution $A + B$ is not possible: A is 3×3 and B is 3×2.

$$A + C = \begin{bmatrix} 1+1 & 1+0 & 1+0 \\ 1+0 & 2+1 & 1+1 \\ 1+0 & 1+0 & 2+1 \end{bmatrix} = \begin{bmatrix} 2 & 1 & 1 \\ 1 & 3 & 2 \\ 1 & 1 & 3 \end{bmatrix}$$

$$C - A = \begin{bmatrix} 1-1 & 0-1 & 0-1 \\ 0-1 & 1-2 & 1-1 \\ 0-1 & 0-1 & 1-2 \end{bmatrix} = \begin{bmatrix} 0 & -1 & -1 \\ -1 & -1 & 0 \\ -1 & -1 & -1 \end{bmatrix}$$

$$B^{\mathrm{T}} + D = \begin{bmatrix} 1 & 1 & 2 \\ 1 & 0 & 1 \end{bmatrix} + \begin{bmatrix} 1 & 2 & 3 \\ 4 & 5 & 6 \end{bmatrix} = \begin{bmatrix} 2 & 3 & 5 \\ 5 & 5 & 7 \end{bmatrix}$$

$$B+D^{\mathrm{T}}=\begin{bmatrix}1&1\\1&0\\2&1\end{bmatrix}+\begin{bmatrix}1&4\\2&5\\3&6\end{bmatrix}=\begin{bmatrix}2&5\\3&5\\5&7\end{bmatrix}=(B^{\mathrm{T}}+D)^{\mathrm{T}}$$

$$3A=\begin{bmatrix}3&3&3\\3&6&3\\3&3&6\end{bmatrix}$$

$$2A-3C=\begin{bmatrix}2&2&2\\2&4&2\\2&2&4\end{bmatrix}-\begin{bmatrix}3&0&0\\0&3&3\\0&0&3\end{bmatrix}=\begin{bmatrix}-1&2&2\\2&1&-1\\2&2&1\end{bmatrix}$$

13.7 In terms of elements, define addition, subtraction and scalar multiplication.

Solution A typical element of A is a_{ij} and of B is b_{ij} and λ is a scalar.
If $C=A+B$ then $c_{ij}=a_{ij}+b_{ij}$.
If $D=A-B$ then $d_{ij}=a_{ij}-b_{ij}$.
If $E=\lambda A$ then $e_{ij}=\lambda a_{ij}$.

13.8 Verify the laws (13.2) to (13.5) for the matrices

$$A=\begin{bmatrix}1&2\\3&4\end{bmatrix},\quad B=\begin{bmatrix}2&-3\\-1&0\end{bmatrix},\quad C=\begin{bmatrix}1&1\\1&1\end{bmatrix}\qquad\text{and for }\lambda=2$$

Solution

$$A+B=\begin{bmatrix}1+2&2-3\\3-1&4+0\end{bmatrix}=\begin{bmatrix}3&-1\\2&4\end{bmatrix}=\begin{bmatrix}2+1&-3+2\\-1+3&0+4\end{bmatrix}=B+A$$

$$A^{\mathrm{T}}+B^{\mathrm{T}}=\begin{bmatrix}1&3\\2&4\end{bmatrix}+\begin{bmatrix}2&-1\\-3&0\end{bmatrix}=\begin{bmatrix}3&2\\-1&4\end{bmatrix}=(A+B)^{\mathrm{T}}$$

$$(A+B)+C=\begin{bmatrix}3&-1\\2&4\end{bmatrix}+\begin{bmatrix}1&1\\1&1\end{bmatrix}=\begin{bmatrix}4&0\\3&5\end{bmatrix}$$

$$B+C=\begin{bmatrix}3&-2\\0&1\end{bmatrix};\quad A+(B+C)=\begin{bmatrix}1&2\\3&4\end{bmatrix}+\begin{bmatrix}3&-2\\0&1\end{bmatrix}=\begin{bmatrix}4&0\\3&5\end{bmatrix}$$

$$2(A+B)=\begin{bmatrix}6&-2\\4&8\end{bmatrix};$$

$$2A+2B=\begin{bmatrix}2&4\\6&8\end{bmatrix}+\begin{bmatrix}4&-6\\-2&0\end{bmatrix}=\begin{bmatrix}6&-2\\4&8\end{bmatrix}$$

13.9 Show that for any square matrix A the matrix $\frac{1}{2}(A+A^{\mathrm{T}})$ is symmetric, the matrix $\frac{1}{2}(A-A^{\mathrm{T}})$ is skew-symmetric and

$$A=\tfrac{1}{2}(A+A^{\mathrm{T}})+\tfrac{1}{2}(A-A^{\mathrm{T}})$$

Solution Let $B = \frac{1}{2}(A + A^T)$. Then $B^T = \frac{1}{2}(A + A^T)^T = \frac{1}{2}(A^T + A) = B$, since $(A^T)^T = A$. Therefore B is symmetric.

Let $C = \frac{1}{2}(A - A^T)$; then $C^T = \frac{1}{2}(A^T - A) = -C$. Hence C is skew-symmetric.

$$\frac{1}{2}(A + A^T) + \frac{1}{2}(A - A^T) = \frac{1}{2}A + \frac{1}{2}A^T + \frac{1}{2}A - \frac{1}{2}A^T = A$$

Multiplication of two matrices

To multiply two matrices A and B we proceed as follows: each element of the product AB is found by taking the *scalar product* of a *row* of A with a *column* of B. This requires there to be as many elements in a column of A as there are in a row of B, that is, the number of columns of A = the number of rows of B.

KEY POINT

> An $m \times n$ matrix A can be multiplied by an $n \times p$ matrix B to give an $m \times p$ matrix C for which the element in row i and column j is the scalar product of the row i of A and column j of B.

As an example,

$$\begin{pmatrix} a_{11} & a_{12} \\ a_{21} & a_{22} \end{pmatrix} \begin{pmatrix} b_{11} & b_{12} \\ b_{21} & b_{22} \end{pmatrix} = \begin{pmatrix} a_{11}b_{11} + a_{12}b_{21} & a_{11}b_{12} + a_{12}b_{22} \\ a_{21}b_{11} + a_{22}b_{21} & a_{21}b_{12} + a_{22}b_{22} \end{pmatrix} \quad 13.6$$

The following properties of multiplication hold:

$$A(BC) = (AB)C \qquad \text{(associative law)} \qquad 13.7$$

$$\lambda(AB) = (\lambda A)B = A(\lambda B) \qquad \text{(distributive law)} \qquad 13.8$$

$$(A + B)C = AC + BC \qquad \text{(distributive law)} \qquad 13.9$$

$$A(B + C) = AB + AC \qquad \text{(distributive law)} \qquad 13.10$$

$$AI = A, \; IA = A \qquad 13.11$$

(provided I is the appropriate size)

$$(AB)^T = B^T A^T \qquad 13.12$$

Rotation of axes

For a geometrical application of matrix multiplication we look at rotation of the coordinate axes. Consider Figure 13.1. Let the point P have coordinates (x,y) relative to the axes OX and OY and coordinates (x',y') relative to the axes OX' and OY'. Now

$$x' = OM + MN + NS = OR\cos\theta + RN\sin\theta + NP\sin\theta$$
$$= x\cos\theta + (RN + NP)\sin\theta$$
$$= x\cos\theta + y\sin\theta$$

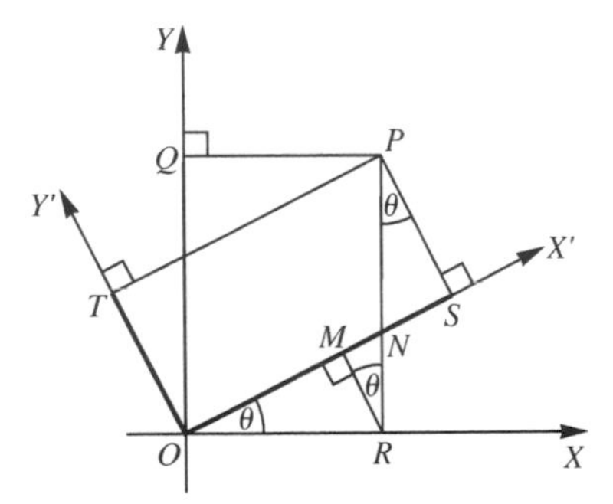

Figure 13.1.
Rotation of axes

Also

$$y' = OT = PS = PN\cos\theta = (PR - NR)\cos\theta = y\cos\theta - MR$$
$$= y\cos\theta - OR\sin\theta = y\cos\theta - x\sin\theta$$

The two equations

$$x' = x\cos\theta + y\sin\theta$$
$$y' = -x\sin\theta + y\cos\theta$$

can be written as a single matrix equation

$$\begin{pmatrix} x' \\ y' \end{pmatrix} = \begin{pmatrix} \cos\theta & \sin\theta \\ -\sin\theta & \cos\theta \end{pmatrix}\begin{pmatrix} x \\ y \end{pmatrix} \qquad 13.13$$

or

$$\boldsymbol{X}' = \boldsymbol{AX}$$

where $\boldsymbol{A}$ is the **matrix of transformation**.

Note that the right-hand side of (13.13) is the product of a 2×2 matrix with a 2×1 matrix, resulting in a 2×1 matrix. The first components of this matrix and the 2×1 matrix $\boldsymbol{X}'$ are equal and the second components are also equal.

Worked examples

13.10 Given the matrices

$$\boldsymbol{A} = \begin{bmatrix} 1 & 0 \\ -1 & 1 \\ 0 & 2 \end{bmatrix}, \quad \boldsymbol{B} = \begin{bmatrix} 1 & 1 & 1 \\ 1 & 2 & 3 \end{bmatrix}, \quad \boldsymbol{C} = \begin{bmatrix} 0 & 1 \\ 1 & 0 \end{bmatrix},$$

$$\boldsymbol{D} = \begin{bmatrix} 1 & 2 & 3 \\ 0 & 1 & 1 \\ 1 & 0 & 1 \end{bmatrix}, \quad \boldsymbol{E} = \begin{bmatrix} 0 & 1 & -2 \\ -1 & 1 & -1 \\ 1 & -1 & 2 \end{bmatrix}$$

find, where possible, the products $\boldsymbol{AC}$, $\boldsymbol{BC}$, $\boldsymbol{AB}$, $\boldsymbol{BA}$, $\boldsymbol{DE}$.

Solution The elements of the second matrix are in bold for emphasis.

$$AC = \begin{bmatrix} 1\times\mathbf{0}+0\times\mathbf{1} & 1\times\mathbf{1}+0\times\mathbf{0} \\ -1\times\mathbf{0}+1\times\mathbf{1} & -1\times\mathbf{1}+1\times\mathbf{0} \\ 0\times\mathbf{0}+2\times\mathbf{1} & 0\times\mathbf{1}+2\times\mathbf{0} \end{bmatrix} = \begin{bmatrix} 0 & 1 \\ 1 & -1 \\ 2 & 0 \end{bmatrix}$$

$\boldsymbol{BC}$ is not possible (you cannot multiply a 2×3 matrix by a 2×2 matrix).

$$AB = \begin{bmatrix} 1\times\mathbf{1}+0\times\mathbf{1} & 1\times\mathbf{1}+0\times\mathbf{2} & 1\times\mathbf{1}+0\times\mathbf{3} \\ -1\times\mathbf{1}+1\times\mathbf{1} & -1\times\mathbf{1}+1\times\mathbf{2} & -1\times\mathbf{1}+1\times\mathbf{3} \\ 0\times\mathbf{1}+2\times\mathbf{1} & 0\times\mathbf{1}+2\times\mathbf{2} & 0\times\mathbf{1}+2\times\mathbf{3} \end{bmatrix} = \begin{bmatrix} 1 & 1 & 1 \\ 0 & 1 & 2 \\ 2 & 4 & 6 \end{bmatrix}$$

$$BA = \begin{bmatrix} 1\times\mathbf{1}+1\times(\mathbf{-1})+1\times\mathbf{0} & 1\times\mathbf{0}+1\times\mathbf{1}+1\times\mathbf{2} \\ 1\times\mathbf{1}+2\times(\mathbf{-1})+3\times\mathbf{0} & 1\times\mathbf{0}+2\times\mathbf{1}+3\times\mathbf{2} \end{bmatrix} = \begin{bmatrix} 0 & 3 \\ -1 & 8 \end{bmatrix}$$

$$DE = \begin{bmatrix} 1\times\mathbf{0}+2\times(\mathbf{-1})+3\times\mathbf{1} & 1\times\mathbf{1}+2\times\mathbf{1}+3\times(\mathbf{-1}) & 1\times(\mathbf{-2})+2\times(\mathbf{-1})+3\times\mathbf{2} \\ 0\times\mathbf{0}+1\times(\mathbf{-1})+1\times\mathbf{1} & 0\times\mathbf{1}+1\times\mathbf{1}+1\times(\mathbf{-1}) & 0\times(\mathbf{-2})+1\times(\mathbf{-1})+1\times\mathbf{2} \\ 1\times\mathbf{0}+0\times(\mathbf{-1})+1\times\mathbf{1} & 1\times\mathbf{1}+0\times\mathbf{1}+1\times(\mathbf{-1}) & 1\times(\mathbf{-2})+0\times(\mathbf{-1})+1\times\mathbf{2} \end{bmatrix}$$

$$= \begin{bmatrix} 1 & 0 & 2 \\ 0 & 0 & 1 \\ 1 & 0 & 0 \end{bmatrix}$$

Note that, unlike matrix addition, matrix multiplication is in general *not* commutative. Sometimes neither $\boldsymbol{AB}$ nor $\boldsymbol{BA}$ exists, sometimes one does but not the other, sometimes both exist but are different sizes and sometimes the sizes are the same but the matrices are not. To be precise when talking about the product $\boldsymbol{AB}$, we say either that $\boldsymbol{A}$ is post-multiplied by $\boldsymbol{B}$ or that $\boldsymbol{B}$ is pre-multiplied by $\boldsymbol{A}$.

If two matrices $\boldsymbol{A}$ and $\boldsymbol{B}$ are such that $\boldsymbol{AB} = \boldsymbol{BA}$ then they are said to **commute**.

13.11 Verify the distributive law (13.9) for the following matrices:

$$A = \begin{bmatrix} 1 & 1 \\ 2 & 2 \end{bmatrix} \quad B = \begin{bmatrix} 1 & 2 \\ 3 & 4 \end{bmatrix} \quad \text{and} \quad C = \begin{bmatrix} 0 & 1 \\ 1 & 0 \end{bmatrix}$$

Solution The elements of the second matrix in the multiplication are in bold for clarity.

$$AC = \begin{bmatrix} 1\times\mathbf{0}+1\times\mathbf{1} & 1\times\mathbf{1}+1\times\mathbf{0} \\ 2\times\mathbf{0}+2\times\mathbf{1} & 2\times\mathbf{1}+2\times\mathbf{0} \end{bmatrix} = \begin{bmatrix} 1 & 1 \\ 2 & 2 \end{bmatrix}$$

$$BC = \begin{bmatrix} 1\times\mathbf{0}+2\times\mathbf{1} & 1\times\mathbf{1}+2\times\mathbf{0} \\ 3\times\mathbf{0}+4\times\mathbf{1} & 3\times\mathbf{1}+4\times\mathbf{0} \end{bmatrix} = \begin{bmatrix} 2 & 1 \\ 4 & 3 \end{bmatrix}$$

$$AC + BC = \begin{bmatrix} 3 & 2 \\ 6 & 5 \end{bmatrix}$$

$$A + B = \begin{bmatrix} 2 & 3 \\ 5 & 6 \end{bmatrix};$$

$$(A + B)C = \begin{bmatrix} 2\times\mathbf{0}+3\times\mathbf{1} & 2\times\mathbf{1}+3\times\mathbf{0} \\ 5\times\mathbf{0}+6\times\mathbf{1} & 5\times\mathbf{1}+6\times\mathbf{0} \end{bmatrix} = \begin{bmatrix} 3 & 2 \\ 6 & 5 \end{bmatrix}$$

13.12 Show that the result of rotating the axes anticlockwise by θ and then by ϕ is the same as a single rotation by $\theta + \phi$.

Solution Using the results of (13.13) the matrix of the first rotation is

$$\begin{pmatrix} \cos\theta & \sin\theta \\ -\sin\theta & \cos\theta \end{pmatrix}$$

and that of the second is

$$\begin{pmatrix} \cos\phi & \sin\phi \\ -\sin\phi & \cos\phi \end{pmatrix}$$

The matrix of the combined rotation is

$$\begin{pmatrix} \cos\phi & \sin\phi \\ -\sin\phi & \cos\phi \end{pmatrix}\begin{pmatrix} \cos\theta & \sin\theta \\ -\sin\theta & \cos\theta \end{pmatrix} = \begin{pmatrix} \cos\phi\cos\theta - \sin\theta\sin\phi & \cos\phi\sin\theta + \sin\phi\cos\theta \\ -\sin\phi\cos\theta - \cos\phi\sin\theta & -\sin\phi\sin\theta + \cos\phi\cos\theta \end{pmatrix}$$

$$= \begin{pmatrix} \cos(\phi+\theta) & \sin(\phi+\theta) \\ -\sin(\phi+\theta) & \cos(\phi+\theta) \end{pmatrix}$$

which represents a rotation of $\phi + \theta$.

Self-assessment questions 13.2

1. Under what conditions can two matrices be (a) added, and (b) multiplied together?
2. Explain how you would add two $m \times n$ matrices.
3. Explain how you would multiply an $m \times n$ matrix by an $n \times p$ matrix.

Exercise 13.2

1. Given the following matrices:

$$A = \begin{bmatrix} 1 & -1 & 1 \\ -1 & 1 & -1 \\ 1 & -1 & 1 \end{bmatrix},$$

$$B = \begin{bmatrix} 1 & 2 & 1 \\ 3 & 4 & 3 \end{bmatrix},$$

$$C = \begin{bmatrix} 1 & -1 \\ 0 & 1 \\ 2 & 3 \end{bmatrix},$$

$$D = \begin{bmatrix} 0 & 1 & 1 \\ 1 & 0 & 1 \\ 1 & 1 & 0 \end{bmatrix}$$

find, where possible, $\boldsymbol{A}+\boldsymbol{B}$, $\boldsymbol{B}+\boldsymbol{A}$, $\boldsymbol{A}+\boldsymbol{C}$, $\boldsymbol{A}+\boldsymbol{D}$, $\boldsymbol{D}-\boldsymbol{A}$, $\boldsymbol{B}^{\mathrm{T}}+\boldsymbol{C}$, $\boldsymbol{B}+\boldsymbol{C}^{\mathrm{T}}$, $4\boldsymbol{A}$, $\frac{1}{2}\boldsymbol{D}$, $3\boldsymbol{A}-2\boldsymbol{D}$.

2. Verify the laws (13.2) to (13.5) for the matrices

$$A = \begin{bmatrix} 2 & -1 \\ -4 & 3 \end{bmatrix},$$

$$B = \begin{bmatrix} 0 & 1 \\ 1 & 1 \end{bmatrix},$$

$$C = \begin{bmatrix} -1 & 2 \\ 3 & -4 \end{bmatrix}$$

and for $\lambda = 3$.

3. Show that for any matrix $\boldsymbol{A}$ the matrix $\boldsymbol{A}\boldsymbol{A}^{\mathrm{T}}$ is defined. For the matrix

$$\boldsymbol{A} = \begin{bmatrix} a & b \\ c & d \end{bmatrix}$$

show that $\boldsymbol{A}\boldsymbol{A}^{\mathrm{T}}$ is symmetric.

4. Write down matrices which rotate the x–y axes anticlockwise by

(a) 30°
(b) 45°
(c) 90°
(d) 180°.

5. For the matrix

$$\boldsymbol{A} = \begin{bmatrix} a & b \\ c & d \end{bmatrix}$$

show that

$$\boldsymbol{A}^2 = \begin{bmatrix} a^2 + bc & ab + bd \\ ac + bd & bc + d^2 \end{bmatrix}$$

What is the restriction if $\boldsymbol{A}^2$ is to be symmetric?

Show that $\boldsymbol{A}$ satisfies the equation

$$\boldsymbol{A}^2 - (a + d)\boldsymbol{A} + (ad + bc)\boldsymbol{I} = \boldsymbol{0}$$

6. Given the following matrices:

$$\boldsymbol{A} = \begin{bmatrix} 1 & 2 & 3 \\ 1 & 0 & 1 \end{bmatrix},$$

$$\boldsymbol{B} = \begin{bmatrix} 1 & 1 \\ 0 & 1 \end{bmatrix},$$

$$\boldsymbol{C} = \begin{bmatrix} 1 & 2 \\ 0 & -1 \\ 1 & 0 \end{bmatrix},$$

$$\boldsymbol{D} = \begin{bmatrix} 1 & 1 & 0 \\ 0 & 1 & 1 \\ 1 & 0 & 1 \end{bmatrix},$$

$$\boldsymbol{E} = \begin{bmatrix} 2 & -1 & 1 \\ 1 & -2 & -1 \\ -1 & 1 & 2 \end{bmatrix}$$

find, where possible, $\boldsymbol{AC}$, $\boldsymbol{BC}$, $\boldsymbol{AB}$, $\boldsymbol{BA}$, $\boldsymbol{DE}$, $\boldsymbol{ED}$.

Also find, where possible, $\boldsymbol{A}^2$, $\boldsymbol{B}^2$, $\boldsymbol{C}^2$, $\boldsymbol{D}^2$.

7. Verify the distributive law (13.11) for the following matrices:

$$\boldsymbol{A} = \begin{bmatrix} 1 & 2 \\ 2 & 1 \end{bmatrix},$$

$$\boldsymbol{B} = \begin{bmatrix} 1 & -1 \\ -1 & 1 \end{bmatrix}$$

and $\boldsymbol{C} = \begin{bmatrix} 1 & 1 \\ 0 & 1 \end{bmatrix}$

13.3 Determinants

The **determinant** of a square matrix is a number produced by combining the elements of the matrix in a prescribed way; it is written $|\boldsymbol{A}|$ or det $\boldsymbol{A}$.

KEY POINT

The determinant of a 2 × 2 matrix is easily calculated.

If $\boldsymbol{A} = \begin{bmatrix} a & b \\ c & d \end{bmatrix}$ then $|\boldsymbol{A}| = \begin{vmatrix} a & b \\ c & d \end{vmatrix} = ad - bc.$ 13.14

The determinant of a 3×3 matrix can be calculated in a number of ways. One way is presented here; the method is known as *expanding the top row*.

If $\quad A = \begin{bmatrix} a_{11} & a_{12} & a_{13} \\ a_{21} & a_{22} & a_{23} \\ a_{31} & a_{32} & a_{33} \end{bmatrix}$ then

$$|A| = a_{11}\begin{vmatrix} a_{22} & a_{23} \\ a_{32} & a_{33} \end{vmatrix} - a_{12}\begin{vmatrix} a_{21} & a_{23} \\ a_{31} & a_{33} \end{vmatrix} + a_{13}\begin{vmatrix} a_{21} & a_{22} \\ a_{31} & a_{32} \end{vmatrix} \qquad 13.15$$

Each position in a square matrix is assigned a place sign which can be used to evaluate the determinant. For example,

$$\begin{bmatrix} + & - \\ - & + \end{bmatrix} \quad \begin{bmatrix} + & - & + \\ - & + & - \\ + & - & + \end{bmatrix} \quad \begin{bmatrix} + & - & + & - \\ - & + & - & + \\ + & - & + & - \\ - & + & - & + \end{bmatrix}$$

Notice how the signs alternate as we go from row to row and from column to column.

The **minor** $\boldsymbol{M}_{ij}$ of the element a_{ij} in a matrix is found by removing row i and column j and taking the determinant of the remaining elements in the order they appear. If the minor $\boldsymbol{M}_{ij}$ is multiplied by the place sign we obtain the **cofactor** of the element a_{ij}, often written $\boldsymbol{A}_{ij}$.

For a 3×3 matrix $\boldsymbol{A}$,

$$|A| = a_{11}\boldsymbol{M}_1 - a_{12}\boldsymbol{M}_{12} + a_{13}\boldsymbol{M}_{13}$$

or, in terms of cofactors,

$$|A| = a_{11}\boldsymbol{A}_{11} + a_{12}\boldsymbol{A}_{12} + a_{13}\boldsymbol{A}_{13} \qquad 13.16$$

Worked examples

13.13 Evaluate the following determinants:

$$\begin{vmatrix} 1 & 2 \\ 3 & 4 \end{vmatrix}, \quad \begin{vmatrix} 1 & 0 \\ 0 & 1 \end{vmatrix}, \quad \begin{vmatrix} 2 & -1 \\ -3 & -3 \end{vmatrix}, \quad \begin{vmatrix} \cos\theta & \sin\theta \\ -\sin\theta & \cos\theta \end{vmatrix},$$

$$\begin{vmatrix} 1 & 0 & 0 \\ 0 & 1 & 0 \\ 0 & 0 & 0 \end{vmatrix}, \quad \begin{vmatrix} 2 & 1 & 2 \\ 0 & 3 & 1 \\ -1 & 0 & 4 \end{vmatrix}$$

Solution

$$\begin{vmatrix} 1 & 2 \\ 3 & 4 \end{vmatrix} = 1 \times 4 - 3 \times 2 = -2$$

$$\begin{vmatrix} 1 & 0 \\ 0 & 1 \end{vmatrix} = 1$$

$$\begin{vmatrix} 2 & -1 \\ -3 & -3 \end{vmatrix} = 2 \times (-3) - (-3) \times (-1) = -6 - 3 = -9$$

$$\begin{vmatrix} \cos\theta & \sin\theta \\ -\sin\theta & \cos\theta \end{vmatrix} = \cos^2\theta - (-\sin^2\theta) = \cos^2\theta + \sin^2\theta = 1$$

$$\begin{vmatrix} 1 & 0 & 0 \\ 0 & 1 & 0 \\ 0 & 0 & 0 \end{vmatrix} = 1 \times \begin{vmatrix} 1 & 0 \\ 0 & 1 \end{vmatrix} - 0 \times \begin{vmatrix} 0 & 0 \\ 0 & 0 \end{vmatrix} + 0 \times \begin{vmatrix} 0 & 1 \\ 0 & 0 \end{vmatrix} = 1$$

$$\begin{vmatrix} 2 & 1 & 2 \\ 0 & 3 & 1 \\ -1 & 0 & 4 \end{vmatrix} = 2 \times \begin{vmatrix} 3 & 1 \\ 0 & 4 \end{vmatrix} - 1 \times \begin{vmatrix} 0 & 1 \\ -1 & 4 \end{vmatrix} + 2 \times \begin{vmatrix} 0 & 3 \\ -1 & 0 \end{vmatrix}$$

$$= 2 \times (12 - 0) - 1 \times (0 + 1) + 2 \times (0 + 3) = 24 - 1 + 6 = 29$$

13.14 Find the minors and cofactors of the second row of the determinant

$$|\boldsymbol{A}| = \begin{vmatrix} 1 & 2 & 3 \\ 6 & 5 & 4 \\ 7 & 8 & 9 \end{vmatrix}$$

Solution Consider the element $a_{21} = 6$. If we cross out the second row and first column,

$$\begin{vmatrix} 1 & 2 & 3 \\ 6 & 5 & 4 \\ 7 & 8 & 9 \end{vmatrix} \quad \text{leads to} \quad \begin{vmatrix} 2 & 3 \\ 8 & 9 \end{vmatrix} = 2 \times 9 - 8 \times 3 = -6$$

The minor $\boldsymbol{M}_{21} = -6$ and the cofactor $\boldsymbol{A}_{21} = 6$ (place sign $-$).

The element a_{22} is 5. Repeating the procedure above,

$$\begin{vmatrix} 1 & 2 & 3 \\ 6 & 5 & 4 \\ 7 & 8 & 9 \end{vmatrix} \quad \text{leads to} \quad \begin{vmatrix} 1 & 3 \\ 7 & 9 \end{vmatrix} = 1 \times 9 - 7 \times 3 = -12$$

The minor is $\boldsymbol{M}_{22} = -12$ and the cofactor $\boldsymbol{A}_{22} = -12$ (place sign $+$).

The element a_{23} is 4. The procedure

$$\begin{vmatrix} 1 & 2 & 3 \\ 6 & 5 & 4 \\ 7 & 8 & 9 \end{vmatrix} \quad \text{leads to} \quad \begin{vmatrix} 1 & 2 \\ 7 & 8 \end{vmatrix} = 1 \times 8 - 7 \times 2 = -6$$

The minor $\boldsymbol{M}_{23} = -6$ and the cofactor $\boldsymbol{A}_{23} = 6$ (place sign $-$).

The following properties of determinants are stated without proof.

1. $|\boldsymbol{A}^{\mathrm{T}}| = |\boldsymbol{A}|$ which means that the properties which apply to rows will also apply to columns.
2. If two rows (or two columns) of a determinant are equal then the determinant is zero.
3. If a row (or a column) of a determinant is multiplied by a scalar then the value of the determinant is multiplied by that scalar.
4. If two rows (or two columns) of a determinant are interchanged then the value of the determinant is multiplied by -1.
5. The determinant of the product of two matrices is the product of the determinants of the matrices:

$$|\boldsymbol{AB}| = |\boldsymbol{A}| \times |\boldsymbol{B}| \qquad 13.17$$

Worked example

13.15 Verify property (4) with the matrix

$$A = \begin{bmatrix} 1 & 2 \\ 3 & 4 \end{bmatrix}$$

and property (5) with matrices A as above and

$$B = \begin{bmatrix} 2 & -1 \\ 1 & -3 \end{bmatrix}$$

Solution First, $|A| = 1 \times 4 - 3 \times 2 = -2$.

Let

$$C = \begin{bmatrix} 3 & 4 \\ 1 & 2 \end{bmatrix} \qquad |C| = 3 \times 2 - 1 \times 4 = 2$$

Therefore interchanging the two rows of A has led to a matrix whose determinant has a value -1 times that of the determinant of A.

Let

$$D = \begin{bmatrix} 2 & 1 \\ 4 & 3 \end{bmatrix} \qquad |D| = 2 \times 3 - 4 \times 1 = 2$$

Therefore interchanging the two columns of A has led to a matrix whose determinant has a value -1 times that of the determinant of A.

$$AB = \begin{bmatrix} 1 & 2 \\ 3 & 4 \end{bmatrix}\begin{bmatrix} 2 & -1 \\ 1 & -3 \end{bmatrix} = \begin{bmatrix} 1 \times 2 + 2 \times 1 & 1 \times (-1) + 2 \times (-3) \\ 3 \times 2 + 1 \times 1 & 3 \times (-1) + 4 \times (-3) \end{bmatrix}$$

$$= \begin{bmatrix} 4 & -7 \\ 10 & -15 \end{bmatrix}$$

$$|AB| = 4 \times (-15) - 10 \times (-7) = 10$$

$$|A| = 4 - 6 = -2, \quad |B| = -6 + 1 = -5; \quad \text{then} \quad (-2) \times (-5) = 10$$

Self-assessment questions 13.3

1. Write down the determinant of the matrix

$$\begin{bmatrix} a & b \\ c & d \end{bmatrix}$$

2. Explain the terms minor and cofactor of an element of a matrix.
3. State four properties of determinants.
4. What is the value of the determinant of an identity matrix?

Exercise 13.3

1. Evaluate the following determinants:

$$\begin{vmatrix} 0 & 1 \\ 1 & 0 \end{vmatrix}, \quad \begin{vmatrix} a & b \\ 0 & d \end{vmatrix}, \quad \begin{vmatrix} a & 0 \\ c & d \end{vmatrix},$$

$$\begin{vmatrix} -3 & 2 \\ 4 & -4 \end{vmatrix}, \quad \begin{vmatrix} \sec\theta & \tan\theta \\ \tan\theta & \sec\theta \end{vmatrix},$$

$$\begin{vmatrix} 1 & 1 & 0 \\ 0 & 1 & 1 \\ 1 & 0 & 1 \end{vmatrix}, \quad \begin{vmatrix} 1 & -2 & 3 \\ 2 & -1 & 0 \\ 3 & 0 & 2 \end{vmatrix}$$

2. Find the minors and cofactors of the first row of the determinant

$$|A| = \begin{vmatrix} 1 & 4 & 3 \\ 2 & -1 & 2 \\ 3 & 2 & -1 \end{vmatrix}$$

Hence evaluate $|A|$.

3. Verify properties (2) and (3) with the matrices

$$A = \begin{bmatrix} 1 & 3 \\ 2 & 4 \end{bmatrix} \quad \text{and} \quad B = \begin{bmatrix} 1 & 1 & 0 \\ 0 & 1 & 1 \\ 1 & 0 & 1 \end{bmatrix}$$

4. The value of a determinant is unchanged if a multiple of any row (or column) is added to any other row (column).

(a) By subtracting the third column from the first column evaluate

$$|D| = \begin{vmatrix} 3 & 6 & 3 \\ 7 & 8 & 2 \\ 6 & 5 & 2 \end{vmatrix}$$

(b) Given

$$D = \begin{vmatrix} 8 & 7 & 9 \\ 3 & 1 & -2 \\ 4 & 1 & 3 \end{vmatrix}$$

first subtract three times the third row from the first row, then add the first column to the second column. Hence evaluate $|D|$.

13.4 Inverse of a matrix

If $AC = CA = I$ then C is the inverse of A, written A^{-1}. If

$$A = \begin{bmatrix} a & b \\ c & d \end{bmatrix}$$

then

$$A^{-1} = \frac{1}{ad - bc}\begin{bmatrix} d & -b \\ -c & a \end{bmatrix} = \frac{1}{|A|}\begin{bmatrix} d & -b \\ -c & a \end{bmatrix} \qquad 13.18$$

For matrices of larger sizes we use the following results. The matrix which is obtained by replacing each element a_{ij} of the transpose of a matrix A by its cofactor A_{ij} is called the **adjoint matrix** of A and is written adj A. It can be shown that:

KEY POINT

$$A^{-1} = \frac{1}{|A|}\,\text{adj}\,A \qquad 13.19$$

A matrix which has no inverse is said to be **singular**. One which has an inverse is said to be **non-singular**.

KEY POINT

> A is non-singular if and only if $|A| = 0$.

The following properties of inverses are stated without proof:

1. If a matrix has an inverse then this inverse is unique (we can therefore speak of *the* inverse of a matrix).
2. $(AB)^{-1} = B^{-1}A^{-1}$ 13.20

Worked examples

13.16 Find, where possible, the inverse of the matrix A where A is

(a) $\begin{bmatrix} 1 & 2 \\ 3 & 4 \end{bmatrix}$, (b) $\begin{bmatrix} 1 & -1 \\ 1 & -1 \end{bmatrix}$ (c) $\begin{bmatrix} 1 & 0 \\ 0 & 1 \end{bmatrix}$

Solution (a)

$$\begin{vmatrix} 1 & 2 \\ 3 & 4 \end{vmatrix} = 4 - 6 = -2$$

$$\begin{bmatrix} 1 & 2 \\ 3 & 4 \end{bmatrix}^{-1} = \frac{1}{(-2)}\begin{bmatrix} 4 & -2 \\ -3 & 1 \end{bmatrix} = \begin{bmatrix} -2 & 1 \\ \frac{3}{2} & -\frac{1}{2} \end{bmatrix}$$

(You can check that $AA^{-1} = A^{-1}A = I$.)

(b)

$$A^{-1} = \frac{1}{ad - bc}\begin{bmatrix} d & -b \\ -c & a \end{bmatrix} = \frac{1}{|A|}\begin{bmatrix} d & -b \\ -c & a \end{bmatrix}$$

But

$$\begin{vmatrix} 1 & -1 \\ 1 & -1 \end{vmatrix} = -1 + 1 = 0$$

hence the matrix has no inverse.

(c)

$$\begin{vmatrix} 1 & 0 \\ 0 & 1 \end{vmatrix} = 1 \text{ so that } A^{-1} = \begin{bmatrix} 1 & 0 \\ 0 & 1 \end{bmatrix}$$

(Hence I is its own inverse.)

13.17 Find, where possible, the inverses of the following matrices:

(a) $A = \begin{bmatrix} 2 & -1 & 4 \\ 1 & 0 & 0 \\ 1 & -2 & 0 \end{bmatrix}$ (b) $A = \begin{bmatrix} 5 & -2 & 9 \\ 3 & 2 & -1 \\ 2 & -1 & 4 \end{bmatrix}$

Solution (a) $|A| = 2 \times \begin{vmatrix} 0 & 0 \\ -2 & 0 \end{vmatrix} - (-1) \times \begin{vmatrix} 1 & 0 \\ 1 & 0 \end{vmatrix} + 4 \times \begin{vmatrix} 1 & 0 \\ 1 & -2 \end{vmatrix} = -8$

$$A^{\mathrm{T}} = \begin{bmatrix} 2 & 1 & 1 \\ -1 & 0 & -2 \\ 4 & 0 & 0 \end{bmatrix}$$

$$\operatorname{adj} A = \begin{bmatrix} 0 & -8 & 0 \\ 0 & -4 & 4 \\ -2 & 3 & 1 \end{bmatrix} \quad \text{and so} \quad A^{-1} = \tfrac{1}{8}\begin{bmatrix} 0 & 8 & 0 \\ 0 & 4 & -4 \\ 2 & -3 & -1 \end{bmatrix}$$

(b) $|A| = 5 \times \begin{vmatrix} 2 & -1 \\ -1 & 4 \end{vmatrix} + 2 \times \begin{vmatrix} 3 & -1 \\ 2 & 4 \end{vmatrix} + 9 \times \begin{vmatrix} 3 & 2 \\ 2 & -1 \end{vmatrix}$

$= 5 \times 7 + 2 \times 14 + 9 \times (-7) = 0$ The matrix has no inverse.

Self-assessment questions 13.4

1. Under what conditions does a matrix have an inverse?
2. Outline the steps by which you would find the inverse of a 3×3 matrix.

Exercise 13.4

1. Find, where possible, the inverse of each of the following matrices:

$$A = \begin{bmatrix} 0 & 1 \\ -1 & 0 \end{bmatrix}, \quad B = \begin{bmatrix} \cos\theta & \sin\theta \\ -\sin\theta & \cos\theta \end{bmatrix},$$

$$C = \begin{bmatrix} 4 & 2 \\ 10 & 5 \end{bmatrix}, \quad D = \begin{bmatrix} 1 & 1 \\ 0 & 1 \end{bmatrix},$$

$$E = \begin{bmatrix} a & b \\ 0 & d \end{bmatrix}, \quad F = \begin{bmatrix} a & 0 \\ c & d \end{bmatrix},$$

$$G = \begin{bmatrix} a & 0 \\ 0 & d \end{bmatrix}, \quad H = \begin{bmatrix} a & -b \\ -a & b \end{bmatrix}$$

Where the inverse exists verify your result by multiplying the matrix by its inverse.

2. A non-singular matrix A for which $A^{-1} = A^{\mathrm{T}}$ is said to be **orthogonal**. Hence, if A is orthogonal then $A^{\mathrm{T}}A = AA^{\mathrm{T}} = I$ and vice versa.

   Verify in two ways that the following matrices are orthogonal:

$$A = \begin{bmatrix} 0 & 1 \\ -1 & 0 \end{bmatrix}, \quad B = \begin{bmatrix} \cos\theta & \sin\theta \\ -\sin\theta & \cos\theta \end{bmatrix}$$

   Is the identity matrix orthogonal?

3. Each of the matrices of Question 2 has a determinant of 1. Show that this is always true for an orthogonal matrix. By considering the matrix

$$A = \begin{bmatrix} 1 & 1 \\ 0 & 1 \end{bmatrix}$$

   show that the converse is not true, that is, not every matrix whose determinant is 1 is orthogonal.

4. If

$$A = \begin{bmatrix} 1 & 3 \\ 0 & 2 \end{bmatrix} \quad \text{and} \quad B = \begin{bmatrix} 3 & 1 \\ 5 & 2 \end{bmatrix}$$

verify (13.20).

5. Find, where possible, the inverses of the following matrices:

$$A = \begin{bmatrix} 1 & 2 & 3 \\ 4 & 5 & 6 \end{bmatrix}, \quad B = \begin{bmatrix} 2 & 5 & 2 \\ -3 & 4 & -2 \\ 1 & 1 & -1 \end{bmatrix},$$

$$C = \begin{bmatrix} 2 & 5 & 3 \\ -1 & -2 & 2 \\ 4 & 9 & -1 \end{bmatrix},$$

$$D = \begin{bmatrix} 3 & 2 & 5 \\ 1 & -1 & 5 \\ 0 & 1 & -7 \end{bmatrix}$$

Test and assignment exercises 13

1. If

$$A = \begin{bmatrix} 1 & 3 \\ 2 & 1 \end{bmatrix}, \quad B = \begin{bmatrix} 3 & 4 \\ -2 & 0 \end{bmatrix}$$

and $C = \begin{bmatrix} 1 & 0 & 4 \\ 3 & 2 & 0 \end{bmatrix}$

find, where possible, the following:

(a) $A + B$ (b) $4A - 2B$
(c) $A^T + B^T$ (d) $C + C^T$

2. Given the matrices

$$A = \begin{bmatrix} 2 \\ 0 \\ 1 \end{bmatrix}, \quad B = \begin{bmatrix} 1 & -1 \\ 0 & 3 \\ 3 & 0 \end{bmatrix}$$

and $C = \begin{bmatrix} 4 & 0 & -1 \\ 0 & 2 & 3 \\ 1 & -1 & 0 \end{bmatrix}$

find, where possible,

(a) AB (b) A^TB (c) BA (d) B^TA
(e) A^2 (f) B^2 (g) C^2 (h) BB^T
(i) B^TB (j) A^TCA

3. Evaluate the following matrix products:

(a) $\begin{bmatrix} 3 & 1 & 1 & 2 \end{bmatrix} \begin{bmatrix} 2 \\ 1 \\ 1 \\ 2 \end{bmatrix}$

(b) $\begin{bmatrix} 1 & 1 & 1 & 0 \\ 1 & 2 & 2 & 2 \\ 0 & 3 & 1 & 2 \end{bmatrix} \begin{bmatrix} 1 & 1 \\ 0 & 0 \\ 1 & -1 \\ 1 & 2 \end{bmatrix}$

4. Given that

$$A = \begin{bmatrix} 9 & -3 & 1 \\ 1 & 4 & 7 \end{bmatrix}, \quad B = \begin{bmatrix} -5 & 6 \\ 3 & 1 \\ -4 & 2 \end{bmatrix}$$

verify that $(AB)^T = B^TA^T$.

5. Evaluate the determinants:

(a) $\begin{vmatrix} 1 & 2 \\ 6 & -3 \end{vmatrix}$ (b) $\begin{vmatrix} 1 & 2 \\ 6 & 3 \end{vmatrix}$

(c) $\begin{vmatrix} -1 & -2 \\ -3 & 4 \end{vmatrix}$ (d) $\begin{vmatrix} 3 & 9 & 4 \\ 1 & 5 & 7 \\ 7 & 18 & 8 \end{vmatrix}$

(e) $\begin{vmatrix} -1 & 2 & 1 \\ 6 & 9 & -2 \\ 3 & 1 & -4 \end{vmatrix}$ (f) $\begin{vmatrix} 6 & 9 & -2 \\ -1 & 2 & 1 \\ 3 & 1 & -4 \end{vmatrix}$

(g) $\begin{vmatrix} 1 & 1 & 1 \\ 1 & 1 & 1 \\ 3 & 4 & -5 \end{vmatrix}$ (h) $\begin{vmatrix} 1 & 1 & 1 \\ 3 & 4 & 5 \\ 4 & 5 & 6 \end{vmatrix}$

(i) $\begin{vmatrix} 6 & 9 & -2 \\ -2 & 4 & 2 \\ 3 & 1 & -4 \end{vmatrix}$

6. Evaluate the determinant

$$\begin{vmatrix} 0 & 4 & 6 & 9 \\ 2 & 4 & 6 & 4 \\ 2 & 1 & 4 & 0 \\ 1 & 2 & 3 & 2 \end{vmatrix}$$

7. (a) Write down the cofactor of the elements 0 and 4 in the determinant

$$\begin{vmatrix} 1 & 2 & 1 \\ 2 & 1 & 4 \\ 1 & 0 & 3 \end{vmatrix}$$

(b) Write down the cofactor matrix and the adjoint matrix of

$$A = \begin{vmatrix} 1 & 2 & 1 \\ 2 & 1 & 4 \\ 1 & 0 & 3 \end{vmatrix}$$

8. Find the inverse of the following matrices, where possible:

(a) $\begin{bmatrix} 4 & 3 \\ 2 & -1 \end{bmatrix}$

(b) $\begin{bmatrix} 2 & 1 \\ 4 & 2 \end{bmatrix}$

(c) $\begin{bmatrix} 3 & 2 \\ 4 & 2 \end{bmatrix}$

9. Using the adjoint matrix find, where possible, the inverse of

(a) $\begin{bmatrix} 0 & 1 & 1 \\ -1 & 0 & 2 \\ 2 & 0 & 3 \end{bmatrix}$

(b) $\begin{bmatrix} 3 & 2 & 1 & 4 \\ 1 & 4 & 0 & 1 \\ 2 & -1 & 1 & 0 \end{bmatrix}$

(c) $\begin{bmatrix} 2 & 0 & 3 \\ 4 & 1 & 4 \\ 0 & 1 & 0 \end{bmatrix}$ (d) $\begin{bmatrix} 2 & 1 & 3 \\ 3 & 2 & 5 \\ 1 & 2 & 3 \end{bmatrix}$

10. (a) Verify that $(AB)^{-1} = B^{-1}A^{-1}$ for the matrices

$$A = \begin{bmatrix} 4 & 3 \\ 2 & -1 \end{bmatrix} \quad \text{and} \quad B = \begin{bmatrix} 3 & 2 \\ 4 & 2 \end{bmatrix}$$

(b) For the matrices above verify that

$$|AB| = |A| \times |B| = |BA|$$

14 Linear equations

Objectives

This chapter

- shows how to express a system of linear equations in matrix form and vice versa
- uses Cramer's rule for solving systems of linear equations
- develops the method of Gaussian elimination
- illustrates the problem of ill-conditioning
- considers two ways of improving the accuracy of the solution

14.1 Solving simultaneous equations

The simultaneous linear equations

$$2x + 5 = 9, \qquad 3x + 6y = 12$$

can be written in matrix form as

$$\begin{bmatrix} 2 & 5 \\ 3 & 6 \end{bmatrix} \begin{bmatrix} x \\ y \end{bmatrix} = \begin{bmatrix} 9 \\ 12 \end{bmatrix} \qquad \text{or} \qquad \boldsymbol{AX} = \boldsymbol{b}$$

where

$$\boldsymbol{A} = \begin{bmatrix} 2 & 5 \\ 3 & 6 \end{bmatrix}, \boldsymbol{X} = \begin{bmatrix} x \\ y \end{bmatrix} \text{ and } \boldsymbol{b} = \begin{bmatrix} 9 \\ 12 \end{bmatrix}$$

Pre-multiplying the matrix equation by $\boldsymbol{A}^{-1}$ we obtain

$$\boldsymbol{A}^{-1}\boldsymbol{AX} = \boldsymbol{A}^{-1}\boldsymbol{b} \qquad \text{that is} \qquad \boldsymbol{X} = \boldsymbol{A}^{-1}\boldsymbol{b} \tag{14.1}$$

From Section 13.4, we can find the inverse of $\boldsymbol{A}$ as

$$\boldsymbol{A}^{-1} = \begin{bmatrix} 2 & 5 \\ 3 & 6 \end{bmatrix}^{-1} = \frac{1}{\begin{vmatrix} 2 & 5 \\ 3 & 6 \end{vmatrix}} \begin{bmatrix} 6 & -5 \\ -3 & 2 \end{bmatrix} = \frac{1}{(-3)} \begin{bmatrix} 6 & -5 \\ -3 & 2 \end{bmatrix}$$

Hence

$$\begin{bmatrix} x \\ y \end{bmatrix} = \frac{-1}{3} \begin{bmatrix} 6 & -5 \\ -3 & 2 \end{bmatrix} \begin{bmatrix} 9 \\ 12 \end{bmatrix} = \frac{-1}{3} \begin{bmatrix} -6 \\ -3 \end{bmatrix} = \begin{bmatrix} 2 \\ 1 \end{bmatrix}$$

That is, $x = 2$, $y = 1$. We can check in the original equations that this is the (unique) solution. If $\boldsymbol{A}^{-1}$ does not exist then no unique solution can be found.

Worked examples

14.1 Find, where possible, the unique solution of the following sets of simultaneous equations:

(a) $\begin{bmatrix} 3 & 1 \\ 4 & 4 \end{bmatrix} \begin{bmatrix} x \\ y \end{bmatrix} = \begin{bmatrix} 1 \\ -4 \end{bmatrix}$ (b) $\begin{bmatrix} 2 & -1 \\ -6 & 3 \end{bmatrix} \begin{bmatrix} x \\ y \end{bmatrix} = \begin{bmatrix} 0 \\ 0 \end{bmatrix}$

(c) $\begin{bmatrix} 2 & -1 \\ -6 & 3 \end{bmatrix} \begin{bmatrix} x \\ y \end{bmatrix} = \begin{bmatrix} 4 \\ 11 \end{bmatrix}$

Solution (a) $\begin{bmatrix} 3 & 1 \\ 4 & 4 \end{bmatrix}^{-1} = \frac{1}{8} \begin{bmatrix} 4 & -1 \\ -4 & 3 \end{bmatrix}$

Hence

$$\begin{bmatrix} x \\ y \end{bmatrix} = \frac{1}{8} \begin{bmatrix} 4 & -1 \\ -4 & 3 \end{bmatrix} \begin{bmatrix} 1 \\ -4 \end{bmatrix} = \frac{1}{8} \begin{bmatrix} 8 \\ -16 \end{bmatrix} \begin{bmatrix} 1 \\ -2 \end{bmatrix}$$

Therefore $x = 1$, $y = -2$.

(b) $\begin{vmatrix} 2 & -1 \\ -6 & 3 \end{vmatrix} = 6 - 6 = 0$

The equations are $2x - y = 0$, $-6x + 3y = 0$.

The second equation is -3 times the first so that there is only one independent equation. The best we can say is that $y = 3x$ and therefore there are **infinitely many solutions**

(c) The equations are $2x - y = 4$ and $-6x + 3y = 11$. Multiplying the first equation by -3 gives $-6x + 3y = -12$ and this conflicts with the second equation. The equations are said to be **inconsistent** and there is no solution.

14.2 Find, where possible, the solutions of the following sets of simultaneous equations:

(a)
$$\begin{aligned} 2x - y + z &= 9 \\ -2x + 7y - z &= -33 \\ 3x + 4y - 2z &= -5 \end{aligned}$$

(b)
$$\begin{aligned} 5x + 7y + 3z &= 1 \\ 7x + 8y + 4z &= 2 \\ 8x + 13y + 5z &= 0 \end{aligned}$$

(c)
$$\begin{aligned} 5x + 7y + 3z &= 1 \\ 7x + 8y + 4z &= 3 \\ 8x + 13y + 5z &= 0 \end{aligned}$$

Solution (a) $\begin{vmatrix} 2 & -1 & 1 \\ -2 & 7 & -1 \\ 3 & 4 & -2 \end{vmatrix} = 2 \times \begin{vmatrix} 7 & -1 \\ 4 & -2 \end{vmatrix} + 1 \times \begin{vmatrix} -2 & -1 \\ 3 & -2 \end{vmatrix} + 1 \times \begin{vmatrix} -2 & 7 \\ 3 & 4 \end{vmatrix}$

$$= 2 \times (-10) + 7 + (-29) = -42$$

$$A^{\mathrm{T}} = \begin{bmatrix} 2 & -2 & 3 \\ -1 & 7 & 4 \\ 1 & -1 & -2 \end{bmatrix} \qquad \operatorname{adj} A = \begin{bmatrix} -10 & 2 & -6 \\ -7 & -7 & 0 \\ -29 & -11 & 12 \end{bmatrix}$$

$$A^{-1} = \frac{-1}{42}\begin{bmatrix} -10 & 2 & -6 \\ -7 & -7 & 0 \\ -29 & -11 & 12 \end{bmatrix}$$

Therefore

$$\begin{bmatrix} x \\ y \\ z \end{bmatrix} = -\frac{1}{42}\begin{bmatrix} -10 & 2 & -6 \\ -7 & -7 & 0 \\ -29 & -11 & 12 \end{bmatrix}\begin{bmatrix} 9 \\ -33 \\ -5 \end{bmatrix} = -\frac{1}{42}\begin{bmatrix} -126 \\ 168 \\ 42 \end{bmatrix} = \begin{bmatrix} 3 \\ -4 \\ -1 \end{bmatrix}$$

so that $x = 3$, $y = -4$, $z = -1$.

You can substitute these values into the equations and check.

(b) $\begin{vmatrix} 5 & 7 & 3 \\ 7 & 8 & 4 \\ 8 & 13 & 5 \end{vmatrix} = 5 \times \begin{vmatrix} 8 & 4 \\ 13 & 5 \end{vmatrix} - 7 \times \begin{vmatrix} 7 & 4 \\ 8 & 5 \end{vmatrix} + 3 \times \begin{vmatrix} 7 & 8 \\ 8 & 13 \end{vmatrix}$

$$= 5 \times (-12) - 7 \times 3 + 3 \times 27 = 0$$

If we add the second and third equations we obtain

$$15x + 21y + 9z = 2$$

Multiplying the first equation by 3 we obtain

$$15x + 21y + 9z = 3$$

This is a conflict. The equations are **inconsistent** and there is no solution.

(c) The left-hand sides of this system are identical to those of (b). On this occasion the sum of the second and third equation is exactly the same as three times the first equation.

The equations are **consistent**. There is no unique solution.

Cramer's rule is a means of finding the solution to a system of linear equations via determinants. It is practical only for sets of two or three equations.

KEY POINT

The solution of the system

$$a_{11}x_1 + a_{12}x_2 = b_1$$
$$a_{21}x_1 + a_{22}x_2 = b_2$$

is

$$x_1 = \frac{\begin{vmatrix} b_1 & a_{12} \\ b_2 & a_{22} \end{vmatrix}}{D} \qquad x_2 = \frac{\begin{vmatrix} a_{11} & b_1 \\ a_{21} & b_2 \end{vmatrix}}{D}$$

where

$$D = \begin{vmatrix} a_{11} & a_{12} \\ a_{21} & a_{22} \end{vmatrix} = |\boldsymbol{A}| \qquad 14.2$$

The solution of the system

$$a_{11}x_1 + a_{12}x_2 + a_{13}x_3 = b_1$$
$$a_{21}x_1 + a_{22}x_2 + a_{23}x_3 = b_2$$
$$a_{31}x_1 + a_{32}x_2 + a_{33}x_3 = b_3$$

is

$$x_1 = \frac{\begin{vmatrix} b_1 & a_{12} & a_{13} \\ b_2 & a_{22} & a_{23} \\ b_3 & a_{32} & a_{33} \end{vmatrix}}{D} \qquad x_2 = \frac{\begin{vmatrix} a_{11} & b_1 & a_{13} \\ a_{21} & b_2 & a_{23} \\ a_{31} & b_3 & a_{33} \end{vmatrix}}{D}$$

$$x_3 = \frac{\begin{vmatrix} a_{11} & a_{12} & b_1 \\ a_{21} & a_{22} & b_2 \\ a_{31} & a_{32} & b_3 \end{vmatrix}}{D}$$

where

$$D = \begin{vmatrix} a_{11} & a_{12} & a_{13} \\ a_{21} & a_{22} & a_{23} \\ a_{31} & a_{32} & a_{33} \end{vmatrix} = |\boldsymbol{A}| \qquad 14.3$$

Worked example

14.3 Using Cramer's rule, find the solution of the systems in (a) Example 14.1(a) and (b) Example 14.2(a).

Solution (a) $\begin{vmatrix} 3 & 1 \\ 4 & 4 \end{vmatrix} = 12 - 4 = 8$

By Cramer's rule

$$x = \frac{\begin{vmatrix} 1 & 1 \\ -4 & 4 \end{vmatrix}}{8} = \frac{8}{8} = 1 \qquad y = \frac{\begin{vmatrix} 3 & 1 \\ 4 & -4 \end{vmatrix}}{8} = -\frac{16}{8} = -2$$

(b) $\begin{vmatrix} 2 & -1 & 1 \\ -2 & 7 & -1 \\ 3 & 4 & -2 \end{vmatrix} = -42$

$$x = -\frac{1}{42}\begin{vmatrix} 9 & -1 & 1 \\ -33 & 7 & -1 \\ -5 & 4 & -2 \end{vmatrix}$$

$$= -\frac{1}{42}\left\{9 \times \begin{vmatrix} 7 & -1 \\ 4 & -2 \end{vmatrix} + 1 \times \begin{vmatrix} -33 & -1 \\ -5 & -2 \end{vmatrix} + 1 \times \begin{vmatrix} -33 & 7 \\ -5 & 4 \end{vmatrix}\right\}$$

$$= -\frac{1}{42}\{9 \times (-10) + 61 - 97\}$$

$$= +\frac{126}{42} = 3$$

Similarly,

$$y = -\frac{1}{42}\begin{vmatrix} 2 & 9 & 1 \\ -2 & -33 & -1 \\ 3 & -5 & -2 \end{vmatrix} = -\frac{168}{42} = -4$$

$$z = -\frac{1}{42}\begin{vmatrix} 2 & -1 & 9 \\ -2 & 7 & -33 \\ 3 & 4 & -5 \end{vmatrix} = -\frac{42}{42} = -1$$

Self-assessment questions 14.1

1. What is meant by an inconsistent system of linear equations?
2. Explain Cramer's rule.

Exercise 14.1

1. (a) Write out the system of equations represented in matrix form as

$$\begin{bmatrix} 2 & 0 & -3 \\ 1 & 3 & 4 \\ 4 & -2 & 1 \end{bmatrix}\begin{bmatrix} x \\ y \\ z \end{bmatrix} = \begin{bmatrix} 5 \\ -1 \\ 3 \end{bmatrix}$$

(b) Express the following system of equations in matrix form:

$$\begin{aligned} x_1 - x_2 + 2x_3 &= 4 \\ x_1 + x_2 + x_3 &= 6 \\ 2x_1 \qquad + x_3 &= 9 \end{aligned}$$

2. Solve, where possible, the following equations:

(a) $\begin{bmatrix} 4 & 2 \\ 5 & 5 \end{bmatrix}\begin{bmatrix} x \\ y \end{bmatrix} = \begin{bmatrix} 2 \\ -3 \end{bmatrix}$

(b) $\begin{bmatrix} 3 & -2 \\ -9 & 6 \end{bmatrix}\begin{bmatrix} x \\ y \end{bmatrix} = \begin{bmatrix} 0 \\ 0 \end{bmatrix}$

(c) $\begin{bmatrix} 3 & -2 \\ -9 & 6 \end{bmatrix}\begin{bmatrix} x \\ y \end{bmatrix} = \begin{bmatrix} 3 \\ 10 \end{bmatrix}$

3. Find, where possible, the solutions of the following equations:

(a) $\begin{aligned} x - 2y + z &= 6 \\ 2x + y - 3z &= -5 \\ 7x - 4y - 3z &= 8 \end{aligned}$

(b) $\begin{aligned} x - 2y + z &= 6 \\ 2x + y - 3z &= -5 \\ 7x - 4y - 3z &= 9 \end{aligned}$

(c) $\begin{aligned} x - 2y + z &= 6 \\ 3x + y - 2z &= -1 \\ -2x + 3y - z &= -9 \end{aligned}$

4. Use Cramer's rule to find the solutions of the following systems: (a) Question 2(a) (b); and (b) Question 3(c).

14.2 Gaussian elimination

Consider the equations

$$\begin{aligned} 2x_1 + x_2 - 3x_3 &= 4 \\ \frac{3}{2}x_2 + \frac{1}{2}x_3 &= 5 \\ 9x_3 &= 9 \end{aligned} \qquad (14.4)$$

These can be solved by **back substitution** to produce successively

$$x_3 = 1$$

$$\frac{3}{2}x_2 + \frac{1}{2} \times 1 = 5$$

so that $x_2 = 3$,

$$2x_1 + 3 - 3 \times 1 = 4$$

so that

$$x_1 = 2$$

In matrix form, equations (14.4) can be written

$$\begin{pmatrix} 2 & 1 & -3 \\ 0 & \frac{3}{2} & \frac{1}{2} \\ 0 & 0 & 9 \end{pmatrix}\begin{pmatrix} x_1 \\ x_2 \\ x_3 \end{pmatrix} = \begin{pmatrix} 4 \\ 5 \\ 9 \end{pmatrix} \qquad (14.5)$$

The matrix of coefficients is said to be in **echelon form**.

Now consider the equations

$$2x_1 + x_2 - 3x_3 = 4 \tag{14.6a}$$

$$x_1 + 2x_2 - x_3 = 7 \tag{14.6b}$$

$$4x_1 - x_2 + 2x_3 = 7 \tag{14.6c}$$

Equation (14.6b) $-$ ($\frac{1}{2}$ $\times$ equation (14.6a)) gives

$$\frac{3}{2}x_2 + \frac{1}{2}x_3 = 5 \tag{14.6d}$$

Equation (14.6c) $-$ (2 $\times$ equation (14.6a)) gives

$$-3x_2 + 8x_3 = -1 \tag{14.6e}$$

Equation (14.6e) + (2 $\times$ equation (14.6d)) gives

$$9x_3 = 9 \tag{14.6f}$$

Equations (14.6a), (14.6d) and (14.6f) are equivalent to (14.4), which we can readily solve.

KEY POINT

Gaussian elimination consists of two phases:

1. Reduce the system of equations to an equivalent system in echelon form.
2. Solve the equivalent system by back substitution.

We could carry out the method on the system (14.6) in matrix form. For example, we use instructions like $R3 \to R3 - 2 \times R1$; this means 'take row 3, subtract twice row 1 and put the result as new row 3'. This is equivalent to subtracting twice the first equation from the third.

Worked examples

14.4 Using Gaussian elimination, solve the following equations:

$$\begin{aligned} 2x_1 + x_2 - 3x_3 &= 4 \\ x_1 + 2x_2 - x_3 &= 7 \\ 4x_1 - x_2 + 2x_3 &= 7 \end{aligned}$$

Solution The matrix form of the equations is

$$\begin{bmatrix} 2 & 1 & -3 \\ 1 & 2 & -1 \\ 4 & -1 & 2 \end{bmatrix} \begin{bmatrix} x_1 \\ x_2 \\ x_3 \end{bmatrix} = \begin{bmatrix} 4 \\ 7 \\ 7 \end{bmatrix}$$

Taking the coefficient matrix and adding the column vector of right-hand sides we obtain the **augmented matrix**

$$\left[\begin{array}{rrr|r} 2 & 1 & -3 & 4 \\ 1 & 2 & -1 & 7 \\ 4 & -1 & 2 & 7 \end{array}\right]$$

Now we carry out the row operations:

$$\left[\begin{array}{rrr|r} 2 & 1 & -3 & 4 \\ 1 & 2 & -1 & 7 \\ 4 & -1 & 2 & 7 \end{array}\right] \begin{array}{l} \\ R2 - \frac{1}{2} \times R1 \\ R3 - 2 \times R1 \end{array}$$

This means replace row 2 by row $2 - \frac{1}{2} \times$ row 1 and then replace row 3 by row $3 - 2 \times$ row 1. Look at the first column: the top element is 2; in row 2 the element is 1 and $1 \div 2$ is the multiple of row 1 we subtract to replace 1 by 0; the element in row 3 is 4 and $4 \div 2$ is the multiple of row 1 we subtract to replace 4 by 0. The result is

$$\left[\begin{array}{rrr|r} 2 & 1 & -3 & 4 \\ 0 & \frac{3}{2} & \frac{1}{2} & 5 \\ 0 & -3 & 8 & -1 \end{array}\right] \begin{array}{l} \\ \\ R3 + 2 \times R2 \end{array}$$

In the second column we move to the element in the second row, $\frac{3}{2}$; the element below it is -3 and $3 \div \frac{3}{2} = 2$. The $-$ sign means that we 'add' rather than 'subtract' and the multiple of row 2 we take is $3 \div \frac{3}{2} = 2$. The result of applying the operation is

$$\left[\begin{array}{rrr|r} 2 & 1 & -3 & 4 \\ 0 & \frac{3}{2} & \frac{1}{2} & 5 \\ 0 & 0 & 9 & 9 \end{array}\right]$$

The matrix is now in echelon form. We now apply back substitution.

The last row is translated as

$$9x_3 = 9 \qquad \text{so that} \qquad x_3 = 1$$

The second row reads

$$\frac{3}{2}x_2 + \frac{1}{2}x_3 = 5, \text{ that is } \frac{3}{2}x_2 + \frac{1}{2} = 5; \text{ hence } x_2 = 3$$

The first row reads

$$2x_1 + x_2 - 3x_3 = 4, \text{ that is } 2x_1 + 3 - 3 = 4; \text{ hence}$$

$$x_1 = 2$$

The solution is therefore $x_1 = 2$, $x_2 = 3$ and $x_3 = 1$.

14.5 Use Gaussian elimination to solve the following equations:

$$\begin{aligned} x_1 - x_2 + 2x_3 &= 1 \\ -x_1 + x_2 + x_3 &= 3 \\ 5x_1 - 5x_2 + x_3 &= 2 \end{aligned}$$

Solution The method proceeds as follows:

$$\left[\begin{array}{rrr|r} 1 & -1 & 2 & 1 \\ -1 & 1 & 1 & 3 \\ 5 & -5 & 1 & 2 \end{array}\right] \begin{array}{l} \\ R2 + R1 \\ R3 - 5 \times R1 \end{array}$$

$$\left[\begin{array}{rrr|r} 1 & -1 & 2 & 1 \\ 0 & 0 & 3 & 4 \\ 0 & 0 & -9 & -3 \end{array}\right] \begin{array}{l} \\ \\ R3 + 3 \times R2 \end{array}$$

$$\left[\begin{array}{rrr|r} 1 & -1 & 2 & 1 \\ 0 & 0 & 3 & 4 \\ 0 & 0 & 0 & 9 \end{array}\right]$$

The last line is interpreted as

$$0x_1 + 0x_2 + 0x_3 = 9$$

that is, $0 = 9$, which is clearly impossible. This means that the system of equations is inconsistent and there are no solutions.

14.6 Use Gaussian elimination to solve the following equations:

$$\begin{aligned} x_1 + x_2 - 2x_3 &= 4 \\ 3x_1 + 2x_2 - 6x_3 &= 9 \\ x_1 - x_2 - 2x_3 &= -2 \end{aligned}$$

Solution The steps in the procedure are:

$$\left[\begin{array}{rrr|r} 1 & 1 & -2 & 4 \\ 3 & 2 & -6 & 9 \\ 1 & -1 & -2 & -2 \end{array}\right] \begin{array}{l} \\ R2 - 3 \times R1 \\ R3 - R1 \end{array}$$

$$\left[\begin{array}{rrr|r} 1 & 1 & -2 & 4 \\ 0 & -1 & 0 & -3 \\ 0 & -2 & 0 & -6 \end{array}\right] \begin{array}{l} \\ \\ R3 - 2R2 \end{array}$$

$$\left[\begin{array}{rrr|r} 1 & 1 & -2 & 4 \\ 0 & -1 & 0 & -3 \\ 0 & 0 & 0 & 0 \end{array}\right]$$

The last line implies that $0 = 0$ which is true but unhelpful. We cannot solve the equations uniquely. There is an infinite number of solutions and to find a recipe for them we proceed as follows.

The two remaining equations are

$$x_1 + x_2 - 2x_3 = 4 \qquad \text{and} \qquad -x_2 = -3$$

Hence $x_2 = 3$ and therefore the first of the equations is $x_1 - 2x_3 = 1$.

Let $x_3 = \lambda$; then $x_1 = 1 + 2\lambda$ giving a general solution formula

$$x_1 = 1 + 2\lambda,\ x_2 = 3,\ x_3 = \lambda$$

Application to networks

Consider the network shown in Figure 14.1. We want to find the currents I_1, I_2 and I_3, which are called **mesh currents**, and we can find them by solving a set of simultaneous equations.

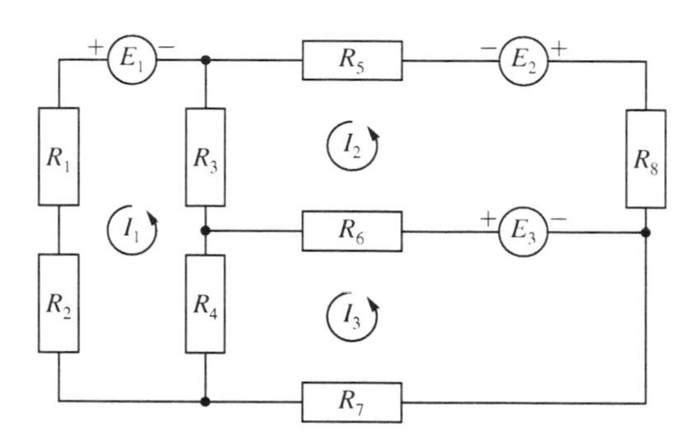

Figure 14.1. Mesh currents

Applying Kirchhoff's laws to each of the subcircuits we eventually obtain the following equations:

$$\begin{bmatrix} R_1 + R_2 + R_3 + R_4 & -R_3 & -R_4 \\ -R_3 & R_3 + R_5 + R_6 + R_8 & -R_6 \\ -R_4 & -R_6 & R_4 + R_6 + R_7 \end{bmatrix} \times \begin{bmatrix} I_1 \\ I_2 \\ I_3 \end{bmatrix}$$

$$= \begin{bmatrix} E_1 \\ -E_2 - E_3 \\ E_3 \end{bmatrix} \qquad 14.7$$

Worked example

14.7 In the case when $R_1 = 4\,\Omega$, $R_2 = 2\,\Omega$, $R_3 = 1\,\Omega$, $R_4 = 3\,\Omega$, $R_5 = 3\,\Omega$, $R_6 = 2\,\Omega$, $R_7 = 4\,\Omega$, $R_8 = 3\,\Omega$; $E_1 = 5\,\text{V}$, $E_2 = -3\,\text{V}$, $E_3 = 17\,\text{V}$, find the mesh currents I_1, I_2 and I_3.

Solution Equation (14.7) becomes

$$\begin{bmatrix} 10 & -1 & -3 \\ -1 & 9 & -2 \\ -3 & -2 & 9 \end{bmatrix} \begin{bmatrix} I_1 \\ I_2 \\ I_3 \end{bmatrix} = \begin{bmatrix} 5 \\ -14 \\ 17 \end{bmatrix}$$

The procedure is

$$\left[\begin{array}{ccc|c} 10 & -1 & -3 & 5 \\ -1 & 9 & -2 & -14 \\ -3 & -2 & 9 & 17 \end{array}\right] \begin{array}{l} \\ R2 + 0.1 \times R1 \\ R3 + 0.3 \times R1 \end{array}$$

$$\left[\begin{array}{ccc|c} 10 & -1 & -3 & 5 \\ 0 & 8.9 & -2.3 & -13.5 \\ 0 & -2.3 & 8.1 & 18.5 \end{array}\right] \begin{array}{l} \\ \\ R3 + \dfrac{2.3}{8.9} \times R2 \end{array}$$

$$\left[\begin{array}{ccc|c} 10 & -1 & -3 & 5 \\ 0 & 8.9 & -2.3 & -13.5 \\ 0 & 0 & \frac{66.8}{8.9} & \frac{133.6}{8.9} \end{array}\right]$$

The last line implies that $I_3 = 2$.

Back substitution gives first

$$8.9I_2 - 2.3 \times I_3 = -13.5$$

that is

$$8.9I_2 - 4.6 = -13.5$$

so that

$$8.9I_2 = -8.9 \text{ and } I_2 = -1$$

Finally,

$$10I_1 - I_2 - 3I_3 = 5$$

that is

$$10I_1 + 1 - 6 = 5$$

so that $I_1 = 1$. The mesh currents are $I_1 = 1\,\text{A}$, $I_2 = -1\,\text{A}$, $I_3 = 2\,\text{A}$.

Self-assessment questions 14.2

1. What are the two phases of Gaussian elimination?
2. What row operations does Gaussian elimination use? How do they relate to the original system of equations?

Exercise 14.2

1. Using Gaussian elimination solve the following equations:

(a) $x_1 + 2x_2 + 2x_3 = 22$
$2x_1 + x_2 + x_3 = 14$
$3x_1 + 4x_2 + x_3 = 28$

(b) $x_1 - x_3 = 2$
$x_2 + 3x_3 = 1$
$x_1 - 2x_2 = 7$

2. Write down the augmented matrix corresponding to the following equations and determine the nature of the solution:

(a) $2x_1 - 5x_2 + 2x_3 = 1$
$3x_1 + 2x_2 - x_3 = 4$
$5x_1 + 16x_2 - 7x_3 = 10$

(b) $x_1 - 4x_2 + 7x_3 = 3$
$3x_1 + 2x_2 - x_3 = 4$
$2x_1 - x_2 + 3x_3 = 1$

3. Use Gaussian elimination to solve the following equations. Work to 4 d.p. throughout and round the solution to 3 d.p.

$5x_1 + 11x_2 + 5x_3 = 9$
$2x_1 + 6x_2 + 7x_3 = 5$
$3x_1 + 4x_2 + 8x_3 = 5$

4. Find the value of k for which the following system of equations has more than one solution:

$$\begin{aligned} 4x_1 - 5x_2 - 7x_3 &= 10 \\ 5x_1 - 3x_2 - 8x_3 &= 14 \\ x_1 - 11x_2 - kx_3 &= -2 \end{aligned}$$

5. Solve the following equations by (a) Gaussian elimination and (b) Cramer's rule. Compare the methods.

$$\begin{aligned} x_1 - x_2 - x_3 &= 0 \\ 2x_1 + 2x_2 + x_3 &= 2 \\ 3x_1 + x_2 + 2x_3 &= 6 \end{aligned}$$

14.3 Ill-conditioning

It is possible to check the accuracy of solutions by substituting them back into the original equations. If the equations balance *exactly* then the solutions are exact. However, what if the equations only nearly balance?

Worked example

14.8 The exact solution of the following system of equations is

$$\{x_1, x_2, x_3, x_4\} = \{1, 1, 1, 1\}$$

$$\begin{aligned} 5x_1 + 5x_2 + 6x_3 + 7x_4 &= 23 \\ 5x_1 + 10x_2 + 9x_3 + 7x_4 &= 31 \\ 7x_1 + 7x_2 + 8x_3 + 10x_4 &= 32 \\ 6x_1 + 9x_2 + 10x_3 + 8x_4 &= 33 \end{aligned}$$

Substitute the following sets of values for $\{x_1, x_2, x_3, x_4\}$.
(a) $\{14.6, 3.1, -2.5, -7.2\}$ (b) $\{2.36, 1.21, 0.65, 0.18\}$
and comment on the results.

Solution Making the substitutions, the left-hand sides of the four equations become, respectively,
(a) 23.1, 31.1, 31.9, 32.9 (b) 23.01, 31.01, 31.99, 32.99
In each case the equations nearly balance; in the second case, very nearly so. Yet the two sets of 'solutions' are very dissimilar from each other and from the exact solution.

The system of equations in Example 14.8 is said to be **ill-conditioned**. Had we obtained set (b) as our supposed solution, we would have checked it out by substituting into the equations and have been satisfied because the approximate balancing of left- and right-hand sides was almost exact.

KEY POINT

A system of equations is ill-conditioned if small changes in the coefficients cause unduly large changes in the solutions.

The following examples illustrate other aspects of ill-conditioning.

It is important to note that ill-conditioning is inherent in a system: it is a function of the system. All we can do is to use as accurate a method as possible to calculate the solution.

Worked examples

14.9 Solve the following equations by Gaussian elimination:

(a) $$\begin{bmatrix} 4 & 1 \\ 1 & 0.2499 \end{bmatrix} \begin{bmatrix} x \\ y \end{bmatrix} = \begin{bmatrix} 0.5 \\ 1.0 \end{bmatrix}$$

(b) $$\begin{bmatrix} 4 & 1 \\ 1 & 0.2501 \end{bmatrix} \begin{bmatrix} x \\ y \end{bmatrix} = \begin{bmatrix} 0.5 \\ 1.0 \end{bmatrix}$$

(c) $$\begin{bmatrix} 4 & 1 \\ 1 & 0.25 \end{bmatrix} \begin{bmatrix} x \\ y \end{bmatrix} = \begin{bmatrix} 0.5 \\ 0.125 \end{bmatrix}$$

(d) $$\begin{bmatrix} 4 & 1 \\ 1 & 0.25 \end{bmatrix} \begin{bmatrix} x \\ y \end{bmatrix} = \begin{bmatrix} 0.5 \\ 0.124 \end{bmatrix}$$

Solution (a) The augmented matrix is

$$\left[\begin{array}{cc|c} 4 & 1 & 0.5 \\ 1 & 0.2499 & 1.0 \end{array}\right] R2 - 0.25 \times R1$$

$$\left[\begin{array}{cc|c} 4 & 1 & 0.5 \\ 1 & -0.0001 & 0.875 \end{array}\right]$$

Hence $y = -8750$, $x = 2187.625$.

(b) $$\left[\begin{array}{cc|c} 4 & 1 & 0.5 \\ 1 & 0.2501 & 1.0 \end{array}\right] R2 - 0.25 \times R1$$

$$\left[\begin{array}{cc|c} 4 & 1 & 0.5 \\ 1 & 0.0001 & 0.875 \end{array}\right]$$

Hence $y = 8750$, $x = -2187.375$.

(c) $$\left[\begin{array}{cc|c} 4 & 1 & 0.5 \\ 1 & 0.25 & 0.125 \end{array}\right] R2 - 0.25 \times R1$$

$$\left[\begin{array}{cc|c} 4 & 1 & 0.5 \\ 0 & 0 & 0 \end{array}\right]$$

Hence there is no unique solution.
If, for example, $x_2 = \lambda$ then $x_1 = \frac{1}{4}(0.5 - \lambda)$.

(d) $$\left[\begin{array}{cc|c} 4 & 1 & 0.5 \\ 1 & 0.25 & 0.124 \end{array}\right] R2 - 0.25 \times R1$$

$$\left[\begin{array}{cc|c} 4 & 1 & 0.5 \\ 0 & 0 & -0.01 \end{array}\right]$$

The equations are inconsistent and there is no solution.

Notice how sensitive the solution is to relatively small changes in the coefficients and in the right-hand sides. This is a symptom of ill-conditioning.

14.10 Find the determinants of the matrices of Example 14.9(a) and (b). The determinant of the coefficients in Example 14.8 is -1. What does this tell us about the systems of linear equations in these cases?

Solution

$$\begin{vmatrix} 4 & 1 \\ 1 & 0.2499 \end{vmatrix} = -0.0004 \qquad \begin{vmatrix} 4 & 1 \\ 1 & 0.2501 \end{vmatrix} = 0.0004$$

We are given that

$$\begin{vmatrix} 5 & 5 & 6 & 7 \\ 5 & 10 & 9 & 7 \\ 7 & 7 & 8 & 10 \\ 6 & 9 & 10 & 8 \end{vmatrix} = -1$$

In each case the determinant is small in magnitude compared to the elements themselves. In the first two cases it is very small. This is an indication of an ill-conditioned system.

Figure 14.2 shows schematically the case of an ill-conditioned system of two equations in two unknowns. The lines which represent the equations are almost coincident. Any small change in the coefficients in either equation could move the point of intersection (i.e. the solution of the equation) by a large amount.

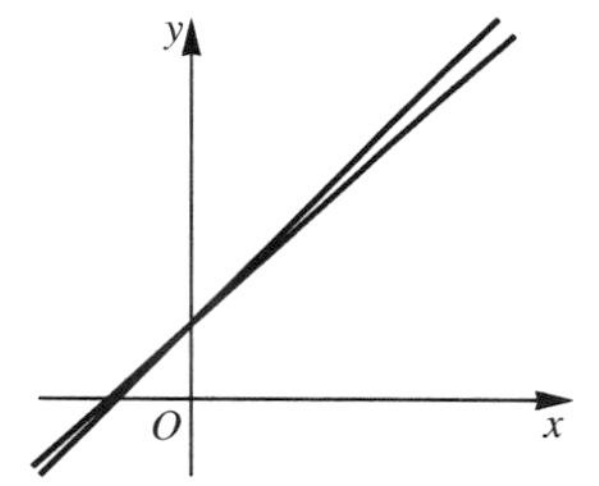

Figure 14.2. Ill-conditioning

Self-assessment questions 14.3

1. Explain the term ill-conditioning.
2. Why can you not check satisfactorily an approximate solution of an ill-conditioned system?

Exercise 14.3

1. Solve the following system of equations:

   $x_1 + 2x_2 = 10, \; 1.1x_1 + 2x_2 = 10.4$

   Now solve the following system:

   $x_1 + 2x_2 = 10, \; 1.05x_1 + 2x_2 = 10.4$

   Comment.

2. Check by substitution that $x_1 = 6.0$, $x_2 = -7.2$, $x_3 = 2.9$ and $x_4 = -0.1$ is apparently a 'good' solution to the following system of equations:

$$\begin{aligned} 10x_1 + 7x_2 + 8x_3 + 7x_4 &= 32 \\ 7x_1 + 5x_2 + 6x_3 + 5x_4 &= 23 \\ 8x_1 + 6x_2 + 10x_3 + 9x_4 &= 33 \\ 7x_1 + 5x_2 + 9x_3 + 10x_4 &= 31 \end{aligned}$$

Given that the exact solution is $x_1 = x_2 = x_3 = x_4 = 1$, comment on your calculations.

3. Find the exact solution of the following system of equations:

$$\begin{aligned} 55x_1 + 89x_2 &= 144 \\ 89x_1 + (144 + a)x_2 &= 233 \end{aligned}$$

Evaluate the solution when $a = 0.02$ and again when $a = 0.018$ and comment on your results. What happens when $a = 1/55$?

4. Working to 6 s.f. solve the following system of equations:

$$\begin{aligned} 200x + 4y + 2z &= 55 \\ 100x + 2y + 4z &= 5 \\ x + 2y + 100z &= 5 \end{aligned}$$

Now change the coefficient 200 to 201 and repeat the calculations. Comment on your results.

14.4 Improving the accuracy of the solution

The examples so far have been deliberately chosen so that exact arithmetic could be used. This will not always be so and round-off error will affect the accuracy of the solution. A method which attempts to reduce the errors propagating through the system as a result of the calculations being performed is that of **partial pivoting**. It ensures that the multiples of rows being added or subtracted are less than 1 by sometimes interchanging two rows.

Worked example

14.11 Use Gaussian elimination with partial pivoting to solve the equations whose augmented matrix is

$$\left[\begin{array}{ccc|c} 2 & 1 & -3 & 4 \\ 4 & -1 & 2 & 7 \\ 1 & 2 & -1 & 7 \end{array}\right]$$

Solution The largest-sized element in column 1 is 4 which is in row 2. We interchange rows 1 and 2 ($R1 \leftrightarrow R2$) to obtain

$$\left[\begin{array}{ccc|c} 4 & -1 & 2 & 7 \\ 2 & 1 & -3 & 4 \\ 1 & 2 & -1 & 7 \end{array}\right] \begin{array}{l} \\ R2 - \frac{2}{4} \times R1 \\ R3 - \frac{1}{4} \times R1 \end{array}$$

$$\left[\begin{array}{ccc|c} 4 & -1 & 2 & 7 \\ 0 & \frac{6}{4} & -\frac{16}{4} & \frac{2}{4} \\ 0 & \frac{9}{4} & -\frac{6}{4} & \frac{21}{4} \end{array}\right] R3 \leftrightarrow R2$$

$$\left[\begin{array}{ccc|c} 4 & -1 & 2 & 7 \\ 0 & \frac{9}{4} & -\frac{6}{4} & \frac{21}{4} \\ 0 & \frac{6}{4} & -\frac{16}{4} & \frac{2}{4} \end{array}\right] R3 - \tfrac{2}{3} R2$$

$$\left[\begin{array}{ccc|c} 4 & -1 & 2 & 7 \\ 0 & \frac{9}{4} & -\frac{6}{4} & \frac{21}{4} \\ 0 & 0 & -\frac{12}{4} & -\frac{12}{4} \end{array}\right]$$

Hence, by back substitution, $x_3 = 1$, $x_2 = 3$, $x_1 = 2$.

In this example the arithmetic is harder than in the straightforward method but it is meant only to illustrate the idea of partial pivoting.

A second method to improve accuracy can be carried out when some coefficients are much larger in size than the others. In essence we multiply each equation by a number so that the size of the largest coefficient is 1. This is called **scaling**.

A word of warning: scaling *does* lead to round-off error and therefore should be employed only if the system of equations contains coefficients of widely different magnitudes. Example 14.12 has been chosen to illustrate the ideas.

Worked example

14.12 Solve the following equations by Gaussian elimination: (a) without scaling, and (b) with scaling.
(Keep three significant figures in your calculations.)

$$\begin{aligned} 2x_1 + 10^6 x_2 &= 10^6 \\ x_1 + \quad x_2 &= 2 \end{aligned}$$

Solution (a)

$$\left[\begin{array}{cc|c} 2 & 1\,000\,000 & 1\,000\,000 \\ 1 & 1 & 2 \end{array}\right] R2 - \tfrac{1}{2} \times R1$$

$$\left[\begin{array}{cc|c} 2 & 1\,000\,000 & 1\,000\,000 \\ 0 & -500\,000 & -500\,000 \end{array}\right] \qquad \text{(3 s.f. retained, remember)}$$

Hence, by back substitution, $x_2 = 1$ and $x_1 = 0$.

(b) Scaling gives

$$\left[\begin{array}{cc|c} 0.000\,002 & 1 & 1 \\ 1 & 1 & 2 \end{array}\right]$$

Applying pivoting, we obtain

$$\left[\begin{array}{cc|c} 1 & 1 & 2 \\ 0.000\,002 & 1 & 1 \end{array}\right] \quad R2 - 0.000\,002 \times R1$$

$$\left[\begin{array}{cc|c} 1 & 1 & 2 \\ 0 & 1 & 1 \end{array}\right]$$

Hence $x_2 = 1$ and $x_1 = 1$.

The solution from (b) is much closer to the exact solution

$$x_1 = 1.000\,002,\ x_2 = 0.999\,998 \qquad (6 \text{ d.p.})$$

and agrees with it exactly to 3 s.f.

Self-assessment questions 14.4

1. What is meant by (a) partial pivoting, and (b) scaling?
2. What is the potential disadvantage of scaling?

Exercise 14.4

1. Use the method of partial pivoting to solve the following systems:
   (a) Exercise 14.1, Question 3(c)
   (b) Exercise 14.2, Question 1(a)
2. Solve the following equations by Gaussian elimination: (a) without scaling, and (b) with scaling. Keep 3 s.f. in your calculations.
   $$2x_1 + 10^5 x_2 = 10^5,\ x_1 + x_2 = 2$$
3. Evaluate the determinant of the left-hand coefficients of each of the following systems:
   (a) $3x_1 + 2x_2 = 18$
   $-x_1 + 2x_2 = 2$
   (b) $x_1 + 2x_2 = 10$
   $1.1x_1 + 2x_2 = 10.4$

   Now write down the scaled version of each system keeping 3 s.f. in your results. Evaluate the determinants of those scaled versions.

Test and assignment exercises 14

1. Using the inverse matrix find, where possible, the solutions of the following equations:

   (a) $\begin{bmatrix} 4 & 1 \\ 1 & -2 \end{bmatrix}\begin{bmatrix} x \\ y \end{bmatrix} = \begin{bmatrix} 2 \\ 5 \end{bmatrix}$

   (b) $\begin{bmatrix} 4 & 1 \\ 1 & -2 \end{bmatrix}\begin{bmatrix} x \\ y \end{bmatrix} = \begin{bmatrix} 0 \\ 0 \end{bmatrix}$

   (c) $\begin{bmatrix} 1 & -3 \\ -2 & 6 \end{bmatrix}\begin{bmatrix} x \\ y \end{bmatrix} = \begin{bmatrix} 0 \\ 0 \end{bmatrix}$

   (d) $\begin{bmatrix} 1 & -3 \\ -2 & 6 \end{bmatrix}\begin{bmatrix} x \\ y \end{bmatrix} = \begin{bmatrix} 5 \\ 2 \end{bmatrix}$

2. Find, where possible, the inverse of the matrix of coefficients and hence the solution of the equations in the following cases:

   (a) $2x_1 - 4x_2 - 3x_3 = 5$
   $x_1 - 7x_2 + 2x_3 = 33$
   $x_1 - x_2 + 2x_3 = 9$

   (b) $13x_1 + 5x_2 + 8x_3 = 0$
   $8x_1 + 4x_2 + 7x_3 = 2$
   $7x_1 + 3x_2 + 5x_3 = 1$

   (c) $13x_1 + 5x_2 + 8x_3 = 0$
   $8x_1 + 4x_2 + 7x_3 = 3$
   $7x_1 + 3x_2 + 5x_3 = 1$

3. Use Cramer's rule to find the solutions of the following systems:
   (a) Question 1(a)
   (b) Question 2(a)

4. Using Gaussian elimination solve the following equations:
   (a) $$\begin{aligned} 4i_1 + i_2 + 2i_3 &= 4 \\ 3i_1 - 3i_2 + i_3 &= 1 \\ i_1 - 2i_2 - 5i_3 &= 1.5 \end{aligned}$$
   (b) $$\begin{aligned} x_1 - 3x_2 + 6x_3 &= 3 \\ 2x_1 + 3x_2 + 4x_3 &= 9 \\ 6x_1 - 3x_2 + 8x_3 &= 12 \end{aligned}$$

5. Solve, as far as possible, the following systems of equations using Gaussian elimination:
   (a) $$\begin{aligned} 2x_1 - x_2 + 3x_3 &= 2 \\ 5x_1 - 3x_2 + 4x_3 &= 0 \\ 4x_1 - x_2 + 13x_3 &= 14 \end{aligned}$$
   (b) $$\begin{aligned} x_1 + x_2 - x_3 &= 1 \\ 3x_1 + x_2 + 2x_3 &= 2 \\ 9x_1 + 5x_2 + x_3 &= 5 \end{aligned}$$

6. Use Gaussian elimination to solve the system
   $$\begin{aligned} x_1 + x_2 + x_3 &= 0 \\ -2x_1 + 2x_2 + 3x_3 &= p \\ -x_1 + 3x_2 + qx_3 &= 2 \end{aligned}$$
   For which values of p and q do the equations
   (a) have no solution?
   (b) have an infinite number of solutions?
   (c) have a unique solution?

7. Solve the equations
   $$\begin{aligned} 0.5x_1 - x_2 &= -19 \\ 0.28x_1 - 0.5x_2 &= -9.44 \end{aligned}$$
   Now solve the equations
   $$\begin{aligned} 0.55x_1 - x_2 &= -19 \\ 0.28x_1 - 0.5x_2 &= -9.44 \end{aligned}$$
   (In each case work to 3 s.f.)
   What conclusion can be drawn?

8. Use Gaussian elimination with partial pivoting to solve the following systems:
   (a) Question 4(b)
   (b) Question 4(a)

15 Non-linear equations

Objectives

This chapter

- uses sketch graphs to locate approximately the root(s) of a non-linear equation
- uses Descartes' rules of signs to help determine the number and nature of the real roots of a polynomial equation
- develops and uses the methods of successive bisection and false position to reduce the intervals containing the roots of a non-linear equation
- uses the method of fixed-point iteration and states a criterion for convergence of the successive approximations
- derives and uses the Newton–Raphson and secant methods of point estimation

15.1 Approximate location of roots

In order to find the load P at which a slender column of length l will buckle, we need to solve the equation

$$\tan\sqrt{\frac{P}{k}}\,l = \sqrt{\frac{P}{k}}\,l$$

where k is a constant.

Even if we simplify the equation by writing

$$\frac{P}{k}\,l^2 = x^2$$

then we have still to solve the **non-linear equation**

$$\tan x = x \qquad 15.1$$

To locate approximately the roots of a non-linear equation we can first draw *suitable* graphs.

Worked examples

15.1 What conclusions can you draw about the roots of the equation $f(x) = 0$ for the following continuous functions $f(x)$?

(a)

x	0	1	2
$f(x)$	1	-2	3

(b)

x	0	1	2
$f(x)$	2	1	3

Solution (a) The function changes sign between $x = 0$ and $x = 1$: there is at least one value of x in $0 < x < 1$ for which $f(x) = 0$. The function also changes sign between $x = 1$ and $x = 2$. There is at least one root of $f(x) = 0$ in $1 < x < 2$.

(b) Since there are no changes of sign in the tabulated values of $f(x)$ there may not be any roots of the equation $f(x) = 0$ between $x = 0$ and $x = 2$. It is, however, possible that the graph of $f(x)$ crosses the x-axis an even number of times between $x = 0$ and $x = 1$ and again between $x = 1$ and $x = 2$.

These conclusions are the consequence of the **intermediate value theorem** which states that a function which has a continuous graph in $a \leqslant x \leqslant b$ takes all values between $f(a)$ and $f(b)$ at least once in $a < x < b$.

15.2 By drawing suitable graphs determine approximately the location of the roots of the following equations:

(a) $x^3 - x + 1 = 0$ (b) $x^4 + x - 1 = 0$ (c) $x^4 - x + 1 = 0$

Solution (a) Figure 15.1(a) shows the graphs of $y = x^3$ and $y = x - 1$ drawn on the same axes. Where these graphs intersect, $x^3 = x - 1$ so that $x^3 - x + 1 = 0$. There appears to be only one root and the following tabulation helps **bracket the root**:

x	-2	-1
$f(x)$	-5	1

There is a root in the interval $(-2, -1)$.

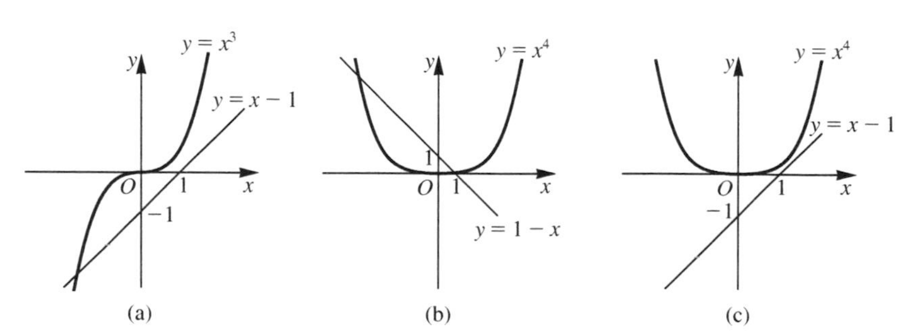

Figure 15.1.
Graphs for Example 15.2

(b) Figure 15.1(b) shows that the graphs of $y = x^4$ and $y = 1 - x$ intersect twice, and therefore the equation $x^4 + x - 1 = 0$ has one positive and one negative root.

The following tabulation shows that the negative root occurs in the interval $(-2,-1)$ and the positive root occurs in the interval $(0,1)$.

x	-2	-1	0	1
$f(x)$	13	-1	-1	1

(c) From Figure 15.1(c) we see that the equation $x^4 = x - 1$ or $x^4 - x + 1 = 0$ has no roots.

15.3 What conclusions can be drawn about the roots of the following equations?

(a) $\tan x = x$ (b) $e^x \sin x = 1$

Solution (a) Figure 15.2 shows the graphs of $y = x$ and $y = \tan x$. Note first that $\tan 0 = 0$ so that $x = 0$. Further, if x^* is a root, $\tan x^* = x^*$, then $\tan(-x^*) = -\tan x^* = -x^*$ so that $-x^*$ is also a root. Each branch of the curve $y = \tan x$ is cut by the line $y = x$ and as we move to the right the intersections occur higher up each branch, that is, closer to where the curve breaks up. Hence the roots are located at $x \simeq 3\pi/2$, $5\pi/2$, $7\pi/2$ and so on, and bearing in mind the earlier result, at $x \simeq -3\pi/2$, $-5\pi/2$, $-7\pi/2$ and so on. The approximations improve as we move further away from the origin. The approximations are called **point estimates**.

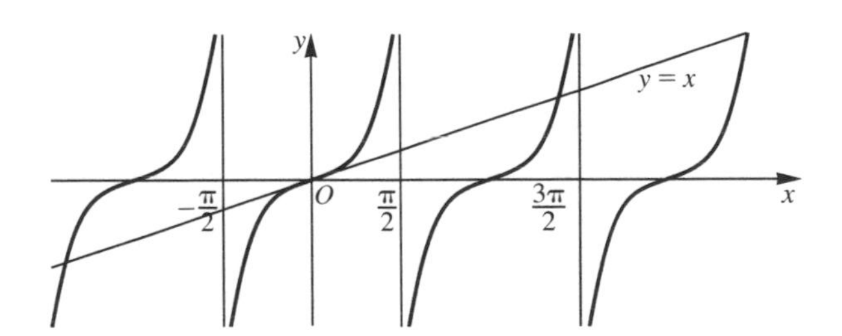

Figure 15.2. $y = x$ and $y = \tan x$

(b) The equation can be rewritten $\sin x = e^{-x}$. Figure 15.3 shows the graphs of $y = \sin x$ and $y = e^{-x}$.

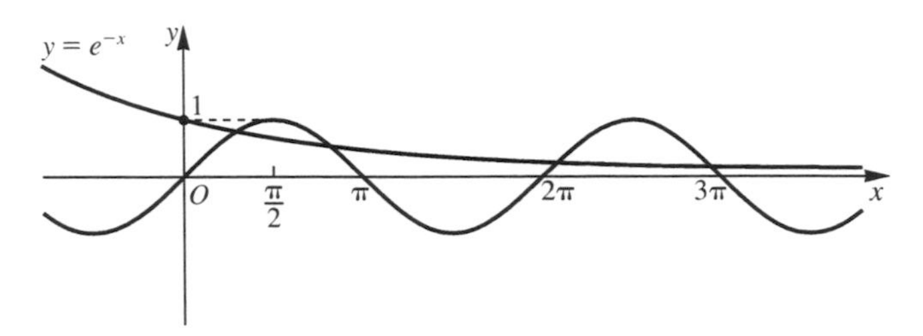

Figure 15.3. $y = \sin x$ and $y = e^{-x}$

It is clear that there are no negative roots. There is one root between $x = 0$ and $x = \pi/2$ and an infinite number of roots whose values are given approximately by $x = n\pi$ where $n = 1,2,3,\ldots$. The approximation improves as n increases.

15.4 Consider the equations (a) $x^{10} - 0$, and (b) $x^{\frac{1}{10}} - 0$. Calculate the left-hand sides for $x = 0.99, 0.999, 1.001, 1.01$ and comment.

Solution

	x	0.99	0.999	1.001	1.01
(a)	$x^{10} - 1$	−0.0956	−0.0100	0.0100	0.1046
(b)	$x^{\frac{1}{10}} - 1$	−0.0010	−0.0001	0.0001	0.0010

Figure 15.4 shows sketch graphs of the two functions for $x \geqslant 0$.

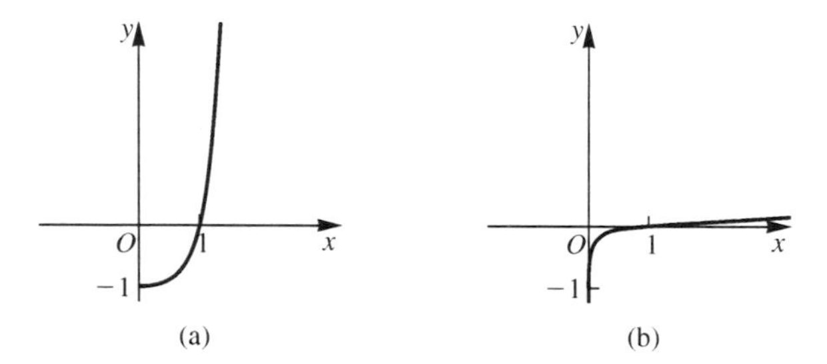

Figure 15.4. Graphs of the functions in Example 15.4

The function $x^{10} - 1$ rises rapidly as it crosses the x-axis (its gradient at that point is 10) whereas the function $x^{\frac{1}{10}} - 1$ rises slowly through the x-axis (its gradient there is 0.1). The absolute value of the second function is smaller when x is in error by ± 0.01 than that of the first function when x is in error by only ± 0.001.

A small value of $f(x)$ does not therefore guarantee an accurate estimate of a root.

If the equation is a polynomial equation then a second method is also possible; it employs the signs of the coefficients of the polynomial.

The equation is written as $p_n(x) = 0$ where $p_n(x)$ is a polynomial of degree n with real coefficients. The first thing to note is that there are at most n real roots and that any complex roots occur in conjugate pairs (see Section 16.1). Hence for a polynomial of degree n the number of real roots will be

1. either n or $n - 2$ or $n - 4$ ... or 0 if n is even,
2. either n or $n - 2$ or $n - 4$ ... or 1 if n is odd.

Descartes' rules of signs give full or partial information about the number of roots of the equation $p_n(x) = 0$. The method is as follows:

1. The coefficients of $p_n(x)$ are written down. The number of changes of sign is the maximum number of positive roots. Other options are found by subtracting 2 until we reach the value 0 or 1.
2. The coefficients of $p_n(-x)$ are written down. The number of changes of sign is the maximum number of negative roots. Other options are found by subtracting 2 until we reach 0 or 1.

Worked example

15.5 Use Descartes' rules of signs to draw conclusions about the roots of the following equations:

(a) $x^3 - x - 1 = 0$ (b) $x^3 + x + 1 = 0$ (c) $x^3 + x - 1 = 0$
(d) $x^3 - x + 1 = 0$ (e) $x^4 + x + 1 = 0$

Solution (a) $p_n(x) = x^3 - x - 1$. The coefficients are 1,−1,−1. There is one sign change and hence one positive root.

$p_n(-x) = -x^3 + x - 1$. The coefficients are −1,1,−1. There are two sign changes and hence two negative roots or no negative roots.

(b) $p_n(x) = x^3 + x + 1$. The coefficients are 1,1,1. No sign changes mean no positive roots.

$p_n(-x) = -x^3 - x + 1$. The coefficients are −1, −1,1. One sign change means one negative root.

(c) $p_n(x) = x^3 + x - 1$. There is one sign change and hence one positive root.

$p_n(-x) = -x^3 - x - 1$. No sign changes mean no negative roots.

(d) $p_n(x) = x^3 - x + 1$. Two sign changes mean either two positive roots or no positive roots.

$p_n(-x) = -x^3 + x + 1$. One sign change means one negative root.

(e) $p_n(x) = x^4 + x + 1$. No sign change means no positive root.

$p_n(-x) = x^4 - x + 1$. Two sign changes mean either two negative roots or no negative roots.

Note that in Example 15.5(b) and (c) we can be certain about the roots. In the other parts there is an element of uncertainty. However, we can make some progress in these other parts. For instance, if we have located one negative root in part (e) we know that there must be a second negative root.

Self-assessment questions 15.1

1. Under what conditions on $f(a)$ and $f(b)$ can you *guarantee* that the equation $f(x) = 0$ has a root in the interval $a < x < b$?
2. When can Descartes' rules of signs give exact information about the number and nature of the real roots of a polynomial equation?

Exercise 15.1

1. What conclusions can be drawn about the real roots of the equation $f(x) = 0$ for the continuous functions tabulated opposite?

(a)

x	0	1	2	3
$f(x)$	2	−3	4	−1

(b)

x	0	1	2	3	4
$f(x)$	2	−1	−2	3	4

2. Use suitable sketch graphs to obtain information as to the number, nature and approximate location of the roots of the following equations. If you have a graphics package available you can use it to verify your sketches.

(a) $x^3 + 3x^2 + 1 = 0$

(b) $x^3 - 3x^2 - 1 = 0$

(c) $100 - x - \frac{1}{x} = 0$

(d) $6 - x - \frac{10}{x} = 0$

(e) $\sin x - e^{-3x} = 0$

(f) $x - 2 - \sin x = 0$

3. Find as far as possible the number and nature of the roots of the following equations:

(a) $2x^3 - 3x^2 + 5x - 1 = 0$

(b) $2x^3 + 4x^2 - 3x - 5 = 0$

(c) $4x^3 - 7x^2 - 6x - 2 = 0$

(d) $x^4 - 6x^3 - 6x^2 + 5x + 4 = 0$

4. Consider the equations

(a) $x^8 - 1 = 0$, and

(b) $x^{\frac{1}{8}} - 1 = 0$.

Calculate the left-hand sides for $x = 0.995$, 0.9995, 1.0005 and 1.005, and comment.

15.2 Interval reduction

We assume that by one or more of the methods of Section 15.1 we have located a root of $f(x) = 0$ in an interval (x_L, x_R). **Interval reduction** is a general process by which we reduce the width of this interval progressively, still keeping the root inside each of the narrower intervals.

The method of **successive bisection** proceeds as follows:

1. Check that $f(x_L) \times f(x_R) < 0$. This ensures that there is a root in (x_L, x_R).

2. Evaluate $x_M = \frac{1}{2}(x_L + x_R)$ and $f(x_M)$. If $f(x_M) = 0$ then x_M is a root of $f(x) = 0$.

3. Check the sign of $f(x_L) \times f(x_M)$.

   If this is negative then replace x_R by x_M.

   If this is positive then replace x_L by x_M.

4. Continue with steps 2 and 3 until $x_R - x_L$ is sufficiently small.

Such a process is known as an **iterative** process and the successive values of x_M are called **iterations**. Note that each time step 3 is applied the interval is halved in length.

Worked examples

15.6 How small should $x_R - x_L$ be in order to estimate the root correct to
(a) 2 d.p.? (b) 3 d.p.? (c) n d.p.?

Solution
(a) If the root is to be estimated to 2 d.p. then the error is at most ± 0.005. Hence if we continue until we obtain $x_R - x_L < 0.01$ and then calculate x_M, this value is accurate to 2 d.p.
(b) A similar argument requires $x_R - x_L < 0.001$.
(c) The general result is $x_R - x_L < 10^{-n}$.

15.7 Find the positive root of the equation $x^4 + x - 1 = 0$ correct to 2 d.p.

Solution From Example 15.2(b) we know that the root lies in the interval (0,1). Hence we write $x_L = 0$, $x_R = 1$. Then $f(x_L) = -1$, $f(x_R) = 1$. We calculate $x_M = \frac{1}{2}(0 + 1) = 0.5$ and $f(x_M) = -0.4375$. Then $f(x_L) \times f(x_M) > 0$ so we put $x_L = 0.5$ and the interval is reduced to (0.5,1). Now we calculate $x_M = \frac{1}{2}(0.5 + 1) = 0.75$ and $f(x_M) = 0.06640625$. Hence $f(x_L) \times f(x_M) < 0$ so we put $x_R = 0.75$ and the interval is reduced to (0.5,0.75). These calculations and subsequent ones are shown in Table 15.1.

Hence the root to 2 d.p. is 0.73 (since x_M and x_R are both 0.73 to 2 d.p.).

Table 15.1.

x_L	x_R	x_M	$f(x_L)$	$f(x_R)$	$f(x_M)$
0	1	0.5	−1	1	−0.437 5
0.5	1	0.75	−0.437 5	1	0.066 406 25
0.5	0.75	0.625	−0.437 5	0.066 406 25	−0.222 412 109
0.625	0.75	0.687 5	−0.222 412 109	0.066 406 25	−0.089 096 069
0.687 5	0.75	0.718 75	−0.089 096 069	0.066 406 25	−0.014 372 825
0.718 75	0.75	0.734 375	−0.014 372 825	0.066 406 25	0.025 226 652
0.718 75	0.734 375	0.726 562 5	−0.014 372 825	0.025 226 652	0.005 233 589 5

The method is very slow. A variation which aims to improve matters by reducing the number of iterations is the method of **false position** (formerly known as ***regula falsi***). It uses a more sophisticated means of calculating the **iterate** x_M based on graphical intuition. Figure 15.5 indicates the idea. The values $f(x_L)$ and $f(x_R)$ are opposite in sign. The straight line cuts the x-axis at x_M which can be obtained using similar triangles.

KEY POINT

$$x_M = x_R - \frac{f(x_R)(x_L - x_R)}{f(x_L) - f(x_R)} \qquad 15.2$$

The false position method is essentially the same as that of successive bisection but with x_M calculated by (15.2).

You may find it helpful when reading the examples or doing the exercises to use a graphics package to sketch the function and mark the successive iterates.

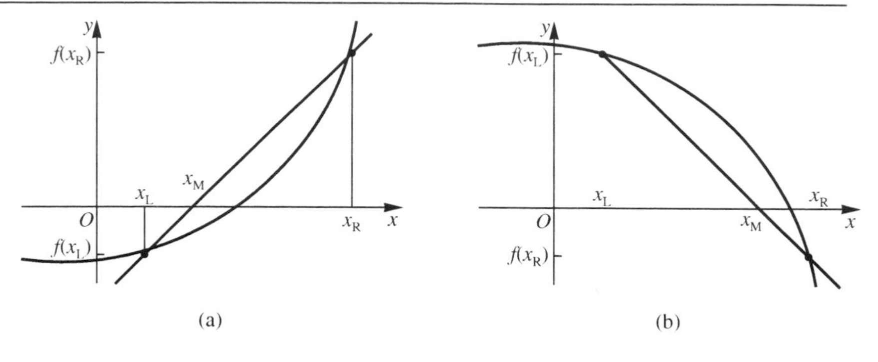

Figure 15.5. The method of false position

Worked examples

15.8 Using similar triangles show that

$$x_M = \frac{x_L f(x_R) - x_R f(x_L)}{f(x_R) - f(x_L)}$$

Hence deduce formula (15.2).

Solution In Figure 15.5(a), using similar triangles we find that

$$\frac{x_M - x_L}{f(x_L)} = -\left(\frac{x_R - x_M}{f(x_R)}\right)$$

The $-$ sign is there because $x_M - x_L$ and $x_R - x_M$ are both positive and $f(x_L), f(x_R)$ are of opposite signs. Hence

$$f(x_R)(x_M - x_L) = -f(x_L)(x_R - x_M)$$

or

$$x_M(f(x_R) - f(x_L)) = x_L f(x_R) - x_R f(x_L)$$

Dividing by $f(x_R) - f(x_L)$ we obtain

$$x_M = \frac{x_L f(x_R) - x_R f(x_L)}{f(x_R) - f(x_L)} = \frac{x_R f(x_L) - x_L f(x_R)}{f(x_L) - f(x_R)}$$

$$= \frac{x_R f(x_L) - x_R f(x_R) + x_R f(x_R) - x_L f(x_R)}{f(x_L) - f(x_R)}$$

$$= x_R + \frac{f(x_R)(x_R - x_L)}{f(x_L) - f(x_R)}$$

$$= x_R - \frac{f(x_R)(x_L - x_R)}{f(x_L) - f(x_R)}$$

15.9 Use the method of false position on the equation $x - \cos x = 0$ to obtain an approximation to the root to 2 d.p. Start with $x_L = 0.5$, $x_R = 1$. Compare with the bisection method.

Solution If $f(x) = x - \cos x$ note that $f(x_L) < 0$ and $f(x_R) > 0$. In fact $f(x_L) = -0.377\,582\,6$ and $f(x_R) = 0.459\,697\,7$. Then

$$x_M = 1 - \frac{0.459\,697\,7(0.5 - 1)}{-0.377\,582\,6 - 0.459\,697\,7} = 0.725\,482\,16$$

and $f(x_M) = -0.022\,698\,4$. Therefore $x_L = 0.725\,482\,6$.

The method is continued in Table 15.2. Results are recorded to 5 d.p. Hence the root is 0.74 to 2 d.p.

Table 15.3 summarizes the steps of the bisection method on the same equation. Again the root is found to be 0.74 to 2 d.p.

Table 15.2.

x_L	x_R	x_M	$f(x_L)$	$f(x_R)$	$f(x_M)$
0.5	1	0.725 48	−0.377 58	0.459 70	−0.022 70
0.725 48	1	0.738 40	−0.022 70	0.459 70	−0.001 47
0.738 40	1	0.739 05	−0.001 47	0.459 70	−0.000 06
0.739 05	1	0.739 08	−0.000 06	0.459 70	−0.000 00

Table 15.3.

x_L	x_R	x_M	$f(x_L)$	$f(x_R)$	$f(x_M)$
0.5	1	0.75	−0.377 58	0.459 70	0.018 31
0.5	0.75	0.625	−0.377 58	0.018 31	−0.185 96
0.625	0.75	0.687 5	−0.185 96	0.018 31	−0.085 33
0.687 5	0.75	0.718 75	−0.085 33	0.018 31	−0.033 88
0.718 75	0.75	0.734 38	−0.033 88	0.018 31	−0.007 87
0.734 38	0.75	0.742 19	−0.007 87	0.018 31	−0.005 20
0.734 38	0.742 19	0.738 28	−0.007 87	0.005 20	−0.001 35
0.738 28	0.742 19	0.740 23	−0.001 35	0.005 20	−0.001 92

The method of false position gets close to the correct value (0.739 085 to 6 d.p.) very quickly, but the value of x_R is stubbornly constant. Were we to continue the process we should find that x_L gradually makes its way towards the correct value from below. The method of successive bisection is much slower but x_R *does* change.

For some equations the method of false position obtains approximations with fewer iterations than that of successive bisection. However, it must be remembered that the calculation of each x_M requires much more effort for false position.

Self-assessment questions 15.2

1. Outline the method of successive bisection.
2. Roughly how many extra steps of successive bisection would it take to improve the accuracy of an approximation to a root from 3 d.p. to 4 d.p.?
3. How does the method of false position differ from that of successive bisection?

Exercise 15.2

1. For each of the following equations verify that there is a root between the two values of x given in parentheses following the equation. Take three steps of the method of successive bisection to improve the estimation of the location of the root.

   (a) $0.874x^2 - 1.75x - 2.627 = 0$ (2.8, 3.2)
   (b) $0.111x^3 - 0.65x^2 + 1.035x - 0.35 = 0$ (0.4, 0.6)
   (c) $0.658x^5 - 8.68x^4 + 41.6x^3 - 88.09x^2 + 79.35x - 23.335 = 0$ (4.5, 5)
   (d) $1.56 - 3.6605x + 2.88275x^2 - 0.617295x^3 = 0$ (2.8, 3.0)
   (e) $\tan x = 1.1x$ (0.4, 0.6)
   (f) $\ln x^2 = 1.2$ (1, 2)

2. Repeat Question 1 using the method of false position.

15.3 Fixed-point iteration

An alternative approach to interval reduction is that of refining a point estimate. First, we make a definition.

The value $x = a$ is a **fixed point** of the function $f(x)$ if $f(a) = a$. The method of **fixed-point iteration** (or **basic iteration**) requires that the equation whose roots we seek be written in the form $x = F(x)$.

An initial estimate x_0 is made. Then, successively, we calculate $x_1 = F(x_0)$, $x_2 = F(x_1)$, $x_3 = F(x_2)$ and so on. Under certain conditions the sequence $\{x_0, x_1, x_2, x_3, \ldots\}$ will converge to a root of the equation.

Again, sensible use of a graphics package can help make clear how the process operates.

Worked examples

15.10 Find the root of the equation $x - \cos x = 0$ correct to 2 d.p. using fixed-point iteration.

Solution First, we write the equation as $x = \cos x$. Figure 15.6 shows the graphs of $y = x$ and $y = \cos x$.

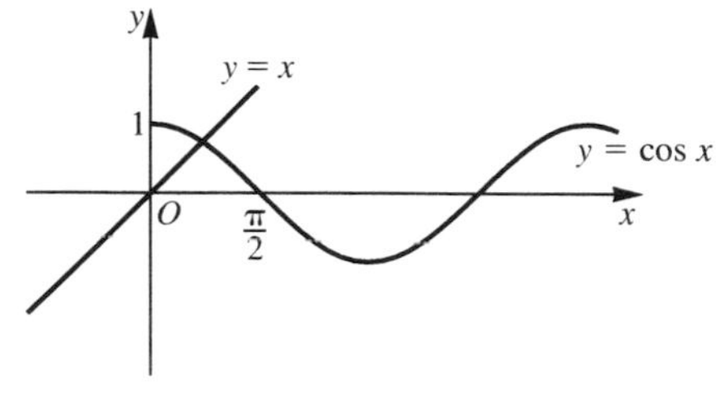

Figure 15.6. $y = x$ and $y = \cos x$

We take as a first estimate $x_0 = 0.78 \simeq \pi/4$. Then $x_1 = \cos x_0$, $x_2 = \cos x_1$ and so on. The first 10 approximations are shown in Table 15.4 where results are recorded to 4 d.p.

Notice that the estimates converge on the correct value but the estimates are alternately above and below it. This is because $F'(x) = -\sin x$ and $F'(x)$ is negative in the neighbourhood of the root.

Table 15.4.

n	x_n
0	0.7800
1	0.7071
2	0.7602
3	0.7247
4	0.7487
5	0.7326
6	0.7435
7	0.7361
8	0.7411
9	0.7377
10	0.7400

15.11 Use fixed-point iteration to find the roots of the equation $x^2 - 4x + 3 = 0$.

Solution First, we rearrange the equation to $x = \frac{1}{4}(x^2 + 3)$. We know, of course, that the equation has roots $x = 1$ and $x = 3$. We try in turn $x_0 = 0.9$, 1.1, 2.9 and 3.1. Table 15.5 shows the first five iterations with results recorded to 4 d.p.

Notice that the approximations $x_0 = 0.9$ and $x_0 = 1.1$ lead to convergence to the root $x = 1$ but we are not as successful in our attempts to locate the root $x = 3$. Table 15.6 shows the result of using the rearrangement $x = (4x - 3)^{\frac{1}{2}}$. Iterations are recorded to 4 d.p.

This time the root $x = 3$ is ever more closely approximated but the root $x = 1$ appears unreachable.

Table 15.5.

n	x_n	x_n	x_n	x_n
0	0.9	1.1	2.9	3.1
1	0.9525	1.0525	2.8525	3.1525
2	0.9768	1.0269	2.7842	3.2346
3	0.988 5	1.0137	2.6879	3.3656
4	0.9943	1.0069	2.5562	3.5818
5	0.9972	1.0034	2.3836	3.9574

Table 15.6.

n	x_n	x_n	x_n	x_n
0	0.9	1.1	2.9	3.1
1	0.7746	1.1832	2.9326	3.0659
2	0.3137	1.3164	2.9547	3.0436
3	cannot take the	1.5052	2.9697	3.0290
4	square root of a	1.7380	2.9797	3.0192
5	negative number	1.9880	2.9864	3.0128

KEY POINT

A **sufficient condition** for the iterative scheme $x_{n+1} = F(x_n)$ to converge to a root of the equation $x = F(x)$ is

$$|F'(x)| < 1 \qquad \text{near} \qquad x = a \qquad (15.3)$$

Note that if the condition is not satisfied then we do not specify whether or not the scheme will converge.

Worked example

15.12 The real roots of the equation $x^4 + x^2 = 90$ are $x = \pm 3$. Three possible rearrangements are shown together with the 'scale factors' $F'(x)$. Predict the behaviour of the corresponding iterative schemes and carry out three iterations with each to find the positive root, starting with $x_0 = 3.1$. How would you find the negative root?

(a) $x = \sqrt[4]{90 - x^2}, \qquad F'(x) = \dfrac{-x}{2(90 - x^2)^{\frac{3}{4}}}$

(b) $x = \sqrt{90 - x^4}, \qquad F'(x) = \dfrac{-2x^3}{\sqrt{90 - x^4}}$

(c) $x = \sqrt{90/(1 + x^2)}, \qquad F'(x) = \dfrac{-\sqrt{90}x}{(1 + x^2)^{\frac{3}{2}}}$

Solution (a) Near $x = 3$, $\quad F'(x) = \dfrac{-3}{2(90 - 9)^{\frac{3}{4}}} = -\dfrac{1}{18}$

(b) Near $x = 3$, $\quad F'(x) = \dfrac{-2 \times 27}{\sqrt{9}} = -18$

(c) Near $x = 3$, $\quad F'(x) = \dfrac{-\sqrt{90} \times 3}{10\sqrt{10}} = -\dfrac{9}{10}$

In case (b), $|F'(x)| = 18 > 1$ and we cannot be sure whether the scheme will converge.

In cases (a) and (c), $|F'(x)| < 1$ so that convergence is guaranteed. The fact that $F'(x) < 0$ implies that the successive iterates are above and below the root.

We would expect (a) to converge more quickly, since the magnitude of $F'(x)$ is smaller than in case (c).

(Note that $(x_{n+1} - a) \simeq F'(a)(x_n - a)$ and if $|F'(a)| < 1$ then the approximation improves as $x_n \to a$.)

Table 15.7 shows the attempts at iteration with the three schemes. Our suggestions are borne out by these results.

Note that in case (a)

$$\frac{x_3 - 3}{x_2 - 3} = -\frac{0.000\,2}{0.000\,31} \simeq -\frac{1}{18}$$

and in case (b)

$$\frac{x_3 - 3}{x_2 - 3} = -\frac{0.071\,06}{0.808\,76} \simeq -0.9$$

Table 15.7.

	a	*b*	*c*
x_0	3.1	3.1	3.1
x_1	2.994 34	cannot take the	2.912 48
x_2	3.000 31	square root of a	3.080 76
x_3	2.999 98	negative number	2.928 94

Self-assessment questions 15.3

1. Explain how to set up a non-linear equation for the application of the fixed-point iteration method.
2. Explain carefully the relevance of the criterion $|F'(x)| < 1$ in relation to convergence of the fixed-point iteration method.

Exercise 15.3

1. Find, correct to 2 d.p., the roots of the following equations using the method of fixed-point iteration:

   (a) $xe^x = 1$

   (b) $x = 2\sin x$

2. Rearrange the equation $x - x^2 = 0$ in two ways in the form $x = F(x)$. Use initial guesses $x_0 = 0.2$, 0.8 and 1.2 in each case. After three iterations in each instance draw conclusions.

3. Use fixed-point iteration to obtain the solution of the equation $xe^x = 4$ correct to 2 d.p. Solve the equations $xe^x = 2$; $xe^x = 10$ to the same accuracy.

4. Show that $x = (x^2 + 2)/(2x + 1)$ is a rearrangement of $x^2 + x - 2 = 0$. Find algebraically the exact roots of the equation. Use three steps of fixed-point iteration with $x_0 = -3, -1, 0$ and 2 in turn. Find the scale factor for the rearrangement given.

15.4 Further methods of point estimates

Once again, the use of a graphics package to sketch suitable functions for the examples and exercises can be a help.

The **Newton–Raphson method** is derived from a geometric approach. Refer to Figure 15.7.

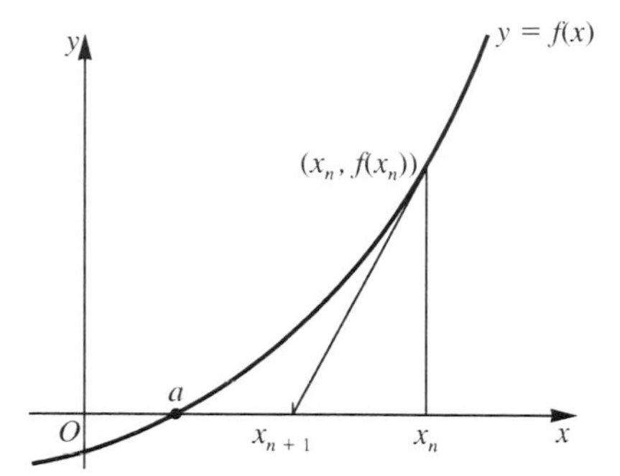

Figure 15.7. Derivation of the Newton–Raphson formula

Let x_n be an approximation to the root a of the equation $f(x) = 0$. Draw the tangent to the curve $y = f(x)$ at the point $(x_n, f(x_n))$. Where it crosses the x-axis is the next approximation, x_{n+1}.

The gradient of this tangent is $f'(x_n)$ and its equation is

$$y - f(x_n) = f'(x_n)(x - x_n)$$

This crosses the x-axis where $y = 0$ and $x = x_{n+1}$. Then the Newton–Raphson formula is

KEY POINT

$$x_{n+1} = x_n - \frac{f(x_n)}{f'(x_n)} \qquad 15.4$$

The problem with the method is that when it works, it works very well, but it can fail badly, especially if there is a point of inflection near the root or if there is a point where $f'(x)$ is very small (for example, a local maximum or minimum).

Worked example

15.13 Find the Newton–Raphson formula for the equation $x - \cos x = 0$. Show that the formula could be used for fixed-point iteration. Find the root correct to 5 d.p.

Solution $f(x) = x - \cos x$ and $f'(x) = 1 + \sin x$. The required formula is

$$x_{n+1} = x_n - \frac{x_n - \cos x_n}{1 + \sin x_n}$$

This can be rearranged as

$$x_{n+1} = \frac{x_n \sin x_n + \cos x_n}{1 + \sin x_n} = F(x_n)$$

(Note that the equation

$$x = \frac{x \sin x + \cos x}{1 + \sin x}$$

becomes

$$x + x \sin x = x \sin x + \cos x$$

(that is, $x = \cos x$ or $x - \cos x = 0$.)

Starting with $x_0 = 0.78$ as in Example 15.10, succeeding iterations are shown in Table 15.8. After three iterations we have obtained the root correct to 5 d.p.

Table 15.8.

n	0	1	2	3
x_n	0.78	0.739 439	0.739 085	0.739 085

A method which does not require the derivative of $f(x)$ is the **secant method**. It is illustrated geometrically in Figure 15.8.

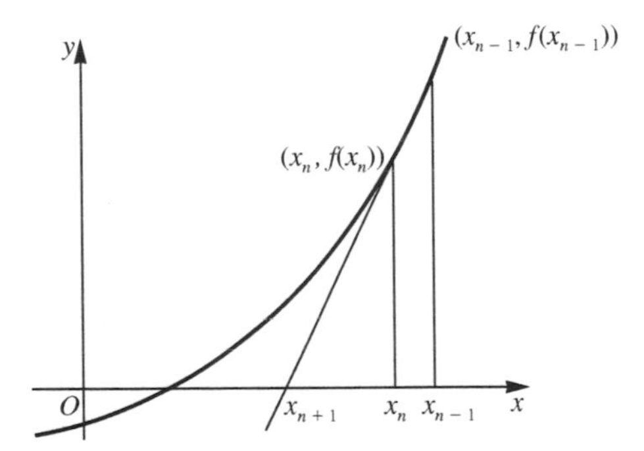

Figure 15.8.
Derivation of the secant formula

Suppose that we have two successive iterations x_{n-1} and x_n. We draw the **secant** to the curve $y = f(x)$ through the points $(x_{n-1}, f(x_{n-1}))$ and $(x_n, f(x_n))$; where this cuts the x-axis is x_{n+1}.

KEY POINT

The secant formula is

$$x_{n+1} = x_n - \frac{f(x_n)(x_{n-1} - x_n)}{f(x_{n-1}) - f(x_n)} \qquad 15.5$$

Notice that we need *two* starting values, x_0 and x_1. This formula is similar to that for false position (15.2). However, the secant method does not guarantee that x_n and x_{n+1} lie on either side of the root and the method may diverge. It is tedious to calculate each step and the method is best suited to a computer-based evaluation.

Worked example

15.14 Starting with $x_0 = 0.5$, $x_1 = 1$, use the secant formula twice to obtain an approximation to the root of the equation $x - \cos x = 0$.

Solution

$$x_2 = x_1 - \frac{(x_1 - \cos x_1)(x_0 - x_1)}{(x_0 - \cos x_0) - (x_1 - \cos x_1)}$$

$$= 1 - \frac{(1 - \cos 1)(0.5 - 1)}{(0.5 - \cos 0.5) - (1 - \cos 1)} = 0.725\,48$$

$$x_3 = x_2 - \frac{(x_2 - \cos x_2)(x_1 - x_2)}{(x_1 - \cos x_1) - (x_2 - \cos x_2)}$$

$$= 0.725\,48 - \frac{(0.725\,48 - \cos 0.725\,48)(1 - 0.725\,48)}{(1 - \cos 1) - (0.725\,48 - \cos 0.725\,48)}$$

$$= 0.738\,40$$

Self-assessment questions 15.4

1. How is the Newton–Raphson formula derived?
2. When might it fail to converge?
3. How does the secant method differ from the Newton–Raphson method?

Exercise 15.4

1. Use the Newton–Raphson method on the equations of Exercise 15.3, Question 1 to obtain the roots correct to 4 d.p.

2. Use the Newton–Raphson method on the equations of Exercise 15.1, Question 2 to obtain the roots correct to 3 d.p. In Question 2(e) find the smallest positive root.

3. Repeat Question 1 using the secant method.

4. A sphere of density 0.4 and radius a floats on water. When it is submerged to a depth $h < a$ the volume submerged is $\frac{\pi}{3}h^2(3a - h)$. Find a suitable approximation to h/a and use the Newton–Raphson method to refine the approximation to 2 d.p.

Test and assignment exercises 15

1. Use suitable sketch graphs to find approximate locations of the roots of the following equations:

   (a) $x^3 + x + 4 = 0$ (b) $x^3 - x + 4 = 0$
   (c) $x^3 - x - 4 = 0$ (d) $x^3 + x - 4 = 0$

2. Find, as far as possible from Descartes' rules of signs, information about the number and nature of the roots of the equation of Exercise 1.

3. Use successive bisection to obtain to 2 d.p. the positive roots of the following equations:

   (a) $x^3 + x + 4 = 0$ (b) $x^3 - x + 4 = 0$
   (c) $x^3 - x - 4 = 0$ (d) $x^3 + x - 4 = 0$

   Use information obtained in Exercise 1.

4. Repeat Exercise 3 using the method of false position.

5. We wish to find the first positive root of the equation $x - \tan x = 0$. Find two rearrangements and check the scale factor of each. Then apply the fixed-point iteration method three times in each case, starting from a suitable initial approximation.

6. Obtain via fixed-point iteration the roots of the following equations correct to 2 d.p.

   (a) $x^2 e^x = 1$
   (b) $x^2 = \cos x$

7. Repeat Exercise 6 using the Newton–Raphson method.

8. Repeat Exercise 6 using the secant method.

9. The displacement $x(t)$ of the centre of mass of a car at time t is given by

$$x = 0.3e^{-nt}\left(\cos pt + \frac{n}{p}\sin pt\right)$$

   where $n = 5.8333$ and $p = 31.743$. Using the methods of bisection, false position and secant, obtain approximations to the values of t at which $x = 0$ on the first three occasions. Compare the results of the three methods.

16 Complex numbers

Objectives

This chapter

- defines a complex number
- shows how to represent complex numbers graphically
- expresses complex numbers in Cartesian, polar and exponential forms
- explains how to add, subtract, multiply and divide two complex numbers
- uses complex numbers to define loci and regions in the plane
- states De Moivre's theorem and uses it to find roots of complex numbers

16.1 Complex numbers and their representation

If a mass suspended from a vertical spring is in equilibrium and is then displaced downwards and released (Figure 16.1a), the resulting motion of the mass is in the form of oscillations with decaying amplitude (Figure 16.1b). If x measures the displacement (downwards positive) and t the time elapsed from the moment of release then the relationship between them is of the form $x = e^{kt}(A\cos nt + B\sin nt)$, where n is the angular frequency of oscillation and k is a decay constant.

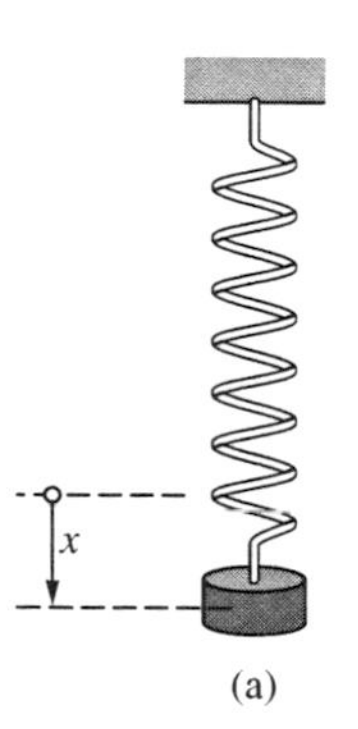

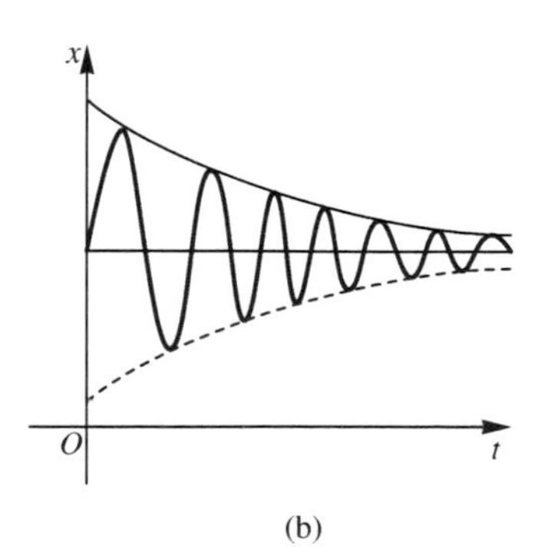

Figure 16.1. Oscillating spring

The values of k and n are found by solving a quadratic equation as will be shown in Chapter 23. A typical example is $x^2 + 2x + 5 = 0$.

Attempting to solve this by the well-known formula method leads to the result $x = -1 \pm \sqrt{-4}$. Since we can write $-4 = -1 \times 2 \times 2$ we might say that $x = -1 \pm 2\sqrt{-1}$, but we can go no further since no real number has a square root of -1. In order to make progress we introduce a quantity j, called the imaginary unit, which obeys the following rule:

KEY POINT

$$\mathrm{j}^2 = -1 \qquad 16.1$$

Notice that $(-\mathrm{j})^2 = +\mathrm{j}^2 = -1$, also. (Mathematicians use the symbol i instead of j.)

The solutions of the quadratic equation above can be written as $x = -1 \pm 2\mathrm{j}$. We define a **complex number** to be a quantity of the form $a + b\mathrm{j}$ where a and b are real numbers. Examples are 2j, $3.2 + 1.5\mathrm{j}$, $-2\mathrm{j}, -4 - 3\mathrm{j}$. When $a = 0$ the number is said to be **purely imaginary**; when $b = 0$ the number is real. Hence real numbers are a special case of complex numbers.

The form $x + y\mathrm{j}$ is used to indicate a general complex number and is usually symbolized as z; x is called the **real part** of z, written Rez and y is the **imaginary part**, written Imz. Related to each complex number $z = x + y\mathrm{j}$ is its (complex) **conjugate**, $\bar{z} = x - y\mathrm{j}$; the real part is not changed but the imaginary part has its sign changed. The solutions of the quadratic equation introduced earlier, namely, $-1 + 2\mathrm{j}$ and $-1 - 2\mathrm{j}$, are a complex conjugate pair. Note that x has a new role; in this context we should write the quadratic equation as

$$z^2 + 2z + 5 = 0$$

Worked examples

16.1 Simplify the following expressions:

(a) $\sqrt{-25}$ (b) j^3 (c) j^4

Solution

(a) $-25 = (-1) \times 25$ so that $\sqrt{-25} = \sqrt{(-1)} \times \sqrt{25} = \pm 5\mathrm{j}$

(b) $\mathrm{j}^3 = \mathrm{j}^2 \times \mathrm{j} = (-1) \times \mathrm{j} = -\mathrm{j}$

(c) $\mathrm{j}^4 = \mathrm{j}^2 \times \mathrm{j}^2 = (-1) \times (-1) = 1$

16.2 Write down the complex conjugates of the following:

(a) $-5 + \mathrm{j}$ (b) $4 - 3\mathrm{j}$ (c) 2 (d) $-2\mathrm{j}$

Solution

(a) $-5 - \mathrm{j}$ (b) $4 + 3\mathrm{j}$ (c) 2 (d) 2j

Note that in part (c) there is no imaginary part to change.

Graphical representation

The complex number $z = x + yj$ can be represented as a point in the x–y plane by treating (x,y) as a Cartesian coordinate pair. In this context, the x-axis is called the **real axis** and the y-axis is called the **imaginary axis**; the whole picture is known as the **Argand diagram** and the x–y plane is referred to as the **complex plane**. Figure 16.2 depicts four complex numbers represented on an Argand diagram. Each complex number corresponds to a unique point and each point to a unique complex number. The numbers z_2 and z_3 are a complex conjugate pair: note that each is a mirror image of the other in the real axis.

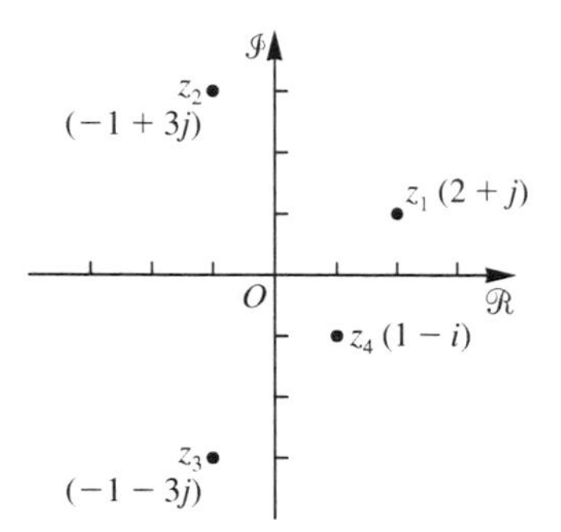

Figure 16.2.
Complex numbers on an Argand diagram

An alternative way of representing points in a plane is to use polar coordinates. Figure 16.3 shows the relationship between the Cartesian form and the **polar form** of a complex number.

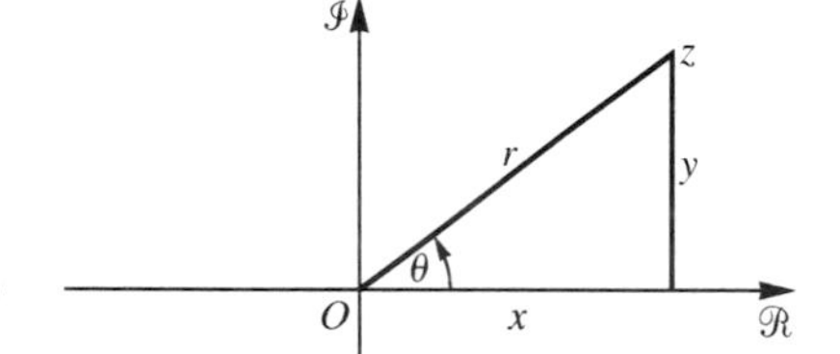

Figure 16.3.
Cartesian and polar forms of a complex number

The distance of the point from the origin, r, is called the **modulus** of the complex number and often written $|z|$. The angle θ, measured anticlockwise from the positive real axis, is called the **argument** of z, often written $\arg(z)$; it is customary to define θ to be in the range $-180° < \theta \leqslant 180°$ or, in radians, $-\pi < \theta \leqslant \pi$. It can be seen from Figure 16.3 that

$$x = r\cos\theta \text{ and } y = r\sin\theta \qquad 16.2$$

Alternatively, we have the relationships

$$r^2 = x^2 + y^2 \text{ and } \tan\theta = \frac{y}{x} \qquad 16.3$$

There is one potential pitfall: in determining the value of θ we must be careful to calculate the correct angle since two angles in the given range share each value of $\tan\theta$.

Using the polar form we can write $z = r\cos\theta + r\sin\theta\mathrm{j}$ or, more succinctly, $z = r\angle\theta$. The complex conjugate of z can be written as $\bar{z} = r\angle(-\theta)$.

Worked examples

16.3 Depict the complex numbers $z_1 = 2 - \mathrm{j}$ and $z_2 = -1 + \mathrm{j}$ on an Argand diagram.

Solution The point corresponding to z_1 has coordinates (2,−1) and that which corresponds to z_2 has coordinates (−1,1); they are plotted in Figure 16.4.

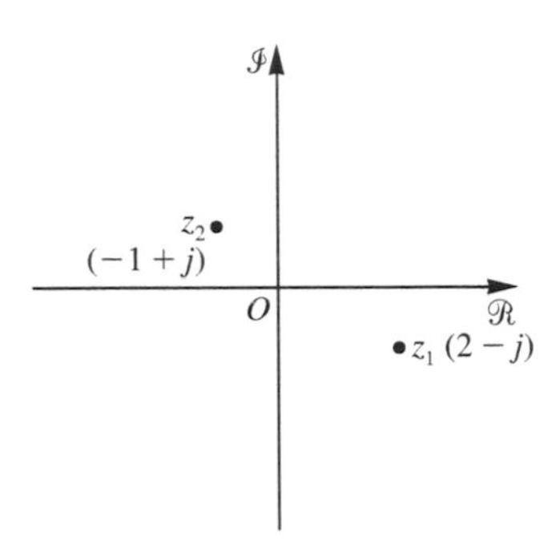

Figure 16.4. Argand diagram for Example 16.3

16.4 Express the complex numbers in Cartesian form:
(a) $z_1 = 4\angle 0$ (b) $z_2 = 3\angle(\pi/2)$ (c) $z_3 = 4\angle\pi$
(d) $z_4 = 1\angle(-\pi/2)$ (e) $z_5 = 2\angle(\pi/3)$

Solution In general, the number $z = r\angle\theta$ can be rewritten in the form $z = r\cos\theta + (r\sin\theta)\mathrm{j}$. Hence:
(a) $z_1 = 4\cos 0 + 4\sin 0\mathrm{j} = 4$
(b) $z_2 = 3\cos(\pi/2) + 3\sin(\pi/2)\mathrm{j} = 3\mathrm{j}$
(c) $z_3 = 4\cos\pi + 4\sin\pi\mathrm{j} = -4$
(d) $z_4 = \cos(-\pi/2) + \sin(-\pi/2)\mathrm{j} = -\mathrm{j}$
(e) $z_5 = 2\cos(\pi/3) + 2\sin(\pi/3)\mathrm{j} = 1 + \sqrt{3}\mathrm{j}$

16.5 Express the complex numbers $z_1 = -2 - \mathrm{j}$ and $z_2 = 2 + \mathrm{j}$ in polar form.

Solution Using the relationship $r^2 = x^2 + y^2$ we see that

$$|z| = r = \sqrt{x^2 + y^2}$$

In the first case this is equal to $\sqrt{((-2)^2 + (-1)^2)} = \sqrt{5}$.
In the second case this is equal to $\sqrt{((2)^2 + (1)^2)} = \sqrt{5}$, also.
Using the relationship $\tan\theta = y/x$ we find that $\tan\theta = \frac{1}{2}$ in both cases; a calculator will give the value of $\theta = 26.6^\circ$. Reference to an Argand diagram quickly establishes that, whereas this is the correct value for z_2, it is not correct for z_1; for the latter we must subtract 180° to obtain $\arg(z_1) = -153.4^\circ$.
Hence $z_1 = \sqrt{5}\angle -153.4^\circ$, $z_2 = \sqrt{5}\angle 26.6^\circ$.

Self-assessment questions 16.1

1. Explain what is meant by (a) a complex number, (b) a purely imaginary number, and (c) a complex conjugate.
2. Explain the use of an Argand diagram to represent complex numbers.
3. Explain the meaning of the modulus and argument of a complex number and show how they are related to the Cartesian form.

Exercise 16.1

1. Solve the equations
   (a) $x^2 + 9 = 0$
   (b) $z^2 - 2z + 4 = 0$
2. Mark on an Argand diagram the complex numbers

   $$z_1 = -2 + 3\mathrm{j},\ z_2 = -2\mathrm{j},$$
   $$z_3 = -1 - 2\mathrm{j},\ z_4 = 2$$

   Find the modulus and argument of each number.
3. Given that $x = 1$ is a root of the equation $x^3 + x^2 + 8x - 10 = 0$ find the other roots.
4. Express the following complex numbers in polar form:
   (a) $4 - \mathrm{j}$
   (b) -3
   (c) $-4\mathrm{j}$
   (d) $-12 + 5\mathrm{j}$

16.2 The arithmetic of complex numbers

First, we note that two complex numbers are equal if and only if: (a) their real parts are equal, and (b) their imaginary parts are equal.

Addition and subtraction

To add two complex numbers we add their real parts and then we add their imaginary parts; the result is a complex number. To subtract one complex number from another we subtract their real parts from each other and then we subtract their imaginary parts from each other; the result is again a complex number.

Worked examples

16.6 Find the values of x and y if $x - 5\mathrm{j}$ and $4 + y\mathrm{j}$ are the same number.

Solution Since $x - 5\mathrm{j} = 4 + y\mathrm{j}$ then, equating the real parts gives $x = 4$ and equating the imaginary parts gives $-5 = y$, or $y = -5$.

16.7 If $z_1 = 2 - 3j$ and $z_2 = 5 + 2j$, find $z_1 + z_2$, $z_1 - z_2$ and $2z_1 - 3z_2$.

Solution

$$z_1 + z_2 = (2 - 3j) + (5 + 2j) = (2 + 5) + (-3j + 2j) = 7 - j$$

$$z_1 - z_2 = (2 - 3j) - (5 + 2j) = (2 - 5) + (-3j - 2j) = -3 - 5j$$

$$2z_1 - 3z_2 = 2(2 - 3j) - 3(5 + 2j) = (2 \times 2 - 3 \times 5) + (-2 \times 3j - 3 \times 2j) = (4 - 15) + (-6j - 6j) = -11 - 12j$$

Multiplication and division

To multiply two complex numbers we carry out the multiplication as in 'ordinary' algebra and use the result that $j^2 = -1$. To divide one complex number by another we multiply the numerator and denominator by the complex conjugate of the denominator to obtain a real number on the denominator; then we simplify the result.

Worked examples

16.8 If $z_1 = 2 - 3j$ and $z_2 = 5 + 2j$ find $z_1 z_2$.

Solution $z_1 z_2 = (2 - 3j)(5 + 2j)$

Multiplying out we obtain

$$\begin{aligned} z_1 z_2 &= 10 - 15j + 4j - 6j^2 \\ &= 10 - 15j + 4j + 6 \qquad \text{using } j^2 = -1 \\ &= 16 - 11j \end{aligned}$$

16.9 If $z_1 = 2 - 3j$ and $z_2 = 5 + 2j$ find z_1/z_2.

Solution The complex conjugate of the denominator, z_2, is $\overline{z_2} = 5 - 2j$ and we multiply both the numerator and the denominator by this quantity. The value of the fraction will, of course, be unchanged. Therefore

$$\frac{z_1}{z_2} = \frac{2 - 3j}{5 + 2j} = \frac{(2 - 3j)(5 - 2j)}{(5 + 2j)(5 - 2j)} = \frac{10 - 15j - 4j + 6j^2}{25 + 10j - 10j - 4j^2} = \frac{10 - 19j - 6}{25 + 4}$$

$$= \frac{4 - 19j}{29} = \frac{4}{29} - \frac{19}{29}j$$

In this case the effect of multiplying both the numerator and the denominator of the fraction by the conjugate of the denominator was to produce a new denominator which is a real number. This will be true in general as the next example shows.

16.10 If $z_1 = x_1 + y_1 j$ and $z_2 = x_2 + y_2 j$ find $z_1 z_2$ and z_1/z_2. Deduce the expression for $z\bar{z}$.

Solution The product of the two complex numbers is given by

$$\begin{aligned} z_1z_2 &= (x_1 + y_1\text{j})(x_2 + y_2\text{j}) = x_1x_2 + x_1y_2\text{j} + x_2y_1\text{j} + y_1y_2\text{j}^2 \\ &= x_1x_2 - y_1y_2 + (x_1y_2 + x_2y_1)\text{j} \end{aligned}$$

The ratio of the two complex numbers is given by

$$\frac{z_1}{z_2} = \frac{(x_1 + y_1\text{j})(x_2 - y_2\text{j})}{(x_2 + y_2\text{j})(x_2 - y_2\text{j})} = \frac{x_1x_2 + x_2y_1\text{j} - x_1y_2\text{j} - y_1y_2\text{j}^2}{x_2x_2 + x_2y_2\text{j} - x_2y_2\text{j} - y_2y_2\text{j}^2}$$

$$= \frac{x_1x_2 + y_1y_2 + (x_2y_1 - x_1y_2)\text{j}}{x_2^2 + y_2^2}$$

and we see that the denominator is a real quantity.

If $z_1 = z$ and $z_2 = \bar{z}$ then $x_2 = x_1 = x$ and $y_1 = y$, $y_2 = -y$ so that

$$z\bar{z} = x^2 + y^2 + (-xy + xy)\text{j} = x^2 + y^2$$

KEY POINT

$$z\bar{z} = x^2 + y^2 \qquad 16.4$$

Multiplication and division in polar form

To multiply two numbers in polar form we simply multiply their moduli and add their arguments: that is, if $z_1 = r_1\angle\theta_1$ and $z_2 = r_2\angle\theta_2$ then:

KEY POINT

$$z_1z_2 = r_1r_2\angle(\theta_1 + \theta_2) \qquad 16.5$$

Similarly, to divide one complex number by another we divide one modulus by the other and subtract one argument from the other, that is:

KEY POINT

$$\frac{z_1}{z_2} = \frac{r_1}{r_2}\angle(\theta_1 - \theta_2) \qquad 16.6$$

Worked examples

16.11 If $z_1 = 6\angle(\pi/3)$ and $z_2 = 2\angle(\pi/4)$ find z_1z_2 and z_1/z_2.

Solution Multiplying the moduli we have $r_1r_2 = 12$ and adding the arguments we have $\theta_1 + \theta_2 = 7\pi/12$. Therefore $z_1z_2 = 12\angle(7\pi/12)$.

Similarly, dividing the moduli we have $r_1/r_2 = 3$ and subtracting the arguments we have $\theta_1 - \theta_2 = \pi/12$. Therefore $z_1/z_2 = 3\angle(\pi/12)$.

16.12 If $z = 2\angle(\pi/4)$, find z^2, z^3 and z^{10}, expressing each answer in both polar and Cartesian forms.

Solution To multiply a complex number by itself we square its modulus and double its argument. Therefore, $z^2 = 4\angle(\pi/2) = 4\text{j}$. In the same way, to cube a complex number we cube its modulus and multiply its argument by three so that

$$\begin{aligned} z^3 &= 8\angle(3\pi/4) = 8\cos(3\pi/4) + 8\sin(3\pi/4)\text{j} \\ &= 8(-1/\sqrt{2}) + 8(1/\sqrt{2})\text{j} = -4\sqrt{2} + 4\sqrt{2}\text{j} \end{aligned}$$

Finally,

$$\begin{aligned} z^{10} &= 2^{10}\angle(10\pi/4) = 1024\{\cos(10\pi/4) + \sin(10\pi/4)\text{j}\} \\ &= 1024\{0 + 1\text{j}\} = 1024\text{j} \end{aligned}$$

Self-assessment questions 16.2

1. Explain how to add, subtract, multiply and divide two complex numbers in Cartesian form.
2. Explain how to multiply and divide two complex numbers in polar form.

Exercise 16.2

1. Express the following in both Cartesian and polar forms:

   (a) $2\text{j}(3 - \text{j})$
   (b) $(3 - 2\text{j})(1 + 3\text{j})$

2. Express the following in Cartesian form:

   (a) $\dfrac{1}{2+\text{j}}$ (b) $\dfrac{-3}{\text{j}}$

   (c) $\dfrac{2}{\text{j}} - \dfrac{1}{1-\text{j}}$ (d) $\dfrac{\text{j}}{1-\text{j}}$

   (e) $\dfrac{2}{2+\text{j}} + \dfrac{1}{3-\text{j}}$

3. Find the real and imaginary parts of the following numbers:

   (a) j^5 (b) $\dfrac{-1}{\text{j}}$ (c) $\dfrac{1}{\text{j}^3 - 2\text{j}}$

4. Express the following numbers in Cartesian form:

   (a) $2\sqrt{2}\angle(\pi/4)$
   (b) $3\angle(\pi/3)$
   (c) $3\angle(-4\pi/3)$
   (d) $1\angle(-\pi/6)$

5. Find the modulus and argument of the following numbers:

   (a) $z_1 = -1 + \sqrt{3}\text{j}$
   (b) $z_2 = 2 + 2\text{j}$

   Express the product $z_1 z_2$ and the quotient z_1/z_2 in polar form.

6. Simplify the expression

$$\frac{(\sqrt{3}\angle(5\pi/6))^2(2\angle(-\pi/4))^3}{4\angle(-\pi/3)}$$

16.3 Regions in the complex plane

Complex numbers and vectors

We have represented a complex number by a pair of coordinates and we have defined the addition of two complex numbers in a way which is analogous to the triangle law of addition for vectors: this allows us to represent complex numbers by vectors in the x–y plane. The line joining the point (x,y) to the origin is the geometric version of this vector. Figure 16.5 depicts the complex numbers z_1 and z_2 represented by the vectors $\overrightarrow{OA}$ and $\overrightarrow{OB}$, respectively. If we complete the parallelogram $OACB$ then the vector $\overrightarrow{OC}$ represents the sum $z_1 + z_2$ and the vector $\overrightarrow{BA}$ represents the difference, $z_1 - z_2$. The latter result follows since $\overrightarrow{OB} + \overrightarrow{BA} = \overrightarrow{OA}$, that is $z_2 + (z_1 - z_2) = z_1$.

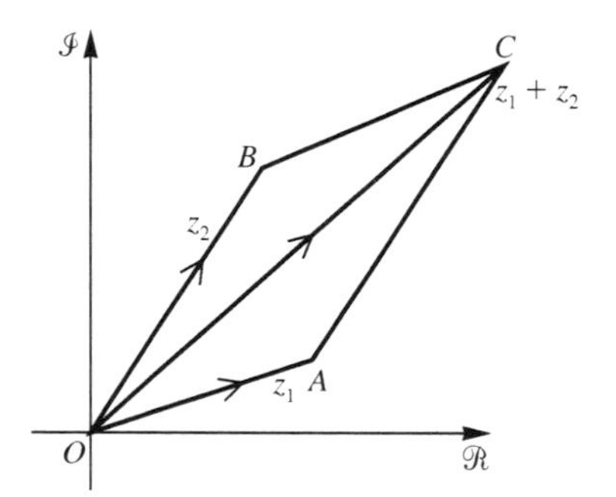

Figure 16.5. Addition of two complex numbers

Worked example

16.13 Represent on an Argand diagram the complex numbers:

$$z_1 = 5 + \mathrm{j},\ z_2 = 2 + 3\mathrm{j}$$

and their sum and difference.

Solution In Figure 16.6 the vectors $\overrightarrow{OA}$ and $\overrightarrow{OB}$ represent z_1 and z_2 respectively. The parallelogram $OACB$ is completed. Then $z_1 + z_2$ is represented by $\overrightarrow{OC}$ and $z_1 - z_2$ by $\overrightarrow{BA}$. We see that

$$\overrightarrow{OC} = \begin{pmatrix} 7 \\ 4 \end{pmatrix}$$

therefore represents the complex number $7 + 4\mathrm{j}$;

$$\overrightarrow{BA} = \begin{pmatrix} 3 \\ -2 \end{pmatrix}$$

represents the complex number $3 - 2\mathrm{j}$. Hence the sum $z_1 + z_2 = 7 + 4\mathrm{j}$ and the difference $z_1 - z_2 = 3 - 2\mathrm{j}$, as can be seen from the algebraic definition of addition and subtraction.

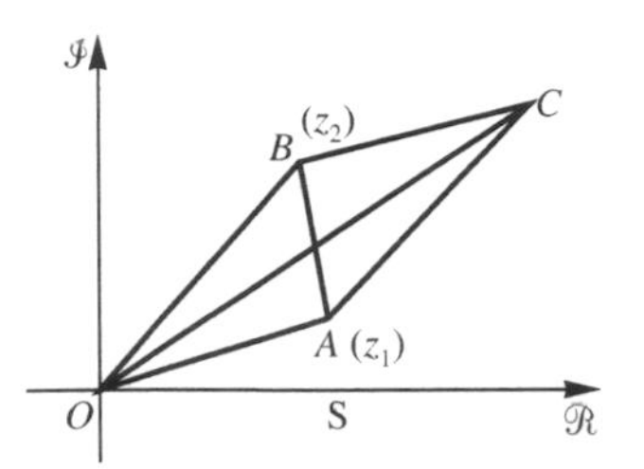

Figure 16.6.
Sum and difference of two complex numbers

Regions in the complex plane

We can describe regions in the (complex) plane by prescribing the locus of a complex number or by placing inequality restrictions on the number. For example, all the complex numbers which lie in the first quadrant of the plane have arguments which have values between 0 and $\pi/2$; this region, shaded in Figure 16.7(a), is therefore described by the inequality $0 \leqslant \arg(z) \leqslant \pi/2$. The equation $|z| = 3$ represents all points which are at a distance of 3 units from the origin. This follows because the equation can also be written as $|z|^2 = 9$ and, since $z = x + y\text{j}$, then $|z|^2 = x^2 + y^2 = 9$ or $r^2 = 9$ and hence $r = 3$. The locus of the complex number z is therefore a circle of radius 3 centred at the origin as shown in Figure 16.7(b). The inequality $|z| < 3$ describes the interior of the circle whereas the inequality $|z| > 3$ represents the exterior.

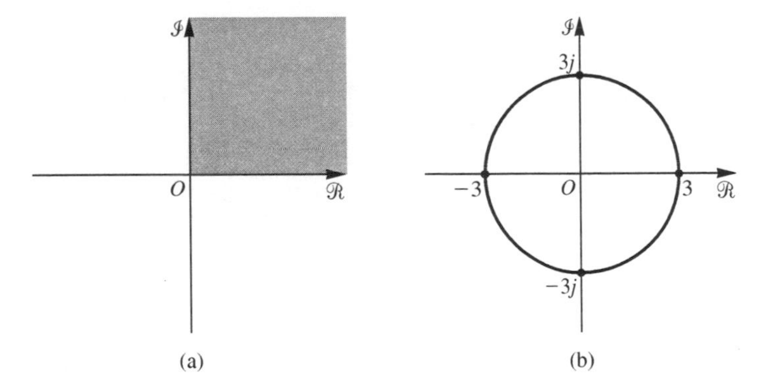

Figure 16.7.
Regions in the complex plane

Worked examples

16.14 Draw the locus of the point z which satisfies $\arg(z) = \pi/3$.

Solution The important point here is that we must be careful not to use the fact that $\tan\theta = \sqrt{3}$ and draw the straight line $y = \sqrt{3}x$ since this would also include points representing those complex numbers whose argument is $-2\pi/3$. The points which we wish to represent lie on a ray starting at the origin as indicated in Figure 16.8.

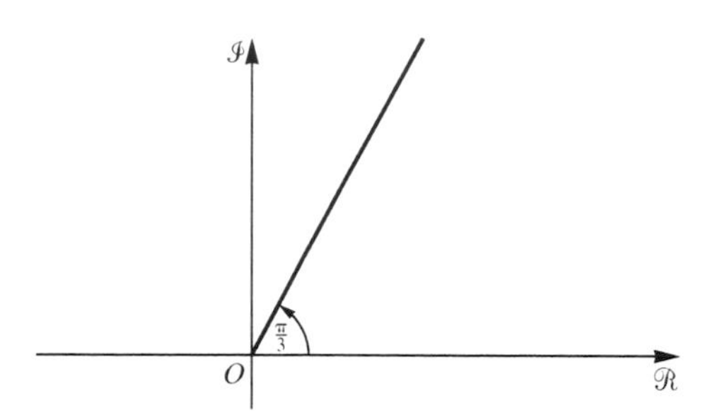

Figure 16.8. Points for which $\arg(z) = \pi/3$

16.15 Draw the locus of the point z which satisfies $|z - 3| = 2$.

Solution Figure 16.9 shows the fixed point C which corresponds to the complex number 3 and the point P which represents the variable number z. The vector $\overrightarrow{CP}$ represents the number $z - 3$ and the distance from C to P represents $|z - 3|$. This distance must be equal to 2 wherever P is placed and therefore the locus of P is a circle centred at O (3,0) and of radius 2. It is of interest to obtain the result algebraically. Since $z = x + y\text{j}$ then

$$z - 3 = (x - 3) + y\text{j}$$

and so

$$|z - 3|^2 = |x - 3 + y\text{j}|^2 = 4$$

that is $(x - 3)^2 + y^2 = 4$. This is the equation of a circle of radius 2 centred at the point (3,0).

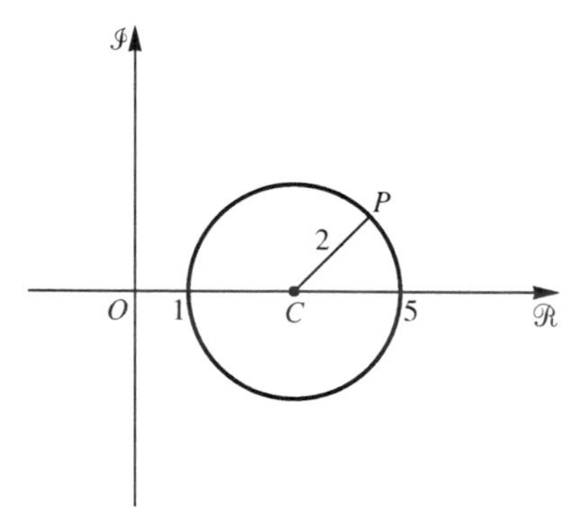

Figure 16.9. Diagram for Example 16.15

16.16 Use algebra to find the locus of the point z which satisfies the equation $|z - 1| = |z - \text{j}|$ and interpret the result geometrically.

Solution Since

$$|z - 1| = |z - \text{j}|,\ |z - 1|^2 = |z - \text{j}|^2, \text{ that is } |(x - 1) + y\text{j}|^2 = |x + (y - 1)\text{j}|^2$$

and therefore

$$(x - 1)^2 + y^2 = x^2 + (y - 1)^2$$

or

$$x^2 - 2x + 1 + y^2 = x^2 + y^2 - 2y + 1$$

from which we deduce that $y = x$. Figure 16.10 shows this result diagrammatically.

In the given equation the left-hand side $|z-1|$ can be interpreted as the distance on the Argand diagram between z and 1. The right-hand side $|z-\mathrm{j}|$ can be interpreted as the distance between z and j. The point we seek is equidistant from 1 and j: the result follows readily.

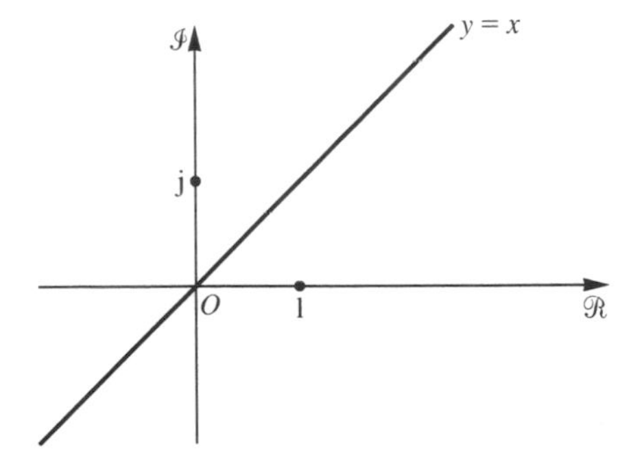

Figure 16.10. Diagram for Example 16.16

Self-assessment questions 16.3

1. Explain geometrically how to represent the sum and the difference of two complex numbers.
2. What is the complex number form of the equation of a circle of radius a, centred at z_0?

Exercise 16.3

1. Sketch the regions defined by the following inequalities:
   (a) $\mathrm{Re}z < 1/2$
   (b) $\mathrm{Im}z > -1/2$
   (c) $|z| > 2$
   (d) $|z+3| \leqslant 1$
   (e) $|z-\mathrm{j}| \geqslant 2$
   (f) $|z-2| < |z+1|$
   (g) $0 \leqslant \arg(z) \leqslant \pi/4$
2. Sketch the loci given by
   (a) $\arg(z) = 3\pi/2$
   (b) $\arg(z-2) = \pi/4$
3. Find graphically the complex number(s) which satisfy the conditions stated; in each case verify your result algebraically.
   (a) $\arg(z) = \pi/4$, $|z-(1+j)| = 1$
   (b) $\arg(z) = \pi/4$, $|z-1| = 2$

16.4 De Moivre's theorem

A result which has widespread applications is **De Moivre's theorem** which we state in two forms.

First, if n is a natural number then:

KEY POINT

$$(\cos\theta + \sin\theta\mathrm{j})^n = \cos(n\theta) + \sin(n\theta)\mathrm{j} \qquad 16.7$$

Worked example

16.17 Verify De Moivre's theorem when $n = 2$.

Solution When $n = 2$ the theorem states that

$$(\cos\theta + \sin\theta \mathrm{j})^2 = \cos(2\theta) + \sin(2\theta)\mathrm{j}$$

Now

$$\begin{aligned}(\cos\theta + \sin\theta \mathrm{j})^2 &= (\cos\theta + \sin\theta \mathrm{j})(\cos\theta + \sin\theta \mathrm{j})\\ &= \cos^2\theta + 2\sin\theta\cos\theta \mathrm{j} + \sin^2\theta \mathrm{j}^2\\ &= \cos^2\theta - \sin^2\theta + 2\sin\theta\cos\theta \mathrm{j}\end{aligned}$$

Making use of the trigonometric identities

$$\cos(2\theta) = \cos^2\theta - \sin^2\theta \qquad \text{and} \qquad \sin(2\theta) = 2\sin\theta\cos\theta$$

we obtain

$$(\cos\theta + \sin\theta \mathrm{j})^2 = \cos(2\theta) + \sin(2\theta)\mathrm{j}$$

The more general statement of the theorem allows the index to be a rational number, p/q, where p and $q \neq 0$ are integers.

KEY POINT

$$(\cos\theta + \sin\theta \mathrm{j})^{p/q} = \cos\left(\frac{p}{q}\theta\right) + \sin\left(\frac{p}{q}\theta\right)\mathrm{j} \qquad 16.8$$

This version is especially useful when we wish to find the roots of a complex number. For example, if we want the square roots of a complex number we use the result

$$(\cos\theta + \sin\theta \mathrm{j})^{1/2} = \cos\left(\frac{1}{2}\theta\right) + \sin\left(\frac{1}{2}\theta\right)\mathrm{j}$$

Whereas this is *one* of the square roots, there is a second root to find. How we find the missing root is demonstrated in Example 16.18.

Worked examples

16.18 Find the complex numbers which satisfy the equation $z^2 = 9\mathrm{j}$.

Solution Let $z = r(\cos\theta + \sin\theta \mathrm{j})$. Then

$$\begin{aligned}z^2 &= r^2(\cos\theta + \sin\theta \mathrm{j})^2\\ &= r^2(\cos 2\theta + \sin 2\theta \mathrm{j})\end{aligned}$$

by De Moivre's theorem. Using the polar form and allowing for the multiple representation of the argument we obtain the result

$$9j = 9\{\cos(\pi/2 + 2n\pi) + \sin(\pi/2 + 2n\pi)j\} \qquad n = 0, 1, 2, \ldots$$

Equating the respective parts of the formulae for z^2 and 9j we see that $r^2 = 9$, so that $r = 3$, and $2\theta = \pi/2 + 2n\pi$ and hence $\theta = \pi/4 + n\pi$. When $n = 0$, $\theta = \pi/4$ and when $n = 1$ then $\theta = 5\pi/4$. The case when $n = 2$ merely repeats the value for $n = 0$, the case when $n = 3$ repeats the value for $n = 1$, and so on. As we can see in Figure 16.11 the solutions are spaced apart by an amount $2\pi/2 = \pi$. The solutions are therefore

$$z_1 = 3\angle(\pi/4), \qquad z_2 = 3\angle(5\pi/4)$$

in polar form and

$$z_1 = \frac{3}{\sqrt{2}} + \frac{3}{\sqrt{2}}j, \qquad z_2 = -\frac{3}{\sqrt{2}} - \frac{3}{\sqrt{2}}j$$

in Cartesian form.

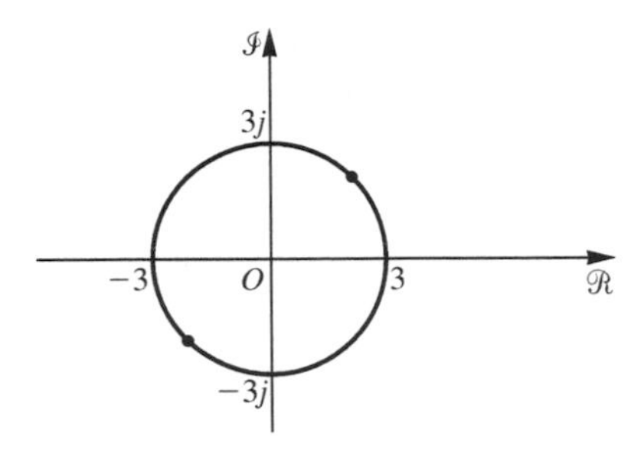

Figure 16.11. Diagram for Example 16.18

16.19 Find all the complex numbers which satisfy the equation $z^3 = 1$.

Solution The given equation is equivalent to the equation $z = 1^{\frac{1}{3}}$.

If $z = r(\cos\theta + \sin\theta j)$ then by De Moivre's theorem it follows that

$$z^3 = r^3(\cos 3\theta + \sin 3\theta j)$$

The number 1 can be written in polar form as

$$1\{\cos 0 + \sin 0j\} \quad \text{or} \quad 1\{\cos 2\pi + \sin 2\pi j\} \quad \text{or}$$
$$1\{\cos 4\pi + \sin 4\pi j\}$$

or, in general, as $1\{\cos 2n\pi + \sin 2n\pi j\}$ where n is any of the numbers 0, 1, 2, ... Hence we may write

$$r^3(\cos 3\theta + \sin 3\theta j) = 1\{\cos 2n\pi + \sin 2n\pi j\}$$

Therefore $r^3 = 1$, from which we conclude that $r = 1$, that is, the solutions of the equation are represented by points lying on the unit circle (centred at the origin and of radius 1).

Also, $3\theta = 2n\pi$ and the first three solutions are found by taking $n = 0$, 1 and 2; this gives $\theta = 0$, $2\pi/3$, $4\pi/3$, respectively. The solutions are $z_1 = 1\angle 0$, $z_2 = 1\angle 2\pi/3$ and $z_3 = 1\angle 4\pi/3$. Any further values of θ merely repeat these points, for example $\theta = 6\pi/3 = 2\pi$ gives the same point as $\theta = 0$. Figure 16.12 shows the solutions on an Argand diagram.

Note that the two non-real solutions are a complex conjugate pair. In Cartesian form the solutions are

$$z_2 = -\frac{1}{2} + \frac{\sqrt{3}}{2}\mathrm{j} \text{ and } z_3 = -\frac{1}{2} - \frac{\sqrt{3}}{2}\mathrm{j}$$

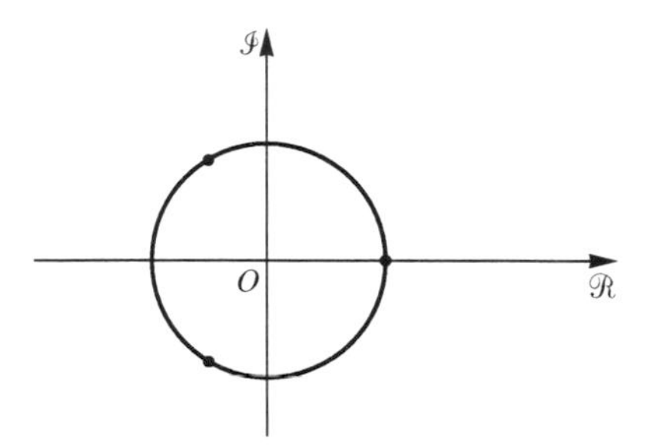

Figure 16.12. Diagram for Example 16.19

16.20 If $z = \cos\theta + \sin\theta\mathrm{j}$ find expressions for $z + (1/z)$ and $z - (1/z)$. Hence derive expressions for

$$z^n + \frac{1}{z^n} \quad \text{and} \quad z^n - \frac{1}{z^n}$$

Solution First, we use De Moivre's theorem to find an expression for $1/z$.

$$\frac{1}{z} = z^{-1} = (\cos\theta + \sin\theta\mathrm{j})^{-1} = \cos(-\theta) + \sin(-\theta)\mathrm{j}$$

Since $\cos(-\theta) = \cos(\theta)$ and $\sin(-\theta) = -\sin(\theta)$ then $1/z = \cos\theta - \sin\theta\mathrm{j}$.

Therefore it follows that

$$z + \frac{1}{z} = 2\cos\theta \quad \text{and} \quad z - \frac{1}{z} = (2\sin\theta)\mathrm{j}$$

Further use of De Moivre's theorem yields the expressions

$$z^n = \cos(n\theta) + \sin(n\theta)\mathrm{j}$$

and

$$z^{-n} = \cos(-n\theta) + \sin(-n\theta)\mathrm{j} = \cos(n\theta) - \sin(n\theta)\mathrm{j}$$

Finally, we obtain the results required.

KEY POINT

$$z^n + \frac{1}{z^n} = 2\cos(n\theta) \text{ and } z^n - \frac{1}{z^n} = 2\sin(n\theta)\mathrm{j} \qquad 16.9$$

The behaviour of the expression $\cos\theta + \sin\theta\mathrm{j}$ when raised to a power is one piece of evidence which was used to suggest the following result, known as **Euler's formula**, which is stated without proof.

KEY POINT

$$\cos\theta + \sin\theta\mathrm{j} \equiv \mathrm{e}^{\theta\mathrm{j}} \qquad 16.10$$

This allows us to write a complex number in a third form, namely $z = r\mathrm{e}^{\theta\mathrm{j}}$; this is known as the **exponential form**.

Worked examples

16.21 (a) Find the Cartesian form of
(i) $2e^{\pi j/2}$ (ii) $3e^{\pi j/4}$ (iii) $e^{\pi j}$
(b) Given that $z_1 = 2e^{\pi j/3}$ and $z_2 = 5e^{\pi j/2}$ find the Cartesian form of z_1^2/z_2.

Solution (a)

$$\text{(i) } 2e^{\pi j/2} = 2\cos\left(\frac{\pi}{2}\right) + 2\sin\left(\frac{\pi}{2}\right)j = 2j$$

$$\text{(ii) } 3e^{\pi j/4} = 3\cos\left(\frac{\pi}{4}\right) + 3\sin\left(\frac{\pi}{4}\right)j = \frac{3\sqrt{2}}{2} + \frac{3\sqrt{2}}{2}j$$

$$\text{(iii) } e^{\pi j} = \cos(\pi) + \sin(\pi)j = -1$$

(b)

$$z_1^2 = (2e^{\pi j/3})^2 = 2^2(e^{\pi j/3})^2 = 4e^{2\pi j/3}$$

therefore

$$\frac{z_1^2}{z_2} = \frac{4e^{2\pi j/3}}{5e^{\pi j/2}} = \frac{4}{5}e^{\pi j/6}$$

$$= \frac{4}{5}\cos\left(\frac{\pi}{6}\right) + \frac{4}{5}\sin\left(\frac{\pi}{6}\right)j = \frac{2\sqrt{3}}{5} + \frac{2}{5}j$$

16.22 Show that

$$\cos\theta = \frac{e^{\theta j} + e^{-\theta j}}{2} \qquad \sin\theta = \frac{e^{\theta j} - e^{-\theta j}}{2j}$$

and deduce that $\cosh\theta j = \cos\theta$, $\sinh\theta j = j\sin\theta$, $\cos\theta j = \cosh\theta$, $\sin\theta j = j\sinh\theta$.

Solution Since $e^{\theta j} = \cos\theta + \sin\theta j$ then $e^{-\theta j} = \cos\theta - \sin\theta j$. Hence

$$\frac{e^{\theta j} + e^{-\theta j}}{2} = \cos\theta \qquad \frac{e^{\theta j} - e^{-\theta j}}{2j} = \sin\theta$$

Also, since

$$\cosh\theta = \frac{e^{\theta} + e^{-\theta}}{2} \qquad \cosh\theta j = \frac{e^{\theta j} + e^{-\theta j}}{2} = \cos\theta$$

and since

$$\sinh\theta = \frac{e^{\theta} - e^{-\theta}}{2} \qquad \sinh\theta j = \frac{e^{\theta j} - e^{-\theta j}}{2} = j\sin\theta$$

Finally,

$$\cos\theta\mathrm{j} = \frac{\mathrm{e}^{\theta\mathrm{j}^2} + \mathrm{e}^{-\theta\mathrm{j}^2}}{2} = \frac{\mathrm{e}^{-\theta} + \mathrm{e}^{\theta}}{2} = \cosh\theta$$

and

$$\sin\theta\mathrm{j} = \frac{\mathrm{e}^{\theta\mathrm{j}^2} - \mathrm{e}^{-\theta\mathrm{j}^2}}{2\mathrm{j}} = \frac{\mathrm{e}^{-\theta} - \mathrm{e}^{\theta}}{2\mathrm{j}} = -\mathrm{j}\left(\frac{\mathrm{e}^{-\theta} - \mathrm{e}^{\theta}}{2}\right) = \mathrm{j}\sinh\theta$$

(These results are the basis for Osborne's rule, met in Chapter 6.)

KEY POINT

$$\mathrm{e}^{\pi\mathrm{j}} = -1 \qquad 16.11$$

Application to network analysis

When an alternating current $i = i_0 \sin\omega t$ flows through a resistor of resistance R, the voltage across the resistor is $i_0 R\sin\omega t$; across a capacitor of capacitance C the corresponding voltage is

$$\frac{i_0}{\omega C}\sin\left(\omega t - \frac{\pi}{2}\right)$$

and across an inductor of inductance L the voltage is

$$\omega L i_0 \sin\left(\omega t + \frac{\pi}{2}\right)$$

Hence the voltage across the resistor is in phase with the current, whereas that across the capacitor lags behind and that across the inductor leads the current. To study the voltage in a circuit comprising a number of these components we make use of the exponential form of complex numbers.

Since $\sin\theta = \mathrm{Im}\,\mathrm{e}^{\theta\mathrm{j}}$, then $\sin\omega t = \mathrm{Im}\,\mathrm{e}^{\omega t\mathrm{j}}$ and therefore

$$\sin\left(\omega t - \frac{\pi}{2}\right) = \mathrm{Im}\,\mathrm{e}^{(\omega t - \pi/2)\mathrm{j}} = (\mathrm{Im}\,\mathrm{e}^{\omega t\mathrm{j}}.(-\mathrm{j}))$$

Similarly,

$$\sin\left(\frac{\omega t + \pi}{2}\right) = \mathrm{Im}\,(\mathrm{e}^{\omega t\mathrm{j}}.\mathrm{j})$$

We can write the voltage as $v = \mathrm{Im}\, i_0 Z\mathrm{e}^{\omega t\mathrm{j}}$ where the **complex impedance** Z is given by $Z = R$ for a resistor, $Z = -\mathrm{j}/\omega C$ for a capacitor and $Z = \mathrm{j}\omega L$ for an inductor.

For a series LCR circuit the voltage across the components is given by $v = i_0 \operatorname{Im} Ze^{\omega t j}$ where

$$Z = j\omega L - \frac{j}{\omega C} Z + R$$

Worked example

16.23 For the series LCR circuit show that $v = i_0|Z|\sin(\omega t + \phi)$ where

$$|Z| = \left[R^2 + \left(L\omega - \frac{1}{\omega C}\right)^2\right]^{1/2}$$

is the **impedance** of the circuit and

$$\phi = \tan^{-1}\left(\frac{L\omega - \dfrac{1}{\omega C}}{R}\right)$$

is the **phase**.

Solution If $Z = |Z|e^{\phi j}$ then $Ze^{\omega t j} = |Z|e^{(\phi + \omega t)j}$ and $\operatorname{Im} Ze^{\omega t j} = |Z|\sin(\phi + \omega t)$, hence the result for v follows. Since

$$Z = R + \left(\omega L - \frac{1}{\omega C}\right)j$$

then $|Z|$ is as quoted and

$$\tan\phi = \frac{L\omega - \dfrac{1}{\omega C}}{R}$$

which leads readily to the formula given for ϕ.

Self-assessment questions 16.4

1. State De Moivre's theorem for an integer index, n, and a fractional index p/q.
2. Explain how to use De Moivre's theorem to find the roots of a complex number.

Exercise 16.4

1. Express $(\cos\theta + \sin\theta j)^5$ and $(\cos\theta + \sin\theta j)^{1/2}$ in the form $\cos n\theta + \sin n\theta j$.
2. Use De Moivre's theorem to simplify the following expressions:

   (a) $(\cos 2\theta + \sin 2\theta j)(\cos 3\theta + \sin 3\theta j)$
   (b) $(\cos 6\theta + \sin 6\theta j)(\cos 3\theta + \sin 3\theta j)$

3. Find $(4-4j)^{1/3}$ and plot the answers on an Argand diagram.
4. Verify the identities $\cos 3\theta \equiv 4\cos^3\theta - 3\cos\theta$ and $\sin 3\theta \equiv 3\sin\theta - 4\sin^3\theta$.
5. Solve the equation $z^4 + 16 = 0$.
6. Find the fifth roots of $-j$, displaying the results on an Argand diagram.
7. (a) Show that $\tanh xj = j\tan x$ and $\tan xj = j\tanh x$.
   (b) If $z = re^{\theta j}$ show that $\ln z = u + vj$ where $u = \ln r$ and $v = \theta + 2n\pi j$, $n = 0, \pm 1, \pm 2, \ldots$. (This means that $\ln z$ has many values: the one corresponding to $n = 0$ is called the **principal value**.)

Test and assignment exercises 16

1. Show that $1/(\cos\theta + \sin\theta j) \equiv \cos\theta - \sin\theta j$.
2. Simplify the following expressions:

   (a) $(5-4j)/(5+4j)$
   (b) $1/(2-3j)$
   (c) $1/(x+yj)$
   (d) $1/(3+2j) + 1/(3-2j)$
3. Find the modulus and argument of the following numbers:

   (a) $-2j$ (b) -4
   (c) $1-j$ (d) $\cos\theta - \sin\theta j$
4. Express the following in Cartesian form:

   (a) $(\cos\theta + \sin\theta j)^7$
   (b) $1/(\cos\theta + \sin\theta j)^2$
   (c) $(\cos\theta - \sin\theta j)/(\cos 2\theta + \sin 2\theta j)^2$
5. Show that

   (a) $z + \bar{z} = 2\mathrm{Re}\, z$ (b) $z - \bar{z} = 2\,\mathrm{Im}\, z$
   (c) $\bar{\bar{z}} = z$ (d) $z\bar{z} = |\bar{z}|^2$
6. Sketch the region in the complex plane defined by $|z - 2j| < 1$.
7. Solve the equation $z^4 = 1 - \sqrt{3}j$.

17 Differentiation II

Objectives

This chapter

- differentiates functions defined implicitly
- develops and uses logarithmic differentiation
- differentiates functions defined parametrically
- introduces the concept of curvature
- differentiates vector quantities

17.1 Implicit and logarithmic differentiation

In the relationship $y = f(x)$, y is expressed **explicitly** in terms of x. The relationship

$$y^2 + \sin y = x^2 + \cos x$$

cannot be so expressed: it is said to be an **implicit** relationship. In some cases it is not possible to rearrange an implicit relationship into an explicit one. In some cases it *is* possible to rearrange an implicit relationship into an explicit one but it is not easy to differentiate the result. As an example the relationship

$$\frac{x^2}{4} + \frac{y^2}{9} = 1$$

becomes

$$y = \pm 3\left(1 - \frac{x^2}{4}\right)^{\frac{1}{2}}$$

We can differentiate the implicit relationship term by term which is often easier. The technique is known as **implicit differentiation**. To differentiate a function of y with respect to x, say, we apply the chain rule; for example

$$\frac{\mathrm{d}}{\mathrm{d}x}(\sin y) = \frac{\mathrm{d}}{\mathrm{d}y}(\sin y) \times \frac{\mathrm{d}y}{\mathrm{d}x} = \cos y \frac{\mathrm{d}y}{\mathrm{d}x}$$

KEY POINT

A second, important example is

$$\frac{d}{dx}(\ln y) = \frac{d}{dy}(\ln y) \times \frac{dy}{dx} = \frac{1}{y}\frac{dy}{dx}$$

Worked examples

17.1 Find dy/dx given the relationship

$$x^2y^3 + e^{2y} = 0$$

Solution First, $d/dx(x^2y^3)$ can be found by using the product rule. This gives

$$x^2\frac{d}{dx}(y^3) + 2xy^3$$

Applying the chain rule gives

$$x^2 \times 3y^2\frac{dy}{dx} + 2xy^3$$

Similarly,

$$\frac{d}{dx}(e^{2y}) = \frac{d}{dy}(e^{2y}) \times \frac{dy}{dx} = 2e^{2y}\frac{dy}{dx}$$

Hence differentiating the given relationship term by term we obtain

$$x^2 \times 3y^2\frac{dy}{dx} + 2xy^3 + 2e^{2y}\frac{dy}{dx} = 0$$

that is

$$(3x^2y^2 + 2e^{2y})\frac{dy}{dx} = -2xy^3$$

Therefore

$$\frac{dy}{dx} = -\frac{2xy^3}{(3x^2y^2 + 2e^{2y})}$$

17.2 Find the gradient of the tangent to the circle

$$x^2 + y^2 - 6x + 2y - 30 = 0$$

at the point (9,1).

Solution Differentiating the equation of the circle with respect to x, we obtain

$$2x + 2y\frac{dy}{dx} - 6 + 2\frac{dy}{dx} = 0$$

that is

$$(2y+2)\frac{dy}{dx} = 6 - 2x$$

so that

$$\frac{dy}{dx} = \frac{6-2x}{2y+2} = \frac{3-x}{y+1}$$

At the point (9,1),

$$\frac{dy}{dx} = \frac{-6}{2} = -3$$

17.3 Differentiate the equation

$$x^2 + y^2 - 6x + 2y - 30 = 0$$

a second time to find d^2y/dx^2.

Solution The first differentiation yields

$$2x + 2y\frac{dy}{dx} - 6 + 2\frac{dy}{dx} = 0$$

A second differentiation gives

$$2 + 2y\frac{d^2y}{dx^2} + 2\frac{dy}{dx}\cdot\frac{dy}{dx} - 0 + 2\frac{d^2y}{dx^2} = 0$$

that is

$$(2y+2)\frac{d^2y}{dx^2} = -2 - 2\left(\frac{dy}{dx}\right)^2$$

or

$$(y+1)\frac{d^2y}{dx^2} = -1 - \left(\frac{dy}{dx}\right)^2$$

$$= -1 - \left(\frac{3-x}{y+1}\right)^2 \qquad \text{using Example 17.2}$$

Hence

$$\frac{d^2y}{dx^2} = -\frac{1}{(y+1)} - \frac{(3-x)^2}{(y+1)^3}$$

We *could* simplify this expression by using the original equation in the form

$$(x-3)^2 + (y+1)^2 = 40$$

to obtain a formula for d^2y/dx^2 in terms of x only. It is unwise to carry out the 'simplification' unless we know the reason why we need d^2y/dx^2.

When we have to differentiate a function $y = f(x)$ in the form of an awkward product or quotient the technique of **logarithmic differentiation** may be useful. The process involves taking the natural logarithm of the function which we wish to differentiate, applying the laws of logarithms, and differentiating the result. We use the result

$$\frac{d}{dx}(\ln y) = \frac{1}{y}\frac{dy}{dx}$$

and multiply through by y before simplifying the resulting expression.

Worked examples

17.4 Find the derived function of $y = x^4(1+x)^6 e^x$.

Solution Taking natural logarithms, we obtain

$$\begin{aligned}\ln y &= \ln\left[x^4(1+x)^6 e^x\right]\\ &= \ln x^4 + \ln\left[(1+x)^6\right] + \ln e^x\\ &= 4\ln x + 6\ln(1+x) + x\end{aligned}$$

We now differentiate this equation:

$$\begin{aligned}\frac{1}{y}\frac{dy}{dx} &= \frac{4}{x} + \frac{6}{1+x} + 1\\ &= \frac{4(1+x) + 6x + x(1+x)}{x(1+x)}\\ &= \frac{x^2 + 11x + 4}{x(1+x)}\end{aligned}$$

Multiplying by y we obtain

$$\begin{aligned}\frac{dy}{dx} &= \frac{(x^2 + 11x + 4)}{x(1+x)} \times x^4(1+x)^6 e^x\\ &= x^3(1+x)^5(x^2 + 11x + 4)e^x\end{aligned}$$

17.5 Find dy/dt for the function

$$y = \frac{e^t(1+t)^{\frac{1}{2}}}{t^3}$$

Solution

$$\ln y = \ln(e^t) + \ln\left[(1+t)^{\frac{1}{2}}\right] - \ln t^3$$
$$= t + \tfrac{1}{2}\ln(1+t) - 3\ln t$$

Hence

$$\frac{1}{y}\frac{dy}{dt} = 1 + \frac{1}{2}\cdot\frac{1}{1+t} - \frac{3}{t}$$
$$= \frac{2t(1+t) + t - 3\times 2(1+t)}{2t(1+t)}$$
$$= \frac{2t^2 - 3t - 6}{2t(1+t)}$$

Therefore

$$\frac{dy}{dt} = \frac{(2t^2 - 3t - 6)}{2t(1+t)} \times \frac{e^t(1+t)^{\frac{1}{2}}}{t^3}$$
$$= \frac{(2t^2 - 3t - 6)\,e^t}{2t^4(1+t)^{\frac{1}{2}}}$$

Self-assessment questions 17.1

1. Explain when and how to use implicit differentiation.
2. Explain when and how to use logarithmic differentiation.

Exercise 17.1

1. Obtain dy/dx from each of the following equations and hence find the equation of the tangent to curves (a), (b) and (c) at the point indicated:

   (a) $xy^4 = 3x^3 - 2xy^2$ at (1,1)
   (b) $y^2 + x^2 + 2x - 4y = 15$ at (1,–2)
   (c) $x^2 - y^2 + x - y = 0$ at (2,2)
   (d) $\cos(x+y) = xy$

2. Find d^2y/dx^2 at the point $(a/\sqrt{2}, b/\sqrt{2})$ on the ellipse

$$\frac{x^2}{a^2} + \frac{y^2}{b^2} = 1$$

   (Note that the equation of the ellipse can be written $b^2x^2 + a^2y^2 = a^2b^2$.)

3. Differentiate the following:

   (a) $x^2(1-x)^5e^{-4x}$

   (b) $\dfrac{e^{2x}\cos x}{x^2}$

   (c) $\dfrac{(1+x^3)^2e^{-2x}}{(3+x)^5}$

   (d) $x^2e^{-3x}\ln x$

17.2 Parametric differentiation

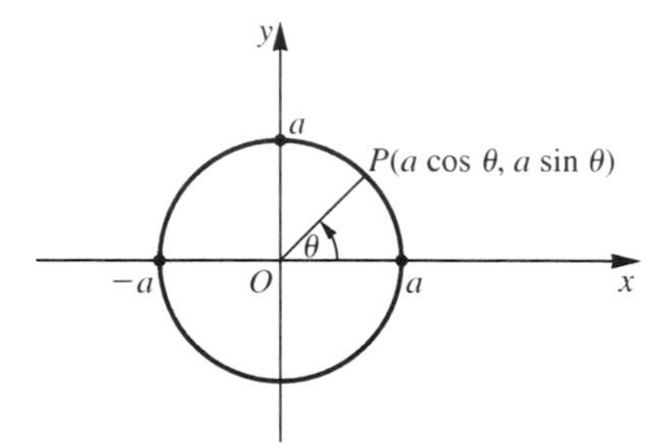

Figure 17.1. Parametric representation of a point on a circle

Referring to Figure 17.1, the position of the point P on the circle $x^2 + y^2 = a^2$ can be specified by the **parametric equations**

$$x = a\cos\theta, \qquad y = a\sin\theta \tag{17.1}$$

The variable θ is a **parameter**. We can eliminate θ to obtain the equation of the circle and in this case the resulting equation is relatively simple. It can be differentiated implicitly to obtain

$$2x + 2y\frac{dy}{dx} = 0$$

that is

$$\frac{dy}{dx} = \frac{-x}{y}$$

Often, however, a curve can only be expressed in parametric form and we need to use the parametric equations to obtain dy/dx. In the current example we use the chain rule to find the following:

KEY POINT

$$\frac{dy}{dx} = \frac{dy}{d\theta} \times \frac{d\theta}{dx} = \frac{dy}{d\theta} \div \frac{dx}{d\theta} \tag{17.2}$$

In this case the rule gives

$$\frac{dy}{dx} = \frac{a\cos\theta}{-a\sin\theta} = -\cot\theta$$

To find d^2y/dx^2 takes more care; refer to Example 17.7.

Worked examples

17.6 For the curve whose equation is $x = 3\cos t$, $y = 4\sin t$, use parametric differentiation to find dy/dx. By eliminating t from the parametric equations and using implicit differentiation check the result.

Solution First,

$$\frac{dx}{dt} = -3\sin t, \quad \frac{dy}{dt} = 4\cos t$$

Hence

$$\frac{dy}{dx} = \frac{dy}{dt} \div \frac{dx}{dt} = \frac{4\cos t}{-3\sin t} = -\frac{4}{3}\cot t$$

Now $\cos t = x/3$ and $\sin t = y/4$. Eliminating t,

$$\frac{x^2}{9} + \frac{y^2}{16} = 1$$

Differentiating this equation implicitly we obtain

$$\frac{2x}{9} + \frac{2y}{16}\frac{dy}{dx} = 0$$

that is

$$\frac{dy}{dx} = \frac{-16x}{9y} = \frac{-16 \times 3\cos t}{9 \times 4\sin t} = \frac{-4\cos t}{3\sin t} = -\frac{4}{3}\cot t \qquad \text{(again)}$$

17.7 Find d^2y/dx^2 given $x = a(\theta - \sin\theta)$ and $y = a(1 - \cos\theta)$.

Solution

$$\frac{dy}{d\theta} = a(\sin\theta) \text{ and } \frac{dx}{d\theta} = a(1 - \cos\theta)$$

Hence

$$\frac{dy}{dx} = \frac{dy}{d\theta} \div \frac{dx}{d\theta} = \frac{a\sin\theta}{a(1-\cos\theta)} = \frac{\sin\theta}{(1-\cos\theta)}$$

Now

$$\frac{d^2y}{dx^2} = \frac{d}{dx}\left(\frac{dy}{dx}\right) = \frac{d}{dx}\left(\frac{\sin\theta}{1-\cos\theta}\right)$$

Remember that in order to obtain dy/dx where y is a function of θ we differentiated with respect to θ and scaled the result by dividing by $dx/d\theta$. The same procedure is applied to finding d^2y/dx^2. Therefore

$$\frac{d^2y}{dx^2} = \frac{d}{d\theta}\left(\frac{\sin\theta}{1-\cos\theta}\right) \div \frac{dx}{d\theta}$$

Now, using the quotient rule,

$$\frac{d}{d\theta}\left(\frac{\sin\theta}{1-\cos\theta}\right) = \frac{(1-\cos\theta).\cos\theta - \sin\theta.\sin\theta}{(1-\cos\theta)^2}$$

$$= \frac{\cos\theta - \cos^2\theta - \sin^2\theta}{(1-\cos\theta)^2} = \frac{\cos\theta - 1}{(1-\cos\theta)^2}$$

$$= -\frac{1}{(1-\cos\theta)}$$

Finally,

$$\frac{d^2y}{dx^2} = -\frac{1}{(1-\cos\theta)} \div a(1-\cos\theta) = -\frac{1}{a(1-\cos\theta)^2}$$

The curve given by the parametric equations of Example 17.7 is called a **cycloid**. It is the curve traced out by a fixed point P on the rim of a circular wheel of radius a as it rolls along a horizontal surface. P is initially on the ground and θ is the angle turned through clockwise as indicated in Figure 17.2(a). The curve is shown in Figure 17.2(b); in practice, the arches of some bridges are constructed in the shape of one of the branches of the cycloid.

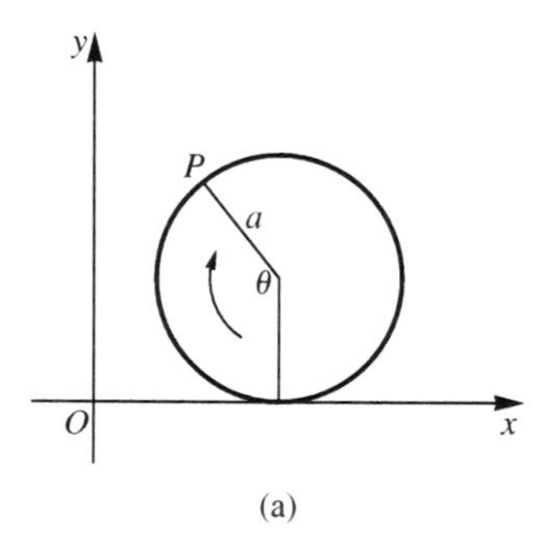

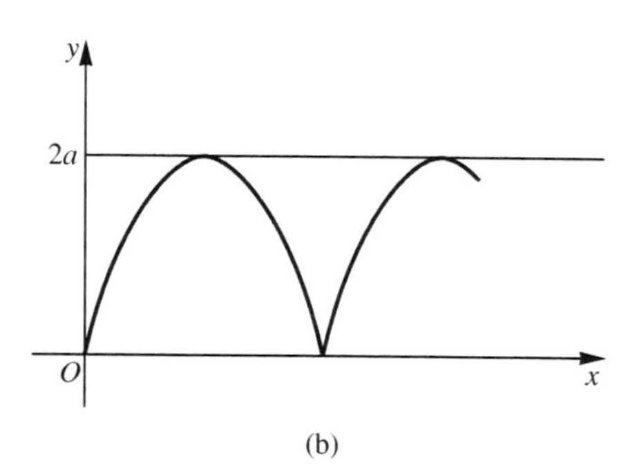

Figure 17.2. Development of the cycloid

Self-assessment questions 17.2

1. Explain what is meant by the phrase: 'a function is defined parametrically'.
2. Explain how to differentiate a function defined parametrically.
3. Explain how to obtain the second derivative of a function defined parametrically.

Exercise 17.2

1. Differentiate the following functions which are defined parametrically:

   (a) $x = t^3$, $y = 1 + t^2$
   (b) $x = (1+t)^2$, $y = 1 + t^2$
   (c) $x = ct$, $y = c/t$
   (d) $x = at^2$, $y = 2at$
   (e) $x = a\cosh\theta$, $y = b\sinh\theta$

2. Obtain the second derivatives of the functions of Question 1.

17.3 Curvature

The curvature, κ, of a plane curve $y = f(x)$ measures the rate at which the curve is bending away from the tangent to the curve at each point on the curve. A straight line therefore has zero curvature at all points. The more rapidly a curve bends the greater its curvature. A more helpful concept is that of radius of curvature. In Figure 17.3 we show a circle which has the same tangent as the curve $y = f(x)$ has at the point of contact P and has the same curvature as the curve at P.

It is called the **circle of curvature**; its centre is the **centre of curvature** at P and its radius is the **radius of curvature** of the curve $y = f(x)$ at P.

The radius of curvature ρ is related to the curvature κ by $\rho = 1/\kappa$.

KEY POINT

The radius of curvature can be expressed as

$$\rho = \frac{\left[1 + \left(\frac{dy}{dx}\right)^2\right]^{\frac{3}{2}}}{\left|\frac{d^2y}{dx^2}\right|} \qquad 17.3$$

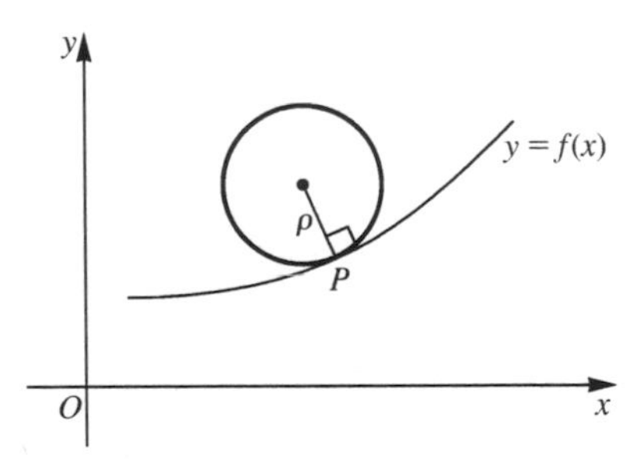

Figure 17.3.
Circle of curvature

Worked examples

17.8 Find the radius of curvature ρ at a point on the hyperbola $xy = c^2$. Where is ρ least?

Evaluate ρ at the points $(c/2, 2c)$, (c, c) and $(2c, c/2)$. Interpret your results in terms of the graph of the hyperbola.

Solution $y = c^2/x$ so that

$$\frac{dy}{dx} = -\frac{c^2}{x^2} \qquad \text{and} \qquad \frac{d^2y}{dx^2} = \frac{2c^2}{x^3}$$

Then

$$\rho = \left(1 + \frac{c^4}{x^4}\right)^{\frac{3}{2}} \div \left|\frac{2c^2}{x^3}\right|$$

$$= \frac{1}{2c^2}\left(1 + \frac{c^4}{x^4}\right)^{\frac{3}{2}} \times |x^3|$$

$$= \frac{1}{2c^2}\left(1 + \frac{c^4}{x^4}\right)^{\frac{3}{2}} (x^2)^{\frac{3}{2}}$$

$$= \frac{1}{2c^2}\left(x^2 + \frac{c^4}{x^2}\right)^{\frac{3}{2}}$$

ρ is minimal when $x^2 + c^4/x^2$ is least.

Using differentiation, this is seen to occur when $x^2 = c^2$, that is $x = \pm c$. This gives the points (c,c) and $(-c,-c)$.

At $\left(\frac{c}{2}, 2c\right)$, $\rho = \frac{1}{2c^2}\left(\frac{17c^2}{4}\right)^{\frac{3}{2}} = \frac{17\sqrt{17}}{16}c$

At (c, c), $\rho = \frac{1}{2c^2}(2c^2)^{\frac{3}{2}} = \sqrt{2}\,c$

At $\left(2c, \frac{c}{2}\right)$, $\rho = \frac{17\sqrt{17}}{16}c$

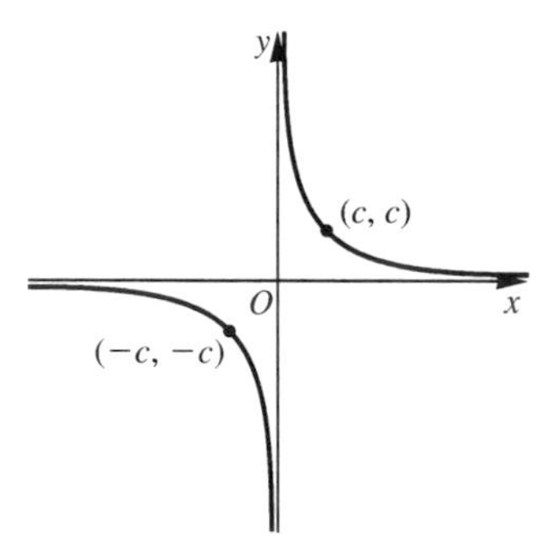

Figure 17.4. $y = \dfrac{c^2}{x}$

Figure 17.4 shows the graph of $y = c^2/x$. The curve is symmetrical about the lines $y = x$ and $y = -x$. As $x \to \pm\infty$ or as $x \to 0$, $\rho \to \infty$ indicating that $\kappa = 1/\rho \to 0$, confirming that the curve is asymptotic to a straight line.

17.9 Find the radius of curvature at a point on the cycloid

$$x = a(\theta - \sin\theta), \qquad y = a(1 - \cos\theta)$$

Solution From Example 17.7 we have

$$\frac{dy}{dx} = \frac{\sin\theta}{(1 - \cos\theta)} \qquad \text{and} \qquad \frac{d^2y}{dx^2} = -\frac{1}{a(1 - \cos\theta)^2}$$

Then

$$\rho = \left[1 + \frac{\sin^2\theta}{(1-\cos\theta)^2}\right]^{\frac{3}{2}} \times \left|\frac{a(1-\cos\theta)^2}{-1}\right|$$

$$= \frac{[(1-\cos\theta)^2 + \sin^2\theta]^{\frac{3}{2}}}{(1-\cos\theta)^3} \times \left|\frac{a(1-\cos\theta)^2}{-1}\right|$$

$$= \frac{(2-2\cos\theta)^{\frac{3}{2}}}{(1-\cos\theta)^3} \times \left|\frac{a(1-\cos\theta)^2}{-1}\right|$$

$$= 2\sqrt{2}a(1-\cos\theta)^{\frac{1}{2}}$$

Note that $\rho = 0$ where $1 - \cos\theta = 0$, that is where $y = 0$; these points are when the curve has a cusp.

Self-assessment questions 17.3

1. Explain the terms curvature and radius of curvature.
2. What does an infinite radius of curvature indicate?

Exercise 17.3

1. Find the radius of curvature of the given curves at the points indicated:

   (a) $y = x^2$ at the points (2,4), (−2,4), (0,0)
   (b) $(x^2/a^2) + (y^2/b^2) = 1$ at the points $(a,0)$, $(0,b)$, $(a/2, \sqrt{3}b/2)$
   (c) $(x+1)^2 + (y-2)^2 = 1$ at the points (−1,3), (−1,1), (0,2), (−2,2)

   Where is the radius of curvature greatest and where is it least? Interpret your results.
2. Find the radius of curvature at a point on the curve given parametrically by $x = at^2, y = 2at$.

17.4 Differentiation of vectors

If the position vector $\boldsymbol{r}$ of an object changes with time t we may write $\boldsymbol{r} = \boldsymbol{r}(t)$. The velocity vector $\boldsymbol{v} = \boldsymbol{v}(t)$ is given by

$\dot{\boldsymbol{r}}(t)$ is read as $\boldsymbol{r}$ dot t.

$$\boldsymbol{v} = \frac{d\boldsymbol{r}}{dt} \equiv \dot{\boldsymbol{r}}(t)$$

and the acceleration vector $\boldsymbol{a} = \boldsymbol{a}(t)$ by

$$\boldsymbol{a} = \frac{d\boldsymbol{v}}{dt} \equiv \dot{\boldsymbol{v}}(t) \quad \text{or} \quad \frac{d^2\boldsymbol{r}}{dt^2} = \ddot{\boldsymbol{r}}(t)$$

In 3D Cartesian components, if $\boldsymbol{r} = x\boldsymbol{i} + y\boldsymbol{j} + x\boldsymbol{k}$ where x, y and z are functions of time then

$$\boldsymbol{v} = \dot{x}\boldsymbol{i} + \dot{y}\boldsymbol{j} + \dot{z}\boldsymbol{k} \qquad 17.4$$

and

$$\boldsymbol{a} = \ddot{x}\boldsymbol{i} + \ddot{y}\boldsymbol{j} + \ddot{z}\boldsymbol{k} \qquad 17.5$$

where $\dot{x} \equiv \mathrm{d}x/\mathrm{d}t$, $\ddot{x} \equiv \mathrm{d}^2x/\mathrm{d}t^2$ and so on.

In 2D the results are

$$\boldsymbol{r} = x\boldsymbol{i} + y\boldsymbol{j}, \quad \boldsymbol{v} = \dot{x}\boldsymbol{i} + \dot{y}\boldsymbol{j}, \quad \boldsymbol{a} = \ddot{x}\boldsymbol{i} + \ddot{y}\boldsymbol{j} \qquad 17.6$$

In general if $\boldsymbol{b} = \boldsymbol{b}(t)$ and $\boldsymbol{c} = \boldsymbol{c}(t)$ are time dependent and $\lambda = \lambda(t)$ is a scalar then the following rules hold:

KEY POINT

$$\frac{\mathrm{d}}{\mathrm{d}t}(\lambda\boldsymbol{b}) = \lambda\frac{\mathrm{d}\boldsymbol{b}}{\mathrm{d}t} + \frac{\mathrm{d}\lambda}{\mathrm{d}t}\boldsymbol{b} \qquad 17.7a$$

$$\frac{\mathrm{d}}{\mathrm{d}t}(\boldsymbol{b} + \boldsymbol{c}) = \frac{\mathrm{d}\boldsymbol{b}}{\mathrm{d}t} + \frac{\mathrm{d}\boldsymbol{c}}{\mathrm{d}t} \qquad 17.7b$$

$$\frac{\mathrm{d}}{\mathrm{d}t}(\boldsymbol{b}.\boldsymbol{c}) = \boldsymbol{b}.\frac{\mathrm{d}\boldsymbol{c}}{\mathrm{d}t} + \frac{\mathrm{d}\boldsymbol{b}}{\mathrm{d}t}.\boldsymbol{c} \qquad 17.7c$$

$$\frac{\mathrm{d}}{\mathrm{d}t}(\boldsymbol{b} \times \boldsymbol{c}) = \boldsymbol{b} \times \frac{\mathrm{d}\boldsymbol{c}}{\mathrm{d}t} + \frac{\mathrm{d}\boldsymbol{b}}{\mathrm{d}t} \times \boldsymbol{c} \qquad 17.7d$$

Worked examples

17.10 If a particle moves according to the relationship $\boldsymbol{r} = \cos t\boldsymbol{i} + \sin t\boldsymbol{j}$ find $\dot{\boldsymbol{r}}$ and $\ddot{\boldsymbol{r}}$ and show that the velocity vector is perpendicular to the position vector. Describe the motion of the particle.

Solution

$$\dot{\boldsymbol{r}} = -\sin t\boldsymbol{i} + \cos t\boldsymbol{j}$$

$$\ddot{\boldsymbol{r}} = -\cos t\boldsymbol{i} - \sin t\boldsymbol{j}$$

Now $\ddot{\boldsymbol{r}} = -\boldsymbol{r}$, so that the direction of the acceleration vector is in the opposite direction to that of the position vector.

Also, $\boldsymbol{r}.\dot{\boldsymbol{r}} = \cos t \times (-\sin t) + \sin t \times \cos t = 0$ and hence the vector $\boldsymbol{r}$ is perpendicular to the vector $\dot{\boldsymbol{r}}$.

The particle is moving in a circle, the velocity vector is tangential to the circle and the acceleration is directed towards the centre of the circle.

Note that the speed of the particle is

$$|\dot{\boldsymbol{r}}| = \sqrt{\sin^2 t + \cos^2 t} = 1$$

17.11 For the vectors

$$\boldsymbol{b} = t\boldsymbol{i} + \boldsymbol{j} + t^2\boldsymbol{k}$$

$$\boldsymbol{c} = t^2\boldsymbol{j} - 2t\boldsymbol{k}$$

verify (17.7c) and (17.7d).

Solution

$$b.c = t \times 0 + 1 \times t^2 - t^2 \times 2t = t^2 - 2t^3$$

$$b \times c = \begin{vmatrix} i & j & k \\ t & 1 & t^2 \\ 0 & t^2 & -2t \end{vmatrix} = (-2t - t^4)i - (-2t^2)j + t^3 k$$

$$\frac{db}{dt} = i + 2tk, \ \frac{dc}{dt} = 2tj - 2k$$

Then

$$\frac{d}{dt}(b.c) = 2t - 6t^2$$

Also,

$$b.\frac{dc}{dt} = t \times 0 + 1 \times 2t - t^2 \times 2 = 2t - 2t^2$$

and

$$\frac{db}{dt}.c = 1 \times 0 + 0 \times t^2 - 2t \times 2t = -4t^2$$

Hence (17.7c) is verified.

Also

$$\frac{d}{dt}(b \times c) = (-2 - 4t^3)i + 4tj + 3t^2 k$$

Now

$$b \times \frac{dc}{dt} = \begin{vmatrix} i & j & k \\ t & 1 & t^2 \\ 0 & 2t & -2 \end{vmatrix} = (-2 - 2t^3)i + 2tj + 2t^2 k$$

and

$$\frac{db}{dt} \times c = \begin{vmatrix} i & j & k \\ 1 & 0 & 2t \\ 0 & t^2 & -2t \end{vmatrix} = -2t^3 i + 2tj + t^2 k$$

Hence (17.7d) is verified.

Self-assessment questions 17.4

1. Explain how to obtain the derivative of a vector specified in Cartesian form.
2. If r is the position vector of an object at time t, what is the interpretation of the vectors $\dot{r}$ and $\ddot{r}$?

Exercise 17.4

1. If $r = \cos \omega t i + \sin \omega t j$ find $\dot{r}$, $\ddot{r}$, $|\dot{r}|$ and $|\ddot{r}|$.
2. Verify (17.7) for the vectors $a = t^2 i + tj$, $b = 2ti - t^2 j$ and the scalar $\lambda = 3$.

Test and assignment exercises 17

1. Find dy/dx and d^2y/dx^2 from the following equations:

   (a) $\sin y = xy$
   (b) $x^2y^2 = x^2 + y^2 + x$

2. Find the tangents to the following curves at the points indicated:

   (a) $x^2y^2 = x^3 + 2y^4$ at $(-1, 1)$ and $(-1, -1)$
   (b) $x^2y^2 = x^2y + x - 1$ at $(1, 1)$

3. Differentiate the following using logarithmic differentiation:

   (a) $xe^{-2x}\cos 3x$ (b) $x^3(2+x)^4e^{-x}$

   (c) $\dfrac{e^{3x}\sin x}{1+x^2}$ (d) $\dfrac{(x^2+1)^2e^{3x}}{1+2x}$

4. Find dy/dx and d^2y/dx^2 for the following functions which are specified parametrically:

   (a) $x = t^2, y = 1 + t^3$
   (b) $x = 1 + t^2, y = (1+t)^2$
   (c) $x = \dfrac{t^2+1}{t}, y = \dfrac{t^2-1}{t}$

5. Find the radius of curvature of a point on the following curves:

   (a) $\dfrac{x^2}{a^2} - \dfrac{y^2}{b^2} = 1$

   (b) $y^2 = 4ax$

   (c) $x = ct, y = \dfrac{c}{t}$

   (d) $x = a\cos\theta, y = a\sin\theta$

6. If a particle moves according to the relationship $\boldsymbol{r} = \cos t\boldsymbol{i} + \sin t\boldsymbol{j} + t\boldsymbol{k}$ describe the motion of the particle; find $\dot{\boldsymbol{r}}$ and $\ddot{\boldsymbol{r}}$.

7. Given the vectors $\boldsymbol{a} = 2t\boldsymbol{i} - \boldsymbol{j}$ and $\boldsymbol{b} = -\boldsymbol{i} + t\boldsymbol{j} + (t^2+1)\boldsymbol{k}$ find

   (a) $\dfrac{d\boldsymbol{a}}{dt}$ (b) $\dfrac{d\boldsymbol{b}}{dt}$

   (c) $\dfrac{d}{dt}(\boldsymbol{a}.\boldsymbol{b})$ (d) $\dfrac{d}{dt}(\boldsymbol{a}\times\boldsymbol{b})$

   Verify (17.7c) and (17.7d) for the vectors $\boldsymbol{a}$ and $\boldsymbol{b}$.

18 Power series

Objectives

This chapter

- defines tangent and quadratic approximations to a function
- develops the Taylor and Maclaurin series for a function
- states the Maclaurin series for standard functions
- uses L'Hôpital's rule for evaluating the limit of a rational function
- applies the Taylor series to the approximation of small errors

18.1 Tangent and quadratic approximations

You may be familiar with the approximation $\sin x \simeq x$ for small values of x (in radians). This is an example of a **tangent approximation**.

Figure 18.1 shows the curve $y = f(x)$ and the tangent to the curve at $x = a$.

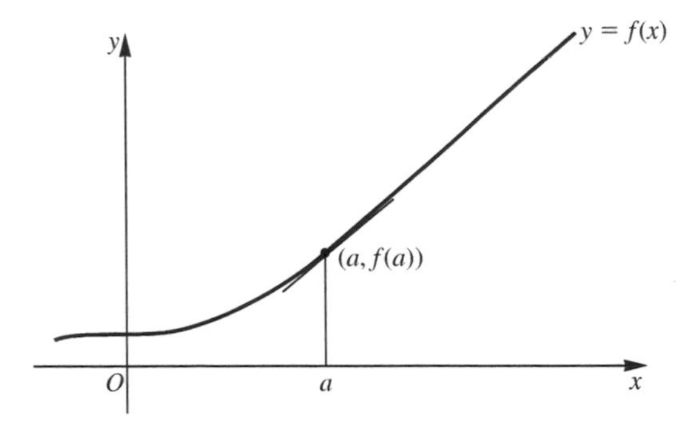

Figure 18.1. Tangent approximation

This tangent has the same gradient as the curve at $x = a$. Let the equation of the tangent be $y = \alpha + \beta x$. Then at $x = a$,

$$y = f(a) = \alpha + \beta a \qquad \text{and} \qquad \frac{\mathrm{d}y}{\mathrm{d}x} = f'(a) = \beta$$

and therefore $\alpha = f(a) - f'(a)a$. We have the following result:

KEY POINT

The tangent approximation to $f(x)$ near $x = a$ is

$$f(x) \simeq p_1(x) = f(a) + (x - a)f'(a) \qquad 18.1$$

The approximation deteriorates as x moves away from a. An improvement is made by replacing the straight line by a quadratic curve which has the same value and same first and second derivatives at $x = a$ as the function itself.

If the quadratic is $y = \alpha + \beta x + \gamma x^2$ then at $x = a$,

$$y = \alpha + \beta a + \gamma a^2 = f(a)$$

$$\frac{dy}{dx} = \beta a + 2\gamma a = f'(a)$$

and

$$\frac{d^2y}{dx^2} = 2\gamma = f''(a)$$

We solve for γ, β and α and obtain the following result:

KEY POINT

The quadratic approximation to $f(x)$ at $x = a$ is

$$f(x) \simeq p_2(x) = f(a) + (x - a)f'(a) + \frac{(x-a)^2}{2} f''(a) \qquad 18.2$$

Worked examples

18.1 Find the tangent approximations to the following functions:
(a) e^x (b) $\sin x$ (c) $\cos x$

Solution (a) $f'(x) = e^x$ and $f'(a) = e^a$. Hence

$$e^x \simeq e^a + (x - a)\,e^a$$

(b) $f'(x) = \cos x$ and $f'(a) = \cos a$. Hence

$$\sin x \simeq \sin a + (x - a)\cos a$$

(c) $f'(x) = -\sin x$ and $f'(a) = -\sin a$. Hence

$$\cos x \simeq \cos a - (x - a)\sin a$$

18.2 Evaluate the tangent approximation to e^x based at $a = 0$ at $x = 0$, ± 0.1, ± 0.5, ± 1.

Compare with the calculator values to 3 d.p.

Table 18.1.

x	$1+x$	e^x (3 d.p.)	Error: $e^x - 1 - x$
−1	0	0.368	0.368
−0.5	0.5	0.607	0.107
−0.1	0.9	0.905	0.005
0	1	1	0
0.1	1.1	1.105	0.005
0.5	1.5	1.649	0.149
1	2	2.718	0.718

Solution $e^a = e^0 = 1$ so that the tangent approximation is $e^x \simeq 1 + x$. The values are compared in Table 18.1.

18.3 Obtain the quadratic approximations to the following functions:
(a) e^x (b) $\sin x$ (c) $\cos x$

Solution (a) $f''(x) = e^x$ and $f''(a) = e^a$. Hence

$$e^x \simeq e^a + (x-a)e^a + \frac{(x-a)^2}{2} e^a$$

(b) $f''(x) = -\sin x$ and $f''(a) = -\sin a$. Hence

$$\sin x \simeq \sin a + (x-a)\cos a - \frac{(x-a)^2}{2} \sin a$$

(c) $f''(a) = -\cos a$. Hence

$$\cos x \simeq \cos a - (x-a)\sin a - \frac{(x-a)^2}{2} \cos a$$

The above approach can be generalized to obtain polynomial approximations of any finite order n.

KEY POINT

> The nth order **Taylor approximation** of $f(x)$ based at $x = a$ is
>
> $$f(x) \simeq p_n(x) = f(a) + (x-a)f'(a) + \frac{(x-a)^2}{2!} f''(a) + \ldots + \frac{(x-a)^n}{n!} f^{(n)}(a) \qquad 18.3$$

Worked examples

18.4 Evaluate the quadratic approximation to e^x, based at $a = 0$, at $x = 0$, ± 0.1, ± 0.5, ± 1. Compare with the calculator values to 3 d.p.

Table 18.2.

x	$1+x+\frac{1}{2}x^2$	e^x (3 d.p.)	$e^x-1-x-\frac{1}{2}x^2$
−1	0.5	0.368	−0.132
−0.5	0.625	0.607	−0.018
−0.1	0.905	0.905	0.000
0	1	1	0
0.1	1.105	1.105	0.000
0.5	1.625	1.649	0.024
1	2.5	2.718	0.218

Table 18.3.

x	$1+x$	$1+x+\frac{1}{2}x^2$	e^x (3 d.p.)
−3	−2	2.5	0.0498
−2	−1	1	0.135
2	3	5	7.389
3	4	8.5	20.086

Solution The quadratic approximation is $1+x+\frac{1}{2}x^2$. The values are compared in Table 18.2.

Notice that in all cases, save $x=0$, the magnitude of the error is reduced when compared to the tangent approximation.

18.5 Find the Taylor approximations to the following functions to the order given in parentheses:
(a) $\sin x$ (6) (b) e^x (n)

Solution (a) $f'''(x)=-\cos x$, $f^{(iv)}(x)=\sin x$, $f^{(v)}(x)=\cos x$, $f^{(vi)}(x)=-\sin x$.
Hence

$$\sin x \simeq \sin a + (x-a)\cos a - \frac{(x-a)^2}{2!}\sin a - \frac{(x-a)^3}{3!}\cos a$$

$$+\frac{(x-a)^4}{4!}\sin a + \frac{(x-a)^5}{5!}\cos a - \frac{(x-a)^6}{6!}\sin a$$

(b) All derivatives of $f(x)$ are equal to e^x. Hence

$$e^x \simeq e^a + (x-a)e^a + \frac{(x-a)^2}{2!}e^a + \ldots + \frac{(x-a)^n}{n!}e^a$$

18.6 Evaluate the linear and quadratic approximations to e^x, based at $a=0$, at $x=\pm 2, \pm 3$. What conclusions can be drawn?

Solution The values are compared in Table 18.3.

Although the quadratic approximation is closer than the linear approximation it is still quite dreadful.

Self-assessment questions 18.1

1. Explain what is meant by the tangent approximation to a function $f(x)$ at the point $x = a$.
2. Suppose that you had the tangent approximation to a given function; how would you obtain the quadratic approximation?

Exercise 18.1

1. Find the tangent approximations to the following functions:

   (a) e^{-x} (b) $\ln(1 + x)$ (c) $\tan x$
   (d) x^2 (e) x^3

2. Use the tangent approximations in Question 1(a), (b) and (c) based at $a = 0$ to estimate to 3 d.p. the values of the functions at $x = 0, \pm 0.1, \pm 0.5, \pm 1$.

3. Obtain the quadratic approximations to the functions in Question 1.

4. Evaluate the quadratic approximations in Question 3, based at $a = 0$, at $x = 0$, $\pm 0.1, \pm 0.5, \pm 1$.

5. Find the fourth order Taylor approximations near $a = 0$ to the following functions:

   (a) $\cos x$ (b) e^{-x} (c) $(1 + x)^4$

18.2 Taylor and Maclaurin series

As n increases, the nth order Taylor approximations become better approximations to the function.

Taylor's theorem guarantees that, under suitable conditions, including the existence of all relevant derivatives,

KEY POINT

$$f(x) = f(a) + (x - a)f'(a) + \frac{(x - a)^2}{2!}f''(a) + \dots + \frac{(x - a)^n}{n!}f^{(n)}(a) + R_n(x) \quad 18.4$$

where

$$R_n(x) = \frac{x^{n+1}}{(n + 1)!}f^{(n+1)}(a + \theta h)\, h = x - a \quad \text{and} \quad 0 < \theta < 1$$

If $R_n(x) \to 0$ as $n \to \infty$ then we may represent $f(x)$ by an infinite series.

KEY POINT

The **Taylor series** for $f(x)$ is

$$f(x) = f(a) + (x - a)f'(a) + \frac{(x - a)^2}{2!}f''(a) + \dots + \frac{(x - a)^n}{n!}f^{(n)}(a) + \dots \quad 18.5$$

Replacing x by $x + a$ we obtain

$$f(x+a) = f(a) + xf'(a) + \frac{x^2}{2!}f''(a) + \ldots + \frac{x^n}{n!}f^{(n)}(a) + \ldots \qquad 18.6$$

Sometimes (18.6) is valid for all values of x, but sometimes for a restricted set of values of x.

To test a series for convergence can often be a complicated process. One useful test which is simple to carry out is the ratio test, which applies for any general value of x. Note that if the Taylor series converges, it does so to $f(x)$.

The **ratio test** for the series $a_0 + a_1x + a_2x^2 + \ldots$ first forms the ratio

$$\left|\frac{a_nx^n}{a_{n-1}x^{n-1}}\right| = \left|\frac{a_n}{a_{n-1}}\right| \cdot |x|$$

then, if this ratio tends to a limit $l < 1$ as $n \to \infty$, the series converges.

If $l > 1$ then the series diverges, whereas if $l = 1$ then the ratio test is inconclusive and we must use other tests to determine whether the series converges or diverges.

Worked example

18.7 For what values of x do the following series converge?

(a) $1 + 2x + 3x^2 + 4x^3 + \ldots$ (b) $1 + x + \frac{1}{2!}x^2 + \frac{1}{3!}x^3 + \ldots$

Solution

(a) The general term is $(n+1)x^n$ and its predecessor is nx^{n-1}. The required ratio is

$$\left|\frac{(n+1)x^n}{nx^{n-1}}\right| = \left|\frac{n+1}{n}\right| \cdot |x|$$

As $n \to \infty$ for a fixed x this ratio $\to 1.|x| = |x|$. Hence for convergence $|x| < 1$, that is $-1 < x < 1$.

(b) The general term is $x^n/n!$, its predecessor is $x^{n-1}/(n-1)!$ and the required ratio is

$$\left|\frac{x^n}{n!}\frac{(n-1)!}{x^{n-1}}\right| = \frac{1}{n} \cdot |x|$$

As $n \to \infty$ for a fixed x, the ratio approaches zero. Hence the series converges for all x.

The **Maclaurin series** for a function is a special case of the Taylor series when $a = 0$. It is given by:

KEY POINT

$$f(x) = f(0) + xf'(0) + \frac{x^2}{2!}f''(0) + \ldots + \frac{x^n}{n!}f^{(n)}(0) + \ldots \qquad 18.7$$

Table 18.4.

Function	Maclaurin series	Convergence
$(1+x)^s$	$1+\frac{s}{1}x+\frac{s(s-1)}{1.2}x^2+\frac{s(s-1)(s-2)}{1.2.3}x^3+\ldots$	$-1<x<1$
e^x	$1+x+\frac{x^2}{2!}+\frac{x^3}{3!}+\frac{x^4}{4!}+\ldots$	All x
$\sin x$	$x-\frac{x^3}{3!}+\frac{x^5}{5!}-\frac{x^7}{7!}+\ldots$	All x
$\cos x$	$1-\frac{x^2}{2!}+\frac{x^4}{4!}-\frac{x^6}{6!}+\ldots$	All x
$\sinh x$	$x+\frac{x^3}{3!}+\frac{x^5}{5!}+\frac{x^7}{7!}+\ldots$	All x
$\cosh x$	$1+\frac{x^2}{2!}+\frac{x^4}{4!}+\frac{x^6}{6!}+\ldots$	All x
$\ln(1+x)$	$x-\frac{x^2}{2}+\frac{x^3}{3}-\frac{x^4}{4}+\ldots$	$-1<x\leqslant 1$

Some standard Maclaurin series are given in Table 18.4.

Note that the series for $\ln(1+x)$ has been quoted since $\ln x$ and its derivatives are not defined at $x=0$.

Worked example

18.8 Obtain the Maclaurin series for the following functions, stating the values of x for which convergence is guaranteed:

(a) $\sin 2x$ (b) $\left(1+\frac{x}{2}\right)^{-1}$ (c) $\ln(1-4x)$

Solution (a) Replace x by $2x$ in the Maclaurin series for $\sin x$:

$$\sin 2x = 2x-\frac{(2x)^3}{3!}+\frac{(2x)^5}{5!}-\ldots = 2x-\frac{8x^3}{3!}+\frac{32x^5}{5!}-\ldots$$

This is valid for all x.

(b) Replacing s by -1 and x by $x/2$ in the standard series:

$$\left(1+\frac{x}{2}\right)^{-1} = 1+\left(\frac{-1}{1}\right)\left(\frac{x}{2}\right)+\frac{(-1)(-2)}{1.2}\left(\frac{x}{2}\right)^2+\frac{(-1)(-2)(-3)}{1.2.3}\left(\frac{x}{2}\right)^3+\ldots$$

$$= 1-\frac{x}{2}+\frac{x^2}{4}-\frac{x^3}{8}+\ldots$$

This is valid for $-1<x/2<1$, that is for $-2<x<2$.

(c) Replace x by $-4x$ in the standard series:

$$\ln(1-4x) = -4x - \frac{(-4x)^2}{2} + \frac{(-4x)^3}{3} - \dots$$
$$= -4x - 8x^2 - \frac{64}{3}x^3 - \dots$$

The series converges for $-1 < -4x \leqslant 1$, that is $-1 \leqslant 4x < 1$, or $-1/4 \leqslant x < 1/4$.

Self-assessment questions 18.2

1. Explain how to use the ratio test to determine the range of convergence of a Taylor series.
2. How are a Taylor series and a Maclaurin series for a function $f(x)$ related?

Exercise 18.2

1. For which values of x do the following series converge?

   (a) $1 + \frac{1}{2}x + \frac{1}{4}x^2 + \frac{1}{8}x^3 + \dots$

   (b) $x - \frac{x^3}{3} + \frac{x^5}{5} - \frac{x^7}{7} + \dots$

2. Using Table 18.4 obtain the Maclaurin series for the following functions. State the values of x for which the series converge.

   (a) $\cos\frac{1}{2}x$ (b) $\left(1 - \frac{x}{2}\right)^{-2}$

   (c) $\ln(1 - x^2)$.

18.3 Obtaining and using Maclaurin series

With a complicated function it would be tedious to find its Maclaurin series by differentiating the function repeatedly. The following examples show some of the devices we can use.

Worked examples

18.9 Find the Maclaurin series as far as the term in x^6 for the following functions:
(a) $\sin^2 x$ (b) $\cos^2 x$

Solution (a) We *could* multiply the series for $\sin x$ by itself and build up the terms. However, we use the identity

$$\cos 2x \equiv 1 - 2\sin^2 x$$

from which we obtain

$$\sin^2 x \equiv \tfrac{1}{2}(1 - \cos 2x)$$

Replacing x by $2x$ in the standard series for $\cos x$ we obtain

$$\cos 2x \equiv 1 - \frac{(2x)^2}{2!} + \frac{(2x)^4}{4!} - \frac{(2x)^6}{6!} + \ldots$$

$$\equiv 1 - 2x^2 + \frac{2}{3}x^4 - \frac{4}{45}x^6 + \ldots$$

Then

$$\sin^2 x = \frac{1}{2}\left(2x^2 - \frac{2}{3}x^4 + \frac{4}{45}x^6 \ldots\right)$$

$$= x^2 - \frac{1}{3}x^4 + \frac{2}{45}x^6 \ldots$$

(b) Having obtained the series for $\sin^2 x$ we should use it. Since $\cos^2 x \equiv 1 - \sin^2 x$ then the series is obtained immediately as

$$\cos^2 x = 1 - x^2 + \frac{1}{3}x^4 - \frac{2}{45}x^6 + \ldots$$

18.10 If $y = \ln \cos x$ show that

$$\frac{d^3y}{dx^3} + 2\frac{d^2y}{dx^2} \cdot \frac{dy}{dx} = 0$$

Hence find the Maclaurin series of $\ln \cos x$ as far as the term in x^4. Hence find

$$\lim_{x \to 0} \frac{\ln \cos x}{x^2}$$

Solution If $y = \ln \cos x$,

$$\frac{dy}{dx} = \frac{-\sin x}{\cos x} = -\tan x$$

$$\frac{d^2y}{dx^2} = -\sec^2 x$$

and

$$\frac{d^3y}{dx^3} = -2 \sec x . \sec x \tan x = -2 \sec^2 x . \tan x$$

Hence

$$\frac{d^3y}{dx^3} + 2\frac{d^2y}{dx^2} \cdot \frac{dy}{dx} = -2 \sec^2 x \tan x + 2(-\sec^2 x)(-\tan x) = 0$$

Differentiating this equation we obtain

$$\frac{d^4y}{dx^4} + 2\frac{d^2y}{dx^2} \cdot \frac{d^2y}{dx^2} + 2\frac{d^3y}{dx^3} \cdot \frac{dy}{dx} = 0 \qquad \text{(i)}$$

At $x = 0$, $y = \ln \cos 0 = \ln 1 = 0$, $dy/dx = 0$, $d^2y/dx^2 = -1$, $d^3y/dx^3 = 0$ and from (i)

$$\frac{d^4y}{dx^4} + 2(-1)^2 + 0 = 0 \quad \text{so that} \quad \frac{d^4y}{dx^4} = -2$$

Hence the Maclaurin series is

It is sometimes possible to obtain a Maclaurin series by differentiating a known series term-by-term, but the resulting series may not converge to the correct function.

$$\ln \cos x = \frac{x^2}{2!}(-1) + \frac{x^4}{4!}(-2) - \ldots$$

$$= -\frac{x^2}{2} - \frac{x^4}{12} - \ldots$$

Now

$$\frac{\ln \cos x}{x^2} = -\frac{1}{2} - \frac{x^2}{12} - \ldots$$

Therefore

$$\lim_{x \to 0} \frac{\ln \cos x}{x^2} = -\frac{1}{2}$$

18.11 A uniform beam of length l is simply supported and subjected to a uniformly distributed load w and a compressive force P. The deflection at the mid-point is

$$w\frac{EI}{P^2}\left(\sec\frac{1}{2}ml - 1\right) - \frac{wl^2}{8P}$$

where $m^2 = P/EI$ and E and I are constants. Show that as $P \to 0$, the deflection $\to 5wl^4/384\,EI$. (Use the series $\sec x = 1 + \frac{1}{2}x^2 + \frac{5}{24}x^4 + \ldots$.)

Solution Using the given series we obtain

$O(l^6)$ means that all remaining terms contain l to a power of at least 6

$O(P^3)$ means that all remaining terms contain P to a power of at least 3

$$\sec\frac{1}{2}ml - 1 = \frac{1}{2}\frac{m^2l^2}{4} + \frac{5}{24}\frac{m^4l^4}{16} + O(l^6) + \ldots$$

$$= \frac{Pl^2}{8\,EI} + \frac{5P^2l^4}{16 \times 24\,E^2I^2} + O(P^3)$$

Then the deflection is

$$\frac{w}{8}\frac{l^2}{Pw} + \frac{5wl^4}{384\,EI} + O(P) - \frac{wl^2}{8P} = \frac{5wl^4}{384\,EI} + O(P) \to \frac{5wl^4}{384\,EI} \text{ as } P \to 0$$

Self-assessment questions 18.3

1. Give two examples of how a Maclaurin series can be obtained by using standard series and trigonometric identities.

Exercise 18.3

1. Give the Maclaurin series as far as the term in x^6 for the following functions:

   (a) $\sinh^2 x$
   (b) $\cosh^2 x$
   (c) $\ln\left(\dfrac{1+x}{1-x}\right)$

2. Obtain the Maclaurin series for $f(x) = (1+x^2)^{-1}$. By integrating the series term by term and noting that $\tan^{-1} 0 = 0$, obtain the Maclaurin series for $\tan^{-1} x$.

3. Show that the following series are the Maclaurin series for the following functions:

   (a) $\tan x = x + \dfrac{x^3}{3} + \dfrac{2x^5}{15} + \ldots$

   (b) $\sin^{-1} x = x + \dfrac{1}{2\times 3}x^3 + \dfrac{1\times 3}{2\times 4\times 5}x^5 + \ldots$

   (c) $\sec x = 1 + \dfrac{1}{2}x^2 + \dfrac{5}{24}x^4 + \ldots$

4. A function $y(x)$ is defined by $dy/dx = y^2 - 4x$ and the condition $y(0) = 1$. By calculating successive derivatives of y at $x = 0$ show that the Maclaurin series for y is $1 + x - x^2 - \frac{1}{3}x^3 - \frac{2}{5}x^4 + \ldots$.

5. It is also possible to obtain a Maclaurin series by integrating a known series. Check that by integrating term by term the series for $\cos x$ that you obtain the series for $\sin x$; to evaluate the constant of integration remember that $\sin 0 = 0$.

18.4 Applications of Taylor series

If we wish to examine the behaviour of a function $f(x)$ at a particular value $x = a$, the natural thing to do is to substitute $x = a$ into the formula for $f(x)$. If this results in $0/0$, an expression which is mathematically indeterminate, then we may employ **L'Hôpital's rule**.

KEY POINT

L'Hôpital's rule states that if $f(a) = 0$, $g(a) = 0$ then

$$\lim_{x\to a} \frac{f(x)}{g(x)} = \frac{f'(a)}{g'(a)} \qquad 18.8$$

Worked example

18.12 Use L'Hôpital's rule to evaluate the limits as $x \to 0$ of the following functions:

(a) $\frac{e^x - 1}{x}$ (b) $\frac{\sec x - 1}{x \sin x}$

Solution (a) Putting $x = 0$ we obtain

$$\frac{e^0 - 1}{0} = \frac{1 - 1}{0} = \frac{0}{0}$$

Differentiating the top and bottom of the function we obtain $e^x/1$. As $x \to 0$ this expression $\to e^0/1 = 1$.

(b) Putting $x = 0$ gives $0/0$. Differentiating top and bottom gives

$$\frac{\sec x \tan x}{x \cos x + \sin x}$$

Unfortunately, putting $x = 0$ gives $0/0$ again.

Repeating the process we differentiate top and bottom to obtain

$$\frac{\sec x \,.\, \sec^2 x + \sec x \tan x \,.\, \tan x}{-x \sin x + 2 \cos x}$$

Putting $x = 0$ gives $1/2$.

We may also use a Taylor series to obtain an estimate of small errors in calculations. Suppose x has been measured as a. If an error δx has been made in the measurement of x then there is an error δf in the value of $f(x)$. If the magnitude of δx is small then the **percentage error** in the value of $f(x)$ is given approximately by:

KEY POINT

$$\left(\frac{\delta f}{f(a)} \times 100\right)\% \simeq \left(\delta x \frac{f'(a)}{f(a)} \times 100\right)\% \qquad 18.9$$

Worked examples

18.13 A tower 30 m high stands on a bank of a river. The elevation of its top from a point across the river is measured as 20°. If the angle is subject to an error of 0.1°, find the approximate percentage error in the measurement of the distance of the observation point from the foot of the tower.

Solution

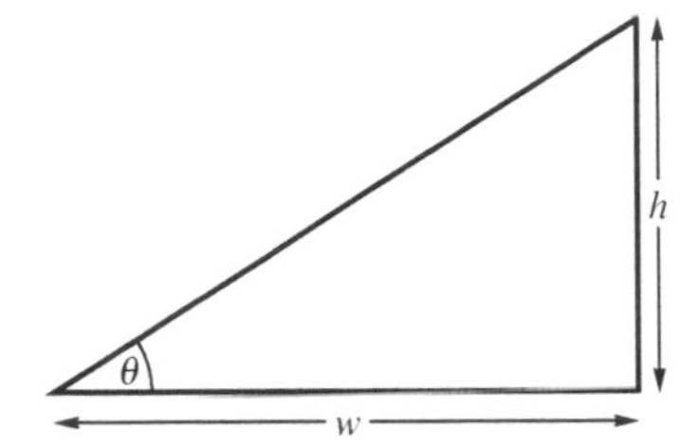

Figure 18.2. Diagram for Example 18.13

In Figure 18.2 the height of the tower is h, the distance of the observation point from the foot of the tower is w and the angle of elevation is θ. Then

$$w = h\cot\theta \qquad \text{and} \qquad \frac{\mathrm{d}w}{\mathrm{d}\theta} = -h\operatorname{cosec}^2\theta$$

If the error in the angle is $\delta\theta$ then the percentage error in w is approximately

$$\left(\frac{\delta\theta}{w}\cdot\frac{\mathrm{d}w}{\mathrm{d}\theta}\times 100\right)\% = \left(\frac{\delta\theta}{h\cot\theta}\times(-h\operatorname{cosec}^2\theta)\times 100\right)\%$$
$$= \left(-\frac{100\delta\theta}{\cos\theta\sin\theta}\right)\%$$

With the given data the approximate percentage error is

$$\pm\left(\frac{-100\times 0.1\times\pi}{\cos 20^\circ\times\sin 20^\circ\times 180}\right)\% = \pm 0.543\% \quad \text{(3 s.f.)}$$

18.14 The current in a series RL circuit is given by

$$i = \frac{E}{R}\left(1 - \mathrm{e}^{\frac{-Rt}{L}}\right)$$

Find a suitable formula for i when R is negligible.

Solution Putting $R = 0$ gives $i = 0/0$. Differentiating $1 - \mathrm{e}^{\frac{-Rt}{L}}$ gives

$$+\frac{t}{L}\,\mathrm{e}^{\frac{-Rt}{L}}$$

Then as $R \to 0$, by L'Hôpital's rule

$$\lim_{R\to 0}\frac{E}{R}\left(L - \mathrm{e}^{\frac{-Rt}{L}}\right) = \frac{E}{1}\,\frac{t}{L}\,\mathrm{e}^{\frac{-Rt}{L}} = \frac{Et}{L}\cdot\mathrm{e}^0 = \frac{Et}{L}$$

Hence $i \to Et/L$.

We can also use a Taylor series to obtain formulae for approximate differentiation.

Putting $x = a + h$ in (18.5) gives

$$f(a+h) = f(a) + hf'(a) + \frac{h^2}{2!}f''(a) + \frac{h^3}{3!}f'''(a) + \ldots$$

Putting $x = a - h$ in (18.5) gives

$$f(a-h) = f(a) - hf'(a)\frac{h^2}{2!}f''(a) - \frac{h^3}{3!}f'''(a) + \ldots \qquad \text{ii}$$

Subtracting (ii) from (i) gives

$$f(a+h) - f(a-h) = 2hf'(a) + O(h^3)$$

Hence

$$f'(a) = \frac{f(a+h) - f(a-h)}{2h} + O(h^2) \qquad 18.10$$

Adding (i) and (ii) gives

$$f(a+h) + f(a-h) = 2f(a) + \frac{2h^2}{2!}f''(a) + O(h^4)$$

and therefore

$$f''(a) = \frac{f(a+h) - 2f(a) + f(a-h)}{h^2} + O(h^2) \qquad 18.11$$

KEY POINT

$$f'(a) = \frac{f(a+h) - f(a-h)}{2h} + O(h^2)$$

$$f''(a) = \frac{f(a+h) - 2f(a) + f(a-h)}{h^2} + O(h^2)$$

Worked example

18.15 The angle in radians turned through by a shaft under the action of a torque is shown in Table 18.5. Estimate the angular speed $\dot{\theta}$ and angular acceleration $\ddot{\theta}$ when $t = 0.4$.

Solution With $h = 0.2$ we use (18.10) to find

$$\dot{\theta}(0.4) \simeq \frac{\theta(0.6) - \theta(0.2)}{0.4} = \frac{0.285 - 0.060}{0.4} = \frac{0.225}{0.4} = 0.55 \text{ rad s}^{-1}$$

Using (18.11) we find

$$\ddot{\theta}(0.4) \simeq \frac{\theta(0.6) - 2\theta(0.4) + \theta(0.2)}{h^2}$$

$$= \frac{0.285 - 2 \times 0.151 + 0.060}{(0.2)^2} = 1.075 \text{ rad s}^{-2}$$

Table 18.5.

t (seconds)	0	0.2	0.4	0.6	0.8
θ (radians)	0	0.060	0.151	0.285	0.472

Self-assessment questions 18.4

1. Explain when and how you would use L'Hôpital's rule.
2. State simple approximations for $f'(a)$ and $f''(a)$.

Exercise 18.4

1. Use L'Hôpital's rule to evaluate the following limits:

(a) $\lim_{x \to 0} \frac{1 - e^x}{x^2}$

(b) $\lim_{x \to 0} \frac{1 - \cos x}{x^2}$

(c) $\lim_{x \to 0} \frac{x - \sin x}{x^2}$

(d) $\lim_{x \to 0} \frac{\sin^2 x}{x}$

(e) $\lim_{x \to 0} \frac{\ln(1 + x)}{x}$

(f) $\lim_{x \to 1} \frac{\ln(\cos 2\pi x)}{1 + \cos \pi x}$

2. The radius of curvature of a cycloid at the origin is given by

$$\lim_{\theta \to 0} \frac{(\theta + \sin \theta)^2}{1 - \cos \theta}$$

Find this radius.

3. A particle moves along a straight line and its distance from a fixed point is recorded at $t = 0,1,2,\ldots,9$ as tabulated below. Estimate its speed and acceleration at $t = 1,2,\ldots,8$.

t	0	1	2	3	4	5	6	7	8	9
x	0	3	9	24	43	50	52	50	43	24

Comment on the results of your calculation.

Test and assignment exercises 18

1. Find the tangent approximations to the following functions:

(a) e^{2x}
(b) $\cosh x$
(c) $\sinh x$

Estimate the value of the tangent approximations, based at $a = 0$, at $x = 0.1, 0.5$ and 1 and compare with the actual values.

2. Repeat Exercise 1 using a quadratic approximation.

3. For what values of x do the following series converge?

(a) $1 + \frac{1}{3}x + \frac{1}{9}x^2 + \frac{1}{27}x^3 + \ldots$

(b) $1 + 2x + 3x^2 + 4x^3 + \ldots$

4. Find the Maclaurin series for

(a) $\ln(1 - x^2)$ (b) $\ln(1 + x^2)$

(c) $\ln\left(\frac{1 + x^2}{1 - x^2}\right)$

State the values of x for which each series converges.

5. Use L'Hôpital's rule to evaluate the following limits:

(a) $\lim_{x \to -1} \frac{a^{x+1} - 1}{x + 1}$ $(a > 0)$

(b) $\lim_{x \to 0} \frac{\ln(1 + x)}{x}$

6. Repeat Question 3 of Exercise 18.4 with the following data:

t	0	1	2	3	4	5	6	7	8	9	10
x	104	102	98	76	55	41	32	29	29	30	34

19 Partial differentiation

Objectives

This chapter

- introduces functions of two or more variables
- represents graphically a function of two variables
- shows how to find first and second partial derivatives
- states and uses the chain rule for partial derivatives
- defines differentials and applies them to the estimation of small errors

19.1 Functions of several variables

The functions which we have studied so far have had one independent variable input. Many systems in engineering have several inputs and we need to extend our ideas to cover **functions of several variables**. For example, van der Waals' equation of state is

$$\left(p+\frac{a}{V^2}\right)(V-b)=RT \qquad 19.1$$

where p, V and T are the pressure, volume and absolute temperature of the substance and R is the gas constant; a and b are constants. We can rearrange (19.1) to the form

$$p=\frac{RT}{(V-b)}-\frac{a}{V^2}$$

or

$$p=f(T,V)$$

This notation indicates that T and V are the independent variables and p is the dependent variable.

A second example is the angular frequency of oscillation of a series LCR circuit, which is given by

$$\omega = \left(\frac{1}{LC} - \frac{R^2}{4L^2}\right)^{\frac{1}{2}} \qquad 19.2$$

or

$$\omega = f(L, C, R)$$

The independent variables are L, C and R and ω is the dependent variable.

A function of two variables can be represented pictorially in two ways. The curves $f(x,y) = c$, where c is a constant, are called **level curves** and correspond to contours on a map. In Figure 19.1(a) the function $f(x,y) = x^2 + y^2$ is depicted by sketching the curve $x^2 + y^2 = c$ for values of $c = 1$, 4 and 9. Note that each level curve is a circle, centre the origin and radius $c^{\frac{1}{2}}$.

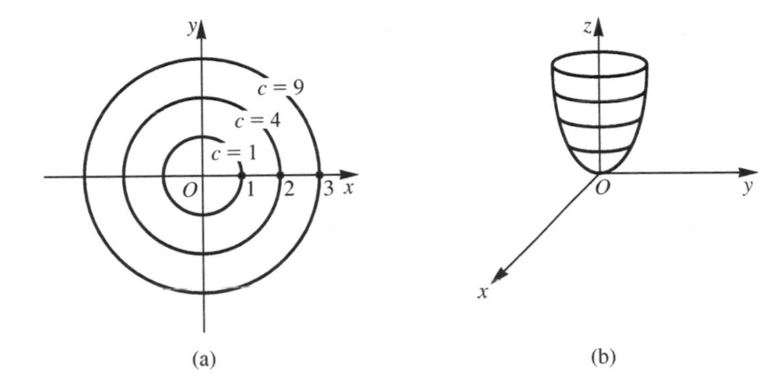

Figure 19.1. (a) Contours of $f(x,y) = x^2 + y^2$, (b) the surface $z = x^2 + y^2$

In Figure 19.1(b) the surface $z = x^2 + y^2$ has been sketched. The height z of the surface at each point (x,y,z) on the surface is the value of the function $f(x,y) = x^2 + y^2$.

We can sketch by hand only relatively simple surfaces; more complicated ones require the use of a computer package.

Note that a **plane** is a simple surface: the plane with equation

$$ax + by + cz = d \quad (c \neq 0)$$

can be described as

$$z = (d - ax - by)/c$$

that is, $z = f(x,y)$ where $f(x,y) = (d - ax - by)/c$.

The plane $z = z_0$, z_0 constant, is a plane parallel to the x–y plane (we sometimes refer to it as a horizontal plane). It cuts the surface $z = f(x,y) = x^2 + y^2$ in a circle, $x^2 + y^2 = z_0$, which corresponds to a contour in Figure 19.1(a).

Worked examples

19.1 Sketch some contours of the surfaces which represent the following functions:

(a) $z = x^2 + 4y^2$ (b) $z = (4 - x^2 - y^2)^{\frac{1}{2}}$

Describe the surface.

Solution (a) The contours are the curves $x^2 + 4y^2 = c$, constant. They are ellipses centred at the origin. Some examples are shown in Figure 19.2(a). The surface is bowl-shaped with its base on the origin and elongated in the x-direction. Note how the values of c increase rapidly. Had we drawn contours where the values of c increase in equal steps they would have bunched together as we moved away from the origin, indicating a more steeply rising surface.

(b) The contours are the curves $(4 - x^2 - y^2)^{\frac{1}{2}} = c$, constant, or $x^2 + y^2 =$ constant, that is, circles centred at the origin, as in Figure 19.2(b). Note that as we move further from the origin the values of c *decrease*. The surface is the upper half of a sphere of radius 2.

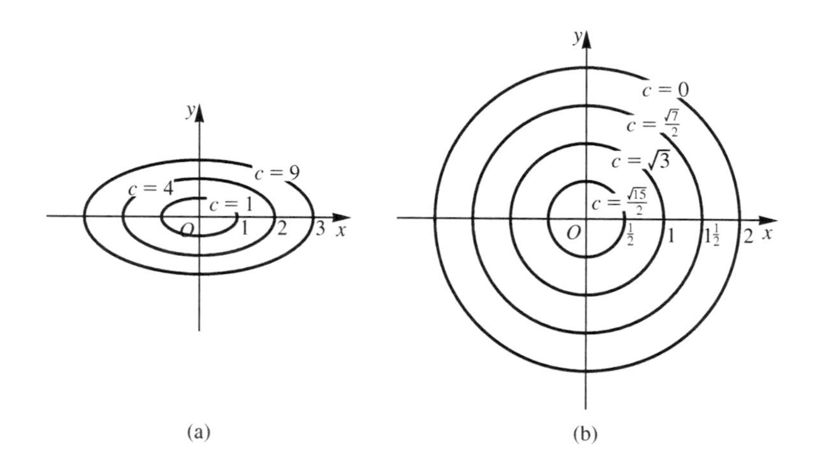

Figure 19.2. Contours for the functions of Example 19.1

19.2 For each of the following functions $f(x,y)$ evaluate $f(0,0)$, $f(1,2)$, $f(-2,1)$:

(a) x^2y (b) $x^2 - 2y^2 + 3x + 2$ (c) $\ln(xy)$ (d) $\sin\left(\frac{\pi}{2}x\right)\cos \pi y$.

Solution (a) $f(0,0)$ means 'substitute $x = 0$, $y = 0$ in the formula'. Hence $f(0,0) = 0, f(1,2) = (1)^2 \times 2 = 2$ and $f(-2,1) = (-2)^2 \times 1 = 4$.

(b) $f(0,0) = 0 - 0 + 0 + 2 = 2, f(1,2) = 1 - 8 + 3 + 2 = -2,$
$f(-2,1) = 4 - 2 - 6 + 2 = -2.$

(c) $f(0,0)$ does not exist. $f(1,2) = \ln 2, f(-2,1)$ does not exist. (We cannot take logs of negative numbers or of zero.)

(d) $f(0,0) = \sin 0 \times \cos 0 = 0 \times 1 = 0$
$f(1,2) = \sin \pi/2 \times \cos 2\pi = 1$
$f(-2,1) = \sin(-\pi) \times \cos \pi = 0$

Self-assessment questions 19.1

1. Explain the meaning of the notation $f(x,y)$.
2. Describe two ways of representing graphically a function of two variables.

Exercise 19.1

1. Draw contours and describe the surfaces in 3D representing the following functions:

   (a) $f(x,y) = x^2 + y^2$
   (b) $f(x,y) = (x^2 + y^2)^{\frac{1}{2}}$
   (c) $f(x,y) = (x^2 + y^2)^{\frac{1}{2}}$
   (d) $z^2 = 4a(x^2 + y^2)$

2. For the following functions $f(x,y)$, find $f(0,0)$, $f(0.5,0)$, and $f(-0.5,1)$:

   (a) x^2y^3
   (b) $x^2 + y^2 + 2$
   (c) $\dfrac{1}{\sqrt{1 + x^2 + y^2}}$
   (d) $\cos \pi x \cos \pi y$

19.2 Partial derivatives

If we move along the surface $z = f(x,y)$ in a direction parallel to the x-axis then y is constant but x varies. If we cut the surface by a vertical plane parallel to the y-axis the 'cliff face' which is exposed has a profile which is a function of x only.

For example, if $z = f(x,y) = x^2 + x + y^2$ then, along the plane $y = 3$, $z = x^2 + x + 9$. See Figure 19.3. The gradient along this curve is found by differentiation with respect to x.

In the same way, we can cut the surface with a plane parallel to the x-axis; the exposed profile is a function of y only.

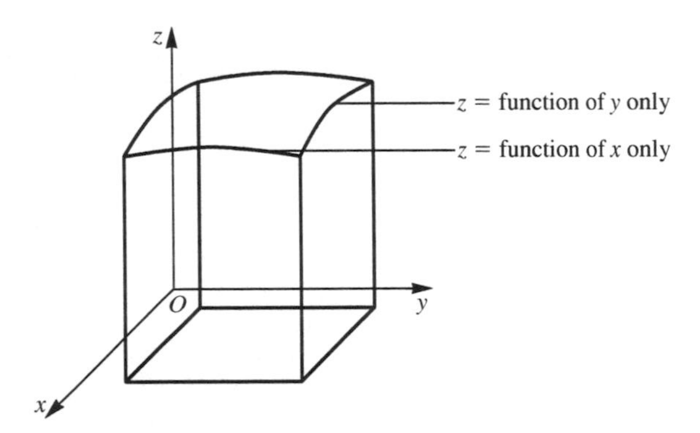

Figure 19.3.
Profiles parallel to the axes

Consider again the examples which introduced Section 19.1, namely $p = f(T, V)$ and $\omega = f(L,C,R)$. In the first example, p changes either if T changes or if V changes. The *rate of change* of p with V when T is held constant is the **partial derivative** of p with respect to V and denoted by

$$\frac{\partial p}{\partial V}$$

Similarly

$$\frac{\partial p}{\partial T}$$

is the partial derivative of p with respect to T, holding V constant. In the second example,

$$\frac{\partial \omega}{\partial L}$$

denotes the partial derivative of ω with L, holding C and R constant, and so on.

In general, if we have a function $f(x,y)$ then $\partial f/\partial x$ is the partial derivative of f with respect to x, holding y constant. It is sometimes read as 'partial dee f by dee x'. In the same way, $\partial f/\partial y$ is the partial derivative of f with respect to y, holding x constant. The symbol ∂ is called 'curly d' and denotes partial differentiation.

The formal definition of partial derivatives follows:

KEY POINT

The first **partial derivatives** of $f(x,y)$ are

$$\frac{\partial f}{\partial x} = \lim_{h \to 0} \frac{f(x+h,y) - f(x,y)}{h}$$

$$\frac{\partial f}{\partial y} = \lim_{k \to 0} \frac{f(x,y+k) - f(x,y)}{k} \qquad 19.3$$

KEYPOINT

To evaluate $\partial f/\partial x$ we treat y as a constant. Hence

1. If an additive term is a function of y only, it differentiates to zero.
2. If a term is a product of a function of x multiplied by a function of y the function of x is differentiated (with respect to x) and multiplied by the function of y.

Worked examples

19.3 Find $\partial f/\partial x$ and $\partial f/\partial y$ for the following functions $f(x,y)$:
(a) $x^2 + y^3$ (b) x^2y^3 (c) $xy^4 + x^3y + x^3 + 2y$

Solution

(a) $\frac{\partial f}{\partial x} = 2x + 0 = 2x \qquad \frac{\partial f}{\partial y} = 0 + 3y^2 = 3y^2$

(b) $\frac{\partial f}{\partial x} = 2xy^3 \qquad \frac{\partial f}{\partial y} = x^2 \times 3y^2 = 3x^2y^2$

(c) $\frac{\partial f}{\partial x} = y^4 + 3x^2y + 3x^2 + 0 \qquad \frac{\partial f}{\partial y} = 4xy^3 + x^3 + 0 + 2$

19.4 Evaluate the first partial derivatives of $f(x,y) = x^3y + xy^3$ at (1,1) and (−2,0).

Solution $\dfrac{\partial f}{\partial x} = 3x^2y + y^3 \qquad \dfrac{\partial f}{\partial y} = x^3 + x \times 3y^2 = x^3 + 3xy^2$

An alternative notation for $\partial f/\partial x$ is f_x and for $\partial f/\partial y$ is f_y. Hence

$$f_x(1,1) = 3 \times 1 \times 1 + 1 = 4 \qquad \text{and} \qquad f_y(1,1) = 1 + 3 = 4$$

Also

$$f_x(-2,0) = 3 \times 4 \times 0 + 0 = 0 \qquad f_y(-2,0) = -8 + 0 = -8$$

19.5 For a gas which obeys the law $pV = RT$ show that

$$\left(\frac{\partial p}{\partial V}\right)_T \left(\frac{\partial V}{\partial T}\right)_p \left(\frac{\partial T}{\partial p}\right)_V = -1$$

Solution The notation $(\partial p/\partial V)_T$ emphasizes that p is a function of V and T and that we are holding T constant and differentiating p with respect to V.

Therefore we write $p = RT/V$ and obtain

$$\left(\frac{\partial p}{\partial V}\right)_T = RT\left(-\frac{1}{V^2}\right) = -\frac{RT}{V^2}$$

Similarly,

$$V = \frac{RT}{p} \qquad \text{and} \qquad \left(\frac{\partial V}{\partial T}\right)_p = \frac{R}{p}$$

Also

$$T = \frac{pV}{R} \qquad \text{and} \qquad \left(\frac{\partial T}{\partial p}\right)_V = \frac{V}{R}$$

Then

$$\left(\frac{\partial p}{\partial V}\right)_T \left(\frac{\partial V}{\partial T}\right)_p \left(\frac{\partial T}{\partial p}\right)_V = -\frac{RT}{V^2} \times \frac{R}{p} \times \frac{V}{R} = -\frac{RT}{pV} = -1$$

(since $pV = RT$)

As with ordinary differentiation we can define higher derivatives.

KEY POINT

For a function $f(x,y)$ the pure second derivatives are

$$\frac{\partial}{\partial x}\left(\frac{\partial f}{\partial x}\right) = \frac{\partial^2 f}{\partial x^2} \equiv f_{xx} \qquad \text{and}$$

$$\frac{\partial}{\partial y}\left(\frac{\partial f}{\partial y}\right) = \frac{\partial^2 f}{\partial y^2} \equiv f_{yy} \qquad (19.4)$$

The mixed second partial derivatives are

$$\frac{\partial}{\partial x}\left(\frac{\partial f}{\partial y}\right)=\frac{\partial^2 f}{\partial x \partial y}\equiv f_{yx} \quad \text{and} \quad \frac{\partial}{\partial y}\left(\frac{\partial f}{\partial x}\right)=\frac{\partial^2 f}{\partial y \partial x}\equiv f_{xy} \qquad 19.5$$

Notice the order of x and y in f_{xy} and f_{yx}.

For all functions of engineering interest,

$$f_{xy} \equiv f_{yx}$$

Many of the rules for ordinary differentiation can be adapted to partial differentiation. For example, if u and v are both functions of x and y then

$$\frac{\partial}{\partial x}(uv)=u\frac{\partial v}{\partial x}+\frac{\partial u}{\partial x}v \quad \text{and} \quad \frac{\partial}{\partial y}\left(\frac{u}{v}\right)=\left(v\frac{\partial u}{\partial y}-u\frac{\partial v}{\partial y}\right)\Big/ v^2$$

Worked examples

19.6 Find the second partial derivatives of the functions $f(x,y)=x^2+y^3$ and $g(x,y)=xy^2+x^2+y^4$ and evaluate them at the point (1,3).

Solution

$$\frac{\partial f}{\partial x}=2x \qquad \frac{\partial f}{\partial y}=3y^2$$

$$\frac{\partial^2 f}{\partial x^2}=2 \qquad \frac{\partial^2 f}{\partial y \partial x}=\frac{\partial}{\partial y}(2x)=0 \qquad \frac{\partial^2 f}{\partial x \partial y}=\frac{\partial}{\partial x}(3y^2)=0$$

$$\frac{\partial^2 f}{\partial y^2}=6y$$

$$f_{xx}(1,3)=2 \qquad f_{xy}=0=f_{yx} \qquad f_{yy}=18$$

$$\frac{\partial g}{\partial x}=y^2+2x \qquad \frac{\partial g}{\partial y}=2xy+4y^3$$

$$\frac{\partial^2 g}{\partial x^2}=2 \qquad \frac{\partial^2 g}{\partial y \partial x}=2y=\frac{\partial^2 g}{\partial x \partial y} \qquad \frac{\partial^2 g}{\partial y^2}=2x+12y^2$$

$$g_{xx}(1,3)=2 \qquad g_{xy}=6=g_{yx} \qquad g_{yy}=2+108=110$$

19.7 The **critical point** of a pure substance occurs where $\partial p/\partial V$ and $\partial^2 p/\partial V^2$ are both zero. Find the values of p, V and T where a substance which obeys van der Waals' law

$$\left(p+\frac{a}{V^2}\right)(V-b)=RT$$

has its critical point.

Solution Rearranging the equation we obtain

$$p = \frac{RT}{V - b} - \frac{a}{V^2}$$

that is, $p = p(V,T)$. Then

$$\frac{\partial p}{\partial V} = -\frac{RT}{(V-b)^2} + \frac{2a}{V^3}$$

and

$$\frac{\partial^2 p}{\partial V^2} = \frac{2RT}{(V-b)^3} - \frac{6a}{V^4}$$

The critical point occurs where both

$$\frac{RT}{(V-b)^2} = \frac{2a}{V^3} \qquad \text{i}$$

and

$$\frac{2RT}{(V-b)^3} = \frac{6a}{V^4} \qquad \text{ii}$$

Dividing (i) by (ii) gives

$$\frac{V-b}{2} = \frac{V}{3}$$

which leads to $V = 3b$. Then (i) becomes

$$\frac{RT}{4b^2} = \frac{2a}{27b^3}$$

so that $T = \dfrac{8a}{27Rb}$. Then

$$p = \frac{R}{2b} \cdot \frac{8a}{27Rb} - \frac{a}{9b^2} = \frac{4a}{27b^2} - \frac{3a}{27b^2} = \frac{a}{27b^2}$$

In summary, at the critical point

$$p = \frac{a}{27b^2} \qquad V = 3b \qquad \text{and} \qquad T = \frac{8a}{27Rb}$$

The chain rule

If the variables x and y are each functions of a third variable t then it is possible to find the rate of change of the function $f(x,y)$ with respect to t by using the **chain rule**.

KEY POINT

$$\frac{df}{dt} = \frac{\partial f}{\partial x} \cdot \frac{dx}{dt} + \frac{\partial f}{\partial y} \cdot \frac{dy}{dt} \qquad 19.6$$

If x and y are each functions of two independent variables s and t then the chain rule takes the form

$$\frac{\partial f}{\partial s}=\frac{\partial f}{\partial x}.\frac{\partial x}{\partial s}+\frac{\partial f}{\partial y}.\frac{\partial y}{\partial s} \quad \text{and} \quad \frac{\partial f}{\partial t}=\frac{\partial f}{\partial x}.\frac{\partial x}{\partial t}+\frac{\partial f}{\partial y}.\frac{\partial y}{\partial t} \qquad 19.7$$

Notice the use of partial derivatives throughout, whereas in (19.6) x and y were assumed to be functions of t only so we used ordinary derivative notation.

Worked example

19.8 The height of a cylinder is increasing at a rate of $2\,\text{mm}\,\text{s}^{-1}$ and its radius decreases at $3\,\text{mm}\,\text{s}^{-1}$. Find the rate of change of the volume when the height is 100 mm and the radius is 80 mm.

Solution The volume of a cylinder with height h and radius r is $V=\pi r^2 h$. Applying (19.6)

$$\begin{aligned}\frac{dV}{dt}&=\frac{\partial V}{\partial h}.\frac{dh}{dt}+\frac{\partial V}{\partial r}.\frac{dr}{dt}\\&=\pi r^2.\frac{dh}{dt}+2\pi rh.\frac{dr}{dt}\\&=\pi\times 6400\times 2+2\pi\times 80\times 100\times(-3)\\&=-35200\pi\end{aligned}$$

Hence V is *decreasing* at a rate of $35\,200\pi\,\text{mm}^3\,\text{s}^{-1}$.

19.9 If $f(x,y)=x^2+xy-y^2$ and $x=r\cos\theta$, $y=r\sin\theta$ find $\partial f/\partial r$ and $\partial f/\partial\theta$.

Solution First,

$$\frac{\partial x}{\partial r}=\cos\theta \qquad \frac{\partial y}{\partial r}=\sin\theta$$

Using the chain rule (19.7)

$$\begin{aligned}\frac{\partial f}{\partial r}&=\frac{\partial f}{\partial x}.\frac{\partial x}{\partial r}+\frac{\partial f}{\partial y}.\frac{\partial y}{\partial r}\\&=(2x+y).\cos\theta+(x-2y).\sin\theta\\&=(2r\cos\theta+r\sin\theta).\cos\theta+(r\cos\theta-2r\sin\theta).\sin\theta\\&=r(2\cos^2\theta+\sin\theta\cos\theta+\cos\theta\sin\theta-2\sin^2\theta)\\&=r(2\cos^2\theta-2\sin^2\theta+2\sin\theta\cos\theta)\\&=r(2\cos 2\theta+\sin 2\theta)\end{aligned}$$

Similarly,

$$\frac{\partial x}{\partial\theta}=-r\sin\theta \qquad \frac{\partial y}{\partial\theta}=r\cos\theta$$

Then

$$\frac{\partial f}{\partial \theta} = \frac{\partial f}{\partial x}\cdot\frac{\partial x}{\partial \theta} + \frac{\partial f}{\partial y}\cdot\frac{\partial y}{\partial \theta}$$
$$= (2x + y).(-r\sin\theta) + (x - 2y).r\cos\theta$$
$$= (2r\cos\theta + r\sin\theta).(-r\sin\theta) + (r\cos\theta - 2r\sin\theta).r\cos\theta$$
$$= r^2(-2\cos\theta\sin\theta - \sin^2\theta + \cos^2\theta - 2\sin\theta\cos\theta)$$
$$= r^2(\cos 2\theta - 2\sin 2\theta)$$

Self-assessment questions 19.2

1. Explain the meaning of $\partial f/\partial x$ and$\partial f/\partial y$ in relation to a function $f(x,y)$.
2. What do the notations z_{xx}, z_{yy}, z_{xy} and z_{yx} mean in relation to a function $z = f(x,y)$?
3. What are the two versions of the chain rule for partial derivatives?

Exercise 19.2

1. Find the first partial derivatives of the following functions at the points (1,2), (−1,−1) and (2,1):
   (a) $x^2 + y^2$
   (b) x^3y^2
   (c) $x^3y + xy^2 + 2x + 3y^2$
   (d) $\sin x + \cos y$
   (e) $e^{-x}\cos y$
2. Find the second partial derivatives of the functions of Question 1.
3. The volume of a cone is increasing at the rate of $20\,\text{mm}^3\,\text{s}^{-1}$. If the base radius increases at the rate of $2\,\text{mm}\,\text{s}^{-1}$, find the rate at which the height is changing when it is 40 mm and the radius is 25 mm.
4. (a) Find dz/dt if $z = x^3y^2$, $x = \cos at$ and $y = \sin at$.
   (b) Find dz/dt if $z^2 = x^2 + y^2$, $y = t^2 + 2$ and $x = 3t + 1$.
   (c) Find $\partial f/\partial u$ and $\partial f/\partial v$ if $f = \cos xe^y$, $x = 2uv$ and $y = u^2 - v^2$.
5. For a gas which obeys the law $pV = RT$, the entropy S is given by the formula $S = c_v \ln T + R\ln V + a$, where c_v, R and a are constants.
   Find $(\partial S/\partial V)_T$, $(\partial S/\partial p)_T$ and $(\partial S/\partial p)_V$.
   Show that
   $$\left(\frac{\partial S}{\partial V}\right)_T = \left(\frac{\partial S}{\partial T}\right)_V \quad \text{and}$$
   $$\left(\frac{\partial S}{\partial p}\right)_T = -\left(\frac{\partial V}{\partial T}\right)_p$$

19.3 Differentials and small errors

Suppose x changes to $x + \delta x$ and, independently, y changes to $y + \delta y$, where δx and δy are small. Then the change in the function $f(x,y)$ is given by

$$\delta f \equiv f(x + \delta x,\ y + \delta y) - f(x,y)$$
$$\equiv [f(x + \delta x,\ y + \delta y) - f(x,y + \delta y)] + [f(x,y + \delta y) - f(x,y)]$$

The term in the first square brackets on the right-hand side is the change in f resulting from the change in x and the term in the second brackets is the change in f resulting from the (independent) change in y. The change can be written

$$\delta f = \frac{[f(x+\delta x, y+\delta y) - f(x,y+\delta y)]}{\delta x}\delta x + \frac{[f(x,y+\delta y) - f(x,y)]}{\delta y}\delta y$$

Remembering the definition of partial derivatives in (19.3) with h and k replaced by δx and δy, respectively, we make the approximation

$$\delta f \simeq \left(\frac{\partial f}{\partial x}\right)\delta x + \left(\frac{\partial f}{\partial y}\right)\delta y \qquad 19.8$$

The **differential**, df, is defined by the formula

$$\mathrm{d}f = \left(\frac{\partial f}{\partial x}\right)\delta x + \left(\frac{\partial f}{\partial y}\right)\delta y \qquad 19.9$$

We see that df is an approximation to δf.

In Example 19.10 we shall see that differentials dx and dy are d$x = \delta x$ and d$y = \delta y$. Then we can write:

KEY POINT

$$\mathrm{d}f = \left(\frac{\partial f}{\partial x}\right)\mathrm{d}x + \left(\frac{\partial f}{\partial y}\right)\mathrm{d}y \qquad 19.10$$

The relationship (19.8) can be used to find the approximate effect of *small* errors in the independent variables on the values of a function.

Worked examples

19.10 Show that d$x = \delta x$ and d$y = \delta y$.

Solution Put $f(x,y) = x$. Then $\partial f/\partial x = 1$ and $\partial f/\partial y = 0$. Therefore (19.9) becomes

$$\mathrm{d}x = \mathrm{d}f = 1.\delta x + 0.\delta y = \delta x$$

Similarly, if $f(x,y) = y$, $\partial f/\partial x = 0$ and $\partial f/\partial y = 1$ so that (19.9) becomes

$$\mathrm{d}y = \mathrm{d}f = 0.\delta x + 1.\delta y = \delta y$$

19.11 Find the differential of each of the following:
(a) $f(x,y) = x^2y + y^2$ (b) $f(x,y,z) = x^2 + y^2 + z^2$

Solution (a) $\dfrac{\partial f}{\partial x} = 2xy, \quad \dfrac{\partial f}{\partial y} = x^2 + 2y$

Therefore

$$\mathrm{d}f = 2xy\,\mathrm{d}x + (x^2 + 2y)\,\mathrm{d}y$$

(b) $$\frac{\partial f}{\partial x} = 2x \qquad \frac{\partial f}{\partial y} = 2y \qquad \frac{\partial f}{\partial z} = 2z$$

$$\mathrm{d}f = \frac{\partial f}{\partial x}.\mathrm{d}x + \frac{\partial f}{\partial y}.\mathrm{d}y + \frac{\partial f}{\partial z}.\mathrm{d}z$$

$$= 2x\,\mathrm{d}x + 2y\,\mathrm{d}y + 2z\,\mathrm{d}z$$

19.12 The volume of a cone of height h and base radius r is given by $V = \frac{1}{3}\pi r^2 h$. The radius and height are measured as 450 mm and 300 mm, respectively, but each measurement could be in error by ± 0.5 mm. Find approximately the maximum error in the calculation of the volume.

Solution

$$\frac{\partial V}{\partial r} = \frac{2}{3}\pi rh \qquad \frac{\partial V}{\partial h} = \frac{1}{3}\pi r^2 \qquad |\delta r| = 0.5 = |\delta h|$$

Now

$$|\delta V| \simeq \left|\frac{\partial V}{\partial r}\right| \times |\delta r| + \left|\frac{\partial V}{\partial h}\right| \times |\delta h|$$

$$= \frac{2}{3}\pi rh \times 0.5 + \frac{1}{3}\pi r^2 \times 0.5$$

$$= \frac{2}{3}\pi \times 450 \times 300 \times 0.5 + \frac{1}{3}\pi \times 450 \times 450 \times 0.5$$

$$= 78\,700\pi \text{ mm}^3 = 2.47 \times 10^5 \text{ mm}^3 \quad (3 \text{ s.f.})$$

Note that the calculated value of V is 63.6×10^7 mm^3 (3 s.f.). Hence the maximum error is approximately 0.4% of the calculated value.

19.13 The coefficient of viscosity η of a liquid is to be calculated via an experiment measuring the rate of flow of the liquid through a tube of length l and radius a, which uses Poiseuille's equation in the form

$$\eta = \frac{p\pi a^4}{8Ql}$$

where Q is the rate of flow and p is the pressure difference between the ends of the tube.

Q can be measured to within 1%, p to within 2%, a to within 0.1% and l to within 0.5%.

Find the maximum percentage error in the calculated value of η.

Solution

$$\frac{\partial \eta}{\partial p} = \frac{\pi a^4}{8Ql} = \frac{\eta}{p} \qquad \frac{\partial \eta}{\partial a} = \frac{4p\pi a^3}{8Ql} = \frac{4\eta}{a}$$

$$\frac{\partial \eta}{\partial Q} = -\frac{p\pi a^4}{8Q^2 l} = -\frac{\eta}{Q} \quad \text{and} \quad \frac{\partial \eta}{\partial l} = -\frac{p\pi a^4}{8Ql^2} = -\frac{\eta}{l}$$

Hence

$$\partial\eta \simeq \frac{\partial\eta}{\partial p}\delta p + \frac{\partial\eta}{\partial a}\delta a + \frac{\partial\eta}{\partial Q}\delta Q + \frac{\partial\eta}{\partial l}\delta l$$

$$= \frac{\eta}{p}\delta p + 4\frac{\eta}{a}\delta a - \frac{\eta}{Q}\delta Q - \frac{\eta}{l}\delta l$$

Hence

$$\frac{\delta\eta}{\eta} = \frac{\delta p}{p} + 4\frac{\delta a}{a} - \frac{\delta Q}{Q} - \frac{\delta l}{l}$$

The maximum percentage error is

$$\left|\frac{\delta\eta}{\eta}\right| \times 100 \simeq \left\{\left|\frac{\delta p}{p}\right| + 4\left|\frac{\delta a}{a}\right| + \left|\frac{\delta Q}{Q}\right| + \left|\frac{\delta l}{l}\right|\right\} \times 100$$

$$= 2 + 4 \times 0.1 + 1 + 0.5$$

$$= 3.9\%$$

This calculation emphasizes how important it is to measure a as accurately as we can since its percentage error is multiplied by 4.

Self-assessment questions 19.3

1. Explain the term differential.
2. Explain how to estimate the error in the value of a function of two independent variables given the (small) errors in their values.

Exercise 19.3

1. Find the differentials of the following functions:

   (a) $x^2 + xy^3$ (b) $x^2y^2z^2$

2. An experiment to estimate the value of g, the acceleration due to gravity, involves timing the oscillations of a simple pendulum. The period of small oscillations is given by

   $$T = 2\pi\sqrt{\frac{l}{g}}$$

   Measurements arc made of the length of the pendulum and the period of the oscillations as $l = 100 \pm 0.15$ and $T = 20 \pm 0.01$, respectively. Estimate the maximum error in the calculated value of g.

3. The frequency of oscillation of the current in a particular LC circuit is given by

   $$f = \frac{1}{2\pi}\sqrt{\frac{1}{LC}}$$

   If L is increased by 1% and C is decreased by 3% find the percentage change in f.

4. The focal length of a lens is given by

   $$\frac{1}{f} = \frac{1}{u} + \frac{1}{v}$$

   If the measurements of the object distance u and the image distance v are each subject to an error of $\pm 1\%$, find the maximum percentage error in the calculated value of the focal length f.

Test and assignment exercises 19

1. For the following functions find $f(0,0), f(1,-1), f(-1,1)$.

   (a) $x^3 + xy^2$
   (b) $x^4y^4 + x^4 + y^4 + 1$
   (c) $(x^2 + y^2 + 2)^{\frac{1}{2}}$
   (d) $\sin\left(\frac{\pi}{2}x\right).\sin\left(\frac{\pi}{2}y\right)$

2. Find the first partial derivatives of the following functions:

   (a) x^4y^2
   (b) $x^2y + xy^2$
   (c) $\cos x \cos y$
   (d) $e^{-x}(\cos y + 2\sin y)$

3. Evaluate the first partial derivatives of Exercise 2 at the points $(1,-2)$, $(2,1)$ and $(-1,-1)$.

4. Find the first partial derivatives of the following functions:

   (a) $\sin(x + y)$
   (b) $e^{x^2+y^2}$
   (c) $\dfrac{1}{(x^2 + y^2 + 1)^{\frac{1}{2}}}$

5. Find the second partial derivatives of the functions of Exercise 2.

6. If $\phi = y/(x^2 + y^2)$ show that

$$\frac{\partial^2\phi}{\partial x^2} + \frac{\partial^2\phi}{\partial y^2} \equiv 0$$

7. Find the differentials of the following functions:

   (a) x^2y^2 (b) $x^2y + z^2$

8. The total surface area of a circular cylinder of radius r and height h is

$$S = 2\pi r^2 + 2\pi rh$$

   The height is increasing at a rate of $3\,\text{mm}\,\text{s}^{-1}$ and the radius is decreasing at $2\,\text{mm}\,\text{s}^{-1}$. Find the rate of change of total surface area when the height is 50 mm and the radius is 20 mm.

9. The breaking weight of a cantilevered beam is given by

$$W = \frac{Cbd^2}{l^2}$$

   where b is the breadth of the beam, d its depth, l its length and C is a constant. If the length is increased by 2% and the breadth by 5%, by how much should the depth be changed in order to keep the breaking weight the same?

20 Further numerical methods

Objectives

This chapter

- defines and uses the trapezium rule for integration
- estimates the maximum error in an estimate provided by the rule
- defines and uses Simpson's rule for integration
- estimates the maximum error in an estimate provided by Simpson's rule
- calculates the least squares line of best fit to a set of data points

20.1 Numerical integration - trapezium rule

An approximation to the definite integral $I = \int_a^b f(x)\,dx$ can be found by dividing the interval $[a,b]$ into n equal subintervals. Let $a = x_0$ and $b = x_n$, and $x_r - x_{r-1} = h$ for $r = 1, 2, ..., n$. Then the subintervals are $[x_0,x_1]$, $[x_1,x_2]$, ..., $[x_{n-1},x_n]$. First, note that

$$nh = b - a \tag{20.1}$$

The **trapezium rule** is based on approximating the function in each interval $[x_{r-1}, x_r]$ by a straight line joining the points $(x_{r-1}, f(x_{r-1}))$ and $(x_r, f(x_r))$ as in Figure 20.1. We often use the notation f_r for $f(x_r)$, that is, the value of the function at $x = x_r$.

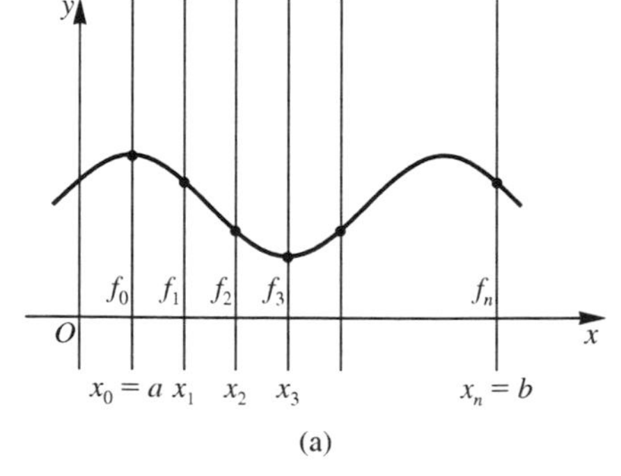

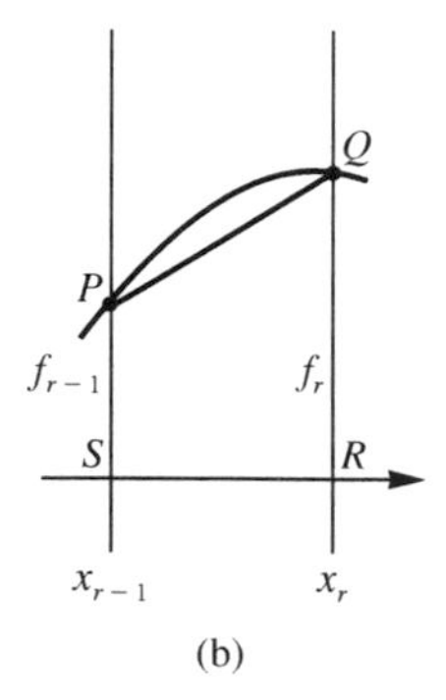

Figure 20.1.
(a) Trapezium rule
(b) typical trapezium

In Figure 20.1(b) a typical trapezium $PQRS$ is shown. Its area is

$$\frac{1}{2}SR(PS + QR) = \frac{1}{2}h(f_{r-1} + f_r)$$

Hence the total area, which is given exactly by $I = \int_a^b f(x)\,dx$, is given approximately by the trapezium rule as:

KEY POINT

$$I_T = \frac{1}{2}h[f_0 + 2f_1 + 2f_2 + \ldots + 2f_{n-1} + f_n] \qquad 20.2$$

The actual value I lies in the interval $I_T \pm \varepsilon_T$ where ε_T is the error in the approximation. The error can be estimated from the formula

$$|\varepsilon_T| \leqslant \frac{(b-a)h^2}{12}M_T \qquad 20.3$$

and M_T is the maximum value that $|f''(x)|$ takes in the interval $a \leqslant x \leqslant b$.

Worked examples

20.1 Use the trapezium rule with four strips and then eight strips to estimate

$$I = \int_1^3 x^2\,dx$$

Solution With $n = 4$, $h = \frac{1}{2}$, and the calculations may be tabulated as in Table 20.1.

The sum of row 4 is 35. Hence

$$I_T = \frac{1}{2} \times \frac{1}{2} \times 35 = 8.75$$

With $n = 8$, $h = \frac{1}{4}$, and the calculations may be tabulated as in Table 20.2.

The sum of row 4 is 69.5. Hence

$$I_T = \frac{1}{2} \times \frac{1}{4} \times 69.5 = 8.6875$$

Table 20.1.

x	1	1.5	2	2.5	3
$f(x)$	1	2.25	4	6.25	9
Coefficients	1	2	2	2	1
Row 2 × Row 3	1	4.5	8	12.5	9

Table 20.2.

x	1	1.25	1.5	1.75	2	2.25	2.5	2.75	3
$f(x)$	1	1.5625	2.25	3.0625	4	5.0625	6.25	7.5625	9
Coefficients	1	2	2	2	2	2	2	2	1
Row 2 × Row 3	1	3.125	4.5	6.125	8	10.125	12.5	15.125	9

20.2 Calculate the errors in the approximations of Example 20.1 and estimate the maximum errors as predicted by (20.3).

Solution The accurate value of I is

$$\left[\frac{1}{3}x^3\right]_1^3 = 9 - \frac{1}{3} = \frac{26}{3} = 8.6667 \quad \text{(4 d.p.)}$$

With four strips,

$$y\varepsilon_T = I_T - I = 8\frac{3}{4} - 8\frac{2}{3} = \frac{1}{12} = 0.0833 \quad \text{(4 d.p.)}$$

With eight strips,

$$\varepsilon_T = 8.6875 - 8.6667 = 0.0208 \quad \text{(4 d.p.)}$$

Doubling the number of strips has divided the error by about 4. Here,

$f''(x) = 2$ and hence $M_T = 2$. Then

$$|\varepsilon_T| \leqslant \frac{(3-1)h^2}{12} \times 2 = \frac{h^2}{3}$$

When $n = 4$, $h = \frac{1}{2}$ and

$$|\varepsilon_T| \leqslant \frac{1}{12} = 0.0833 \quad \text{(4 d.p.)}$$

When $n = 8$, $h = \frac{1}{4}$ and

$$|\varepsilon_T| \leqslant \frac{1}{48} = 0.0208 \quad \text{(4 d.p.)}$$

The actual errors are equal to the maximum predicted errors; this happens because M_T is constant.

20.3 Estimate $I = \int_0^{\frac{\pi}{2}} \cos x \, dx$ by the trapezium rule with four strips.

Calculate the error to 5 d.p. and estimate the maximum error from (20.3).

Solution With $n = 4$, $h = \pi/8$. Table 20.3 shows the calculations.

The total of row 4 is 5.027 339 and therefore

$$I_T = \frac{1}{2} \times \frac{\pi}{8} \times 5.027\,339 = 0.987\,116$$

Now $I = \left[\sin x\right]_0^{\frac{\pi}{2}} = 1$. Hence

$$\varepsilon_T = I_T - I = -0.012\,88 \quad \text{(5 d.p.)}$$

Now $f''(x) = -\cos x$ and hence $M_T = 1$.

This is so because $-1 \leqslant -\cos x \leqslant 1$, hence $-1 \leqslant \cos x \leqslant 1$ and therefore $|\cos x| \leqslant 1$.

Then formula (20.3) becomes

$$|\varepsilon_T| \leqslant \frac{\left(\frac{\pi}{2} - 0\right)}{12} \times \left(\frac{\pi}{8}\right)^2 \times 1 = 0.020\,186 \quad \text{(6 d.p.)}$$

Hence the actual error is within that predicted by (20.3).

Table 20.3.

x	0	$\frac{\pi}{8}$	$\frac{\pi}{4}$	$\frac{3\pi}{8}$	$\frac{\pi}{2}$
$f(x)$	1	0.923 880	0.707 107	0.382 683	0
Coefficients	1	2	2	2	1
Row 2 × Row 3	1	1.847 759	1.414 214	0.765 367	0

20.4 How many strips will guarantee that the trapezium approximation to

$$I = \int_0^{\frac{\pi}{2}} \cos x \, dx$$

is correct to 2 d.p.?

Solution For the approximation to be correct to 2 d.p. we require that

$$|\varepsilon_T| < 0.005 = \frac{1}{200}$$

But

$$|\varepsilon_T| \leqslant \frac{\frac{\pi}{2} h^2}{12} \times 1 = \frac{\pi}{24} h^2$$

Hence we require that $\pi h^2/24 < 1/200$.

Given that $nh = b - a = \pi/2$ then $h = \pi/2n$ and the requirement becomes

$$\frac{\pi}{24} \times \left(\frac{\pi}{2n}\right)^2 < \frac{1}{200}$$

that is,

$$n^2 > \frac{200\pi^3}{24 \times 4}$$

This gives $n > 8.037$ (3 d.p.). Hence we choose nine strips.

(It may well be that we *could* obtain the required accuracy with fewer strips but the problem says 'guarantee'.)

Self-assessment questions 20.1

1. Explain how the trapezium rule can be used to estimate the value of a definite integral.
2. How would you estimate the maximum error in a trapezium approximation to the value of a definite integral?

Exercise 20.1

1. Use the trapezium rule with 5 strips and then 10 strips to evaluate the following integrals:

   (a) $\int_0^1 e^x \, dx$ (b) $\int_1^3 \frac{1}{x} dx$ (c) $\int_0^1 x^4 \, dx$

2. For the examples of Question 1, estimate the maximum error via (20.2). Compare these estimates with the actual errors.

3. For the examples of Question 1, determine how many strips will guarantee an accuracy of:

   (a) 1 d.p.
   (b) 2 d.p.

4. The following data shows the velocity v of an object at time t. Use the trapezium rule to find the approximate distance travelled between $t = 0$ and $t = 8$.

t	0	1	2	3	4	5	6	7	8
v	0	4.2	7.3	9.3	10.2	10.1	9.0	6.9	3.7

20.2 Numerical integration - Simpson's rule

In an attempt to improve the accuracy of numerical integration, we fit a quadratic curve to three consecutive values of $f(x)$. It is necessary to divide the interval $[a,b]$ into an *even* number of strips. Figure 20.2 shows the quadratic through the points (x_{r-1},f_{r-1}), (x_r,f_r) and (x_{r+1},f_{r+1}).

It can be shown that the area under this quadratic curve is

$$\frac{1}{3}h(f_{r-1} + 4f_r + f_{r+1})$$

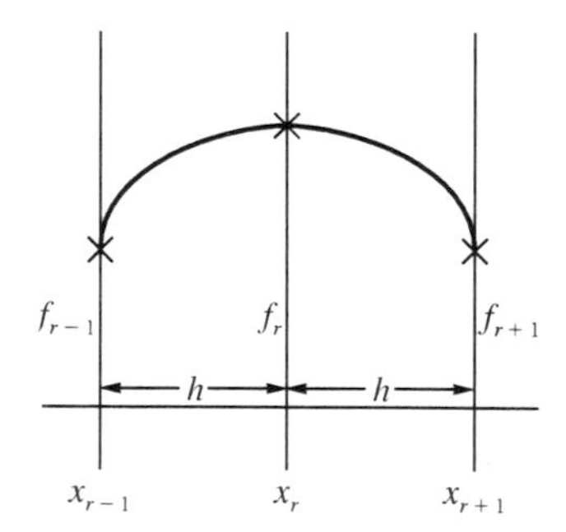

Figure 20.2. Quadratic approximation

Hence if the interval $[a,b]$ is divided into n equal strips of width h, where n is even, then the integral $I = \int_a^b f(x)\,dx$ is given approximately by **Simpson's rule**:

KEY POINT

$$I_S = \frac{1}{3}h[f_0 + 4f_1 + 2f_2 + 4f_3 + 2f_4 + \ldots + 4f_{n-1} + f_n] \qquad 20.4$$

The actual value I lies in the interval $I_S \pm \varepsilon_S$ where

$$|\varepsilon_S| \leqslant \frac{(b-a)h^4}{180} M_S \qquad 20.5$$

and M_S is the maximum value that $|f^{(iv)}(x)|$ takes in the interval $a \leqslant x \leqslant b$.

Table 20.4.

x	0	$\frac{\pi}{8}$	$\frac{\pi}{4}$	$\frac{3\pi}{8}$	$\frac{\pi}{2}$
$f(x)$	1	0.923 880	0.707 107	0.382 683	0
Coefficients	1	4	2	4	1
Row 2 × Row 3	1	3.695 52	1.414 214	1.530 734	0

Worked examples

20.5 Estimate $I = \int_0^{\frac{\pi}{2}} \cos x \, dx$ by Simpson's rule with four strips. Calculate the error to 4 d.p. and estimate the maximum error by (20.5). What would you expect to happen to the error if you doubled the number of strips?

Solution It is helpful to tabulate the calculations as in Table 20.4.

The total of row 4 is 7.640 465 and therefore, since $h = \pi/8$,

$$I_S = \frac{1}{3} \times \frac{\pi}{8} \times 7.640\,465 = 1.000\,135$$

The error $\varepsilon_S = I_S - I = 0.000\,135 = 0.0001$ (4 d.p.).

$$|\varepsilon_S| \leqslant \frac{\left(\frac{\pi}{2} - 0\right)\left(\frac{\pi}{8}\right)^4}{180} \times 1 = 0.000\,208 \text{ (6 d.p.)} = 0.0002 \text{ (4 d.p.)}$$

The actual error lies within the maximum predicted by (20.5).

Compare this answer with that of Example 20.3. We have made the same number of calculations and yet we have achieved a much more accurate result. Since (20.5) shows that the maximum error is approximately proportional to h^4, doubling the number of strips, that is, halving the strip width, will reduce the error by a factor of approximately 16. This compares with the trapezium reduction of approximately 4.

20.6 How many strips will guarantee that the Simpson's approximation to

$$I = \int_0^{\frac{\pi}{2}} \cos x \, dx$$

is correct to 3 d.p.?

Solution We require $|\varepsilon_S| < 0.0005 = 1/2000$. But

$$|\varepsilon_S| \leqslant \frac{\frac{\pi}{2} h^4}{180} \times 1 = \frac{\pi}{360} h^4$$

Hence the requirement is $\pi h^4/360 < 1/2000$.

In terms of n this is

$$\frac{\pi}{360}\left(\frac{\pi}{2n}\right)^4 < \frac{1}{2000}$$

that is

$$n^4 > \frac{2000\pi^5}{360 \times 16}$$

This gives $n > 3.210\,622$ (6 d.p.). We therefore take $n = 4$.

(Note that n is taken as the next *even* integer above the value produced by (20.5).)

20.7 Use Simpson's rule with four strips to estimate $I = \int_1^2 x^3\,dx$.

Solution Table 20.5 is used to help the calculations.

The sum of row 4 = 45 and therefore, since $h = 1/4$,

$$I_S = \frac{1}{3} \times \frac{1}{4} \times 45 = 3.75$$

This is an unexpected bonus. We would anticipate that Simpson's rule would give an exact result for a quadratic function but not for a cubic function. However, (20.5) uses $f^{(iv)}(x)$ and for a cubic function,

$$f(x) = ax^3 + bx^2 + cx + d, \quad f^{(iv)}(x) \equiv 0$$

so that $|\varepsilon_S| \leqslant 0$. Hence $|\varepsilon_S| = 0$ and $\varepsilon_S = 0$.

20.8 The following data shows readings of the rate of flow of water into a reservoir in litres per hour taken over a period of 8 hours. Estimate the total volume flowing in the 8-hour period.

Time	0	1	2	3	4	5	6	7	8
Flow rate	2000	2100	2200	2400	2600	2500	2350	2200	2050

Solution Table 20.6 displays the calculations; note that $h = 1$.

The sum of row 4 is 55 150 and therefore

$$I_S = \frac{1}{3} \times 1 \times 55\,150 = 18\,383 \text{ litres}$$

Table 20.5.

x	1	1.25	1.5	1.75	2
$f(x)$	1	1.953 125	3.375	5.354 375	8
Coefficients	1	4	2	4	1
Row 2 × Row 3	1	7.812 5	6.75	21.437 5	8

Table 20.6.

x	0	1	2	3	4	5	6	7	8
$f(x)$	2000	2100	2200	2400	2600	2500	2350	2200	2050
Coefficients	1	4	2	4	2	4	2	4	1
Row 2 × Row 3	20 000	8400	4400	9600	5200	10 000	4700	8800	2050

Self-assessment questions 20.2

1. Explain how Simpson's rule can be used to estimate the value of a definite integral.
2. How would you estimate the maximum error in a Simpson approximation to the value of a definite integral?

Exercise 20.2

1. Use Simpson's rule with four strips to estimate the values of the integrals in Question 1 of the Exercise 20.1.
2. For the integrals in Exercise 20.1, Question 1, estimate the maximum error via (20.5). Compare these estimates with the actual errors.
3. For the examples of Question 1 determine the number of strips which will guarantee an accuracy of:

   (a) 1 d.p. (b) 2 d.p.
4. Use Simpson's rule to estimate the distance specified in Exercise 20.1, Question 4.

20.3 Least squares curve of best fit

Engineers often fit straight lines to data. In order to obtain consistent results we provide a rule to replace the *ad hoc* approach of using a transparent ruler. This section gives a more consistent method of finding the straight line and then shows how other curves can be fitted using this approach.

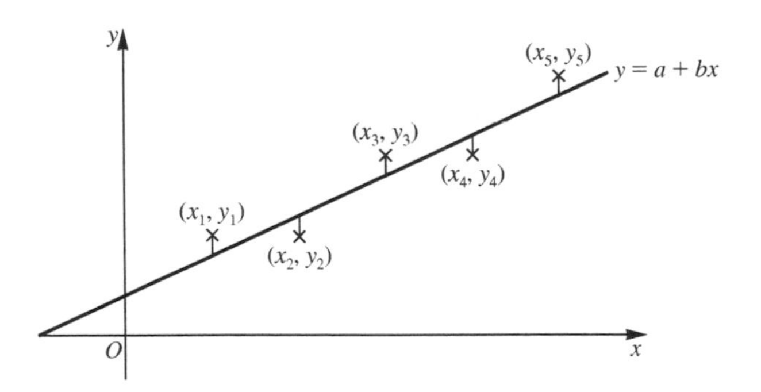

Figure 20.3. Fitting a straight line to data

Figure 20.3 shows some data points and a tentative straight line fit to the data. We assume that the values of y are subject to more error than those of x. At $x = x_1$ the straight line predicts $y = a + bx_1$ and the discrepancy between this value and the observed value of y is $y_1 - a - bx_1$. This will be positive but $y_2 - a - bx_2$ will be negative. To find the total discrepancy over all the data points we do not add the individual discrepancies since cancellation would make things appear much better than they are. Instead, we square first. Then the total discrepancy for n data points (x_i, y_i) is

KEY POINT

$$S = \sum_{i=1}^{n} (y_i - a - bx_i)^2 \qquad 20.6$$

It can be shown that S is a function of a and b.

The minimum value of S can be shown to occur when a and b satisfy the **normal equations**:

KEY POINT

$$na + b\sum_{i=1}^{n} x_i = \sum_{i=1}^{n} y_i$$
$$a\sum_{i=1}^{n} x_i + b\sum_{i=1}^{n} x_i^2 = \sum_{i=1}^{n} x_i y_i \qquad 20.7$$

These are the **least squares** equations for a line of best fit. The solutions are

$$b = \frac{n\sum_{i=1}^{n} x_i y_i - \left(\sum_{i=1}^{n} x_i\right)\left(\sum_{i=1}^{n} y_i\right)}{n\sum_{i=1}^{n} x_i^2 - \left(\sum_{i=1}^{n} x_i\right)^2}, \quad a = \frac{\sum_{i=1}^{n} y_i - b\left(\sum_{i=1}^{n} x_i\right)}{n} \qquad 20.8$$

Similar normal equations can be obtained for other curves. Alternatively, we can sometimes rewrite the equation of the curve in a linear form to make use of (20.7).

Worked examples

20.9 For the following data find S:

x	1	2	3
y	4	5	7

Solution We fit the line $y = a + bx$.

At $x = 1$, $y = a + b$ and the discrepancy is $4 - a - b$

At $x = 2$, $y = a + 2b$ and the discrepancy is $5 - a - 2b$

At $x = 3$, $y = a + 3b$ and the discrepancy is $7 - a - 3b$

Then $S = (4 - a - b)^2 + (5 - a - 2b)^2 + (7 - a - 3b)^2$. This is a function of a and b.

20.10 The following data was collected on the percentage of sand in soil at various depths:

Depth (m)	0	1.5	3	4.5	6	7.5	9	10.5	12
% sand	72.1	54.5	55.8	54.0	49.0	50.7	32.7	38.4	29.1

Table 20.7.

	x	y	xy	x^2
	0	72.1	0	0
	1.5	54.5	81.75	2.25
	3	55.8	167.4	9
	4.5	54.0	243	20.25
	6	49.0	294	36
	7.5	50.7	380.25	56.25
	9	32.7	294.3	81
	10.5	38.4	403.2	110.25
	12	29.1	349.2	144
Total	54	436.3	2213.1	459

Solution Let x be depth and y be percentage of sand. Table 20.7 shows the calculations of the terms in the normal equations.

The number of data points is $n = 9$. The normal equations are therefore

$$9a + 54b = 436.3 \quad \text{and} \quad 54a + 459b = 2213.1$$

Solving these equations via (20.8) gives $b = -2.998$, $a = 66.46$ (4 s.f.). Then the least squares line of best fit is

$$y = 66.46 - 2.998x$$

20.11 Find suitable linearized forms for the following equations:

(a) $y = \dfrac{a}{x^2} + bx$ (b) $y = a\,e^{bx}$

Solution (a) Multiplying by x^2 we obtain the equation

$$x^2y = a + bx^3$$

If we replace x^2y by Y and x^3 by X we obtain the equation

$$Y = a + bX$$

and we can fit a least squares straight line to the 'new' data to find a and b.

(b) Taking natural logarithms we obtain $\ln y = \ln a + bx$ or

$$Y = A + bx$$

Fitting a least squares straight line to this new data we obtain the value of b directly and $A = \ln a$ from which we can obtain a.

It must be stressed that these 'linearized estimates' of a and b are not likely to be the same as if we had applied the least squares method directly to the original curve, but in the latter case the algebra can become very awkward.

Self-assessment questions 20.3

1. Explain the derivation of the formula (20.6).
2. Explain how linearization can be used to fit a least squares straight line to a set of data points.

Exercise 20.3

1. Fit a least squares straight line to the following data:

(a)

x	1	2	3	4
y	2.8	1.1	−1.3	−2.7

(b)

x	0	10	20	30	40	50
y	5.068	5.454	5.662	5.874	6.208	6.633

(c)

x	1	2	3	4	5
y	10	8.1	5.9	3.9	2.1

In each case plot the points on a graph and draw in the least squares line.

2. Linearize the following equations so that a straight line can be fitted to the new data:

(a) $y = ax + \dfrac{b}{x}$ (b) $y = \dfrac{a}{x^2}$

(c) $y = a\mathrm{e}^{b/x}$ (d) $y = ax^k$

3. Fit by least squares a curve of the form $y = ax^b$ to the following data. (First, you will need to linearize the equation of the curve.)

x	1	2	3	4
y	3.1	11.8	27.5	47.7

Test and assignment exercises 20

1. Use the trapezium rule with (i) four strips (ii) eight strips to estimate the values of the following integrals:

(a) $\displaystyle\int_0^1 \mathrm{e}^{-x}\,\mathrm{d}x$ (b) $\displaystyle\int_0^{\frac{\pi}{2}} \sin x\,\mathrm{d}x$

(c) $\displaystyle\int_1^2 \frac{1}{x^2}\,\mathrm{d}x$ (d) $\displaystyle\int_0^2 \mathrm{e}^{-x^2}\,\mathrm{d}x$

2. Estimate the maximum errors in the estimates of (a), (b) and (c) of Exercise 1.

3. How many strips would guarantee an accuracy of (a) 1 d.p. and (b) 2 d.p. in the examples of Exercise 2?

4. Repeat Exercise 1 using Simpson's rule.

5. Repeat Exercise 2 using Simpson's rule.

6. Repeat Exercise 3 using Simpson's rule.

7. Fit the least squares straight line to the following data:

x	100	200	300	400	500
y	4.00	14.02	23.99	34.06	43.97

8. Linearize the equation $y = a\mathrm{e}^{-bx}$.

21 Ordinary differential equations

Objectives

This chapter

- derives some differential equations which model practical engineering phenomena
- classifies ordinary differential equations
- obtains particular solutions of ODEs by applying initial or boundary conditions to the general solution
- introduces a simple numerical method of solution
- uses the method of isoclines to sketch solution curves

21.1 Introduction

First, we consider three models of engineering systems.

Current in an *RL* circuit

Figure 21.1 shows an *RL* circuit to which a step voltage of magnitude E is applied at $t = 0$.

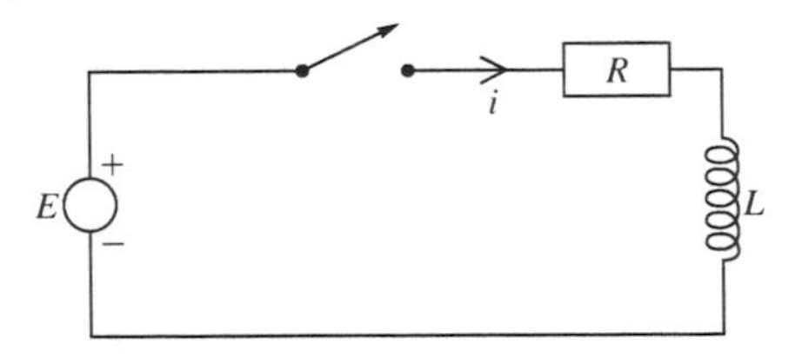

Figure 21.1.
An *RL* circuit

Applying Kirchhoff's law of voltages we obtain

$$V_R + V_L = E \tag{21.1}$$

where V_R is the voltage across the resistor and V_L is the voltage across the inductor.

Ohm's law for the resistor gives

$$V_R = iR \tag{21.2}$$

and Henry's law for the inductor gives

$$V_L = L\frac{di}{dt} \qquad 21.3$$

Combining the equations (21.1), (21.2) and (21.3) we obtain the **differential equation** for the current in the circuit:

$$iR + L\frac{di}{dt} = E \qquad t \geqslant 0 \qquad 21.4$$

Heat transfer in a cooling fin

Newton's law of cooling states that the rate at which a hot body loses heat to its surroundings is proportional to the temperature difference between the body and its surroundings.

If the instantaneous temperature of the body is $\theta(t)$ and the temperature of the surroundings is θ_s (assumed constant) then the law may be stated:

$$\frac{d\theta}{dt} = -k(\theta - \theta_s) \qquad 21.5$$

where k is a positive constant.

The heat developed in an engine can be transferred away by the use of cooling fins. Figure 21.2(a) shows a schematic representation of a cooling fin and part (b) examines the heat flow in a thin section of the fin. The fin is assumed to have constant cross-sectional area, A, and the surrounding air has constant temperature, θ_s.

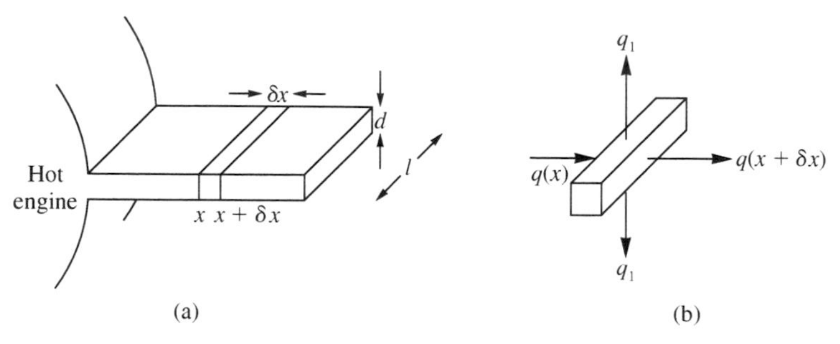

Figure 21.2. A cooling fin

We assume that the heat lost from the sides of the fin is negligible; this is reasonable since $d \ll l$. Heat is transferred from the top and bottom faces by convection. For the strip shown in Figure 21.2(b) this is given by

$$q_1 = h \times l\delta x \times (\theta - \theta_s) \qquad 21.6$$

where q_1 is the rate of heat transfer, h is a heat transfer coefficient and θ is the temperature of the strip at time t; we write $\theta = \theta(x,t)$ to show the dependence on location and time. In a **steady-state** situation $\partial\theta/\partial t = 0$ and we can write $\theta = \theta(x)$.

Heat is transferred along the fin by conduction at a rate

$$q = -kA\frac{d\theta}{dx} \tag{21.7}$$

according to **Fourier's law** of conduction, where k is a heat constant. For the section shown in Figure 21.2(b), the rate of flow of heat in steady state is zero so that

$$q(x) = q(x + \delta x) + 2q_1$$

that is

$$q(x + \delta x) - q(x) = -2q_1 = -2hl\delta x(\theta - \theta_s)$$

from (21.6). Therefore

$$\frac{q(x + \delta x) - q(x)}{\delta x} = -2hl(\theta - \theta_s)$$

As $\delta x \to 0$ we obtain

$$\frac{dq}{dx} = -2hl(\theta - \theta_s) \tag{21.8}$$

Differentiating (21.7) with respect to x and substituting in (21.8), we obtain the equation

$$-kA\frac{d^2\theta}{dx^2} = -2hl(\theta - \theta_s) \tag{21.9}$$

Oscillation of a spring–mass system

For the final example consider the spring–mass system of Figure 21.3. The spring exerts a restoring force which is assumed to be directly proportional to the displacement of the mass; we write this force as $-n^2y$ when y is the displacement from equilibrium and n is a constant. The damping effect is assumed to be proportional to the rate of change of displacement and we write the damping force as $k\dot{y}$ where $\dot{y} = dy/dt$ and k is constant.

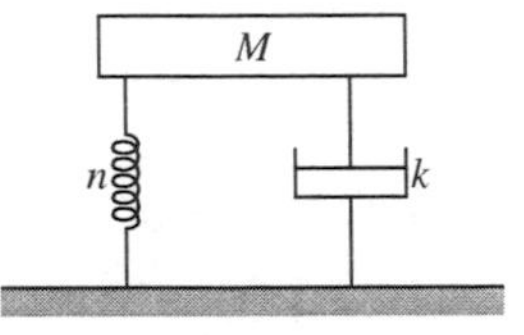

Figure 21.3.
Spring–mass system

Applying Newton's second law of motion to the forces on the mass we obtain the equation

$$M\ddot{y} = -n^2y - k\dot{y} \tag{21.10}$$

where $\ddot{y} = d^2y/dt^2$ is the acceleration of the mass. (M is defined in Figure 21.3.)

If, in addition, a time-varying force $f(t)$ is applied to the mass then the equation is modified to

$$M\ddot{y} = -n^2 y - k\dot{y} + f(t) \qquad 21.11$$

Since (21.4), (21.5), (21.10) and (21.11) are equations which involve ordinary derivatives (as opposed to partial derivatives) they are known as **ordinary differential equations** (or ODEs for short).

The variables t or x are independent variables and i, θ, and y are dependent variables.

Definitions

The **order** of a differential equation is the order of the highest derivative in the equation. The **degree** of an ODE is the power to which the highest derivative is raised.

A **linear** ODE is one in which the dependent variable and its derivatives occur only to the first power and are not combined as products or involved in non-linear functions such as sine or logarithm. Equations which are not linear are **non-linear**. The three ODEs derived earlier are all linear.

If we rearrange (21.10) so that all terms involving the dependent variables and its derivatives are on the left-hand side then the right-hand side is zero, that is

$$M\ddot{y} + k\dot{y} + n^2 y = 0 \qquad 21.12$$

This is referred to as a **homogeneous equation**. A similar rearrangement applied to (21.11) produces

$$M\ddot{y} + k\dot{y} + n^2 y = f(t) \qquad 21.13$$

where the right-hand side is non-zero. This is an example of a **non-homogeneous equation**.

Worked examples

21.1 Find the order and degree of the following ODEs:

(a) $\dfrac{d^2y}{dx^2} + \dfrac{dy}{dx} = x$ (b) $\dfrac{d^2y}{dx^2} + \left(\dfrac{dy}{dx}\right)^3 = x$

(c) $\dfrac{dy}{dx} + \sin x = x^2$ (d) $\left(\dfrac{dx}{dt}\right)^3 + y = \cos t$

Solution
(a) Order 2, degree 1.
(b) Order 2, degree 1 (it is the power to which d^2y/dx^2 is raised).
(c) Order 1, degree 1.
(d) Order 1, degree 3.

21.2 Classify the following ODEs as linear homogeneous, linear non-homogeneous or non-linear. What are the independent and dependent variables in each case?

(a) $iR + L\frac{\mathrm{d}i}{\mathrm{d}t} = E$ (b) $\frac{\mathrm{d}\theta}{\mathrm{d}t} = -k(\theta - \theta_s)$

(c) $-kA\frac{\mathrm{d}^2\theta}{\mathrm{d}x^2} = -2hl(\theta - \theta_s)$ (d) $\frac{\mathrm{d}y}{\mathrm{d}x} = -\cos y$

(e) $\cos x\frac{\mathrm{d}y}{\mathrm{d}x} + y = x^2$ (f) $\frac{\mathrm{d}^2x}{\mathrm{d}y^2} = \frac{x^2+1}{y}$

(g) $\frac{\mathrm{d}^2u}{\mathrm{d}y^2}.\frac{\mathrm{d}u}{\mathrm{d}y} + \cos yu = 3$ (h) $\frac{\mathrm{d}^4y}{\mathrm{d}t^4} = -16y$

Solution (a) Rearranging,

$$L\frac{\mathrm{d}i}{\mathrm{d}t} + Ri = E$$

Linear, non-homogeneous. Independent variable t, dependent variable i.

(b)
$$\frac{\mathrm{d}\theta}{\mathrm{d}t} + k\theta = k\theta_s$$

Linear, non-homogeneous. Independent variable t, dependent variable θ.

(c)
$$\frac{\mathrm{d}^2\theta}{\mathrm{d}x^2} - \frac{2hl\theta}{kA} = -\frac{2hl\theta_s}{kA}$$

Linear, non-homogeneous. Independent variable x, dependent variable θ.

(d)
$$\frac{\mathrm{d}y}{\mathrm{d}x} + \cos x\, y = 0$$

Linear, homogeneous. Independent variable x, dependent variable y.

(e) Linear, non-homogeneous. Independent variable x, dependent variable y.

(f) Non-linear. Independent variable y, dependent variable x.

(g) Non-linear. Independent variable y, dependent variable u.

(h)
$$\frac{\mathrm{d}^4y}{\mathrm{d}t^4} + 16y = 0$$

Linear, homogeneous. Independent variable t, dependent variable y.

Self-assessment questions 21.1

1. Explain the meaning of the order and degree of an ODE.
2. What is the link between a non-homogeneous linear ODE and its associated homogeneous equation?

Exercise 21.1

1. Classify the following ODEs as far as possible:

(a) $\dfrac{d^2y}{dx^2} + 3\dfrac{dy}{dx} = 2$

(b) $\left(\dfrac{d^2y}{dx^2}\right)^3 + \left(\dfrac{dy}{dx}\right)^4 = 0$

(c) $x^2\dfrac{dy}{dx} + xy = 1$

(d) $\dfrac{d^2y}{dx^2} = -25y$

(e) $\dfrac{dm}{dt} = -km$

(f) $v\dfrac{dv}{dx} = -kv - n^2x$

21.2 Solution of an ODE

A **solution** of an ODE is a relationship between the dependent variables and the independent variable such that the ODE is satisfied.

There are usually many solutions of an ODE. The **general solution** comprises all such solutions as special cases. The general solution includes one or more **arbitrary constants**; for an mth order linear equation there are m such constants.

Worked examples

21.3 Show that $y = e^{-x}$ is a solution of the ODE $dy/dx + y = 0$.

Show that $y = Ce^{-x}$ is also a solution for any constant C.

Solution The first formula:

$$\frac{dy}{dx} = -e^{-x} = -y$$

The second formula:

$$\frac{dy}{dx} = C(-e^{-x}) = -Ce^{-x} = -y$$

21.4 Find an ODE satisfied by the following:
(a) $y = x^2$ (b) $y = e^{2x}$
(c) $y = e^{2t} + e^{-2t}$ (d) $y = 4\cos x - 3\sin x$

Solution (a) $\dfrac{dy}{dx} = 2x$ (b) $\dfrac{dy}{dx} = 2e^{2x} = 2y$

(c) $\dfrac{dy}{dt} = 2e^{2t} - 2e^{-2t}$ $\dfrac{d^2y}{dt^2} = 4e^{2t} + 4e^{-2t}$

Hence

$$\frac{d^2y}{dx^2} = 4y$$

(d) $\dfrac{dy}{dx} = -4\sin x - 3\cos x$ $\dfrac{d^2y}{dx^2} = -4\cos x + 3\sin x$

Hence

$$\frac{d^2y}{dt^2} = -y$$

21.5 Show that $y = A\cos\lambda t + B\sin\lambda t$ satisfies the ODE

$$\frac{d^2y}{dt^2} + \lambda^2 y = 0$$

Solution

$$\frac{dy}{dt} = -\lambda A\sin\lambda t + \lambda B\cos\lambda t$$

$$\frac{d^2y}{dt^2} = -\lambda^2 A\cos\lambda t - \lambda^2 B\sin\lambda t = -\lambda^2 y$$

A **particular solution** is obtained from the general solution by giving each of the arbitrary constants a numerical value. These values are determined by imposing conditions on the solution. They are known, in general, as **boundary conditions**.

When time is the independent variable and all the boundary conditions are specified for the same value of the independent variable they are known as **initial conditions**; in this case the ODE and the conditions are known as an **initial-value problem**. If the conditions are specified at both ends of the interval over which the solution is required, for example the deflection of a strut is zero at each end, then the ODE and its condition are known as a **boundary-value problem**.

Worked examples

21.6 The general solution of the ODE $\mathrm{d}y/\mathrm{d}x = -5y$ is $y = Ce^{-5x}$ where C is an arbitrary constant. Find the particular solution which satisfies the condition $y(0) = 2$, that is $y = 2$ when $x = 0$.

Solution Applying the initial condition to the general solution gives $2 = Ce^0 = C$ so that $C = 2$ and the particular solution is $y = 2e^{-5x}$.

21.7 (a) The general solution of the ODE

$$\frac{\mathrm{d}^2y}{\mathrm{d}t^2} = -4y$$

is

$$y = A\cos 2t + B\sin 2t$$

where A and B are arbitrary constants. Find the particular solution which satisfies the conditions $y = 1$, $\mathrm{d}y/\mathrm{d}t = -4$ at $t = 0$.

(b) Find the solution of the ODE

$$\frac{\mathrm{d}^2y}{\mathrm{d}x^2} = -4y$$

which satisfies the condition $y = 0$ at $x = 0$ and at $x = l$, where l is not a multiple of $\pi/2$.

Solution (a) $\dfrac{\mathrm{d}y}{\mathrm{d}t} = -2A\sin 2t + 2B\cos 2t$

$$t = 0,\ y = 1 \text{ gives } A.1 + B.0 = 1, \text{ that is } A = 1$$

$$t = 0,\ \frac{\mathrm{d}y}{\mathrm{d}t} = -4 \text{ gives } 2B = -4, \text{ that is } B = -2$$

The required solution is therefore $y = \cos 2t - 2\sin 2t$.

(b) The general solution is $y = A\cos 2x + \sin 2x$ and hence

$$\frac{\mathrm{d}y}{\mathrm{d}x} = -2A\sin 2x + 2B\cos 2x$$

Now,

$$x = 0,\ y = 0 \text{ gives } A = 0 \text{ so that } y = B\sin 2x$$
$$x = l,\ y = 0 \text{ gives } 2B\sin 2l = 0$$

so that $B = 0$ and $y \equiv 0$.

Self-assessment questions 21.2

1. In a linear ODE of degree 6 how many arbitrary constants should there be in the general solution?
2. What are the two types of condition which allow us to deduce a particular solution from the general solution?

Exercise 21.2

1. Find an ODE, which does not involve the independent variable explicitly, satisfied by each of the following:
   (a) $y = x^3$
   (b) $y = e^{-x}$
   (c) $y = e^{3t} + e^{-3t}$
   (d) $y = 2\cos 2x + 5\sin 2x$
2. Show that $y = A\cosh \lambda t + B\sinh \lambda t$ satisfies the ODE $d^2y/dt^2 = \lambda^2 y$.
3. In each case find the particular solution of an ODE given the general solution and initial conditions:
   (a) $y = Ce^{2t}$; $y(0) = 2$
   (b) $y = A\cos 3t + B\sin 3t$; $y(0) = 1$, $\dot{y}(0) = 0$
   (c) $y = Ae^{-2t} + Be^{2t}$; $y(0) = 0$, $\dot{y}(0) = 1$
4. Find the solution of the ODE $d^2y/dx^2 = -\lambda^2 y$ which satisfies the conditions $y(0) = 0$, $y'(l) = 0$.

21.3 Simple numerical solution for first-order ODEs

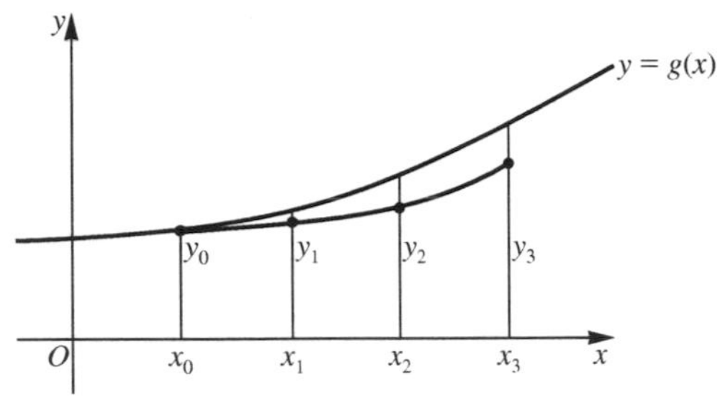

Figure 21.4. Step-by-step solution method

Figure 21.4 shows the solution curve $y = g(x)$ for the differential equation

$$\frac{dy}{dx} = f(x,y) \qquad 21.14$$

Suppose that we are given the condition $y = y_0$ at $x = x_0$. A numerical method of solution of (21.14) estimates the values of y at predetermined values of x. Typically, these values are equally spaced so that $x_1 = x_0 + h$, $x_2 = x_1 + h$, $x_3 = x_2 + h$, and so on, where h is the constant **step size** between the x-values. The estimated value of y at $x = x_1$ is y_1, at $x = x_2$ is y_2, at $x = x_3$ is y_3, and so on.

Euler's method proceeds as follows; refer to Figure 21.5.

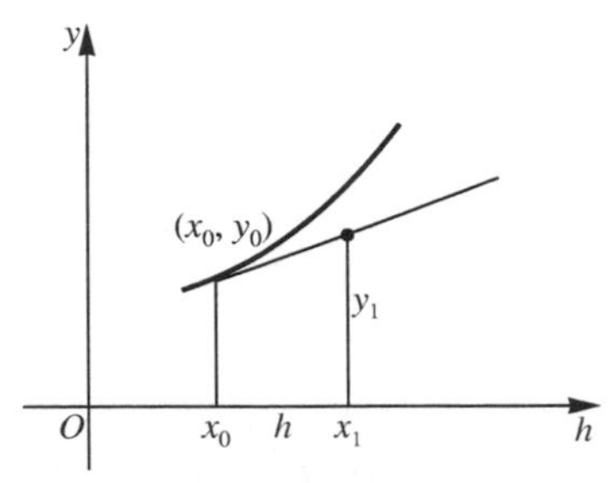

Figure 21.5. Euler's method

From the given point (x_0,y_0) on the solution curve we draw the tangent. This has as gradient dy/dx at $x = x_0$, that is $f(x_0, y_0)$.

The equation of the tangent is

$$y - y_0 = f(x_0,y_0)(x - x_0)$$

At $x = x_1$ this gives

$$y_1 - y_0 = f(x_0, y_0)(x_1 - x_0) = hf(x_0, y_0)$$

that is

$$y_1 = y_0 + hf(x_0,y_0)$$

Now we have a point (x_1,y_1) which we hope is close to the solution curve. The process is repeated to give

$$y_2 = y_1 + hf(x_1,y_1)$$
$$y_3 = y_2 + hf(x_2,y_2)$$

and so on. For an increasing function the estimates become progressively more inaccurate, whereas for a decreasing function they may become progressively more accurate.

KEY POINT

The Euler formula is

$$y_{n+1} = y_n + hf(x_n,y_n) \qquad 21.15$$

Worked examples

21.8 Use the Euler method with $h = 0.2$ to estimate the values of $y = (0.2)$, $y(0.4), \dots, y(1)$ from the equation

$$\frac{dy}{dx} = x + y$$

with the condition $y(0) = 1$.

Compare your results with the analytical solution $y = 2e^x - x - 1$.

Solution We start from $(x_0,y_0) = (0,1)$. Then $f(x_0,y_0) = x_0 + y_0 = 1$.

Applying (21.15) gives $y_1 = 1 + 0.2 \times 1 = 1.2$ which is an estimate of $y(0.2)$. Now

$$(x_1,y_1) = (0.2,1.2) \qquad \text{and} \qquad f(x_1,y_1) = x_1 + y_1 = 0.2 + 1.2 = 1.4$$

Applying (21.15), $y_2 = 1.2 + 0.2 \times 1.4 = 1.48$.

The results are best tabulated as in Table 21.1; the analytical value is quoted to 5 d.p. The error is the Euler value – the analytical value.

Notice how the magnitude of the error becomes progressively worse as the solution proceeds. The error is negative because the Euler values are underestimates. Notice how the number of d.p. increases at each step. In practice we have to round off the estimates.

21.9 With a step size $h = 0.1$, use the Euler method to obtain estimates of $y(0,1), y(0.2), \dots, y(0.6)$ for the equation of Example 21.8.

Compare these results with those of Example 21.8 and comment.

Table 21.1.

n	x_n	y_n	$f(x_n,y_n)$	y_{n+1}	Analytical value	Error
0	0	1	1	1.2	1.242 81	−0.042 8
1	0.2	1.2	1.4	1.48	1.583 65	−0.103 7
2	0.4	1.48	1.88	1.856	2.044 24	−0.188 2
3	0.6	1.856	2.456	2.347 2	2.651 08	−0.303 9
4	0.8	2.347 2	3.147 2	2.976 64	3.436 56	−0.459 9
5	1	2.976 64				

Table 21.2.

n	x_n	y_n	$f(x_n,y_n)$	y_{n+1}	Analytical value	Error
0	0	1	1	1.1	1.110 34	−0.010 3
1	0.1	1.1	1.2	1.22	1.242 81	−0.022 8
2	0.2	1.22	1.42	1.362	1.399 72	−0.037 7
3	0.3	1.362	1.662	1.528 2	1.581 65	−0.053 5
4	0.4	1.528 2	1.928 2	1.721 02	1.797 44	−0.076 4
5	0.5	1.721 02	2.221 02	1.943 122	2.044 24	−0.1011
6	0.6	1.943 122				

Solution The results are given in Table 21.2

The errors are rounded to 4 d.p. The Euler estimates are still underestimates but those at $x = 0.2$, 0.4, 0.6 are better than they were when $h = 0.2$. The error there has been approximately halved.

In practice we would not use the Euler method if we had the means to obtain the analytical solution. There is no easy way to choose a value of h which will deliver estimates to a prescribed accuracy.

Self-assessment questions 21.3

1. What assumption does the Euler method make about the behaviour of the solution curve between $x = x_n$ and $x = x_{n+1}$?
2. Write down the Euler formula and explain the meaning of each of its terms.

Exercise 21.3

1. Use the Euler method with the step size shown to advance four steps from the given initial condition with the given differential equation:

(a) $\dfrac{dy}{dx} = 2$, $y(0) = 1$; $h = 0.25$

(b) $\dfrac{dy}{dx} = 2x$, $y(0) = 1$; $h = 0.25$

(c) $\dfrac{dy}{dx} = y$, $y(0) = 3$; $h = 0.2$

(d) $\dfrac{dy}{dx} = y$, $y(0) = 3$; $h = 0.1$

(e) $\dfrac{dy}{dx} = x + 2y$, $y(0) = 1$; $h = 0.2$

(f) $\dfrac{dy}{dx} = x + 2y$, $y(0) = 1$; $h = 0.1$

(g) $\dfrac{dy}{dx} = xy$, $y(0) = 1$; $h = 0.1$

(h) $\dfrac{dy}{dx} = x^2 + y^2$, $y(0) = 1$; $h = 0.1$

2. The analytical solutions to the equations in Question 1(a)–(g) follow. Comment on the accuracy of the Euler method in each case.

(a) $y = 2x + 1$
(b) $y = x^2 + 1$
(c) and (d) $y = 3e^x$
(e) and (f) $y = -\frac{1}{2}x - \frac{1}{4} + \frac{5}{4}e^{2x}$
(g) $y = e^{x^2/2}$

21.4 Graphical solution

A rough idea of the shape of a solution curve can be obtained by indicating its slope locally. The method is best illustrated by examples.

Worked Examples

21.10 Sketch the solution curves for the equation $dy/dx = x$.

Solution Figure 21.6(a) shows the gradients indicated at $x = -2, -1, 0, 1$ and 2 by short line segments with gradients equal to x. There are five sets of **isoclines** (lines of equal slope).

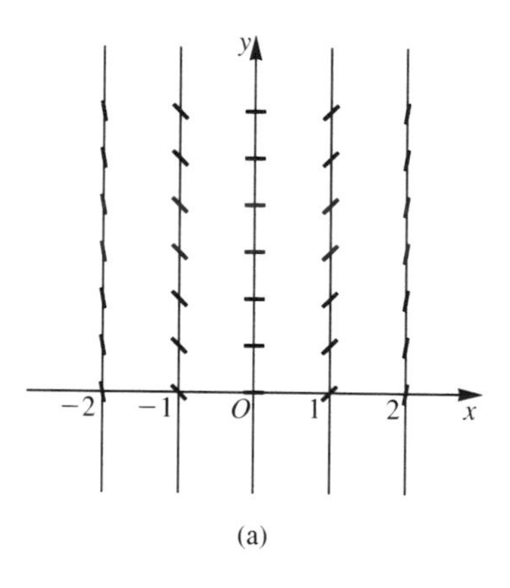

(a)

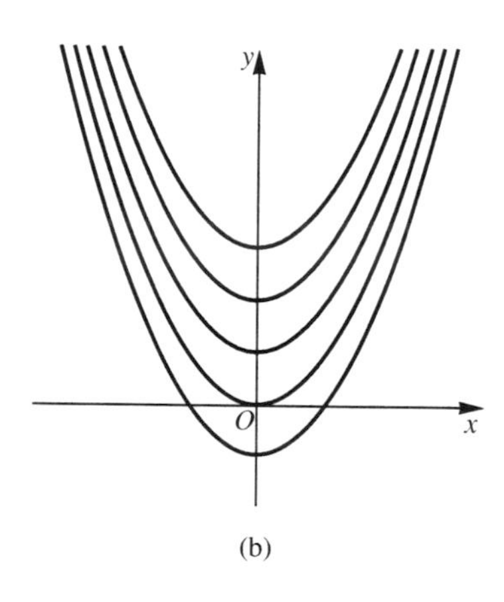

(b)

Figure 21.6. Isoclines and solution curves for Example 21.10

In Figure 21.6(b) five of the solution curves are sketched in using the isoclines. The solution curves are $y = \frac{1}{2}x^2 + C$ where C is an arbitrary constant.

21.11 Sketch the solution curves for the equation $dy/dx = x^2 + y^2$.

Solution Figure 21.7(a) shows some isoclines. Note that dy/dx is constant on the curves $x^2 + y^2 =$ constant. Figure 21.7(b) shows some of the solution curves. This time there is no simple analytical solution.

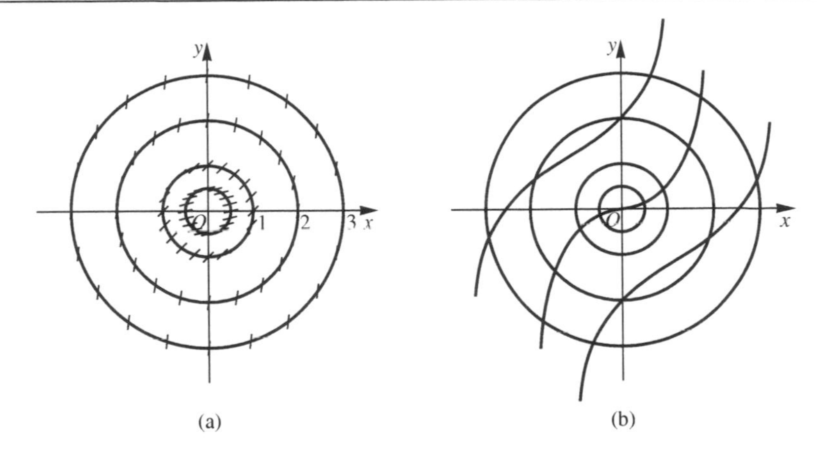

Figure 21.7
Isoclines and solution curves for Example 21.11

Self-assessment questions 21.4

1. Explain the meaning of the term isoclines.
2. Explain how to use isoclines to sketch the solution answer of an ODE.

Exercise 21.4

1. What are the isoclines for each of the following ODEs?

   (a) $\dfrac{dy}{dx} = 2x$ (b) $\dfrac{dy}{dx} = y$

   (c) $\dfrac{dy}{dx} = x^2 + 4y^2$

2. Sketch the solution curves for the ODEs of Question 1.

Test and assignment exercises 21

1. State the order and degree of the following ODEs:

   (a) $\dfrac{d^2y}{dx^2} + \dfrac{dy}{dx}.\dfrac{d^3y}{dx^3} = x^2$

   (b) $\dfrac{dy}{dx} = y\dfrac{d^2y}{dx^2}$

2. Classify the following ODEs:

   (a) $L\dfrac{d^2q}{dt^2} + R\dfrac{dq}{dt} + \dfrac{1}{C}q = e(t)$

   (b) $\left(\dfrac{dy}{dt}\right)^2 = 2\left(1 - \dfrac{y^2}{\sin^2 t}\right)$

   (c) $\dfrac{d^2\theta}{dt^2} + 2k\left(\dfrac{d\theta}{dt}\right)^2 + \dfrac{g}{l}\sin\theta = 0$

3. Find an ODE satisfied by

   (a) $y = \sin 2x$
   (b) $y = e^{ax}$
   (c) $y = \cos\omega t$

4. Verify the following:

   (a) $y = 3e^{2x} + 4e^{-2x}$ is a solution of the equation

   $$\frac{d^2y}{dx^2} = 4y$$

   (b) $y = Ae^{3x} + Be^{-2x}$ is a solution of the equation

   $$\frac{d^2y}{dx^2} - \frac{dy}{dx} - 6y = 0$$

5. (a) Given that $y = Ce^{-x}$ is the general solution of the ODE $dy/dx = -y$, find the particular solution whose graph passes through the following points:

   (i) (0,2) (ii) (1,2) (iii) (0,−1)

   (b) Given that $y = Ae^{2x} + Be^{-3x}$ is the general solution of the ODE

   $$\frac{d^2y}{dx^2} + \frac{dy}{dx} - 6y = 0$$

   find the particular solution which satisfies the conditions $y = 2$ and $dy/dx = 0$ at $x = 0$.

6. With a step size of $h = 0.1$, find the solution of the equation $di/dt = 1 - 2i$ if $i(0) = 0$. Estimate the values of i at $t = 0.1, 0.2, \ldots, 0.5$.

7. Sketch the solution curves for the following equations:

   (a) $\dfrac{dy}{dx} = x + y$

   (b) $\dfrac{dy}{dx} = x - y$

22 First-order ordinary differential equations

Objectives

This chapter

- develops and uses the method of separation of variables
- states a criterion for deciding whether an ODE is exact
- introduces the idea of an integrating factor
- solves a class of equations using the integrating factor method
- uses the improved Euler and fourth-order Runge–Kutta methods to solve equations numerically

22.1 Separation of variables

If a differential equation has the form

$$\frac{dy}{dx} = f(x) \cdot g(y) \qquad 22.1$$

then we can divide the equation by $g(y)$ to obtain

$$\frac{1}{g(y)} \frac{dy}{dx} = f(x)$$

Integrating both sides of the equation with respect to x gives

$$\int \frac{1}{g(y)} \frac{dy}{dx} \cdot dx = \int f(x)\, dx$$

that is

$$\int \frac{1}{g(y)} dy = \int f(x)\, dx \qquad 22.2$$

We have **separated the variables** since each side can be integrated separately. This allows us to find a relationship between y and x which will involve an arbitrary constant.

Although in principle each side being an indefinite integral would produce its own arbitrary constant, only one is necessary for the relationship between y and x.

(Note that, if it helps, you *can* write

$$\frac{1}{g(y)}\,\mathrm{d}y = f(x)\,\mathrm{d}x$$

before putting integral signs in front of each side. If you do, remember that this is *not* a valid equation, merely a means to an end.)

KEY POINT

The solution of the differential equation

$$\frac{\mathrm{d}y}{\mathrm{d}x} = f(x)\cdot g(y)$$

is found as

$$\int \frac{1}{g(y)}\,\mathrm{d}y = \int f(x)\,\mathrm{d}x$$

Worked examples

22.1 Which of the following differential equations can be solved by separation of variables? Obtain the separated form where appropriate.

(a) $\frac{\mathrm{d}y}{\mathrm{d}x} = \frac{x^2}{y^2}$ (b) $\frac{\mathrm{d}y}{\mathrm{d}x} = \frac{y^2}{x^2}$ (c) $\frac{\mathrm{d}y}{\mathrm{d}x} = x^2 + y^2$

(d) $x\frac{\mathrm{d}y}{\mathrm{d}x} + y = 0$ (e) $x\frac{\mathrm{d}y}{\mathrm{d}x} + y = 2$ (f) $y\frac{\mathrm{d}y}{\mathrm{d}x} + x = 2$

Solution

(a) $\int y^2\,\mathrm{d}y = \int x^2\,\mathrm{d}x$

(b) $\int \frac{1}{y^2}\,\mathrm{d}y = \int \frac{1}{x^2}\,\mathrm{d}x$

(c) Not separable

(d) $x\frac{\mathrm{d}y}{\mathrm{d}x} = -y$ so that $\int \frac{1}{y}\,\mathrm{d}y = -\int \frac{1}{x}\,\mathrm{d}x$

(e) $x\frac{\mathrm{d}y}{\mathrm{d}x} = 2 - y$ so that $\int \frac{1}{2-y}\,\mathrm{d}y = \int \frac{1}{x}\,\mathrm{d}x$

(f) $y\frac{\mathrm{d}y}{\mathrm{d}x} = 2 - x$ so that $\int y\,\mathrm{d}y = \int (2-x)\,\mathrm{d}x$

22.2 Find the general solutions of the following equations:

(a) $\dfrac{dy}{dx} = 2y$ (b) $\dfrac{dy}{dx} = \dfrac{x}{y}$

(c) $\dfrac{dy}{dx} = \dfrac{y}{x}$ (d) $\dfrac{dy}{dx} = \dfrac{x^2+4}{y^2+1}$

Solution (a) $\dfrac{1}{y}\dfrac{dy}{dx} = 2$ so that $\displaystyle\int \frac{1}{y}\,dy = \int 2\,dx$

Carrying out the integrations we obtain

$$\ln y = 2x + C$$

(only one '$+C$' is necessary). Then, applying exponentials,

$$y = e^{2x+C} = e^{2x} \cdot e^{C} = Ae^{2x}$$

where we have replaced e^C by A.

(b) $\int y\,dy = \int x\,dx$ and hence $\frac{1}{2}y^2 = \frac{1}{2}x^2 + C$

However, if we multiply by 2 we obtain

$$y^2 = x^2 + 2C$$

and '$2C$' is a strange way to write an unknown number. We therefore write simply $y^2 = x^2 + C$.

(c) $\displaystyle\int \frac{1}{y}\,dy = \int \frac{1}{x}\,dx$

Hence $\ln y = \ln x + C$. For simplification we write the constant not as C but as $\ln C$. This is permissible since $\ln C$ can take all real values. Then

$$\ln y = \ln x + \ln C = \ln(Cx)$$

so that $y = Cx$.

(d) $\int (y^2+1)\,dy = \int (x^2+4)\,dx$

so that

$$\frac{1}{3}y^3 + y = \frac{1}{3}x^3 + 4x + C$$

It is not possible to obtain the relationship in the form $y = f(x)$.

22.3 A heated object is allowed to cool in a room which has a constant temperature of 16 °C. It cools from 80 °C to 48 °C in 30 minutes. How much longer will it take for the object to cool (a) to 20 °C? (b) 16 °C? (Newton's law of cooling states that the rate at which the object cools is proportional to the temperature difference between the object and the air in the room.)

Solution Newton's law may be stated

$$\frac{d\theta}{dt} = -k(\theta - \theta_a)$$

where θ is the temperature of the object, θ_a the air temperature and k is a positive constant related to the thermal capacity of the object. Note the minus sign: if $\theta - \theta_a$ is positive then θ is decreasing with time so that $d\theta/dt$ is negative. Conversely, if $\theta < \theta_a$ then the object would be warming up and $d\theta/dt$ would be positive.

Separating the variables and integrating we obtain

$$\int \frac{1}{\theta - \theta_a}\, d\theta = -\int k\, dt$$

and therefore

$$\ln(\theta - \theta_a) = -kt + C \qquad \text{i}$$

If the initial temperature is θ_0, that is $\theta = \theta_0$ when $t = 0$ then

$$\ln(\theta_0 - \theta_a) = 0 + C \qquad \text{ii}$$

Subtracting (i) and (ii) gives:

$$\ln(\theta - \theta_a) - \ln(\theta_0 - \theta_a) = -kt$$

that is

$$\ln\frac{\theta - \theta_a}{\theta_0 - \theta_a} = -kt$$

so that

$$\frac{\theta - \theta_a}{\theta_0 - \theta_a} = e^{-kt}$$

and

$$\theta = \theta_a + (\theta_0 - \theta_a)e^{-kt} \qquad 22.3$$

Note that as $t \to \infty$, $e^{-kt} \to 0$ and $\theta \to \theta_a$. The value $\theta = \theta_a$ is the **steady- state solution** and the term $(\theta_0 - \theta_a)e^{-kt}$ is called the **transient** since it decays to zero. The graph of θ against t is shown in Figure 22.1.

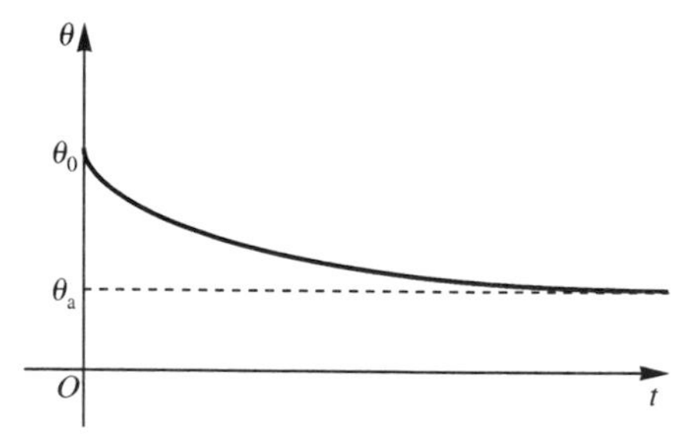

Figure 22.1. Temperature of a cooling object

In this example, $\theta_a = 16$, $\theta_0 = 80$, $\theta = 48$ and $t = 30$. Therefore

$$48 = 16 + 64\,e^{-30k}$$

Hence

$$32 = 64\,e^{-30k} \qquad \text{and} \qquad e^{-30k} = \frac{32}{64} = \frac{1}{2}$$

so that $e^{30k} = 2$ and $30k = \ln 2$. Then $k = \frac{1}{30}\ln 2$.

(a) When $\theta = 20$, $20 = 16 + 64\,e^{-kt}$, that is $4 = 64\,e^{-kt}$, so that

$$e^{-kt} = \frac{1}{16} \qquad \text{and} \qquad t = \frac{1}{k}\ln 16 = 30\frac{\ln 16}{\ln 2}$$

Since $\ln 16 = \ln 2^4 = 4\ln 2$,

$$t = 30 \times \frac{4\ln 2}{\ln 2} = 120 \text{ minutes}$$

Hence the time taken for the object to cool from 48 °C to 20 °C is 90 minutes.

(b) From formula (22.3) we see that in theory the object would take an infinite time to cool to 16 °C. (In practice the object will cool to almost 16 °C in a finite time; for example it will cool from 80 °C to 16.1 °C in approximately 280 minutes.)

22.4 The current in an RL circuit to which a constant voltage of magnitude E is applied from $t \geqslant 0$ is governed by the equation

$$L\frac{di}{dt} + Ri = E$$

Find the current $i(t)$ if $i(0) = 0$.

Solution First,

$$L\frac{di}{dt} = E - Ri$$

so that

$$\frac{1}{E - Ri}\frac{di}{dt} = \frac{1}{L}$$

and hence

$$\int \frac{1}{E - Ri}\,di = \int \frac{1}{L}\,dt$$

Therefore

$$-\frac{1}{R}\ln(E - Ri) = \frac{1}{L}t + C, \quad C \text{ constant}$$

Then

$$\ln(E - Ri) = -\frac{Rt}{L} - RC$$

and

$$E - Ri = e^{-Rt/L - RC} = A\,e^{-Rt/L}, \quad A \text{ constant}$$

Therefore,

$$i = \frac{E}{R} - \frac{A}{R}e^{-Rt/L}$$

When $t = 0$, $i = 0$ so that

$$0 = \frac{E}{R} - \frac{A}{R}$$

Therefore $A = E$ and so, finally,

$$i = \frac{E}{R}(1 - e^{-Rt/L})$$

The graph of $i(t)$ is shown in Figure 22.2.

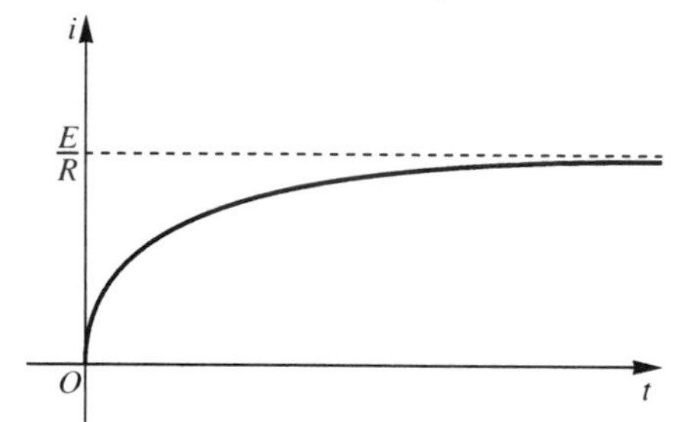

Figure 22.2.
Current against time for the circuit of Example 22.4

As $t \to \infty$, $i \to E/R$, the steady-state value. The larger the fraction R/L, the more rapidly the current approaches the steady-state value.

Self-assessment questions 22.1

1. Explain the method of separation of variables.
2. Why might you write the arbitrary constant of integration in a form other than '$+C$'?

Exercise 22.1

1. Which of the following equations are separable? Obtain the separated form where relevant.

(a) $\frac{dy}{dx} = \frac{x+3}{y-2}$ (b) $\frac{dy}{dx} = \frac{y-2}{x+3}$

(c) $\frac{dy}{dx} = 2x^2 + 4 - y$

(d) $x\frac{dy}{dx} + 2xy = 1$ (e) $y\frac{dy}{dx} + x - 3 = 0$

2. Find the solutions of those equations of Question 1 which are separable.

3. Find the solutions of the following differential equations subject to the given initial condition:

(a) $\frac{dy}{dx} = -2y$, $y(0) = 3$

(b) $\frac{dy}{dt} = -\cos 2\omega t$, $y(0) = 1$

(c) $\frac{du}{dt} = -2t(1 + u^2)$, $u(0) = 1$

4. The velocity v of a rocket attempting to escape from the earth's gravitational field is given by

$$v\frac{dv}{dr} = -g\frac{R^2}{r^2}$$

where r is its distance from the centre of the earth and R is the mean radius of the earth. Find a formula for $v(r)$ and determine the minimum launch velocity v_0 in order that the rocket escapes.

5. At time $t = 0$, water begins to leak from a tank of constant cross-sectional area A. The rate of outflow is proportional to h, the depth of water in the tank at time t. Write the constant of proportion as kA where k is a constant and obtain a differential equation for $h(t)$. How long would it take for the tank to go from full to half full?

6. Repeat Question 5 when the rate of outflow is proportional to $\sqrt{h}$.

22.2 Exact and linear equations

The equation

$$f(x, y) = C, \quad \text{constant} \qquad 22.4$$

can be differentiated with respect to x as follows:

$$\frac{d}{dx}\{f(x, y)\} = \frac{dC}{dx} = 0$$

that is

$$\frac{\partial f}{\partial x} + \frac{\partial f}{\partial y} \cdot \frac{dy}{dx} = 0 \qquad 22.5$$

A differential equation such as (22.5) is called an **exact equation**; it can be solved by integration with respect to x.

Worked examples

22.5 Show that the equation

$$\cos x \frac{dy}{dx} - \sin x \, y = 2$$

is exact and find its general solution.

Solution First, we try to rearrange the equation to the form (22.5). This gives

$$-2 - \sin x \, y + \cos x \frac{dy}{dx} = 0$$

Comparing this with (22.5) then

$$\frac{\partial f}{\partial y} = \cos x \quad \text{and} \quad \frac{\partial f}{\partial x} = -2 - \sin x \, y$$

Now for many functions

$$\frac{\partial^2 f}{\partial x \partial y} \equiv \frac{\partial^2 f}{\partial y \partial x}$$

In this case

$$\frac{\partial^2 f}{\partial x \partial y} = -\sin x \qquad \text{and} \qquad \frac{\partial^2 f}{\partial y \partial x} = 0 - \sin x = -\sin x$$

again. The equation is therefore exact.

Since

$$\frac{\partial f}{\partial y} = \cos x$$

then

$$f = \cos x \, y + \text{a possible function of } x + \text{a possible constant}$$

Since

$$\frac{\partial f}{\partial x} = -2 - \sin x \, y$$

then

$$f = -2x + \cos x \, y + \text{a possible function of } y + \text{a possible constant}$$

Matching these results we obtain

$$f(x,y) = -2x + \cos x \, y + \text{a possible constant}$$

Since the solution is $f(x,y) = \text{constant}$ (from (22.4)) then it can be written as

$$-2x + \cos xy = C, \quad \text{constant}$$

22.6 Show that the equation

$$2x\frac{dy}{dx} + y - \frac{2x}{y} = 0$$

is not exact but by multiplying it by y it becomes exact. (For this equation, y is called the **integrating factor**.)

Solution If the equation is exact then there is a function $f(x,y)$ such that

$$\frac{\partial f}{\partial y} = 2x \quad \text{and} \quad \frac{\partial f}{\partial x} = y - \frac{2x}{y}$$

The first equation gives

$$\frac{\partial^2 f}{\partial x \partial y} = 2$$

and the second gives

$$\frac{\partial^2 f}{\partial y \partial x} = 1 + \frac{2x}{y^2}$$

Hence the ODE is not exact.

Multiplying the ODE by y it becomes

$$2xy\frac{dy}{dx} + y^2 - 2x = 0$$

Here,

$$\frac{\partial f}{\partial y} = 2xy \quad \text{so that} \quad \frac{\partial^2 f}{\partial x \partial y} = 2y$$

and

$$\frac{\partial f}{\partial x} = y^2 - 2x \quad \text{so that} \quad \frac{\partial^2 f}{\partial y \partial x} = 2y$$

The ODE is therefore exact.

In general, the integrating factor is not easy to find. For one class of equations it is possible to find it using the following set of rules.

KEY POINT

The standard form for the ODE is

$$\frac{dy}{dx} + P(x).y = Q(x)$$

where $P(x)$ and $Q(x)$ are functions of x.

The rules are as follows:

1. Take the coefficient of y: $P(x)$.
2. Integrate it, ignoring the arbitrary constant: $\int P(x)\,dx$.
3. Take the exponential: $e^{\int P(x)dx}$.
   This is the integrating factor, referred to as IF.
4. Multiply the ODE by the IF.
5. Check that the left-hand side of the resulting equation is equivalent to

$$\frac{d}{dx}(\text{IF} \times y)$$

6. Integrate both sides of the equation.
7. Divide the resulting equation by the IF.

Worked examples

22.7 Find the general solution of the equation

$$\frac{dy}{dx} + \frac{1}{x}y = x$$

Solution We apply the seven steps outlined above:

1. $\frac{1}{x}$

2. $\int \frac{1}{x}dx = \ln x$

3. $e^{\ln x} = x = \text{IF}$

4. Multiplying the equation by x we obtain

$$x\frac{dy}{dx} + y = x^2$$

5. $\frac{d}{dx}(xy) \equiv x\frac{dy}{dx} + 1.y = x\frac{dy}{dx} + y$

   which checks. Hence

$$\frac{d}{dx}(xy) = x^2$$

6. $xy = \int x^2\,dx = \frac{x^3}{3} + C$

7. $y = \frac{x^2}{3} + \frac{C}{x}$

22.8 Find the solution of the equation

$$\sin x\frac{dy}{dx} + \cos x\,y = 1$$

which satisfies the condition $y(\pi/2) = \pi$.

Solution First, we divide the equation by $\sin x$ to get it in standard form:

$$\frac{dy}{dx} + \cot x\,y = \operatorname{cosec} x$$

1. $\cot x \equiv \frac{\cos x}{\sin x}$

2. $\int \frac{\cos x}{\sin x}\,dx = \ln \sin x$

3. $e^{\ln \sin x} \equiv \sin x = \text{IF}$

4. Multiplying by $\sin x$ gives

$$\sin x\frac{dy}{dx} + \cos x\,y = 1$$

(This is the original equation: as it stood it was exact, but we did not spot this – too bad.)

5. $\frac{d}{dx}(\sin x \times y) \equiv \sin x\frac{dy}{dx} + \cos x\,y$

which checks. Hence

$$\frac{d}{dx}(\sin x\,y) = 1$$

6. $\sin xy = \int 1\,dx = x + C$

7. $y = (x + C)\operatorname{cosec} x$

When $x = \pi/2$, $\operatorname{cosec} x = 1$ and $y = (\pi/2) + C = \pi$. Hence

$$C = \frac{\pi}{2} \quad \text{and} \quad y = \left(x + \frac{\pi}{2}\right)\operatorname{cosec} x$$

22.9 A body of mass m falls in a medium where the resistance to motion is proportional to the speed of the body. The equation of motion is

$$m\frac{dv}{dt} = mg - kv$$

where v is the velocity of the body at time t and k is a constant. Find the velocity $v(t)$ if the body is released from rest.

Solution We first write the equation as

$$\frac{\mathrm{d}v}{\mathrm{d}t} + \frac{k}{m}v = g$$

Then we obtain the integrating factor as

$$\mathrm{e}^{\int \frac{k}{m}\mathrm{d}t} = \mathrm{e}^{kt/m}$$

Multiplying the equation by this expression we obtain

$$\mathrm{e}^{kt/m}\frac{\mathrm{d}v}{\mathrm{d}t} + \frac{k}{m}\mathrm{e}^{kt/m}v = g\mathrm{e}^{kt/m}$$

that is

$$\frac{\mathrm{d}}{\mathrm{d}t}(\mathrm{e}^{kt/m}v) = g\mathrm{e}^{kt/m}$$

Integrating gives

$$\mathrm{e}^{kt/m}v = \int g\mathrm{e}^{kt/m}\,\mathrm{d}t = \frac{mg}{k}\mathrm{e}^{kt/m} + C$$

Then

$$v = \frac{mg}{k} + C\mathrm{e}^{-kt/m}$$

(Note that as $t \to \infty$, $v \to mg/k$.)

When $t = 0$, $v = 0$ so that

$$0 = \frac{mg}{k} + C$$

giving $C = -mg/k$ and, finally,

$$v = \frac{mg}{k}(1 - \mathrm{e}^{-kt/m})$$

(The original ODE could also be solved by separation of variables: carry out the process and verify the result.)

Self-assessment questions 22.2

1. What is meant by the term 'exact equation'?
2. Explain the idea behind the use of an integrating factor.
3. Outline the steps in using the integrating factor method to solve an ODE.

Exercise 22.2

1. Show that the following ODEs are exact:

   (a) $2xy + (x^2 + 1)\dfrac{dy}{dx} = 0$

   (b) $(x - y\cos x) - \sin x\dfrac{dy}{dx} = 0$

2. Show that the equation

   $(3x^2 - y^2)\dfrac{dy}{dx} - 2xy = 0$ is not exact

   but when it is multiplied term by term by $1/y^4$ it *is* exact. Show further that multiplying the equation $(x^2 + y^2)\,dx + xy\,dy = 0$ by the factor $2x$ renders it exact.

3. Solve the following differential equations with the given conditions:

   (a) $\dfrac{dy}{dx} = y + e^x$; $x = 0$, $y = 1$

   (b) $\dfrac{dy}{dx} = xe^{-x^2} - 2xy$; $x = 0$, $y = 2$

   (c) $\dfrac{dy}{dx} - \dfrac{3}{x}y = x^3$; $x = 1$, $y = 4$

4. An RL circuit which includes a constant e.m.f. E is modelled by the equation

   $$L\frac{di}{dt} + Ri = E$$

   If the current i is initially zero and E is constant, find the current at time t and describe what happens to the current as $t \to \infty$. Solve the equation in the case $L = 2$, $R = 10$, $E = 100$. If, now, $E = 100\sin 60t$, find the current at time t.

22.3 Numerical methods

In Section 21.3 the numerical solution of the first-order equation $dy/dx = f(x,y)$ was introduced via the Euler method. This section introduces two further methods.

The **improved Euler** method is an example of a **predictor–corrector** method: at each step we use the Euler method to make a prediction of the next value of y, namely y_{n+1}. Then we evaluate $f(x,y)$ at the point (x_{n+1}, y_{n+1}) and take the average of this value and $f(x_n, y_n)$ to make a second estimate of y_{n+1}; we refer to the estimates as y^{p}_{n+1} and y^{c}_{n+1} (predicted and corrected). The formulae are:

KEY POINT

Predictor:

$$y^{\mathrm{p}}_{n+1} = y_n + hf(x_n, y_n) \qquad 22.6$$

Corrector:

$$y^{\mathrm{c}}_{n+1} = y_n + \frac{1}{2}h\left[f(x_n, y_n) + f(x_n, y^{\mathrm{p}}_{n+1})\right] \qquad 22.7$$

Worked example

22.10 Use the improved Euler method with $h = 0.2$ to estimate $y(0.2)$ and $y(0.4)$ for the ODE $\mathrm{d}y/\mathrm{d}x = x + y$ where $y(0) = 1$.

Solution Note that $f(x, y) = x + y$. We start with $(x_0, y_0) = (0,1)$; then $f(x_0, y_0) = f(0,1) = 1$.

First, we predict $y(0.2)$ by (22.6) as

$$y_1^{\mathrm{p}} = 1 + 0.2 \times 1 = 1.2$$

Next,

$$f(x_1, y_1^{\mathrm{p}}) = f(0.2,1.2) = 1.4$$

The average slope is $\frac{1}{2}(1 + 1.4) = 1.2$. Then

$$y_1^{\mathrm{c}} = 1 + 0.2 \times 1.2 = 1.22$$

We now start from $(0.2,1.22)$; $f(0.2,1.22) = 1.42$. The prediction gives

$$y_2^{\mathrm{p}} = 1.22 + 0.2 \times 1.42 = 1.504$$

Next,

$$f(x_2, y_2^{\mathrm{p}}) = 1.904$$

The average slope is $\frac{1}{2}(1.42 + 1.904) = 1.662$. Then

$$y_2^{\mathrm{c}} = 1.22 + 0.2 \times 1.662 = 1.5524$$

These estimates are closer to the correct answers than the simple Euler values but we have had to do more work to get each one.

The **fourth-order Runge–Kutta method** makes four estimates of the change k we make to go from y_n to y_{n+1}. The formulae are somewhat strange at first sight, but they can be applied straightforwardly with care. They can be summarized as follows:

KEY POINT

$$y_{n+1} = y_n + k$$

where

$$k = \tfrac{1}{6}(k_1 + 2k_2 + 2k_3 + k_4)$$

and

$$\begin{aligned} k_1 &= hf(x_n, y_n) \\ k_2 &= hf\left(x_n + \tfrac{1}{2}h, y_n + \tfrac{1}{2}k_1\right) \\ k_3 &= hf\left(x_n + \tfrac{1}{2}h, y_n + \tfrac{1}{2}k_2\right) \\ k_4 &= hf(x_n + h, y_n + k_3) \end{aligned} \qquad 22.8$$

Worked example

22.11 Use the fourth-order Runge–Kutta method to estimate $y(0.2)$ and $y(0.4)$ for the equation $dy/dx = x + y$ where $y(0) = 1$.

Solution First, $h = 0.2, f(x, y) = x + y, x_0 + \frac{1}{2}h = 0.1$. Then

$$k_1 = 0.2f(0,1) = 0.2$$

$$k_2 = 0.2f(0.1,1 + 0.1) = 0.2f(0.1,1.1) = 0.24$$

$$k_3 = 0.2f(0.1,1 + 0.12) = 0.2f(0.1,1.12) = 0.244$$

$$k_4 = 0.2f(0.2,1 + 0.244) = 0.2 \times (0.2 + 1.244) = 0.2888$$

Then

$$k = \tfrac{1}{6}(0.2 + 2 \times 0.24 + 2 \times 0.244 + 0.2888) = 0.2428$$

and $y = 1.2428$.

We now start from $(0.2,1.2428)$;

$$x_1 + \tfrac{1}{2}h = 0.3$$

Then

$$k_1 = 0.2f(0.2,1.242\,8) = 0.2 \times 1.442\,8 = 0.288\,56$$

$$k_2 = 0.2f(0.3,1.242\,8 + 0.144\,28) = 0.337\,416$$

$$k_3 = 0.2f(0.3,1.242\,8 + 0.168\,708) = 0.342\,301\,6$$

$$k_4 = 0.2f(0.4,1.242\,8 + 0.342\,301\,6) = 0.397\,020\,32$$

$$k = 0.340\,835\,92$$

and therefore $y_2 = 1.242\,8 + 0.340\,835\,92 = 1.583\,6$.

These values are accurate to 4 d.p. (i.e. $0(h^4)$). They required much more effort than Euler or improved Euler but if high accuracy is wanted then this is the price we must pay.

Self-assessment questions 22.3

1. Explain the idea of a predictor–corrector method for solving ODEs.
2. What are the advantages and disadvantages of using the fourth-order Runge–Kutta method?

Exercise 22.3

1. Use the improved Euler method to estimate $y(0.5)$ and $y(1)$ for the following equations, given that $y(0) = 2$. Take $h = 0.5$.

   (a) $\frac{dy}{dx} = 1 - x$ (b) $\frac{dy}{dx} = 1 - y$

   (c) $\frac{dy}{dx} = 1 - x - y$ (d) $\frac{dy}{dx} = 1 - y^2$

2. Use the Runge–Kutta method on the equations of Questions 1(a) and (b). Again, take $h = 0.5$. Then estimate $y(1)$ taking $h = 1$.

Test and assignment exercises 22

1. Obtain, where possible, the separated form of the following differential equations:

   (a) $\frac{dy}{dx} = \frac{y}{x}$

   (b) $\frac{dy}{dx} = x(y - 2)$

   (c) $x\frac{dy}{dx} + x = 2$

   (d) $x\frac{dy}{dx} + y = 2$

   (e) $\frac{dy}{dx} = x + y$

2. Find the general solutions of the following equations:

   (a) $\frac{dy}{dx} = x^2y^3$

   (b) $\frac{dy}{dx} = \frac{x^2}{y^3}$

   (c) $\frac{dy}{dx} = \frac{y+3}{x+1}$

3. Find the solutions of the equations in Exercise 2 which satisfy the condition $y = 1$ when $x = 1$.

4. Which of the following equations are exact?

   (a) $(x^2 + y^2) + 2xy\frac{dy}{dx} = 0$

   (b) $(2 + \sin y) + (1 - x\cos y)\frac{dy}{dx} = 0$

   (c) $\sin x \cos y + \cos x \sin y\frac{dy}{dx} = 0$

5. Solve the following differential equations with the given conditions:

   (a) $\frac{dy}{dx} + 2y = e^{-x}$; $x = 0, y = 2$

   (b) $\frac{dy}{dx} + \tan x\, y = \cos^2 x$; $x = 0, y = 2$

   (c) $\frac{dy}{dx} + \frac{2}{x}y = x^2$; $x = 1, y = 2$

6. Use the improved Euler and fourth-order Runge–Kutta methods with the step sizes shown to solve the following differential equations by estimating the value of y when $x = 0.4$:

   (a) $\frac{dy}{dx} = x + y + 1$; $y(0) = 1, h = 0.1$

   (b) $\frac{dy}{dx} = x^2 + y^2$; $y(0) = 2, h = 0.2$

   (c) $\frac{dy}{dx} = x + y^2$; $y(0) = -1, h = 0.2$

23 Second-order ordinary differential equations

Objectives

This chapter

- explains the strategy for finding the solution to linear second-order ordinary differential equations with constant coefficients
- presents the three types of complementary function
- demonstrates how to find a particular integral for the more common forcing functions
- finds the solution to problems in electrical and mechanical oscillations which are modelled by the equations mentioned above

23.1 Introduction

In this chapter we consider equations of the form

$$\frac{d^2y}{dx^2} + a\frac{dy}{dx} + by = f(x) \qquad 23.1$$

where a and b are constants.

The method of solution is in four stages:

1. Find the most general solution of the associated **homogeneous** equation:

$$\frac{d^2y}{dx^2} + a\frac{dy}{dx} + by = 0 \qquad 23.2$$

This solution is called the **complementary function** of equation (23.1) and we shall sometimes denote it by y_{CF}. It contains two arbitrary constants.

2. Find *any* solution of equation (23.1). This is called a **particular integral** and we shall sometimes denote it by y_{PI}. It contains no arbitrary constants.

3. Add the complementary function and particular integral to obtain the **general solution** of equation (23.1). It contains two arbitrary constants.

4. Apply the given boundary conditions or initial conditions to evaluate the arbitrary constants and obtain the **particular solution**.

The first three steps are made plausible by the following two worked examples.

Worked examples

23.1 If $L\{y\}$ denotes the expression

$$\frac{d^2y}{dx^2}+4\frac{dy}{dx}+3y$$

show that:
(a) $L\{\alpha y_1\}=\alpha L(y_1)$, α constant
(b) $L\{y_1+y_2\}=L\{y_1\}+L\{y_2\}$
Hence deduce that $L\{\alpha y_1+\beta y_2\}=\alpha L\{y_1\}+\beta L\{y_2\}$, for α,β constants. (This result shows that L is a **linear operator**.)

Solution (a) From the constant multiple rule for differentiation

$$\frac{d}{dx}(\alpha y_1)=\alpha\frac{dy_1}{dx}$$

and therefore

$$\frac{d^2}{dx^2}(\alpha y_1)=\alpha\frac{d^2y_1}{dx^2}$$

Hence

$$L\{\alpha y_1\}=\alpha\frac{d^2y_1}{dx^2}+4\alpha\frac{dy_1}{dx}+3\alpha y_1=\alpha L\{y_1\}$$

(b) From the sum rule for differentiation,

$$\frac{d}{dx}(y_1+y_2)=\frac{dy_1}{dx}+\frac{dy_2}{dx}$$

and therefore

$$\frac{d^2}{dx^2}(y_1+y_2)=\frac{d^2y_1}{dx^2}+\frac{d^2y_2}{dx^2}$$

Hence

$$L\{y_1 + y_2\} = \frac{d^2y_1}{dx^2} + \frac{d^2y_2}{dx^2} + 4\frac{dy_1}{dx} + 4\frac{dy_2}{dx} + 3y_1 + 3y_2$$
$$= \left(\frac{d^2y_1}{dx^2} + 4\frac{dy_1}{dx} + 3y_1\right) + \left(\frac{d^2y_2}{dx^2} + 4\frac{dy_2}{dx} + 3y_2\right)$$
$$= L\{y_1\} + L\{y_2\}$$

From part (b) and then part (a),

$$L\{\alpha y_1 + \beta y_1\} = L\{\alpha y_1\} + L\{\beta y_2\}$$
$$= \alpha L\{y_1\} + \beta L\{y_2\}$$

23.2 If y_1 is a solution of the equation

$$\frac{d^2y}{dx^2} + 4\frac{dy}{dx} + 3y = e^{-2x}$$

and y_2 is a solution of the associated equation

$$\frac{d^2y}{dx^2} + 4\frac{dy}{dx} + 3y = 0$$

show that $y_1 + y_2$ is a solution of the original equation.

Solution We are given that

$$\frac{d^2y_1}{dx^2} + 4\frac{dy_1}{dx} + 3y_1 = e^{-2x}, \text{ that is } L\{y_1\} = e^{-2x}$$

and that

$$\frac{d^2y_2}{dx^2} + 4\frac{dy_2}{dx} + 3y_2 = 0, \text{ that is } L\{y_2\} = 0$$

Hence

$$L\{y_1 + y_2\} = L\{y_1\} + L\{y_2\} = e^{-2x} + 0 = e^{-2x}$$

Therefore $y_1 + y_2$ is a solution of the original equation.

Self-assessment questions 23.1

1. Explain the terms homogeneous equation, complementary function, particular integral and general solution in relation to linear second-order ODEs with constant coefficients.
2. Give an example of a linear operator.

Exercise 23.1

1. If $L(y)$ denotes the expression

$$\frac{d^2y}{dx^2} + a\frac{dy}{dx} + by$$

for a and b constants show that

$$L(y_1 + y_2) = L(y_1) + L(y_2) \text{ and}$$
$$L(\alpha y_1) = \alpha L(y_1)$$

where α is a constant.

23.2 Complementary functions

Consider the ODE

$$\frac{d^2y}{dx^2} + a\frac{dy}{dx} + by = 0 \qquad 23.3$$

where a and b are constants.

Suppose $y = e^{mx}$; then

$$\frac{dy}{dx} = me^{mx} \quad \text{and} \quad \frac{d^2y}{dx^2} = m^2e^{mx}$$

Substituting these expressions into the ODE we obtain

$$m^2e^{mx} + ame^{mx} + be^{mx} = 0$$

that is

$$(m^2 + am + b)e^{mx} = 0$$

Now e^{mx} is never zero. Therefore it follows that

$$m^2 + am + b = 0$$

This is the **auxiliary equation** of the ODE.

If this equation has two real roots, m_1 and m_2, then $y = e^{m_1x}$ and $y = e^{m_2x}$ are two independent solutions of (23.3). The general solution of (23.3) is

$$y = Ae^{m_1x} + Be^{m_2x} \qquad A \text{ and } B \text{ constants} \qquad 23.4$$

If the auxiliary equation has a repeated real root m_1 then $y = e^{m_1x}$ and $y = xe^{m_1x}$ are two solutions of (23.4) and the general solution is

$$y = (A + Bx)e^{m_1x} \qquad 23.5$$

If the auxiliary equation has complex conjugate roots $\alpha \pm \beta j$ then the general solution of (23.3) is

$$\begin{aligned} y &= Ce^{(\alpha+\beta j)x} + De^{(\alpha-\beta j)x} \qquad C \text{ and } D \text{ constants} \\ &= e^{\alpha x}(Ce^{\beta jx} + De^{-\beta jx}) \end{aligned}$$

Using the results $e^{\beta jx} \equiv \cos\beta x + (\sin\beta x)j$, $e^{-\beta jx} \equiv \cos\beta x - (\sin\beta x)j$ we may write the general solution as

$$y = e^{\alpha x}(A\cos\beta x + B\sin\beta x) \qquad 23.6$$

where A and B are constants.

KEY POINT

The complementary function is

$y = Ae^{m_1x} + Be^{m_2x}$	(auxiliary equation has two real roots m_1, m_2)
$y = (A + Bx)e^{m_1x}$	(auxiliary equation has one repeated root m_1)
$y = e^{\alpha x}(A\cos\beta x + \beta\sin\beta x)$	(auxiliary equation has complex conjugate roots $\alpha \pm \beta j$)

Worked examples

23.3 Find the general solution of the following ODEs:

(a) $\dfrac{d^2y}{dx^2} + 3\dfrac{dy}{dx} + 2y = 0$ (b) $\dfrac{d^2y}{dx^2} + 4\dfrac{dy}{dx} + 4y = 0$

(c) $\dfrac{d^2y}{dx^2} + 4\dfrac{dy}{dx} + 7y = 0$

Solution (a) Putting $y = e^{mx}$ and substituting in the ODE we obtain the auxiliary equation

$$m^2 + 3m + 2 = 0$$

(The quick way of obtaining this equation is to replace d^2y/dx^2 by m^2, dy/dx by m and y by 1.)

The equation can be written

$$(m + 2)(m + 1) = 0$$

so that $m = -2$ or -1. The general solution is

$$y = Ae^{-2x} + Be^{-x}$$

(b) Putting $y = e^{mx}$ we obtain the auxiliary equation

$$m^2 + 4m + 4 = 0$$

that is, $(m + 2)^2 = 0$ so that $m = -2$, repeated. Hence the general solution is

$$y = (A + Bx)e^{-2x}$$

(c) Putting $y = e^{mx}$ the auxiliary equation is

$$m^2 + 4m + 7 = 0$$

which has solution

$$m = \frac{-4 \pm \sqrt{16 - 28}}{2} = -2 \pm \sqrt{3}j$$

The general solution is

$$y = e^{-2x}(A\cos\sqrt{3}x + B\sin\sqrt{3}x)$$

23.4 The following equations each represent a damped elastic system subjected to an initial disturbance of 1 unit and released from rest, that is, at $t = 0$, $y = 1$, $\mathrm{d}y/\mathrm{d}t = 0$.

(a) $\dfrac{\mathrm{d}^2y}{\mathrm{d}t^2} + 3\dfrac{\mathrm{d}y}{\mathrm{d}t} + 2y = 0$ (b) $\dfrac{\mathrm{d}^2y}{\mathrm{d}t^2} + 4\dfrac{\mathrm{d}y}{\mathrm{d}t} + 4y = 0$

(c) $\dfrac{\mathrm{d}^2y}{\mathrm{d}t^2} + 4\dfrac{\mathrm{d}y}{\mathrm{d}t} + 7y = 0$

Describe the subsequent motion of the systems, illustrating your answer by graphs. (Use the results of Example 23.2 with x replaced by t.)

Solution (a) The general solution is $y = A\mathrm{e}^{-2t} + B\mathrm{e}^{-t}$, from which we deduce that

$$\frac{\mathrm{d}y}{\mathrm{d}t} = -2A\mathrm{e}^{-2t} - B\mathrm{e}^{-t}$$

$$\text{At } t = 0,\ y = 1 \text{ therefore } A + B = 1 \qquad \text{i}$$

$$\text{At } t = 0,\ \frac{\mathrm{d}y}{\mathrm{d}t} = 0 \text{ therefore } -2A - B = 0 \qquad \text{ii}$$

Adding (i) and (ii) gives $-A = 1$ so that $A = -1$.

Using (i) we obtain $B = 2$. Hence the subsequent motion is described by

$$y = -\mathrm{e}^{-2t} + 2\mathrm{e}^{-t}$$

This represents a **heavily damped** or **overdamped system**.

A graph of y against t is shown in Figure 23.1(a). The damping is so strong that the system does not return to equilibrium (the undisturbed position), approaching it asymptotically. There are no oscillations about equilibrium.

(b) The solution is $y = (A + Bt)\mathrm{e}^{-2t}$ and hence

$$\frac{\mathrm{d}y}{\mathrm{d}t} = (B - 2A - 2Bt)\mathrm{e}^{-2t}$$

Applying the initial conditions gives, in turn, $A = 1$, $B = 2$, and therefore the motion is described by

$$y = (1 + 2t)\mathrm{e}^{-2t}$$

This represents a **critically damped system**; the graph is shown in Figure 23.1(b). Again, equilibrium is approached asymptotically; but this case is on the borderline, since if the damping were any weaker then oscillations about equilibrium would occur.

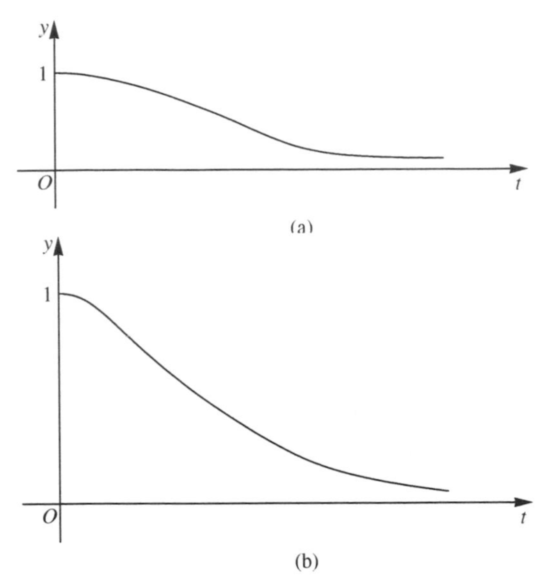

Figure 23.1.
(a) Overdamped system
(b) critically damped system

(c) The general solution is

$$y = e^{-2t}(A\cos\sqrt{3}t + B\sin\sqrt{3}t)$$

so that

$$\frac{dy}{dt} = e^{-2t}(-2A\cos\sqrt{3}t - 2B\sin\sqrt{3}t - \sqrt{3}A\sin\sqrt{3}t + \sqrt{3}B\cos\sqrt{3}t)$$

Applying the initial conditions we find

$$A = 1, \qquad -2A + \sqrt{3}B = 0 \qquad \text{so that} \qquad B = \frac{2\sqrt{3}}{3}$$

giving

$$y = e^{-2t}\left(\cos\sqrt{3}t + \frac{2\sqrt{3}}{3}\sin\sqrt{3}t\right)$$

This represents a **lightly damped** or **underdamped system** with the graph of Figure 23.2: the dashed curves $y = e^{-2t}$ and $y = -e^{-2t}$ represent an envelope within which the oscillations take place with decaying amplitude. The weaker the damping the slower will be the decay.

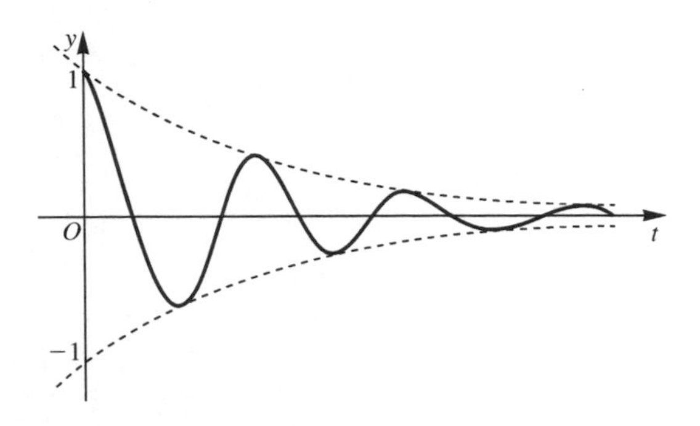

Figure 23.2.
Underdamped system

23.5 Find the general solution of the equation

$$\frac{d^2y}{dt^2} + 4y = 0$$

Find the particular solution for the following initial conditions:

(a) $y = 1, \frac{dy}{dt} = 0$ (b) $y = 0, \frac{dy}{dt} = 1$ (c) $y = 1, \frac{dy}{dt} = 1$

Solution There is no damping term and this case represents **simple harmonic motion** in which the oscillations continue indefinitely with the same amplitude. The auxiliary equation is $m^2 + 4 = 0$ so that $m^2 = -4$ and therefore $m = \pm 2j$.

The general solution is

$$y = A\cos 2t + B\sin 2t$$

Hence

$$\frac{dy}{dt} = -2A\sin 2t + 2B\cos 2t$$

(a) Applying the initial conditions, $A = 1$, $B = 0$. Then the solution is $y = \cos 2t$.
(b) Here $A = 0$, $B = \frac{1}{2}$ and $y = \frac{1}{2}\sin 2t$.
(c) In this case $A = 1$, $B = \frac{1}{2}$ and $y = \cos 2t + \frac{1}{2}\sin 2t$.

Self-assessment questions 23.2

1. Explain the term auxiliary equation.
2. What are the three forms of the general solution to the associated homogeneous equation?

Exercise 23.2

1. The auxiliary equations for some second-order ODEs, where y is the dependent variable and x is the independent variable, are given. Find the general solution of the associated homogeneous equation in each case:

(a) $m^2 - m - 2 = 0$
(b) $m^2 - 10m + 25 = 0$
(c) $m^2 - m + 2 = 0$
(d) $m^2 - 2 = 0$
(e) $m^2 + 2 = 0$

2. Find the general solution of the equation

$$a\frac{d^2y}{dx^2} + b\frac{dy}{dx} + cy = 0$$

where a, b and c are given as follows:

(a) $a = 1, b = -5, c = 4$
(b) $a = 1, b = 6, c = 9$
(c) $a = 1, b = 4, c = 5$
(d) $a = 1, b = -2, c = 1$
(e) $a = 1, b = 0, c = 9$
(f) $a = 1, b = -2, c = 0$
(g) $a = L, b = R, c = 1/C, R^2 = 4L/C$

23.3 Particular integral

The general form of the equation we now consider is

$$\frac{d^2y}{dx^2} + a\frac{dy}{dt} + by = f(t)$$

Table 23.1 specifies common forms of the **forcing function** $f(t)$ and the kinds of solution to try. For example, if $f(t)$ is $\cos 3t$ we are advised to try $y = C\cos 3t + D\sin 3t$. Substituting for y and its derivatives we match the coefficients of each similar term on the left- and right-hand sides of the ODE and hence determine C and D.

In addition, if $f(t)$ is the sum of two functions we try the sum of two expressions from the right-hand column of Table 23.1. For example, if $f(t) = e^{-2t} + 3t$ we would try $y = Ce^{-2t} + Dt + E$. This approach is known as the method of undetermined coefficients.

Table 23.1.

$f(t)$	Trial solution
constant	$y = C$, constant
t	$y = Ct + D$
$\alpha t + \beta$	$y = Ct + D$
polynomial of degree n	$y = C_n t^n + C_{n-1}t^{n-1} + \ldots + C_0$
e^{at}	$y = Ce^{at}$
$\cos \omega t$, $\sin \omega t$, $\alpha\cos\omega t + \beta\sin\omega t$	$y = C\cos\omega t + D\sin\omega t$

Worked examples

23.6 Find particular integrals for the following equations:

(a) $\dfrac{d^2y}{dt^2} + 3\dfrac{dy}{dt} + 2y = 4$ (b) $\dfrac{d^2y}{dt^2} + 3\dfrac{dy}{dt} + 2y = 4t^2 - 4$

(c) $\dfrac{d^2y}{dt^2} + 3\dfrac{dy}{dt} + 2y = e^{-3t}$ (d) $\dfrac{d^2y}{dt^2} + 3\dfrac{dy}{dt} + 2y = \cos 2t$

Solution (a) We try $y = C$, constant. Then $dy/dt = 0$, $d^2y/dt^2 = 0$. Substituting into the ODE we find that $2C = 4$; hence $C = 2$ and $y = 2$ is a particular integral.

(b) $f(t)$ is a quadratic polynomial. We therefore try $y = Ct^2 + Dt + E$ so that

$$\frac{dy}{dt} = 2Ct + D \quad \text{and} \quad \frac{d^2y}{dt^2} = 2C$$

Substitution gives

$$2C + 6Ct + 3D + 2Ct^2 + 2Dt + 2E \equiv 4t^2 - 4$$

Comparing coefficients of t^2 we find $2C = 4$ so that $C = 2$.
Comparing coefficients of t we find $6C + 2D = 0$ so that $D = -6$.
Comparing the constant terms gives $2C + 3D + 2E = -4$ so that $E = 5$.

A particular integral is therefore $y = 2t^2 - 6t + 5$.

(c) We try $y = Ce^{-3t}$ so that

$$\frac{dy}{dt} = -3Ce^{-3t} \quad \text{and} \quad \frac{d^2y}{dt^2} = 9Ce^{-3t}$$

Substitution produces

$$9Ce^{-3t} - 9Ce^{-3t} + 2Ce^{-3t} \equiv e^{-3t}$$

Hence $C = \frac{1}{2}$ and $y = \frac{1}{2}e^{-3t}$ is a particular integral.

(d) We try $y = C\cos 2t + D\sin 2t$ and then

$$\frac{dy}{dt} = -2C\sin 2t + 2D\cos 2t \quad \text{and}$$

$$\frac{d^2y}{dt^2} = -4C\cos 2t - 4D\sin 2t$$

Substituting, we obtain

$$-4C\cos 2t - 4D\sin 2t - 6C\sin 2t + 6D\cos 2t + 2C\cos 2t + 2D\sin 2t \equiv \cos 2t$$

that is

$$(-2C + 6D)\cos 2t + (-2D - 6C)\sin 2t \equiv \cos 2t \equiv 1.\cos 2t + 0.\sin 2t$$

Hence $-2C + 6D = 1$ and $-2D - 6C = 0$, from which we find that $C = -\frac{1}{20}$ and $D = \frac{3}{20}$.

A particular integral is $y = -\frac{1}{20}\cos 2t + \frac{3}{20}\sin 2t$.

23.7 The equation governing the current $i(t)$ in a series LC circuit subjected to a sinusoidal applied voltage $V\sin\omega t$ is

$$L\frac{d^2i}{dt^2} + \frac{i}{C} = V\omega\cos\omega t$$

Find a particular integral for this equation.

Solution To avoid using the symbol C in two contexts we try a solution in the form

$$i = D\cos\omega t + E\sin\omega t$$

Then

$$\frac{di}{dt} = -\omega D\sin\omega t + \omega E\cos\omega t \quad \text{and}$$

$$\frac{d^2i}{dt^2} = -\omega^2 D\cos\omega t - \omega^2 E\sin\omega t$$

Substituting into the ODE gives

$$-L\omega^2 D\cos\omega t - L\omega^2 E\sin\omega t + \frac{1}{C}D\cos\omega t + \frac{1}{C}E\sin\omega t \equiv V\omega\cos\omega t$$

so that

$$-L\omega^2 D + \frac{1}{C}D = V\omega$$

and

$$-L\omega^2 E + \frac{1}{C}E = 0$$

Hence

$$D = \frac{V\omega}{\frac{1}{C} - L\omega^2} = \frac{V\omega C}{1 - \omega^2 LC} \quad \text{and} \quad E = 0$$

Finally a solution of the ODE is

$$i = \frac{V\omega C}{1 - \omega^2 LC}\cos\omega t$$

(Note that if $\omega^2 = 1/LC$ then we have a problem: we cannot find values for D and E. This phenomenon is known as **resonance**. The way out of the difficulty is dealt with in Section 23.4.)

Self-assessment questions 23.3

1. Explain the type of particular integral to look for if the right-hand side of the ODE is a constant, a polynomial of degree n or an exponential function.
2. Explain the type of particular integral to look for if the right-hand side of the ODE is a cosine, a sine or a combination of sine and cosine with the same frequency.

Exercise 23.3

1. The following equations are all of the form

$$\frac{d^2y}{dt^2} + b\frac{dy}{dt} + cy = f(t)$$

Find a particular integral in each case.

(a) $b = 0, c = 4, f(t) = \sin t$
(b) $b = -3, c = 2, f(t) = 6e^{3t}$
(c) $b = -7, c = 10, f(t) = 100t$
(d) $b = 1, c = 0, f(t) = t^3 - t^2$
(e) $b = 5, c = 4, f(t) = t^2$
(f) $b = 4, c = 4, f(t) = e^t$
(g) $b = 3, c = 4, f(t) = \cos t$

2. For the following differential equations, show that the suggested particular integral is valid by finding a value for the constant C.

(a) $\frac{d^2y}{dt^2} + 3\frac{dy}{dt} + 2y = e^{-3t}$ try $y = Ce^{-3t}$

(b) $\frac{d^2y}{dt^2} + 4\frac{dy}{dt} + 3y = e^{-3t}$ try $y = Cte^{-3t}$

(c) $\frac{d^2y}{dt^2} + 6\frac{dy}{dt} + 9y = e^{-3t}$ try $y = Ct^2e^{-3t}$

3. Show that

$$y = \frac{A\sin(\omega t - \alpha)}{\sqrt{\{[(b^2 - \omega^2)^2] + 4\omega^2 a^2\}}}$$

where $\tan\alpha = 2a\omega/(b^2 - \omega^2)$, is a particular integral of the equation

$$\frac{d^2y}{dt^2} + 2a\frac{dy}{dt} + b^2 y = A\sin\omega t$$

23.4 General and particular solutions

Consider the equation

$$\frac{d^2y}{dt^2} + 3\frac{dy}{dt} + 2y = e^{-3t}$$

From Example 23.3(a), replacing x by t we know that the *complementary function* for the equation is

$$y_{cF} = Ae^{-2t} + Be^{-t}$$

From Example 23.6(c) we know that a *particular integral* is

$$y_{PI} = \tfrac{1}{2}e^{-3t}$$

The *general solution*, y_{GS}, is found by

$$y_{GS} = y_{PI} + y_{CF} = \tfrac{1}{2}e^{-3t} + Ae^{-2t} + Be^{-t}$$

A *particular solution* is found by applying the boundary or initial conditions.

KEY POINT

> The general solution is found by adding the complementary function to a particular integral; a particular solution is found by then applying the given conditions.

Worked examples

23.8 Find the solution of the equation

$$\frac{d^2y}{dt^2} + 3\frac{dy}{dt} + 2y = e^{-3t}$$

if $y(0) = 3$ and $y'(0) = 0$.

Remember that $y'(0)$ means the value of dy/dt when $t = 0$.

Solution The general solution is

$$y = \tfrac{1}{2}e^{-3t} + Ae^{-2t} + Be^{-t}$$

Hence

$$\frac{dy}{dt} = -\frac{3}{2}e^{-3t} - 2Ae^{-2t} - Be^{-t}$$

When $t = 0$, $y = 3$ so that

$$\tfrac{1}{2} + A + B = 3 \quad \text{that is} \quad A + B = \tfrac{5}{2} \qquad \text{i}$$

When $t = 0$, $dy/dt = 0$ so that

$$-\tfrac{3}{2} - 2A - B = 0 \quad \text{that is} \quad 2A + B = -\tfrac{3}{2} \qquad \text{ii}$$

Solving (i) and (ii) we obtain $A = -4$, $B = \frac{13}{2}$, and therefore

$$y = \tfrac{1}{2}e^{-3t} - 4e^{-2t} + \tfrac{13}{2}e^{-t}$$

23.9 In a simple model of a shock absorber (Figure 23.3) the vertical motion of the mass can be described by the equation

$$m\frac{d^2x}{dt^2} + k\frac{dx}{dt} + \lambda x = f(t)$$

where x is the vertical displacement of the mass m at time t, $k\,dx/dt$ is the damping force applied by the dashpot, λx is the restoring force applied by the spring and $f(t)$ is the external force applied to the wheel by the undulations in the road surface.

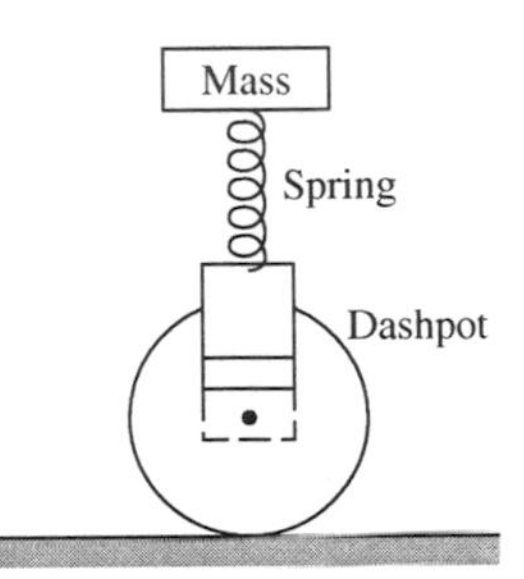

Figure 23.3. Model of a shock absorber

For convenience we divide the equation by m, replace λ/m by ω^2 and k/m by $2\xi\omega$, relabelling $(1/f(t)m)$ as $f(t)$. The equation becomes

$$\frac{d^2x}{dt^2} + 2\xi\omega\frac{dx}{dt} + \omega^2 x = f(t)$$

This is an oscillation with angular frequency Ω (omega).

(a) Find the general solution in the case where $k = 0$, that is $\xi = 0$, and $f(t) = F\cos\Omega t$. Assume that at $t = 0$, x and dx/dt are zero.
(b) What happens to the solution as $\Omega \to \omega$?

Solution (a) The equation is

$$\frac{d^2x}{dt^2} + \omega^2 x = F\cos\Omega t$$

The complementary function is $x_{CF} = A\cos\omega t + B\sin\omega t$. (Compare with Example 23.5.)

For the particular integral we try $x = C\cos\Omega t + D\sin\Omega t$. Then

$$\frac{d^2x}{dt^2} = -\Omega^2 A\cos\Omega t - \Omega^2 B\sin\Omega t$$

and substitution into the ODE produces

$$(\omega^2 - \Omega^2)C\cos\Omega t + (\omega^2 - \Omega^2)D\sin\Omega t \equiv F\cos\Omega t$$

Hence $D = 0$ and $(\omega^2 - \Omega^2)C = F$.

The particular integral is

$$x_{PI} = \frac{F}{\omega^2 - \Omega^2}\cos\Omega t$$

and the general solution is

$$x = \frac{F}{\omega^2 - \Omega^2}\cos \Omega t + A \cos \omega t + B \sin \omega t$$

Then

$$\frac{dx}{dt} = -\frac{\Omega F}{\omega^2 - \Omega^2}\sin \Omega t - \omega A \sin \omega t + \omega B \cos \omega t$$

At $t = 0$, $dx/dt = 0$ so that $B = 0$, and at $t = 0$, $x = 0$ so that

$$A = -\frac{F}{\omega^2 - \Omega^2}$$

Then

$$x = \frac{F}{\omega^2 - \Omega^2}(\cos \Omega t - \cos \omega t)$$

(b) As $\Omega \to \omega$, that is, as the frequency of the applied force approaches the **natural frequency** ω of the system, we can replace Ω by $\omega + \varepsilon$ where ε is small.

Now

$$\cos \Omega t - \cos \omega t \equiv 2 \cos\left(\frac{\Omega + \omega}{2}\right) t \sin\left(\frac{\Omega - \omega}{2}\right) t$$

which represents the product of a slowly varying oscillation

$$\sin\left(\frac{\Omega - \omega}{2}\right) t$$

and a more rapidly varying one

$$\cos\left(\frac{\Omega + \omega}{2}\right) t$$

The phenomenon is known as **beats** and is illustrated in Figure 23.4. The dashed lines indicate the more slowly varying oscillation.

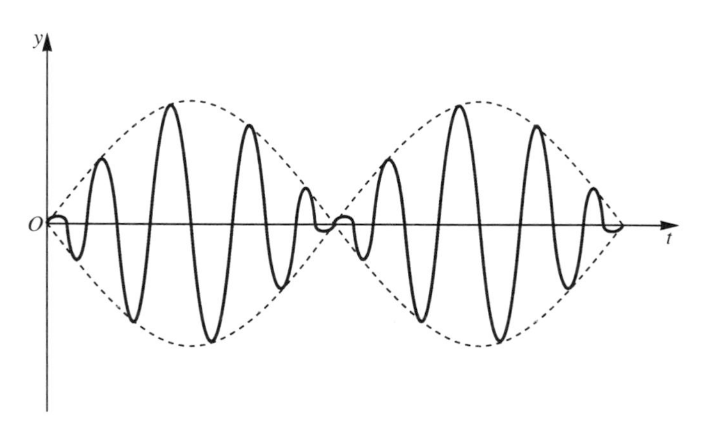

Figure 23.4.
The phenomenon of beats

Notice that the amplitude of x, $F/(\omega^2 - \Omega^2)$, increases as $\Omega \to \omega$. In the limit it becomes infinitely large.

23.10 (a) Show that $y = \frac{1}{4}t \sin 2t$ satisfies the equation

$$\frac{d^2y}{dt^2} + 4y = \cos 2t$$

Hence find the general solution of this equation and deduce the solution which satisfies $y(0) = 0$, $y'(0) = 0$. What happens as t increases?

(b) Find the solution of the equation

$$\frac{d^2y}{dt^2} + 2\frac{dy}{dt} + 4y = \cos 2t$$

which satisfies $y(0) = 0$, $y'(0) = 0$.

Explain the effect of the term $2dy/dt$ on the solution and interpret in terms of a vibrating system.

Solution (a) If $y = \frac{1}{4}t \sin 2t$ then

$$\frac{dy}{dt} = \frac{1}{4}t.2\cos 2t + \frac{1}{4}\sin 2t$$

and

$$\frac{d^2y}{dt^2} = \frac{1}{4}t(-4)\sin 2t + \frac{1}{4}.2\cos 2t + \frac{1}{4}.2\cos 2t = -t\sin 2t + \cos 2t$$

Hence

$$\frac{d^2y}{dt^2} + 4y = -t\sin 2t + \cos 2t + 4.\frac{1}{4}t\sin 2t = \cos 2t$$

The complementary function is

$$y_{CF} = A\cos 2t + B\sin 2t$$

and the general solution is

$$y = \frac{1}{4}t\sin 2t + A\cos 2t + B\sin 2t$$

Hence

$$\frac{dy}{dt} = \frac{1}{2}t\cos 2t + \frac{1}{4}\sin 2t - 2A\sin 2t + 2B\cos 2t$$

When $t = 0$, $y = 0$, that is $A = 0$; when $t = 0$, $y' = 0$, that is $+2B = 0$ so that $B = 0$.

Hence the required solution is

$$y = \frac{1}{4}t\sin 2t$$

As $t \to \infty$, y oscillates with increasing amplitude; see Figure 23.5. This is an example of **resonance**.

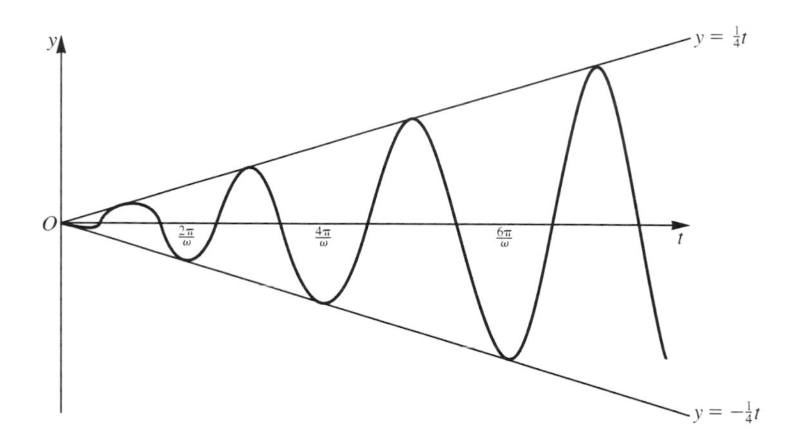

Figure 23.5. The phenomenon of resonance

(b) The auxiliary equation is

$$m^2 + 2m + 4 = 0$$

so that

$$m = \frac{-2 \pm \sqrt{4 - 16}}{2} = -1 \pm \sqrt{3}$$

The complementary function is

$$y_{CF} = e^{-t}(A \cos \sqrt{3}t + B \sin \sqrt{3}t)$$

For a particular integral we try $y = C \cos 2t + D \sin 2t$ so that

$$\frac{dy}{dt} = -2C \sin 2t + 2D \cos 2t \qquad \text{and}$$

$$\frac{d^2y}{dt^2} = -4C \cos 2t - 4D \sin 2t$$

Substituting into the ODE we obtain

$$-4C \cos 2t - 4D \sin 2t - 4C \sin 2t + 4D \cos 2t + 4C \cos 2t$$
$$+ 4D \sin 2t \equiv \cos 2t$$

Hence $4D = 1$ and $-4C = 0$. Therefore $y_{PI} = \frac{1}{4} \sin 2t$.
The general solution is

$$y = \frac{1}{4} \sin 2t + e^{-t}(A \cos 2t + B \sin 2t)$$

and

$$\frac{dy}{dt} = \frac{1}{2} \cos 2t + e^{-t}(-2A \sin 2t + 2B \cos 2t - A \cos 2t - B \sin 2t)$$

At $t = 0$, $y = 0$ so that $A = 0$.

At $t = 0$, $\mathrm{d}y/\mathrm{d}t = 0$, that is $\frac{1}{2} + 2B - A = 0$ so that $B = -\frac{1}{4}$.

Finally, $y = \frac{1}{4}\sin 2t - \frac{1}{4}\mathrm{e}^{-t}\sin 2t$.

The second term decays to zero as $t \to \infty$; it is called the **transient part** of the solution. The first term, which does not decay, is the **steady-state** solution.

The presence of the damping term $2\mathrm{d}y/\mathrm{d}t$ has prevented resonance.

23.11 A pin-ended uniform strut of length l is subject to an axial load P. E is Young's modulus. I is second moment of area. The deflected profile $y(x)$ satisfies approximately the ODE

$$\frac{\mathrm{d}^2y}{\mathrm{d}x^2} + \frac{P}{EI}y = 0$$

with boundary conditions $y(0) = y(l) = 0$.

The load P is increased from zero. Find the value of P at which the strut first buckles into a deflected profile from its original straight profile.

Solution Put $P/EI = m^2$. Then

$$\frac{\mathrm{d}^2y}{\mathrm{d}x^2} + m^2y = 0$$

which has general solution

$$y = A\cos mx + B\sin mx$$

Applying the boundary conditions gives

$$A.1 + B.0 = 0$$

that is, $A = 0$ and $A\cos ml + B\sin ml = 0$, that is $B\sin ml = 0$.

We reject the option $B = 0$ or else $y \equiv 0$. Hence $\sin ml = 0$. The options for ml are $0, \pm\pi, \pm 2\pi, \ldots$.

We ignore $ml = 0$ since, again, this gives $y \equiv 0$. If $ml = \pi$ we obtain

$$y = B\sin\frac{\pi x}{l}$$

and if $ml = -\pi$ we obtain

$$y = B\sin\left(-\frac{\pi x}{l}\right) = -B\sin\frac{\pi x}{l}$$

which is essentially the same solution since B is still arbitrary. Therefore we can ignore negative values of m.

The smallest value of m gives the smallest value of P at which the profile is deflected. Then

$$\frac{P}{EI} = m^2 = \frac{m^2 l^2}{l^2} = \frac{\pi^2}{l^2}$$

so that the critical buckling load is $P = \pi^2 EI/l^2$.

The deflected profile is

$$y = B \sin \frac{\pi x}{l}$$

(Note that we cannot determine B and therefore we do not know the magnitude or direction of the buckle.)

Self-assessment questions 23.4

1. Explain how to combine the complementary function, the particular integral and the initial conditions to obtain the solution to a linear second-order ODE.
2. Explain the phenomena of beats and resonance. How can we prevent resonance?

Exercise 23.4

1. The charge q in a series LCR circuit is governed by the equation

$$L\frac{d^2q}{dt^2} + R\frac{dq}{dt} + \frac{1}{C}q = E$$

where E is the applied e.m.f. Find the current $i = dq/dt$ flowing in the circuit as a function of time in the following cases assuming zero initial conditions, that is, $q = 0$ and $i = 0$ at $t = 0$. Deduce the steady-state current.

(a) $L = 1, R = 100, C = 10^{-4}, E = 1000$
(b) $L = 10, R = 250, C = 10^{-3}, E = 900$
(c) $L = 5, R = 10, C = 0.1, E = 25 \sin t$
(d) $L = 1, R = 7, C = 0.1, E = 100 \sin 10t$

2. Determine the motion of a mass m attached to a spring with spring constant k, displaced a distance x_0 from equilibrium and released with initial velocity v_0. There is assumed to be no damping and there are no external forces. The equation for the displacement is $m\ddot{x} + kx = 0$. Consider the following cases:

(a) $m = 10, k = 1000, x_0 = 1, v_0 = 0$
(b) $m = 10, k = 10, x_0 = 3, v_0 = 4$
(c) $m = 1, k = 25, x_0 = 0, v_0 = 3$

3. The system of Question 2 is subjected to damping with a damping constant c; the equation of motion is $m\ddot{x} + c\dot{x} + kx = 0$. Solve the following cases:

(a) $m = 10, c = 200, k = 1000, x_0 = 1, v_0 = 0$
(b) $m = 10, c = 10\sqrt{5}, k = 10, x_0 = 3, v_0 = 4$
(c) $m = 1, c = 8, k = 25, x_0 = 0, v_0 = 3$

4. The system of Question 3 is now subjected to an external force $\sin \omega t$ where ω is a constant. Find the subsequent motion in the following cases.

(a) $m = 10, c = 200, k = 1000, \omega = 10, x_0 = 1, v_0 = 0$
(b) $m = 1, c = 8, k = 25, \omega = 3, x_0 = 0, v_0 = 3$

Test and assignment exercises 23

1. Find the general solutions of the following equations:

(a) $2\frac{d^2y}{dx^2} + 3\frac{dy}{dx} + y = 0$

(b) $3\frac{d^2y}{dx^2} - 2\frac{dy}{dx} - y = 0$

(c) $\frac{d^2y}{dx^2} + 4\frac{dy}{dx} + 9y = 0$

(d) $\frac{d^2y}{dx^2} - \frac{dy}{dx} + \frac{1}{4}y = 0$

2. Find a particular integral for the following equations:

(a) $\frac{d^2y}{dt^2} + y = \cos 2t$

(b) $\frac{d^2y}{dt^2} + 4\frac{dy}{dt} - 3y = e^{2t}$

(c) $\frac{d^2y}{dt^2} + 2\frac{dy}{dt} + y = t^2 + 1$

(d) $\frac{d^2y}{dt^2} + 6\frac{dy}{dt} + 9y = 5$

3. Find a particular integral for the following equations using the trial function suggested:

(a) $\frac{d^2y}{dx^2} + 4\frac{dy}{dx} + 3y = 2e^x$

try $y = Ce^x$

(b) $\frac{d^2y}{dx^2} - 4\frac{dy}{dx} + 3y = 2e^x$

try $y = Cxe^x$

(c) $\frac{d^2y}{dx^2} - 6\frac{dy}{dx} + 9y = e^{3x}$

try $y = Cx^2 e^{3x}$

4. Find the solution of the following ODEs which satisfy the given initial conditions:

(a) $\ddot{y} + 3\dot{y} + 2y = e^t$;
$y(0) = 1, \dot{y}(0) = 0$

(b) $\ddot{y} + 4\dot{y} + 3y = \cos t$;
$y(0) = 0, \dot{y}(0) = 0$

(c) $\ddot{y} + 2\dot{y} + y = t^2 - 1$;
$y(0) = 1, \dot{y}(0) = -1$

(d) $\ddot{y} + 2\dot{y} + 5y = 4$;
$y(0) = 2, \dot{y}(0) = 1$

24 Laplace transforms

Objectives

This chapter

- defines the Laplace transform of a function
- gives a table of the transforms of common functions
- shows how to extend the table by using four theorems
- uses Laplace transforms to solve initial-value problems
- determines simple transfer functions

24.1 Definitions and simple transforms

An alternative method for solving initial-value problems which involve linear ODEs with constant coefficients is often used in the analysis of signals and systems. It hinges on transforming the ODE, incorporating the initial conditions, to an algebraic equation. This equation is solved and the solution transformed back into the original variables. This chapter looks at one particular transform.

The **Laplace transform** of a function $f(t)$ is a function $F(s)$ defined by

KEY POINT

$$F(s) = \int_0^\infty f(t)e^{-st}dt \qquad 24.1$$

Sometimes we write the Laplace transform as $\mathcal{L}\{f(t)\}$. Note that because the range of the integral is semi-infinite we have to impose the restriction that

$$\lim_{t \to \infty} \{|f(t)|e^{-kt}\} = 0$$

where k is the real part of s. For convenience in practical applications we usually assume that $f(t) = 0$ for $t < 0$.

Worked examples

24.1 Find the Laplace transforms of (a) 1, (b) t, (c) e^{-at}.

Solution (a)

$$\mathcal{L}\{1\} = \int_0^\infty 1e^{-st}dt = \left[\frac{e^{-st}}{-s}\right]_0^\infty = \left[\frac{e^{-st}}{s}\right]_\infty^0 = \frac{1}{s}$$

(since $e^{-st} \to 0$ as $t \to \infty$).

(b)

$$\mathcal{L}\{t\} = \int_0^\infty te^{-st}dt = \left[t\frac{e^{-st}}{-s}\right]_0^\infty - \int_0^\infty 1\frac{e^{-st}}{-s}dt$$

Now

$$t\frac{e^{-st}}{-s} = 0 \text{ at } t = 0 \qquad \text{and} \qquad t\frac{e^{-st}}{-s} \to 0 \text{ as } t \to \infty.$$

Hence

$$\mathcal{L}\{t\} = 0 + \frac{1}{s}\int_0^\infty e^{-st}dt = \frac{1}{s}.\frac{1}{s} = \frac{1}{s^2}$$

(c)

$$\mathcal{L}\{e^{-at}\} = \int_0^\infty e^{-at}e^{-st}dt = \int_0^\infty e^{-(s+a)t}dt = \left[\frac{e^{-(s+a)t}}{-(s+a)}\right]_0^\infty$$

$$= \left[\frac{e^{-(s+a)t}}{s+a}\right]_\infty^0 = \frac{1}{s+a}$$

24.2 If a and b are constants find $\mathcal{L}\{at + be^{-3t}\}$.

Solution

$$\mathcal{L}\{at + be^{-3t}\} = \int_0^\infty (at + be^{-3t})e^{-st}dt$$

$$= a\int_0^\infty te^{-st}dt + b\int_0^\infty e^{-3t}.e^{-st}dt$$

$$= a\mathcal{L}\{t\} + b\mathcal{L}\{e^{-3t}\}$$

$$= \frac{a}{s^2} + \frac{b}{s+3}$$

Table 24.1 shows common functions and their Laplace transforms.

Although Example 24.2 showed how to transform linear combinations of functions, there are transforms of other functions that require other results which are now presented.

The **first shift theorem** states that we can obtain the transform of $e^{-at}f(t)$ from that of $f(t)$ by replacing each s by $s + a$. In symbols:

Table 24.1.

$f(t)$	$F(s)$	$f(t)$	$F(s)$
1	$\frac{1}{s}$	$\sin \omega t$	$\frac{\omega}{s^2+\omega^2}$
t	$\frac{1}{s^2}$	$\cos \omega t$	$\frac{s}{s^2+\omega^2}$
t^2	$\frac{2}{s^3}$	$\sinh \omega t$	$\frac{\omega}{s^2-\omega^2}$
t^n	$\frac{n!}{s^{n+1}}$	$\cosh \omega t$	$\frac{s}{s^2-\omega^2}$
e^{-at}	$\frac{1}{s+a}$	$u(t)$	$\frac{1}{s}$
		$\delta(t)$	1

KEY POINT

If $\mathcal{L}\{f(t)\}$ is $F(s)$ then for a, constant

$$\mathcal{L}\{e^{-at}f(t)\} = F(s+a) \qquad 24.2$$

This represents a shift in the s-domain.

The **second shift theorem** states that if the function $f(t)$ has transform $F(s)$ then the transform of $u(t-d)f(t-d)$ has transform $e^{-ds}F(s)$. In symbols:

KEY POINT

If $\mathcal{L}\{f(t)\} = F(s)$ then for $d > 0$

$$\mathcal{L}\{u(t-d)f(t-d)\} = e^{-ds}F(s) \qquad 24.3$$

Since we have specified that $f(t) = 0$, $t < 0$, we can rewrite $f(t)$ as $u(t)f(t)$. The function $u(t-d)f(t-d)$ represents a shift to the right in the t-domain; see Figure 24.1.

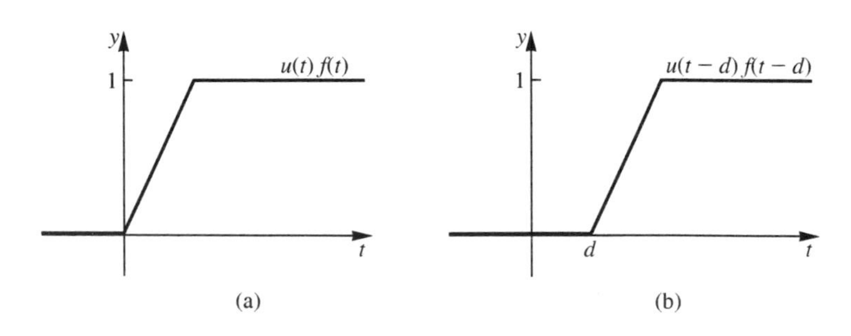

Figure 24.1.
Shift in the t-domain

A result which can be described as the 'multiplication by t' theorem states that:

KEY POINT

If $\mathcal{L}\{f(t)\} = F(s)$ then

$$\mathcal{L}\{tf(t)\} = -\frac{d}{ds}F(s) \qquad 24.4$$

Worked examples

24.3 Find the Laplace transforms of the following functions:
(a) t^2e^{3t} (b) $e^{-3t}\sin 2t$ (c) $e^{2t}\cos 3t$

Solution (a) The transform of t^2 is $2/s^3$ (Table 24.1). Using the first shift theorem and replacing s by $s-3$ we find that the transform of t^2e^{3t} is $2/(s-3)^3$.

(b) The transform of $\sin 2t$ is

$$\frac{2}{s^2+2^2} = \frac{2}{s^2+4}$$

(Table 24.1). Replacing s by $s-(-3)$, that is $s+3$, we obtain the expression

$$\frac{2}{(s+3)^2+4} \equiv \frac{2}{s^2+6s+13}$$

(c) The transform of $\cos 3t$ is $s/(s^2+9)$. Replacing s by $s-2$ we obtain

$$\frac{s-2}{(s-2)^2+9} \equiv \frac{s-2}{s^2-4s+13}$$

24.4 Find the Laplace transform of $t^2u(t-1)$.

Solution

$$t^2 \equiv (t-1)^2 + 2t - 1 \equiv (t-1)^2 + 2(t-1) + 1$$

Then

$$\mathcal{L}\{t^2u(t-1)\} = \mathcal{L}\{(t-1)^2u(t-1)\} + 2\mathcal{L}\{(t-1)u(t-1)\} + \mathcal{L}\{u(t-1)\}$$

Using the second shift theorem and the transforms of t^2, t and 1 which are $2/s^3$, $1/s^2$ and $1/s$, respectively, we find that

$$\mathcal{L}\{(t-1)^2u(t-1)\} = e^{-s}.\frac{2}{s^3}$$

that

$$\mathcal{L}\{(t-1)u(t-1)\} = \frac{e^{-s}}{s^2}$$

and that

$$\mathcal{L}\{u(t-1)\} = \frac{e^{-s}}{s}$$

Hence

$$\mathcal{L}\{t^2u(t-1)\} = e^{-s}.\frac{2}{s^3} + 2\frac{e^{-s}}{s^2} + \frac{e^{-s}}{s} = e^{-s}\left(\frac{2}{s^3} + \frac{2}{s^2} + \frac{1}{s}\right)$$

24.5 Using (24.4), find the Laplace transforms of (a) $t \sin \omega t$, (b) $t^2 e^{3t}$.

Solution (a) The transform of $\sin \omega t$ is $\omega/(s^2 + \omega^2)$. Applying (24.4) the required transform is

$$-\frac{d}{ds}\left(\frac{\omega}{s^2+\omega^2}\right) = -\omega\left(\frac{-2s}{(s^2+\omega^2)^2}\right) = \frac{2\omega s}{(s^2+\omega^2)^2}$$

(b) Starting from the result that the transform of e^{3t} is $1/(s-3)$ we find first the transform of te^{3t} as

$$-\frac{d}{ds}\left(\frac{1}{s-3}\right) = \frac{1}{(s-3)^2}$$

and then the transform of t^2e^{3t} as

$$-\frac{d}{ds}\left(\frac{1}{(s-3)^2}\right) = \frac{2}{(s-3)^3}$$

(Note the easier method shown in Example 24.3a.)

Self-assessment questions 24.1

1. Define the Laplace transform of a function.
2. Explain how to use the shift theorems to obtain further transforms.

Exercise 24.1

1. Find the Laplace transforms of the following functions:

   (a) t^4 (b) t^5
   (c) $\sin 3t$ (d) e^{-3t}
   (e) e^{4t} (f) $\cos(t/3)$
   (g) $\sinh 2t$ (h) $\cosh t$
   (i) $t^2 + 4t + 4$ (j) $3 \sin 2t - 4 \cos 2t$
   (k) $\frac{1}{2}(e^{3t} + e^{-3t})$

2. Find the Laplace transforms of the following:

   (a) $e^{4t} \cos 2t$ (b) $e^{-t} \sin 2t$
   (c) t^2e^{2t} (d) $e^{3t}(t^2 + t)$
   (e) $e^{-t} \cos 2t$ (f) $e^{-t} \sinh 2t$
   (g) $e^{-t}(\sin t + \cos t)$ (h) $t \sin t$
   (i) $t^2 \cos 3t$ (j) $te^t \cos t$

3. Find the Laplace transforms of the following:

   (a) $tu(t-1)$
   (b) $e^tu(t-1)$
   (c) $\sin tu(t - \frac{\pi}{2})$

   (*Hint*: in (b) write e^t as $e.e^{t-1}$; in (c) note that $\cos(t - \frac{\pi}{2}) = \sin t$.)

24.2 Further results, derivatives and integrals

The **final-value theorem** is applied to functions which possess a limit as $t \to \infty$. Note that s is assumed to take real values. The theorem is

KEY POINT

$$\lim_{s \to 0}\{sF(s)\} = \lim_{t \to \infty} f(t) \qquad 24.5$$

The **initial-value theorem** is applied when both $f(t)$ and its derivative $f'(t)$ are of the same exponential order. A function $f(t)$ is of **exponential order** m if positive constants T and M exist and $|f(t)| \leqslant Me^{mt}$ when $t \geqslant T$. The theorem states that if $f(t)$ and $f'(t)$ are continuous for $t \geqslant 0$ and both are of the same exponential order then

KEY POINT

$$\lim_{s \to \infty}\{sF(s)\} = f(0) \qquad 24.6$$

The transforms of the derivatives of a function are required when applying Laplace transforms to the solution of differential equations. The transform of a derivative will be an expression which involves the transform of the function itself. The results for the first and second derivatives are as follows:

KEY POINT

If $\mathcal{L}\{f(t)\} = F(s)$ then

$$\begin{aligned} \mathcal{L}\{f'(t)\} &= sF(s) - f(0) \\ \mathcal{L}\{f''(t)\} &= s^2F(s) - sf(0) - f'(0) \end{aligned} \qquad 24.7$$

where $f(0)$ and $f'(0)$ are the initial values of $f(t)$ and $f'(t)$, respectively.

The transform of the nth derivative, $f^{(n)}(t)$, is given by

$$\mathcal{L}\{f^{(n)}(t)\} = s^nF(s) - s^{n-1}f(0) - s^{n-2}f'(0) - \ldots - f^{(n-1)}(0)$$

We also have a useful result concerning integrals:

KEY POINT

$$\mathcal{L}\left\{\int_0^t f(t)\mathrm{d}t\right\} = \frac{1}{s}F(s) \qquad 24.8$$

For example, the voltage $v(t)$ across a capacitor whose capacitance is C is

$$v(t) = \frac{1}{C}\int_0^t i(t)\mathrm{d}t$$

where $i(t)$ is the current at time t. If the transform of the current is $I(s)$ then the transform of the voltage is

$$\frac{1}{Cs}I(s)$$

Worked examples

24.6 (a) Verify the final-value theorem for $f(t) = e^{-t}$.
(b) Verify the initial-value theorem for $f(t) = \sin \omega t$.

Solution (a) $F(s) = 1/(s+1)$ and $\int_0^\infty e^{-t}e^{-st}dt$ exists provided that $s > -1$; we can therefore safely let $s \to 0$.

Now

$$\lim_{t\to\infty} f(t) = 0$$

Further,

$$sF(s) = \frac{s}{s+1} \qquad \text{and} \qquad \lim_{s\to 0}\{sF(s)\} = 0$$

Hence the theorem is verified.

(b) $f(t)$ and $f'(t)$ are continuous functions, $|f(t)| = |\sin \omega t| \leqslant 1$ and $|f'(t)| = |\omega \cos \omega t| \leqslant \omega$ for all t. We can therefore, for example, take $m = 1$ and $M = \omega$ so that both $f(t)$ and $f'(t)$ are of exponential order 1. The prerequisites of the theorem are satisfied.

Now

$$F(s) = \frac{\omega}{s^2+\omega^2} \qquad \text{so} \qquad sF(s) = \frac{s\omega}{s^2+\omega^2}$$

and

$$\lim_{s\to\infty}\{sF(s)\} = \lim_{s\to\infty}\left\{\frac{s\omega}{s^2+\omega^2}\right\} = 0$$

Since $f(0) = \sin 0 = 0$, the theorem is verified.

24.7 The Laplace transform of $y(t)$ is $Y(s)$. If $y(0) = 1$, $y'(0) = -2$ find an expression for the Laplace transform of

$$\frac{d^2y}{dt^2} + 3\frac{dy}{dt} + 2y$$

Solution Using (24.7)

$$\mathcal{L}\left\{\frac{dy}{dt}\right\} \equiv sY(s) - 1$$

and

$$\mathcal{L}\left\{\frac{d^2y}{dt^2}\right\} \equiv s^2Y(s) - s(1) - (-2) = s^2Y(s) - s + 2$$

Hence the Laplace transform of the given expression is

$$s^2Y(s) - s + 2 + 3(sY(s) - 1) + 2Y(s) \equiv (s^2 + 3s + 2)Y(s) - s - 1$$

Self-assessment questions 24.2

1. State the initial-value and final-value theorems.
2. Express the transforms of the first and second derivatives of a function in terms of the transform of the function itself.

Exercise 24.2

1. Verify the initial-value theorem for the following functions:

   (a) t^3 (b) e^{-2t} (c) $\sin 2t$

2. Verify the final-value theorem for the following functions:

   (a) e^{-3t} (b) te^{-3t} (c) $e^{-2t}\cos t$

3. If $y(0) = 2$, $\dot{y}(0) = 1$ find the Laplace transform of

   (a) $\dot{y}$
   (b) $\ddot{y}$
   (c) $\ddot{y} + 2\dot{y} + 5y$
   (d) $\ddot{y} - 4\dot{y} + 3y$

24.3 Inverting Laplace transforms

We can use Table 24.1 in reverse, together with some imagination, to invert a Laplace transform. This section illustrates some of the main methods.

Worked example

24.8 Find the functions $f(t)$ which have the following Laplace transforms:

(a) $\dfrac{6}{s^3}$ (b) $\dfrac{s+3}{s^2+9}$ (c) $\dfrac{2s+7}{s^2+4}$

(d) $\dfrac{s+2}{s^2+4s+5}$ (e) $\dfrac{s-2}{2s^2+4s+10}$

Solution (a) From Table 24.1 we see that $\mathcal{L}\{t^2\} = 2/s^3$, hence $\mathcal{L}\{3t^2\} = 6/s^3$.

(b) We use the table with $\omega = 3$ to obtain the results

$$\mathcal{L}\{\cos 3t\} = \frac{s}{s^2+9} \qquad \text{and} \qquad \mathcal{L}\{\sin 3t\} = \frac{3}{s^2+9}$$

Hence

$$\mathcal{L}\{\cos 3t + \sin 3t\} = \frac{s+3}{s^2+9}$$

(c) From the table $s/(s^2+4)$ is the transform of $\cos 2t$ and $2/(s^2+4)$ is the transform of $\cos 2t$. Hence $(2s+7)/(s^2+4)$ is the transform of $2\cos 2t + \frac{7}{2}\sin 2t$.

(d) First, note that $s^2+4s+5 \equiv (s+2)^2+1$. Now $s/(s^2+1)$ is the transform of $\cos t$. Applying the first shift theorem,

$$\frac{s+2}{s^2+4s+5} \equiv \frac{s+2}{(s+2)^2+1}$$

is the transform of $e^{-2t}\cos t$.

(e) $2s^2+4s+10 \equiv 2[s^2+2s+5] \equiv 2[(s+1)^2+4]$
Since $s/(s^2+4)$ is the transform of $\cos 2t$,

$$\frac{s+2}{s^2+2s+5}$$

is the transform of $e^{-t}\cos 2t$.

Also, $2/(s^2+4)$ is the transform of $\sin 2t$ and

$$\frac{2}{s^2+2s+5}$$

is the transform of $e^{-t}\sin 2t$. Hence

$$\frac{s-2}{2s^2+4s+10} \equiv \frac{1}{2}\left(\frac{s-2}{s^2+2s+5}\right) \equiv \frac{1}{2}\left(\frac{s+2-4}{s^2+2s+5}\right)$$

is the transform of

$$\frac{1}{2}(e^{-t}\cos 2t - 2e^{-t}\sin 2t) = e^{-t}\left(\frac{1}{2}\cos 2t - \sin 2t\right)$$

Sometimes we have to use partial fractions to carry out the inversion.

Worked example

24.9 Find the inverse Laplace transforms of the following functions:

(a) $\dfrac{2s+2}{(s+2)(s-1)}$ (b) $\dfrac{s^2+1}{(s^2+2s+2)(s+1)}$

Solution (a) Let

$$\frac{2s+2}{(s+2)(s-1)} \equiv \frac{A}{s+2} + \frac{B}{s-1} \equiv \frac{A(s-1)+B(s+2)}{(s+2)(s-1)}$$

Hence $A(s-1)+B(s+2) \equiv 2s+2$.

Putting $s=1$, $B(3)=4$, hence $B=\frac{4}{3}$.

Putting $s=-2$, $A(-3)=-2$, hence $A=\frac{2}{3}$.

Then

$$\frac{2s+2}{(s+2)(s-1)} \equiv \frac{2}{3}.\frac{1}{s+2} + \frac{4}{3}.\frac{1}{s-1}$$

From Table 24.1 the inverse Laplace transforms of $1/(s+2)$ and $1/(s-1)$ are e^{-2t} and e^{t}, respectively. The inverse transform of the given expression is

$$\frac{2}{3}e^{-2t} + \frac{4}{3}e^{t}$$

(b)
$$\frac{s^2+1}{(s^2+2s+2)(s+1)} \equiv \frac{As+B}{s^2+2s+2} + \frac{C}{s+1}$$

$$\equiv \frac{(As+B)(s+1)+C(s^2+2s+2)}{(s^2+2s+2)(s+1)}$$

Hence $(As+B)(s+1)+C(s^2+2s+2) \equiv s^2+1$.

Putting $s=-1$, $C(1-2+2)=2$, hence $C=2$.

Comparing the coefficients of s^2, $A+C=1$, hence $A=-1$.

Comparing the constant terms, $B+2C=1$, hence $B=-3$.

Therefore

$$\frac{s^2+1}{(s^2+2s+2)(s+1)} \equiv \frac{-s-3}{s^2+2s+2} + \frac{2}{s+1}$$

$$\equiv \frac{-(s+1)-2}{s^2+2s+2} + \frac{2}{s+1}$$

Now $s^2+2s+2 \equiv (s+1)^2+1$. Hence

$$\frac{s+1}{s^2+2s+2}$$

is the transform of $e^{-t}\cos t$ and

$$\frac{1}{s^2 + 2s + 2}$$

is the transform of $e^{-t} \sin t$.

Therefore the given expression is the transform of

$$-e^{-t} \cos t - 2e^{-t} \sin t + 2e^{-t}$$

Sometimes we can use complex numbers to help invert transforms.

Worked example

24.10 Find the inverse Laplace transform of

$$\frac{s+2}{s^2 + 4s + 13}$$

Solution The equation $s^2 + 4s + 13 = 0$ has solution $s = -2 \pm 3j$.

Then let

$$\frac{s+2}{s^2 + 4s + 13} \equiv \frac{A}{s-a} + \frac{B}{s-b} \equiv \frac{A(s-b) + B(s-a)}{(s-a)(s-b)}$$

where $a = -2 + 3j$ and $b = -2 - 3j$.

Hence

$$A(s-b) + B(s-a) \equiv s + 2$$

Putting $s = a$,

$$A(a-b) \equiv a + 2$$

that is, $A(6j) = 3j$ so that $A = \frac{1}{2}$.

Putting $s = b$,

$$B(b-a) \equiv b + 2$$

that is, $B(-6j) = -3j$ so that $B = \frac{1}{2}$.

Hence, inverting the transform we obtain

$$\frac{1}{2}e^{(-2+3j)t} + \frac{1}{2}e^{(-2-3j)t} = \frac{1}{2}e^{-2t}(e^{3jt} + e^{-3jt})$$

$$= \frac{1}{2}c^{-2t}(\cos 3t + j \sin 3t + \cos 3t - j \sin 3t)$$

$$= e^{-2t} \cos 3t$$

Compare this method with that used to solve Example 24.8(d).

Self-assessment questions 24.3

1. Explain when you would use partial fractions to invert a Laplace transform.
2. Explain when you would use complex numbers to invert a Laplace transform.

Exercise 24.3

1. Find the inverse Laplace transforms of the following.

(a) $\dfrac{30}{(s+3)^4}$ (b) $\dfrac{s+2}{(s+2)^2+16}$

(c) $\dfrac{8}{(s+2)^2-16}$ (d) $\dfrac{s+5}{s^2+10s+29}$

(e) $\dfrac{2s+4}{s^2+10s+29}$ (f) $\dfrac{5s+8.5}{s^2+s+9.25}$

2. Show that

$$\frac{s+1}{(s-1)(s+2)} \equiv \frac{2}{3}\frac{1}{(s-1)} + \frac{1}{3}\frac{1}{(s+2)}$$

and hence find its inverse Laplace transform.

3. Find the inverse Laplace transforms of the following:

(a) $\dfrac{10s}{(s+1)(2s-3)}$

(b) $\dfrac{s^2-7s+11}{(s-1)(s-2)(s-4)}$

(c) $\dfrac{s+2}{s^2+4}$ (d) $\dfrac{s^2+2s-2}{s^2(s-1)}$

(e) $\dfrac{s^3+3s^2+1}{s^2(s^2+2s+2)}$ (f) $\dfrac{s+1}{(s+1)^2+1}$

(g) $\dfrac{s+5}{s^2+6s+9}$ (h) $\dfrac{2s-1}{(s+1)(s^2+2s+2)}$

(i) $\dfrac{6s-2}{s^2+6s+13}$ (j) $\dfrac{1+2s}{s^2-2s+2}$

24.4 Solving differential equations

By applying the Laplace transform to the terms of an ODE we effectively remove the derivatives and obtain an algebraic equation in the transform of the dependent variable. By inverting this transform we obtain the required solution. There is no general solution to which initial conditions are applied: the solution is tailor-made to fit the given initial conditions only.

Worked examples

24.11 A series *RL* circuit with a sinusoidal applied voltage is modelled by the equation

$$Ri + L\frac{\mathrm{d}i}{\mathrm{d}t} = V_0 \sin \omega t$$

Find $i(t)$, the current in the circuit, if $i(0) = 0$.

Solution Let $\mathcal{L}\{i(t)\} = I(s)$. Then

$$\mathcal{L}\left\{\frac{\mathrm{d}i}{\mathrm{d}t}\right\} = sI(s) - i(0) = sI(s)$$

and

$$\mathcal{L}\{V_0 \sin \omega t\} = \frac{V_0\omega}{s^2 + \omega^2}$$

Transforming the ODE (which we can do term by term) gives

$$RI(s) + LsI(s) = \frac{V_0\omega}{s^2 + \omega^2}$$

Hence

$$I(s) = \frac{V_0\omega}{(s^2 + \omega^2)(Ls + R)} \equiv \frac{As + B}{s^2 + \omega^2} + \frac{C}{Ls + R}$$

Evaluating the constants A, B and C, we obtain

$$A = \frac{-V_0\omega L}{R^2 + \omega^2 L^2}, B = \frac{V_0\omega R}{R^2 + \omega^2 L^2}, C = \frac{V_0\omega L^2}{R^2 + \omega^2 L^2}$$

Then

$$I(s) = \frac{V_0\omega}{R^2 + \omega^2 L^2}\left\{\frac{-Ls}{s^2 + \omega^2} + \frac{R}{s^2 + \omega^2} + \frac{L^2}{Ls + R}\right\}$$

Noting that

$$\frac{L^2}{Ls + R} \equiv \frac{L}{s + \dfrac{R}{L}}$$

we invert the transform to obtain

$$i(t) = \frac{V_0\omega}{R^2 + \omega^2 L^2}\left\{-L\cos \omega t + \frac{R}{\omega}\sin \omega t + L\mathrm{e}^{-Rt/l}\right\}$$

24.12 The temperature $\theta(t)$ of a metal block immersed in a cooling fluid is governed by the equation

$$\frac{\mathrm{d}\theta}{\mathrm{d}t} + 0.1\theta = 1 - 0.2t \qquad 0 \leqslant t \leqslant 10$$

The temperature of the block is initially 70 °C. Find the temperature subsequently.

Solution Let $\mathcal{L}\{\theta(t)\} = \Theta(s)$. Then

$$\mathcal{L}\left\{\frac{\mathrm{d}\theta}{\mathrm{d}t}\right\} = s\Theta(s) - \theta(0) = s\Theta(s) - 70$$

and

$$\mathcal{L}\{1 - 0.2t\} = \frac{1}{s} - \frac{0.2}{s^2}$$

We transform the ODE to obtain

$$s\Theta(s) - 70 + 0.1\Theta(s) = \frac{1}{s} - \frac{0.2}{s^2} = \frac{s - 0.2}{s^2}$$

that is

$$(s + 0.1)\Theta(s) = 70 + \frac{s - 0.2}{s^2} = \frac{70s^2 + s - 0.2}{s^2}$$

Then

$$\Theta(s) = \frac{70s^2 + s - 0.2}{s^2(s + 0.1)} \equiv \frac{A}{s} + \frac{B}{s^2} + \frac{C}{s + 0.1}$$

Evaluating the constants A, B and C, we obtain

$$A = 30, B = -2, C = 40$$

Hence

$$\Theta(s) = \frac{30}{s} - \frac{2}{s^2} + \frac{40}{s + 0.1}$$

Therefore $\theta(t) = 30 - 2t + 40\mathrm{e}^{-0.1t}$.

24.13 The free oscillations of a lightly damped elastic system are governed by the equation

$$\frac{\mathrm{d}^2 y}{\mathrm{d}t^2} + 2\frac{\mathrm{d}y}{\mathrm{d}t} + 5y = 0$$

where $y(t)$ is the displacement from equilibrium.

Find the displacement if the initial values of y and $\mathrm{d}y/\mathrm{d}t$ are 2 and 0, respectively.

Solution Let $\mathcal{L}\{y(t)\} = Y(s)$. Then

$$\mathcal{L}\left\{\frac{\mathrm{d}y}{\mathrm{d}t}\right\} = sY(s) - y(0) = sY(s) - 2$$

and

$$\mathcal{L}\left\{\frac{\mathrm{d}^2 y}{\mathrm{d}t^2}\right\} = s^2 Y(s) - sy(0) - y'(0) = s^2 Y(s) - 2s$$

Transforming the ODE we obtain

$$s^2 Y(s) - 2s + 2(sY(s) - 2) + 5Y(s) = 0$$

that is

$$(s^2 + 2s + 5)Y(s) = 2s + 4$$

Hence

$$Y(s) = \frac{2s+4}{s^2+2s+5}$$

$$\equiv \frac{2s+4}{(s+1)^2+2^2}$$

$$\equiv \frac{2(s+1)}{(s+1)^2+2^2} + \frac{2}{(s+1)^2+2^2}$$

Finally, $y(t) = 2e^{-t}\cos 2t + e^{-t}\sin 2t$.

24.14 The **transfer function** for a system is the ratio of the Laplace transform of the output from the system to the Laplace transform of the input. Find the transfer function for the series RC network modelled by the equation

$$RC\frac{dv}{dt} + v = e(t)$$

where e(t) is an applied time-varying voltage. Assume zero initial input, that is $v(0) = 0$. Hence find the response of the system to the input $e(t) = \delta(t)$, where $\delta(t)$ is the impulse function.

Solution Let $\mathcal{L}\{v(t)\} = V(s)$ and $\mathcal{L}\{e(t)\} = E(s)$. Then

$$\mathcal{L}\left\{\frac{dv}{dt}\right\} = sV(s)$$

Transforming the ODE gives

$$RCsV(s) + V(s) = E(s)$$

that is

$$(RCs+1)V(s) = E(s)$$

and hence the transfer function for the system is

$$\frac{V(s)}{E(s)} = \frac{1}{RCs+1}$$

Since $e(t) = \delta(t)$ then $E(s) = 1$, therefore

$$V(s) = \frac{1}{RCs+1} = \frac{1}{RC}\left(\frac{1}{s+\frac{1}{RC}}\right)$$

and

$$v(t) = \frac{1}{RC}e^{-t/RC}$$

Self-assessment questions 24.4

1. Explain why the method of solving an ODE by Laplace transforms does not produce a general solution.
2. Explain what is meant by the transfer function of a system.

Exercise 24.4

1. The voltage across a capacitor in an RC circuit is given by

$$\frac{dv}{dt} + RCv = 0$$

where $v(0) = v_0$. Show that the solution of the equation is $v = v_0 e^{-t/RC}$.

2. The charge in a series LCR circuit is modelled by the equation

$$L\ddot{q} + R\dot{q} + \frac{q}{C} = e(t)$$

Solve this equation in the case when $L = 2$, $R = 20$, $C = 0.02$, $e(t) = \sin t$ and $\dot{q}(0) = q(0) = 0$.

3. Assuming zero initial conditions, find the transfer functions for the systems with input $f(t)$ which are governed by the following equations:

(a) $\dot{x} - 5x = f(t)$
(b) $\ddot{x} + 4\dot{x} + x = f(t)$

Test and assignment exercises 24

1. Find the Laplace transforms of the following:

(a) t^5 (b) t^9 (c) $\sin \frac{1}{2}t$
(d) e^{-4t} (e) e^{2t}
(f) $2 \sinh 3t + \frac{1}{2}\cosh 3t$
(g) $\frac{1}{2}\sin 4t - \frac{2}{3}\cos 4t$
(h) $t^3 + t$ (i) $tu(t-2)$

2. Find the Laplace transforms of the following:

(a) $t^3 e^{2t}$ (b) $t^3 e^{-2t}$
(c) $e^{-1.5t}\cos 2t$
(d) $e^{-2t}(2\sin 3t + 3\cos 3t)$
(e) $t^2 \sin t$ (f) $te^{-2t}\sin t$

3. Find the inverse Laplace transforms of the following:

(a) $\dfrac{21}{(s-3)^2}$ (b) $\dfrac{3s-2}{s^2+16}$

(c) $\dfrac{3}{(s-1)^2-9}$ (d) $\dfrac{s+6}{s^2+12s+45}$

(e) $\dfrac{3s+7}{s^2-s+4.25}$

4. Find the inverse Laplace transforms of the following:

(a) $\dfrac{s}{(s-1)(2s+3)}$

(b) $\dfrac{s^2+5s+7}{(s+1)(s+2)(s+4)}$

(c) $\dfrac{4s^2-s-2}{s^2(s+1)}$ (d) $\dfrac{s-3}{(s-3)^2+2}$

(e) $\dfrac{s-3}{s^2-8s+16}$ (f) $\dfrac{5s-4}{s^2+8s+20}$

5. Using Laplace transforms, solve the following ODEs with the given initial conditions:

(a) $\ddot{y} + 3\dot{y} + 2y = e^t$; $y(0) = 1$, $\dot{y}(0) = 0$
(b) $\ddot{y} + 4\dot{y} + 3y = \cos t$; $y(0) = 0$, $\dot{y}(0) = 0$
(c) $\ddot{y} + 2\dot{y} + y = t^2 - 1$; $y(0) = 1$, $\dot{y}(0) = -1$
(d) $\ddot{y} + 2\dot{y} + 5y = 4$; $y(0) = 2$, $\dot{y}(0) = 1$

Solutions

Exercise 1.1

1. $A = \{0,1,2,\ldots,11\}$, $B = \{-5,5\}$,
   $C = \{3,6,9,12,15,18\}$, $D = \{4,6,8,10\}$
   $A \cup C = \{0,1,2,\ldots,11,12,15,18\}$,
   $A \cup D = \{0,1,2,\ldots,11\} = A$,
   $C \cup D = \{3,4,6,8,9,10,12,15,18\}$,
   $A \cup C \cup D = \{0,1,2,\ldots,11,12,15,18\}$
   $= A \cup C$,
   $A \cap B = \{5\}$, $A \cap C = \{3,6,9\}$,
   $A \cap C \cap D = \{6\}$,
   $A \cap (C \cup D) = \{3,4,6,8,9,10\}$
   D is a proper subset of A.

2. $\overline{A} = \{12,13,\ldots,18\}$,
   $\overline{C} = \{0,1,2,4,5,7,8,10,11,13,14,16,17\}$,
   $\overline{D} = \{0,1,2,3,5,7,9,11,12,\ldots,18\}$,
   $\overline{A \cup C} = \{13,14,16,17\}$,
   $\overline{A \cup D} = \{12,13,\ldots,18\}$, $\overline{A \cap D} = \overline{D}$,
   $\overline{A \cap C \cap D} = \{0,1,\ldots,5,7,8,\ldots,18\}$

4. (a) A
   (b) A
   (c) $\overline{A} \cup \overline{B}$
   (d) U
   (e) C
   (f) $(A \cap B) \cup C$

5. 103 had faulty exhausts and 37 had faulty brakes only.

Exercise 1.2

3. (a) p
   (b) $p + q + \overline{r}$
   (c) 1
   (d) $q \cdot (p + r)$

4. (a)

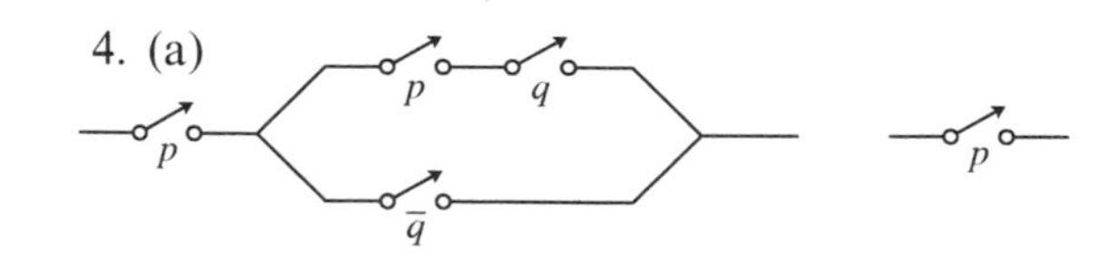

(b)

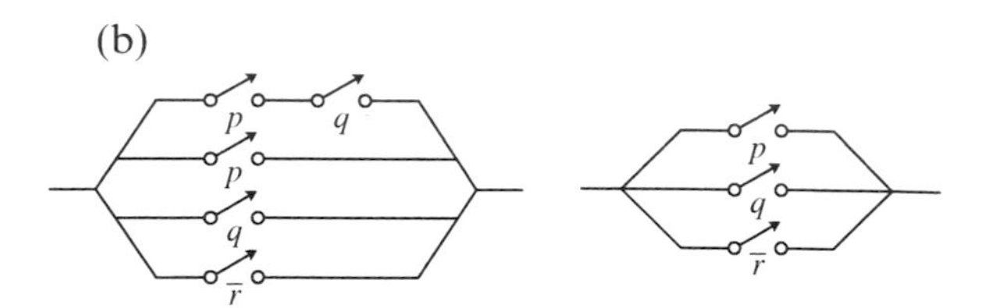

(d)

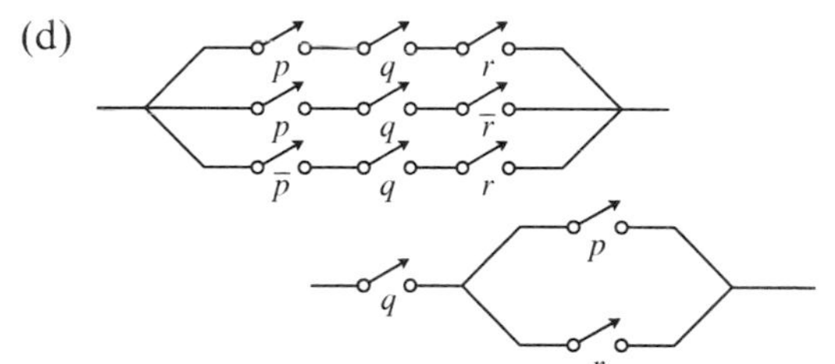

5.

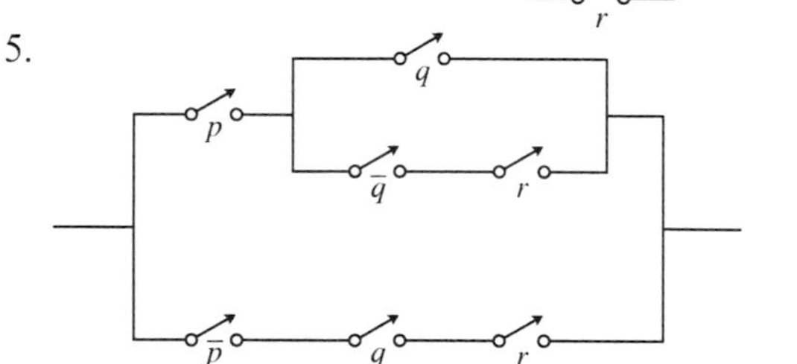

6. (a) $(p + q).(\overline{p} + r)$ (b) $(p + q).r + p.q$

Exercise 1.3

1.

p	q	f
1	1	0
1	0	1
0	1	1
0	0	0

2. $(p + q).r + p.q$

3.

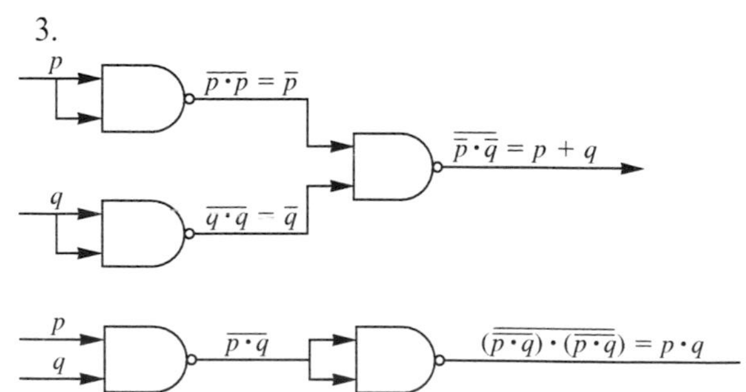

4.

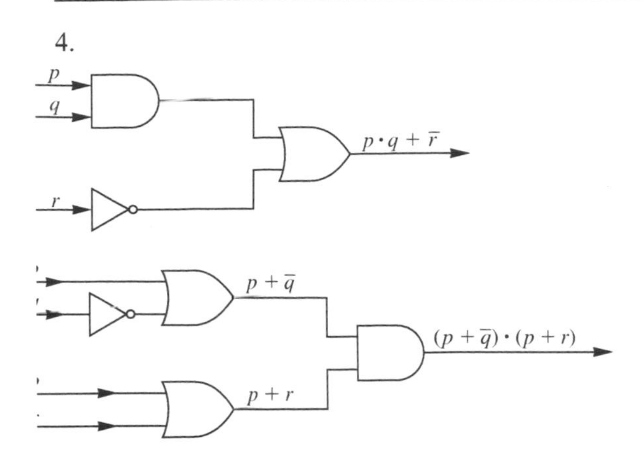

Exercise 1.4

1.

p	q	$p \vee q$	$p \wedge (p \vee q)$
T	T	T	T
T	F	T	T
F	T	T	F
F	F	F	F

2.

p	q	$p \to q$	$p \vee (p \to q)$
T	T	T	T
T	F	F	T
F	T	T	T
F	F	T	T

$\bar{q}$	$\bar{q} \wedge p$	$\bar{q} \wedge p \wedge (p \to q)$
F	F	F
T	T	F
F	F	F
T	F	F

3. (a) Necessary, not sufficient; $(-3)^2 = (3)^2$, for example.
   (b) Necessary and sufficient.
   (c) Necessary, not sufficient; take $p = 1$, $q = -1$, for example.

4. (a) Contrapositive: if a triangle is not isosceles then it is not equilateral.
   Converse: if a triangle is isosceles then it is equilateral.
   (b) Contrapositive: if $x + y > 2$ then $x^2 + y^2 > 4$.
   Converse: if $x + y \leqslant 2$ then $x^2 + y^2 \leqslant 4$.

5. $x = 1\frac{1}{2}$ gives $x^2 - 3x + 2 = -\frac{1}{4}$.

6. Let the consecutive odd stages be $(2n + 1)$, $(2n + 3)$, $(2n + 5)$ and $(2n + 7)$. Their sum is $8n + 16 = 8(n + 2)$.

Exercise 2.1

1. 0, 3, 8, 15, 24

2. (a) $1, \frac{1}{2}, \frac{1}{3}, \frac{1}{4}, \frac{1}{5}, \frac{1}{6}$
   (b) $2, 4\frac{1}{4}, 9\frac{1}{9}, 16\frac{1}{16}, 25\frac{1}{25}, 36\frac{1}{36}$
   (c) $10, -9, 10, -9, 10, -9$
   (d) $1024, 256, 64, 16, 4, 1$

3. (a) $u_k = \dfrac{2}{3^{k-1}}, k = 1,2, \ldots$

   (b) $u_{k+1} = \dfrac{1}{3}u_k, k = 1,2, \ldots$

4. (a) 31, 64 (b) $-6, -39$

5. $a = 10, d = -8$

6. $a = 3, r = 2$

7. (a) decreases, limit is 1
   (b) increases, limit is 1
   (c) decreases, limit is 0
   (d) oscillates, limit is 0
   (e) oscillates and diverges
   (f) decreases, limit is 0

Exercise 2.2

1. (a) 860 (b) 19 553

2. (a) 210 (b) -150 (c) -210 (d) 150

3. 11.75

4. 725

5. $\frac{1}{3}$ or $-\frac{1}{3}$

6. 13, 4

7. 40 m, 8 m from A

Exercise 2.3

1. (a) $a^8 + 8a^7b + 28a^6b^2 + 56a^5b^3 + 70a^4b^4 + 56a^3b^5 + 28a^2b^6 + 8ab^7 + b^8$
   (b) $729 - 7290x + 30\,375x^2 - 67\,500x^3 + 84\,375x^4 - 11\,250x^5 + 15\,625x^6$

2. (a) $1 - \frac{1}{4}x - \frac{3}{32}x^2 - \frac{7}{128}x^3 - \frac{77}{768}x^4 - \frac{231}{3072}x^5$
   (b) $1 + 3x + 6x^2 + 10x^3 + 15x^4 + 21x^5$
   (c) $1 + x + \frac{3}{2}x^2 + \frac{5}{2}x^3 + \frac{35}{8}x^4 + \frac{63}{8}x^5$
3. (a) all x (b) $|x| < \frac{2}{5}$
   (c) $|x| > \sqrt{2}$ (d) $|x| < 3$
4. $2 + 5x + 5x^2 + 5x^3, \quad |x| < 1$
5. $1 - 2x - 2x^2 - 4x^3 - 10x^4$;
   $875/835 \simeq 2.245$ (cf. 2.236)
6. $1 + \dfrac{1}{2x} - \dfrac{1}{8x^2} + \dfrac{1}{16x^3}, \ |x| > 1$;

   $x^{-\frac{1}{2}} + \dfrac{1}{2}x^{\frac{1}{2}} - \dfrac{1}{4}x^{\frac{3}{2}} + \dfrac{3}{8}x^{\frac{5}{2}}, \ |x| < 1$

Exercise 3.1

1. (a) x → Divide by 3 → $x/3$
   (b) x → Multiply by 5 → $5x$
   (c) x → Raise to fifth power → x^5
   (d) x → Take fourth root → $x^{1/4}$

   (a), (b), (c) domain and range all real numbers
   (d) domain and range all non-negative real numbers
2. (a) $\frac{1}{8}$, -8, 226.981
   (b) 2, $-\frac{1}{2}$, 0.1639 (4 d.p.)
   (c) 1, 0, 1
3. (a)

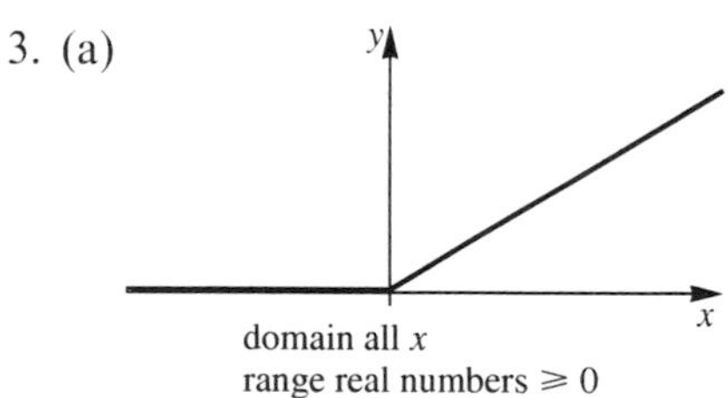

   (b)

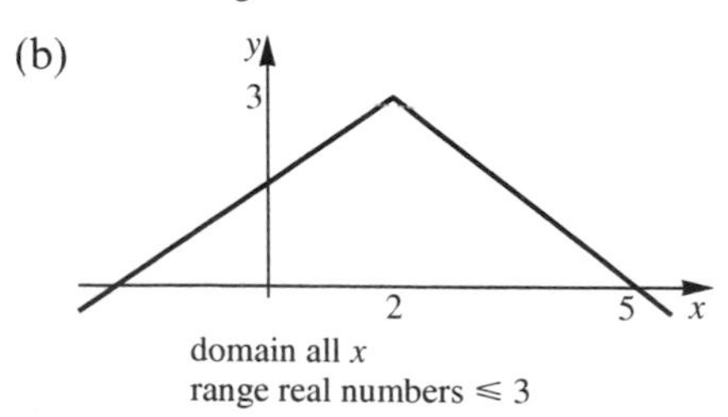

4. Domain $0 \leqslant t \leqslant 60$, range $0 \leqslant v(t) \leqslant 10$

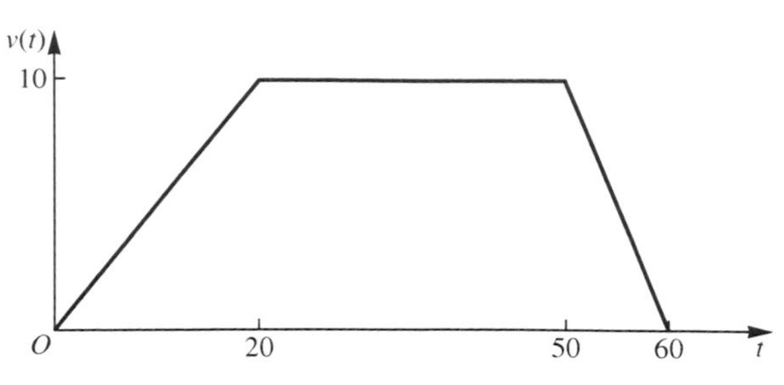

5. (a) $y = -x^3$
   (b) $y = -x$
   (c) $y = \pm\sqrt{x}$ not a function

Exercise 3.2

1. (a) Sum $x^4 + x$, domain all real numbers; product x^5, domain all real numbers.
   (b) Sum $x^{\frac{1}{2}} + x^{\frac{1}{3}}$, domain real numbers $\geqslant 0$; product $x^{\frac{5}{6}}$, domain real numbers $\geqslant 0$.
   (c) Sum $\dfrac{1}{x} + \dfrac{1}{x^3}$, domain real numbers $\neq 0$;
   product $\dfrac{1}{x^4}$, domain real numbers $\neq 0$.
   (d) Sum $x^{\frac{1}{2}} + x^2$, domain real numbers $\geqslant 0$; product $x^{2\frac{1}{2}}$, domain all real numbers $\geqslant 0$.
2. (a) if $g(x) \equiv 0$ for all x, the zero function
   (b) if $g(x) = 1$ for all x
3. (a) $g(x) = x$ for all x
   (b) $f(x) = x$ for all x
4. (a) $f(g(x)) = x^4$, $g(f(x)) = x^4$; domain all real numbers, range real numbers $\geqslant 0$.
   (b) $f(g(x)) = x^{\frac{1}{6}}$, $g(f(x)) = x^{\frac{1}{6}}$; domain and range real numbers $\geqslant 0$.
   (c) $f(g(x)) = x^3$, $g(f(x)) = x^3$; domain and range all real numbers.
   (d) $f(g(x)) = x$; domain and range all real numbers
   $g(f(x)) = x$; domain and range all real numbers $\geqslant 0$.
5. $f(g(1)) = 1$, $g(f(1)) = 1$, $f(g(0)) = 0$, $g(f(0)) = 0$ in all cases.

6. (a)

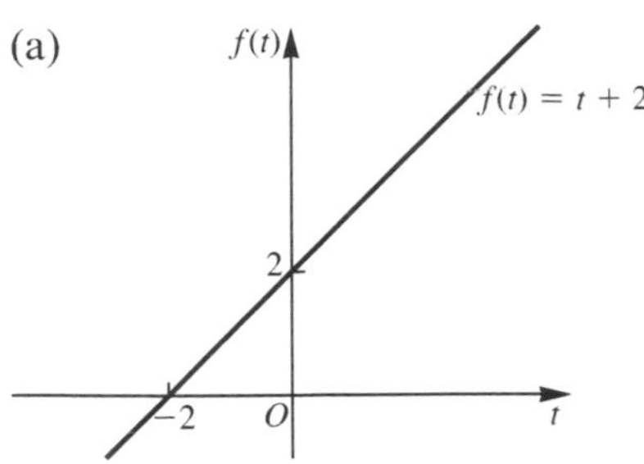

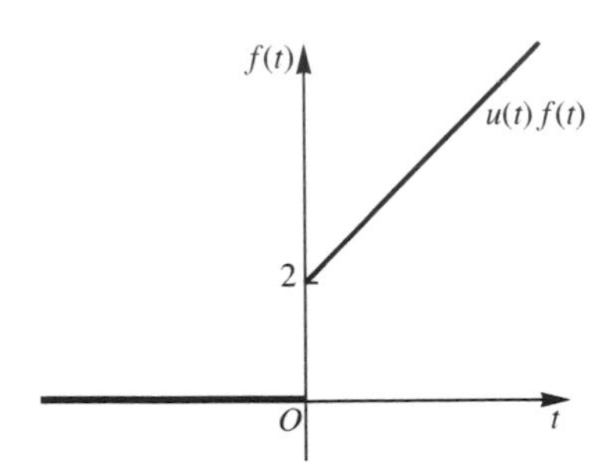

(b)

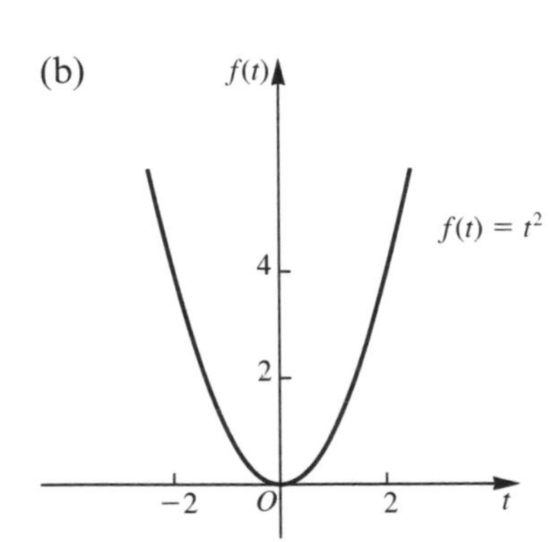

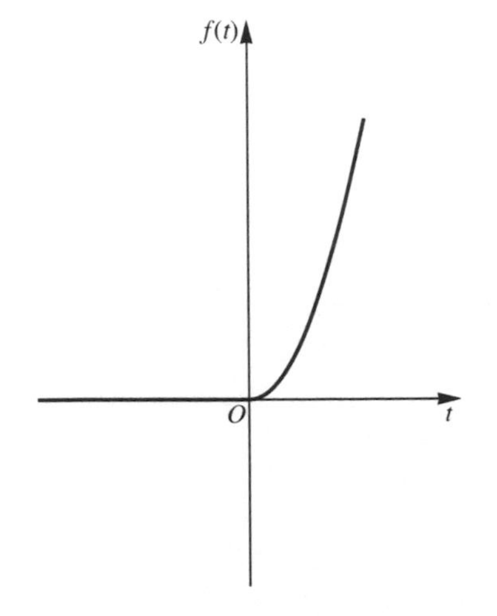

Exercise 3.3

1. (a)

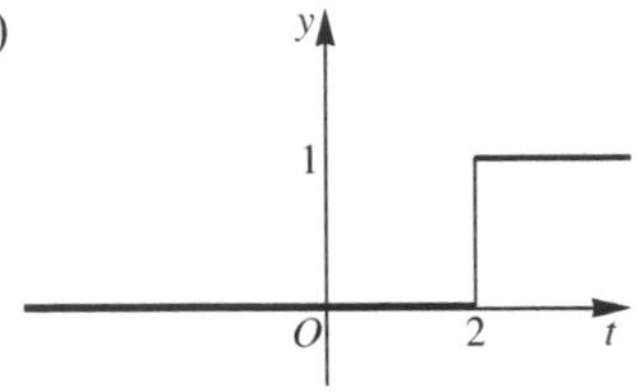

(b)

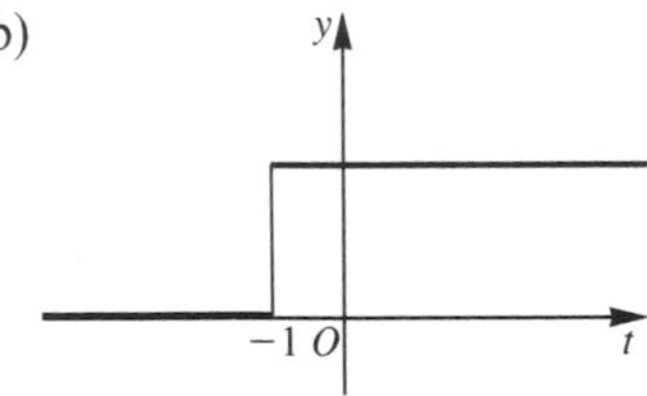

(c)

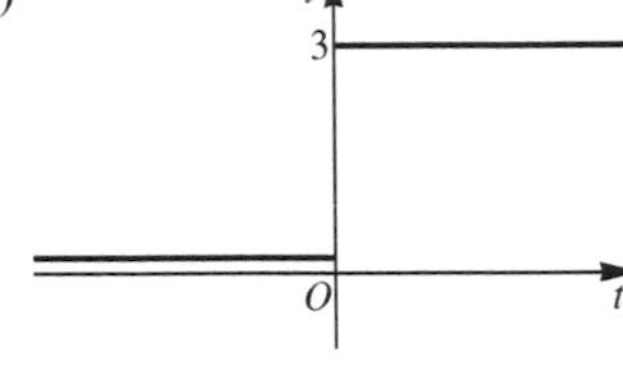

(d)

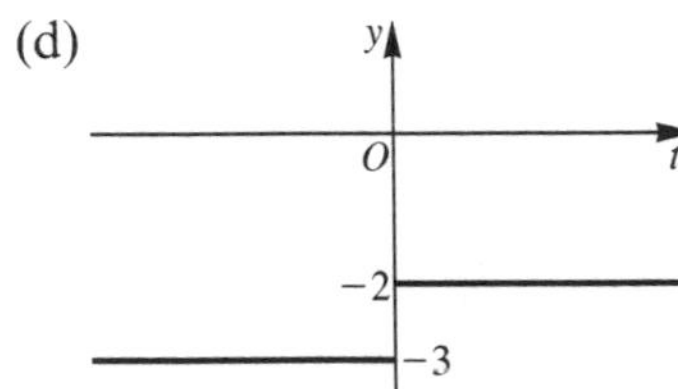

(e)

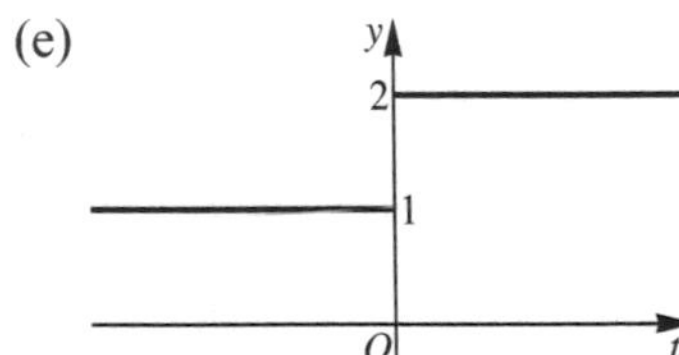

(f)

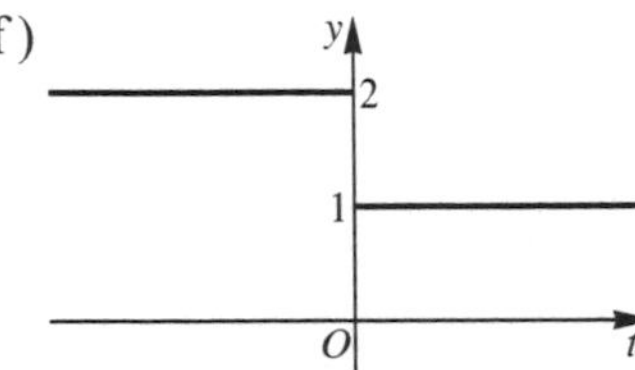

(g)

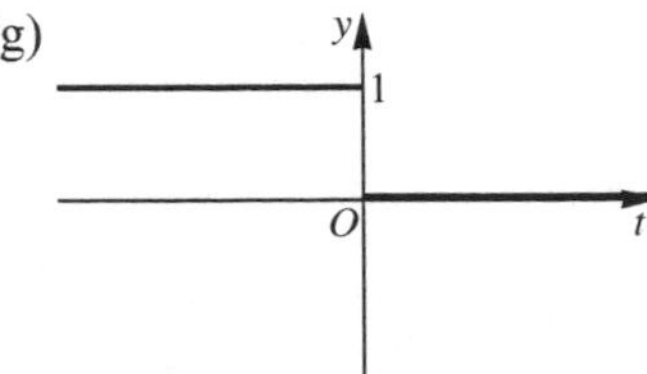

(h)

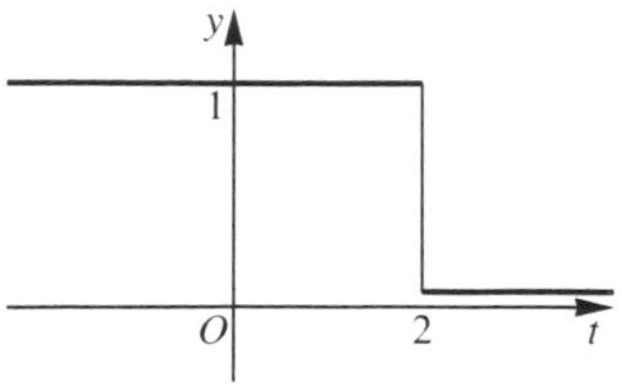
y
1
O
2
t

2. (a)

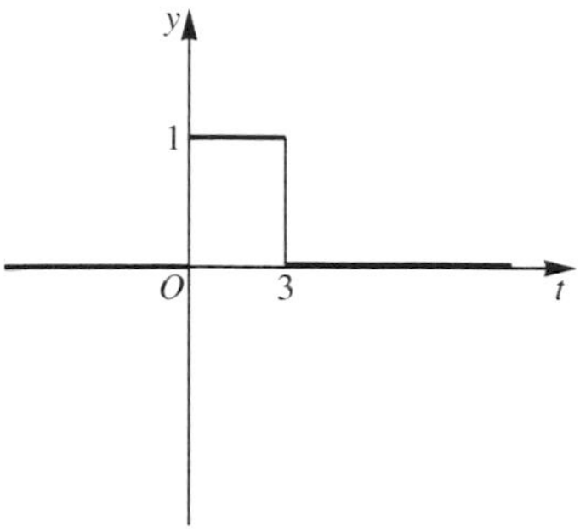
y
1
O
3
t

(b)

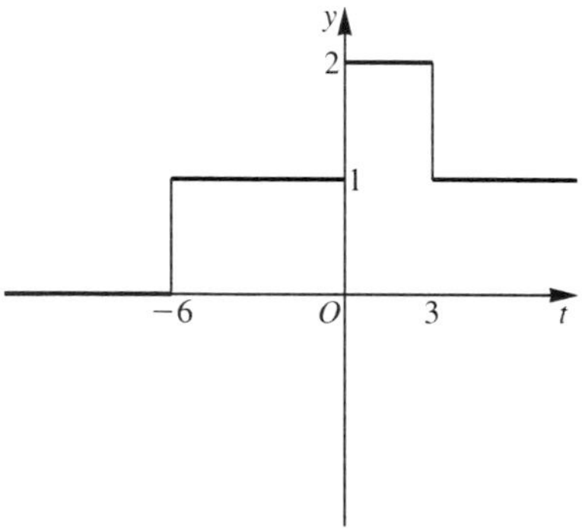
y
2
1
−6
O
3
t

3.

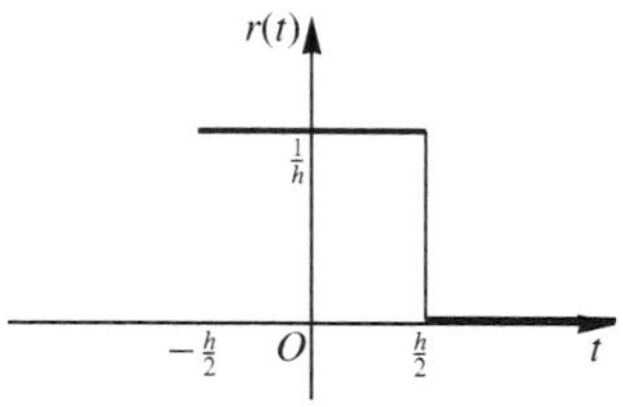
r(t)
1/h
−h/2
O
h/2
t

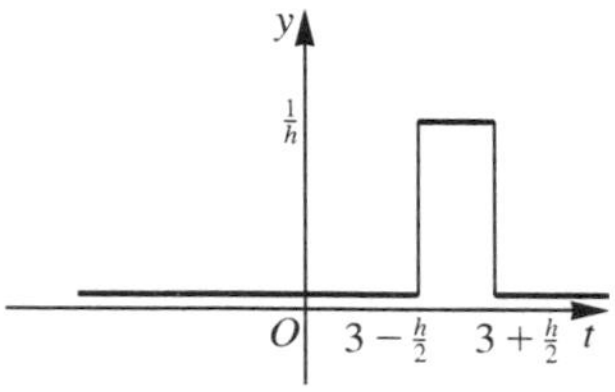
y
1/h
O
3 − h/2
3 + h/2
t

4.

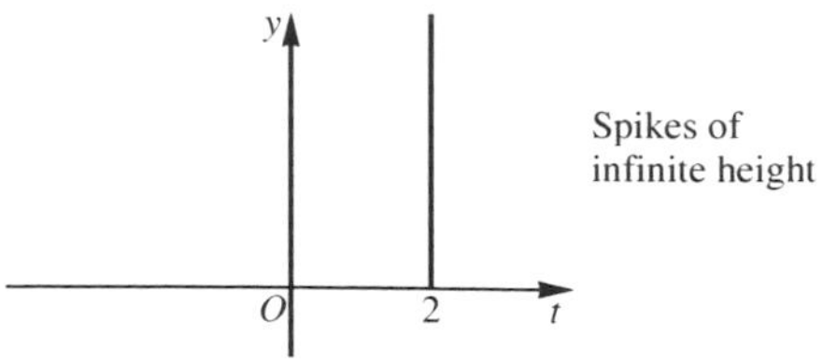
y
Spikes of infinite height
O
2
t

5. $f(t) = \begin{cases} t^2, & t_1 < t < t_2 \\ 0, & \text{elsewhere} \end{cases}$

Exercise 3.4

1. (a) $f^{-1}(x) = x - 2$ (b) $f^{-1}(x) = 2 - x$
   (c) $f^{-1}(t) = \frac{1}{4}t$ (d) $g^{-1}(t) = t^{-\frac{1}{3}}, t \neq 0$

2. (a) $\left(\dfrac{x-2}{6}\right)^{\frac{1}{3}}$
   (b) inverse not a function
   (c) $1 - x^3$
   (d) inverse not a function
   (e) $\dfrac{1}{x-1}, x \neq 1$ (f) $\dfrac{x}{x-1}, x \neq 1$

3. No two values of x give rise to the same value $f(x)$.

4. (a) $h(t) = 6t + 2$, $h^{-1}(t) = \frac{1}{6}(t - 2)$
   (b) $h(t) = (t + 1)^3$, $h^{-1}(t) = t^{\frac{1}{3}} - 1$

Exercise 4.1

1. $A = 1$, $B = -3$, $C = 1$, $D = 2$

2. (a) $(((5x - 2)x + 1)x - 4)x + 2$
   (b) $((5x^2 - 3)x^2 + 2)x$

3. (a) $(x - 4)(x + 4)(x^2 + 16)$
   (b) no factors (c) $(x + 6)(x - 1)$
   (d) $(x + 2)(x^2 - 2x + 4)$
   (e) $(x - 1)(x^2 + x + 1)$ (f) $(x + 2)^3$

4. $(x - 1)^3(x + 2)$

Exercise 4.2

1. $R = \dfrac{R_1 R_2}{R_1 + R_2}$

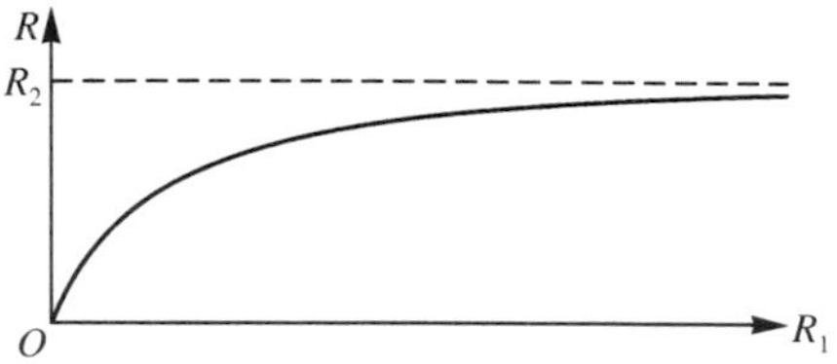
R
R2
O
R1

2. (a)

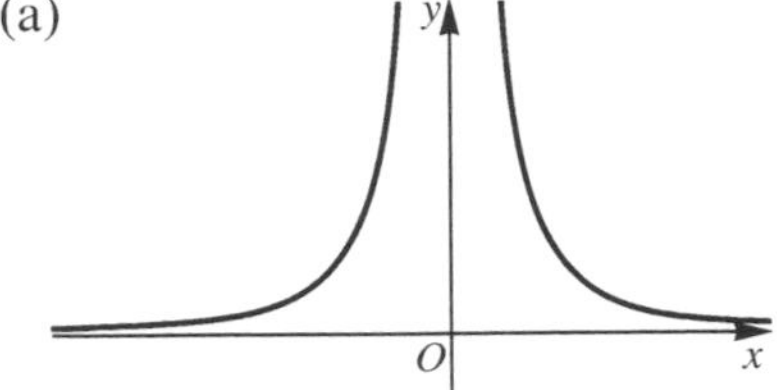

(b)

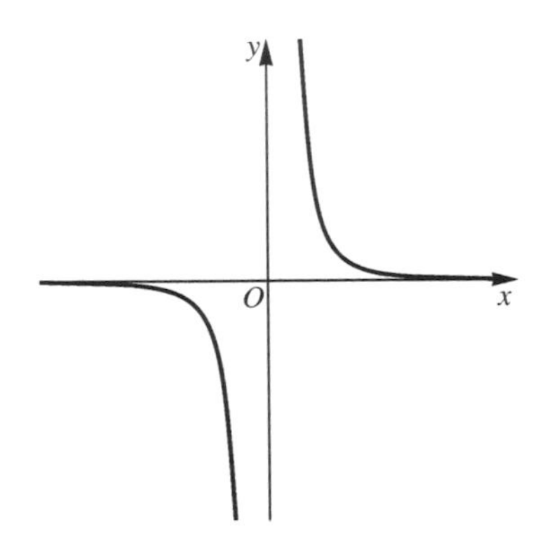

(c)

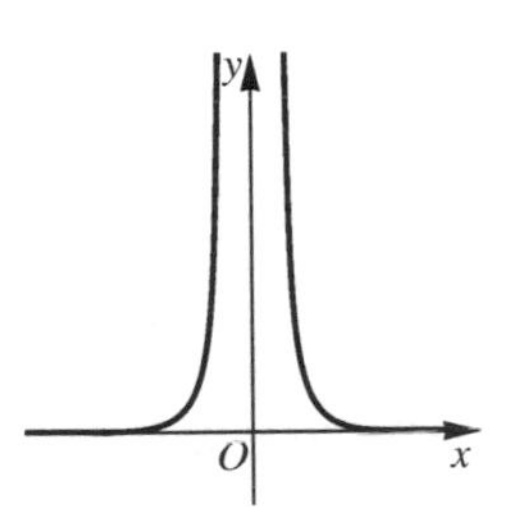

(d)

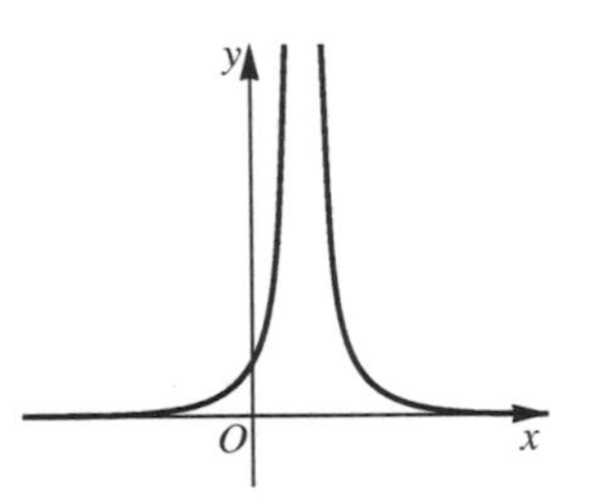

(e)

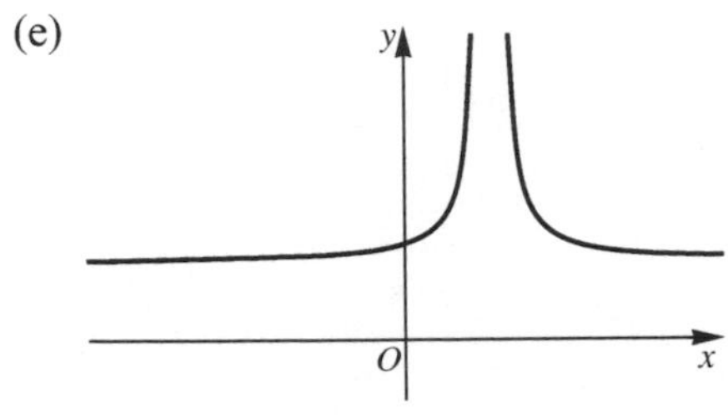

(f)

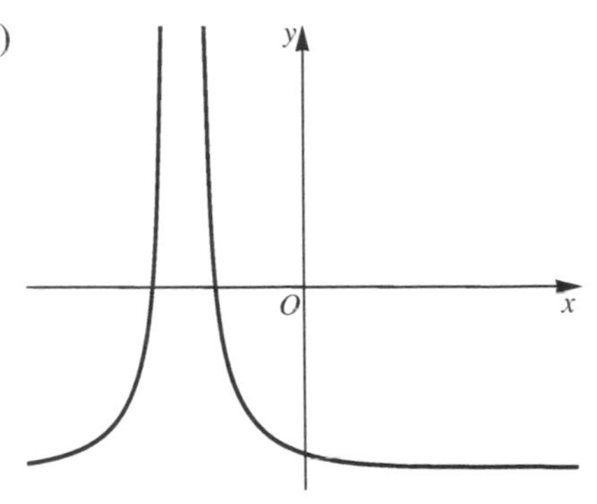

3. (a) Pole $x = 0$, zero $x = -1$; asymptotes $y = 1$, $x = 0$.

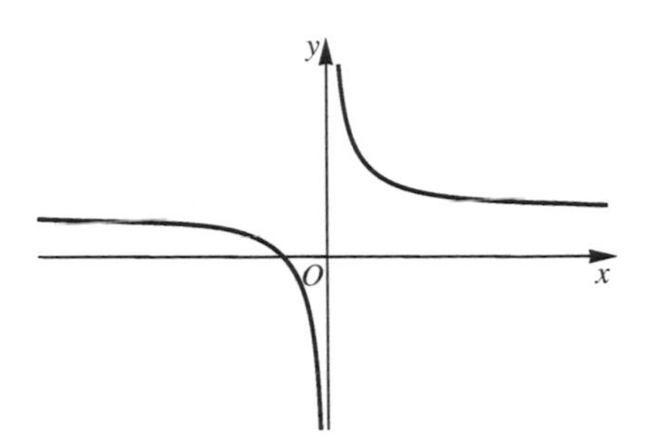

(b) Pole $x = -1$, zero $x = 0$; asymptotes $y = 1$, $x = -1$.

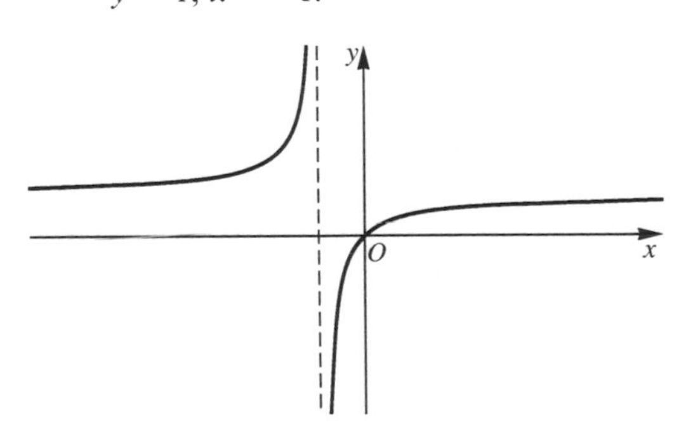

(c) Zero $x = 0$; asymptote $y = 0$.

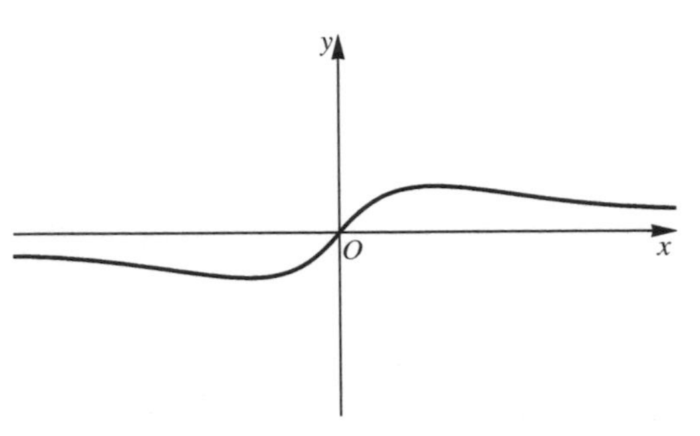

(d) Pole $x = 0$; asymptotes $x = 0$, $y = x$.

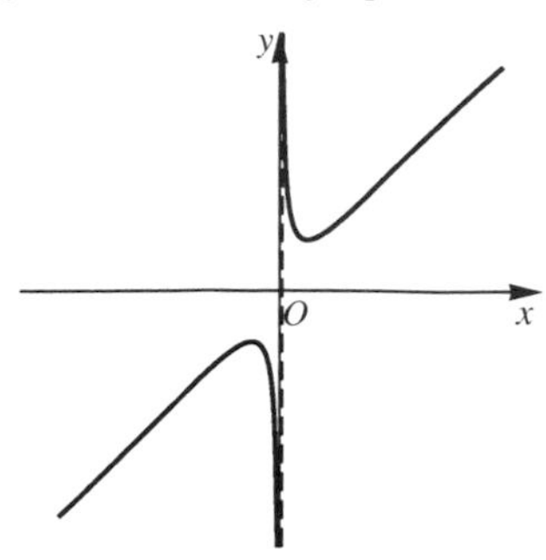

(e) Poles $x = -2, 3$, zero $x = 1$; asymptotes $y = 0$, $x = -2, 3$.

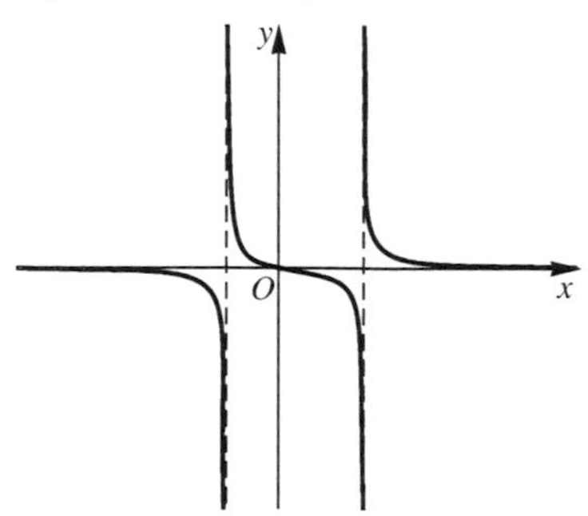

(f) Poles $x = 1, 3$, zero $x = -2$; asymptotes $y = 0$, $x = 1, 3$.

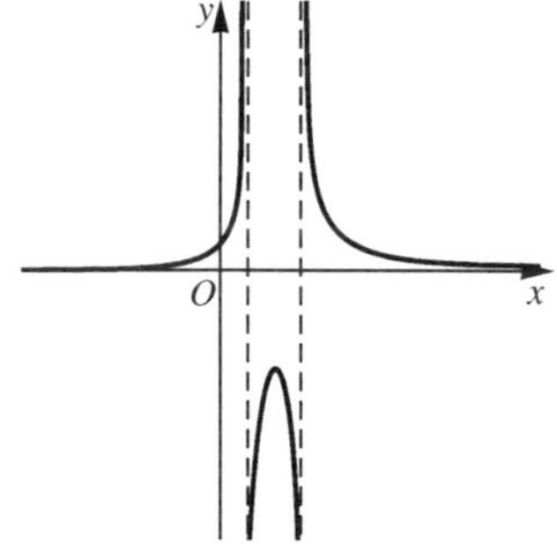

(g) Poles $x = -2, 1$, zero $x = 3$; asymptotes $y = 0$, $x = -2, -1$.

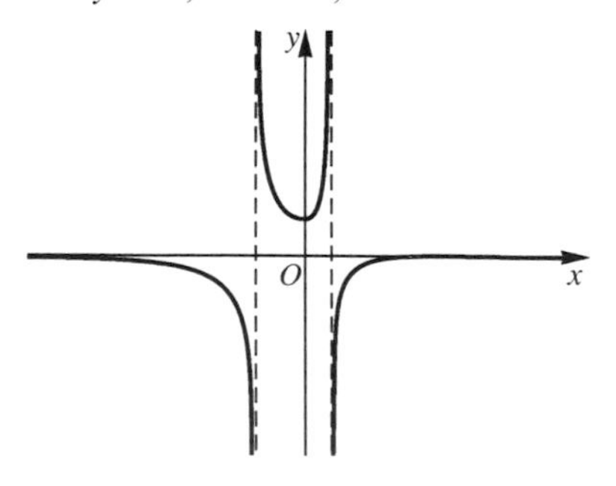

4. (a) $\frac{3}{5}\cdot\frac{1}{2x-3}+\frac{1}{5}\cdot\frac{1}{x+1}$

(b) $-\frac{16}{x-2}+\frac{22}{x-3}$

(c) $\frac{32}{7}\cdot\frac{1}{x+6}+\frac{10}{7}\cdot\frac{1}{x-1}$

(d) $1-\frac{4}{x^2+4}$

(e) $x-\frac{4x}{x^2+4}$

(f) $\frac{14}{9}\cdot\frac{1}{x_{-2}}+\frac{16}{3}\cdot\frac{1}{(x-2)^2}+\frac{4}{9}\cdot\frac{1}{x+1}$

(g) $\frac{3}{4}\cdot\frac{1}{x-1}+\frac{1}{4}\cdot\frac{1}{(x-1)^2}-\frac{1}{4}\cdot\frac{1}{x+1}$
$-\frac{1}{4}\cdot\frac{1}{(x+1)^2}$

(h) $\frac{1}{3}\cdot\frac{1}{x+1}-\frac{2}{3}\frac{(x-2)}{x^2-x+1}$

Exercise 4.3

1. (a)

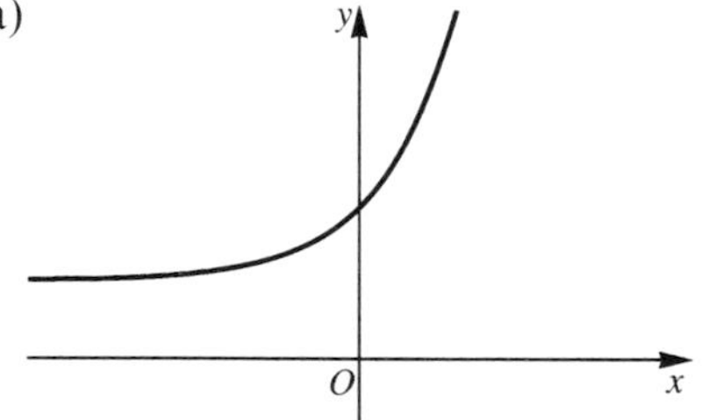

(b)

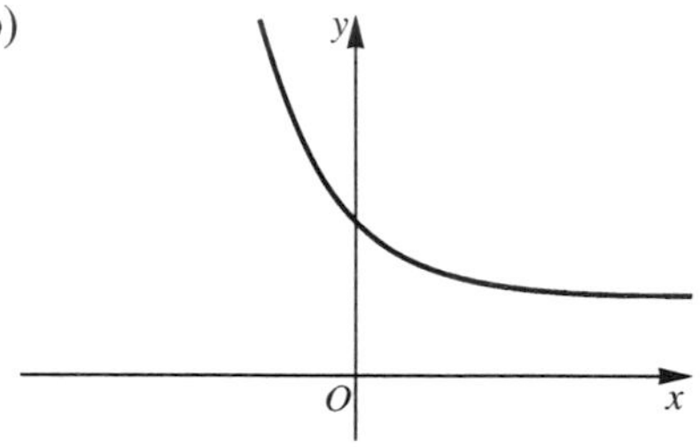

(c)

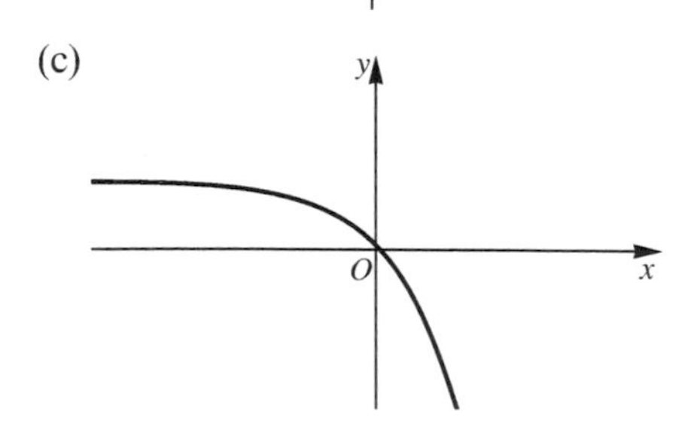

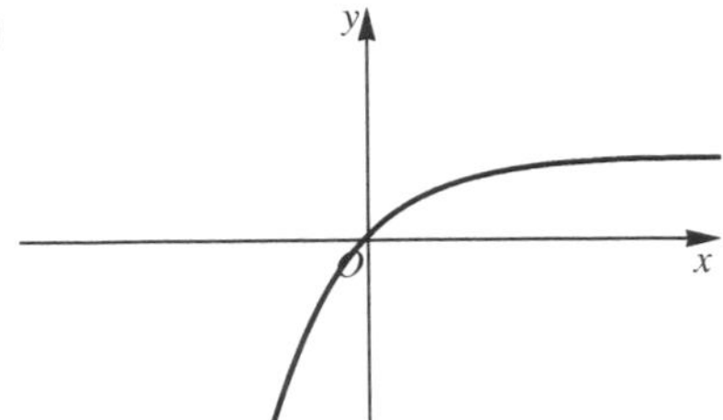

2. (a)

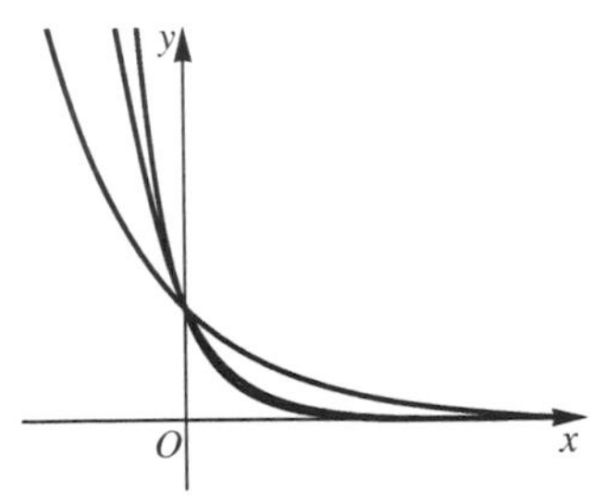

(b)

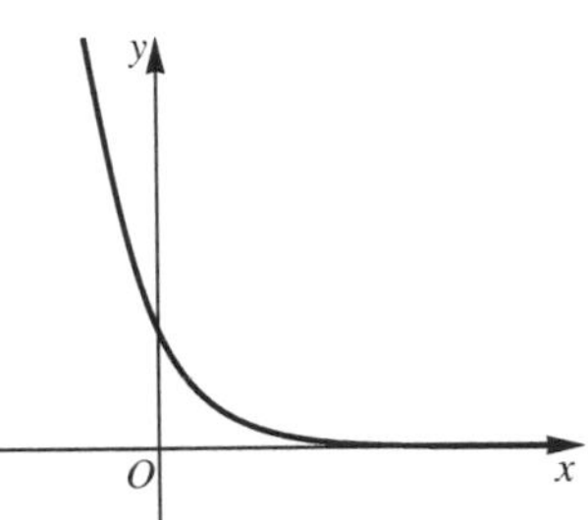

(c)

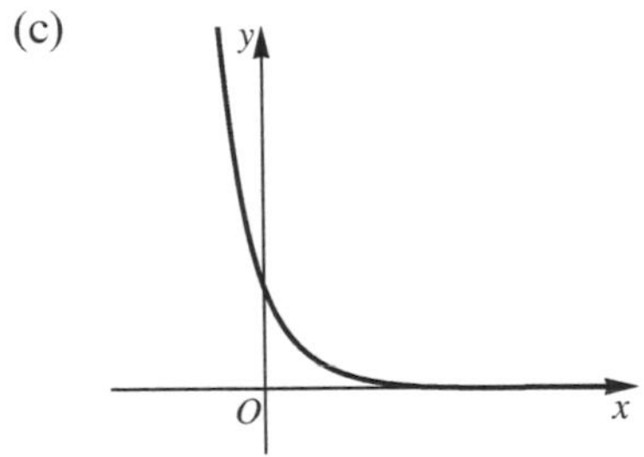

3. 90 dB, $0.01\,\mathrm{W\,m^{-2}}$

4. 17.8 kPa, 5.61 kPa, 1.77 kPa

5. (a) 77.7% (b) 95.0% (c) 98.9%

6. 99.995 46%

Exercise 4.4

2. (a) $2\log_a x$ (b) $5\log_a x$
(c) $2\log_a x + 3\log_a y$ (d) -2

3. (a) $\ln\left(\frac{5+\sqrt{29}}{2}\right)$ (b) $\ln\left(\frac{3\pm\sqrt{5}}{2}\right)$

(c) $\left(\frac{1}{2}\mathrm{e}^{16}\right)^{\frac{1}{6}}$ (d) $-\frac{\log 20}{\log 6}$

4. 17.6

Exercise 5.1

1. (a) $\frac{\sqrt{3}}{2}$ (b) $-\frac{1}{2}$ (c) -1

(d) $-\frac{1}{\sqrt{2}}$ (e) $-\frac{1}{2}$ (f) $-\sqrt{3}$

2. $\frac{5\pi}{4}$

Exercise 5.2

1. (a) amplitude 5, angular frequency 3, period $2\pi/3$, frequency $3/2\pi$, phase angle $\pi/2$

(b) $1, \frac{1}{2}, 4\pi, \frac{1}{4\pi}, \pi$

(c) $2, 5, \frac{2\pi}{5}, \frac{5}{2\pi}, -\frac{\pi}{6}$

(d) $3, \frac{1}{3}, 6\pi, \frac{1}{6\pi}, \frac{\pi}{4}$

2. (a)

(b)

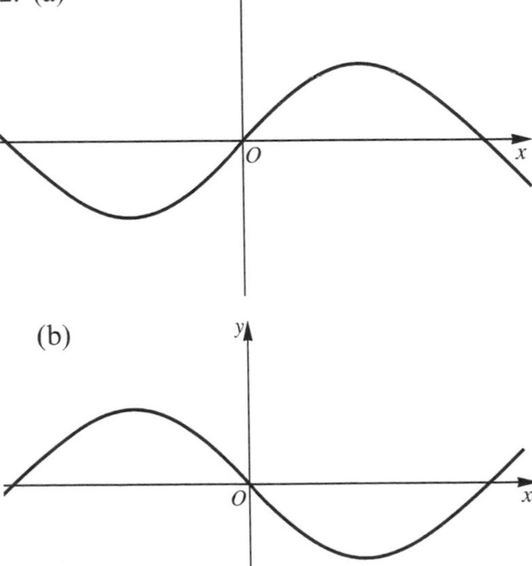

(c)

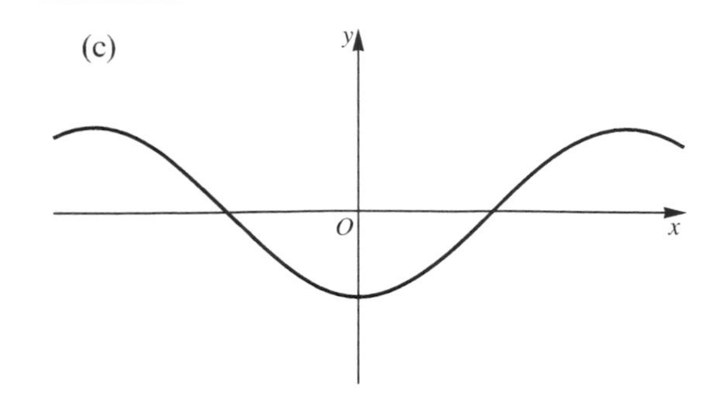

3. (a) $30^\circ = \dfrac{\pi^c}{6}$ (b) $60^\circ = \dfrac{\pi^c}{5}$
   (c) $63.4^\circ = 1.107^c$ (d) No such value
   (e) $-30^\circ = -\dfrac{\pi^c}{6}$ (f) $-120^\circ = -\dfrac{2\pi^c}{3}$
   (g) $-63.4^\circ = -1.107^c$

4. $x = 4\sin\left(\dfrac{\pi}{2}t + \dfrac{\pi}{6}\right)$

5. (a) $\dfrac{2}{\sqrt{3}}$ (b) $-\dfrac{2}{\sqrt{3}}$ (c) $\dfrac{2}{\sqrt{3}}$
   (d) $-\sqrt{2}$ (e) -1 (f) $\sqrt{3}$

Exercise 5.3

1. (a) 0.664, 2.498 (b) 3.785, 5.640
   (c) 0.644, 5.640 (d) −0.644, −5.640
   (e) 1.249, 4.391 (f) −1.249, −4.391

2. Add to each solution $\pm 2n\pi$ where $n = 1,2,3,\ldots$

3. (a) $\dfrac{\pi}{3}, \dfrac{5\pi}{3}$ (b) no solutions
   (c) 0.675, 2.466 (d) 1.190, 4.332
   (e) −1.190, −4.332 (f) −0.675, −1.466

4. (a) 1.231, π, 5.052 (b) 0, 1.318, 4.965, 2π

5. (a) $\dfrac{\pi}{9}, \dfrac{5\pi}{9}, \dfrac{7\pi}{9}, \dfrac{11\pi}{9}, \dfrac{13\pi}{9}, \dfrac{17\pi}{9}$
   (b) $\dfrac{\pi}{8}, \dfrac{5\pi}{8}, \dfrac{9\pi}{8}, \dfrac{13\pi}{8}$
   (c) $\dfrac{7\pi}{6}, \dfrac{11\pi}{6}$ (d) $\dfrac{\pi}{6}, \dfrac{5\pi}{6}$

6. $\dfrac{\pi}{4}, \dfrac{\pi}{2}, \dfrac{3\pi}{4}$

Exercise 5.4

1. (a) 27.3 (b) 0.637 (c) 0.762

2. (a)

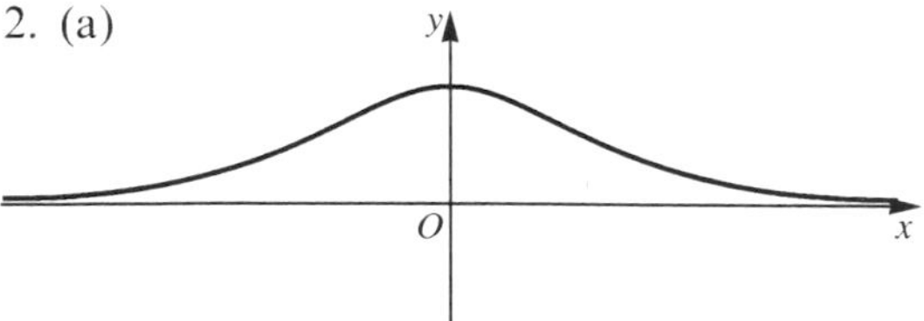

(b)

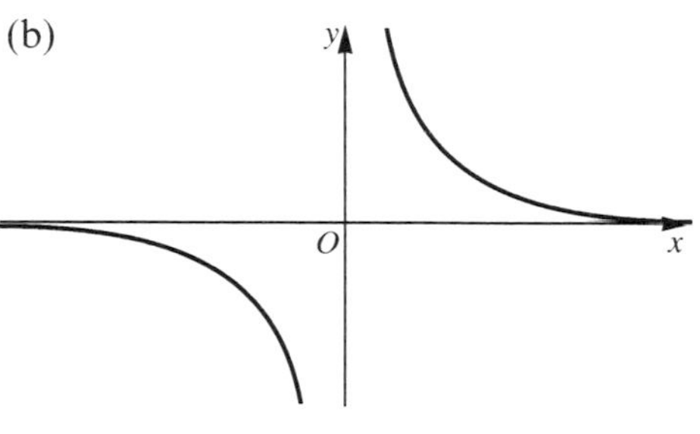

(c)

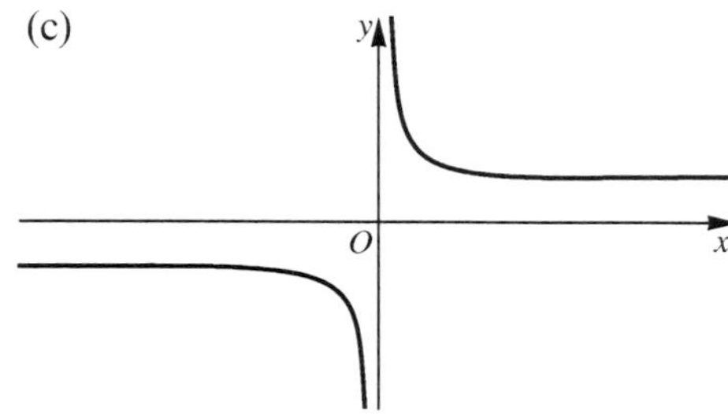

3. (a) $\ln(2 + \sqrt{3})$
   (b) 0.733 (c) $\frac{1}{2}\ln 3$

4. $c\left(\cosh\dfrac{25}{c} - 1\right)$

5. (a) $-\ln 2$ (b) 0, $\ln 7$

Exercise 6.1

1. (a) $\sin\theta\tan\theta$
   (b) $\cot^4\theta$
   (c) $\operatorname{cosec}^3\theta$

2. (a) $y^2 = x^2 + 9$
   (b) $x^2 + y^2 - 2x - 2y + 1 = 0$
   (c) $\left(\dfrac{x}{2}\right)^2 + \left(\dfrac{3}{y}\right)^2 = 1$

4. (a) 0.905, 5.38
   (b) 1.32, 4.97
   (c) $\dfrac{3\pi}{2}$

Exercise 6.2

1. $\frac{44}{125}, \frac{4}{5}$

2. (a) $\sin 3\theta$ (b) $\cot\theta$ (c) $\tan 2\theta$

3. (a) $\frac{\pi}{6}, \frac{\pi}{2}, \frac{5\pi}{6}, \frac{7\pi}{6}, \frac{3\pi}{2}, \frac{11\pi}{6}$

   (b) $0, \frac{\pi}{4}, \frac{3\pi}{4}, \pi, \frac{5\pi}{4}, \frac{7\pi}{4}, 2\pi$

   (c) $\frac{\pi}{6}, \frac{\pi}{4}, \frac{3\pi}{4}, \frac{5\pi}{6}, \frac{5\pi}{4}, \frac{7\pi}{6}$

   (d) $\frac{\pi}{12}, \frac{3\pi}{12}, \frac{5\pi}{12}, \frac{9\pi}{12}, \frac{13\pi}{12}, \frac{15\pi}{12}, \frac{21\pi}{12}, \frac{23\pi}{12},$

5. (a) $\frac{2t}{1-t^2}$ (b) $\frac{2t}{1+t^2}$ (c) $\frac{1-t^2}{1+t^2}$

6. (a) $\frac{\pi}{6}, \frac{5\pi}{6}, \frac{3\pi}{2}$ (b) $\frac{\pi}{3}, \frac{4\pi}{3}$

   (c) $0, \frac{\pi}{4}, \pi, \frac{5\pi}{4}, 2\pi$

Exercise 6.3

1. $5, -5, \theta = 53.1°$

2. (a) $-24.3° \pm 360n°, -114.3° \pm 360n°$
   (b) $360n°, -90 \pm 360n°$
   (c) $31.4° \pm 360n°, -121.4° \pm 360n°$

3. (a) $\pm 120n°, 42.1° \pm 120n°$
   (b) $45° + 360n$
   (c) $\pm 360n°, -141.3° \pm 360n°$
   (d) $90° \pm 360n°, = 30° \pm 360n°$
   (e) $\pm 360n°, -53.1° \pm 360n°$

Exercise 7.1

1. $-3, 0, 3$

2. (a) $x = n\pi$ (b) none (c) none

   (d) $x = (2n+1)\frac{\pi}{2}$

3. $-\frac{2}{x^3}$

4. Not at $x = 0$

5. $40 - 19.6t, t = 2.04$

Exercise 7.2

1. $\frac{dI}{dV} = I + I_s$

2. Tangents: $y = x, y - \frac{1}{\sqrt{2}} = \frac{1}{\sqrt{2}}\left(x - \frac{\pi}{4}\right),$

   $y = 1$

   Normals: $y = -x, y - \frac{1}{\sqrt{2}}$

   $= -\sqrt{2}\left(x - \frac{\pi}{4}\right), x = \frac{\pi}{4}$

3. (a) $42x^5$ (b) $3\cos x + 4x^3$
   (c) $\cos^2 x - \sin^2 x \equiv \cos 2x$

   (d) $\frac{2x}{(1+x^2)^2}$ (e) $e^x(\cos x - \sin x)$

   (f) $1 + \ln x$ (g) $\frac{1 - 3\ln x}{x^4}$

   (h) $\frac{x^2 + 2x\sin x}{1 + \cos x}$

Exercise 7.3

1. (a) $210x^4$ (b) $-3\sin x + 12x^2$
   (c) $-4\sin x\cos x \equiv 2\sin 2x$

   (d) $\frac{2 - 6x^2}{(1+x^2)^3}$ (e) $-2e^x \sin x$

   (f) $\frac{1}{x}$ (g) $\frac{-7 + 12\ln x}{x^5}$

   (h) $\frac{(1+\cos x)(2x + \sin x) + x^2 \sin x}{(1+\cos x)^2}$

2. $\rho = c\cosh^2 \frac{x}{c}$, least value of c at $x = 0$.

4. $\dot\theta = \omega = -2\sin t + 3\cos t,$
   $\ddot\theta = -2\cos t - 3\sin t, t = 0.983$

Exercise 7.4

1. (a) $-4(2-x)^3$ (b) $20(5x-4)^3$
   (c) $8x(x^2+1)^3$ (d) $-2x(x^2+4)^{-2}$
   (e) $4\cos(4x+3)$ (f) $4x^3/(1+x^4)$

   (g) $\frac{1}{x}\cos\ln x$ (h) $-2x\sin(x^2+1)$

   (i) $6e^{3x}(1 + e^{3x})$

2. $\dfrac{1}{\sqrt{1+x^2}}, \dfrac{1}{\sqrt{x^2-1}}, \dfrac{1}{1-x^2}$

3. Tangent $y - \dfrac{1}{2} = -2\sqrt{3}\left(t - \dfrac{\pi}{4\omega}\right)$;

   normal $y - \dfrac{1}{2} = \dfrac{1}{2\sqrt{3}}\left(t - \dfrac{\pi}{4\omega}\right)$

4. $\dfrac{2}{x} \ln x \cos(1 + (\ln x)^2)$

Exercise 8.1

1. (a) 0,4 (b) e^{-1}, e^2 (c) -1, 1
   (d) no maximum, minimum values

2. (a) $\frac{1}{4}$; no minimum
   (b) no maximum and no minimum

3.

	Increasing	Decreasing	Stationary point
(a)	$x < -2$	$x > -2$	$x = -2$
(b)	$x > -2$	$x < -2$	$x = -2$
(c)	$x \neq -1$	–	$x = -1$
(d)	$-\sqrt{2} < x < 0$, $x > \sqrt{2}$	$x < -\sqrt{2}$, $0 < x < \sqrt{2}$	$0, \pm\sqrt{2}$
(e)	all x	–	–
(f)	$x > 0$ ($\ln x$ defined only for $x > 0$)		

Exercise 8.2

1. (a) $x = 0$, local maximum
   (b) $x = 0$, local maximum; $x = \pi$, local minimum

2.

	Concave upwards	Concave downwards	Point of inflection
(a)	$x < 0$	$x > 0$	$x = 0$
(b)	$x > 2$	$x < 2$	$x = 2$
(c)	$x < -1, x > 0$	$-1 < x < 0$	$x = 0, 1$
(d)	$x < -3 - \sqrt{3}$, $x > 3 + \sqrt{3}$	$-3 - \sqrt{3} < x$ $< 3 + \sqrt{3}$	$x = -3 \pm \sqrt{3}$
(e)	$0 \leqslant x < \dfrac{\pi}{2}$	$\dfrac{\pi}{2} < x \leqslant \pi$	$x = \dfrac{\pi}{2}$
(f)	$x < 1$	$x > 1$	$x = 1$

Exercise 8.3

1. (a) $x = 0$, local maximum; $x = \frac{2}{3}$, local minimum; $x = \frac{1}{3}$, inflection
   (b) $x = -1$, local maximum; $x = 1$, local minimum; $x = 0$, inflection
   (c) $x = -1$, inflection; $x = 2$, local minimum; inflection at $x = 1$
   (d) $x = 0$, local minimum, $x = \pm\sqrt{2}$, local maximum; inflection at $x = \pm\sqrt{2/3}$.

2. $\theta \equiv 45^\circ$

3. $v = (T/3)^{\frac{1}{2}}$

4. 14.7 cm square

Exercise 8.4

1. (a) $x = \pm\sqrt{1/6}$ (b) $x = 2$
   (c) $x = 2 \pm \sqrt{2}$ (d) $x = n \pm \sqrt{n}$

2. $t_1 = t_* + \dfrac{\pi}{\omega}$, $t_2 = t_* + \dfrac{2\pi}{\omega}$;

   $t_* = \dfrac{1}{\omega} \tan^{-1}\left(-\dfrac{\alpha}{\omega}\right)$

3. $(2A/B)^{\frac{1}{6}}$

Exercise 9.1

1. (a) $x^6 + C$ (b) $\frac{1}{6}u^6 + C$ (c) $-\dfrac{1}{x} + C$

2. (a) 80 (b) 20

Exercise 9.2

1. (a) $\tan x + C$ (b) 1 (c) $2\ln 2 - 1$

2. (a) 2 (b) 1 (c) 1
   (d) 0 (e) -1 (f) 1

4. (a) $\dfrac{3}{10}x^{\frac{10}{3}} + \dfrac{3}{2}x^{\frac{4}{3}} + C$ (b) $\ln|v| - \dfrac{1}{v} + C$

   (c) $3 - 2e^{\frac{\pi}{4}} + 2$ (d) $\dfrac{1}{4} + \dfrac{\pi}{8}$

5. (a) 4.07 (b) $\frac{1}{2}\ln 2$ (c) $2\ln 2$
   (d) $2\ln 2 - 1$

Exercise 9.3

1. (a) $\dfrac{2}{9}(x^3 - 8)^{\frac{3}{2}} + C$ (b) $\dfrac{1}{2}\ln(1 + 2e^x) + C$

   (c) $-\dfrac{1}{4(4 + u^2)^2} + C$ (d) $\dfrac{2}{5}(1 + \tan x)^{\frac{5}{2}} + C$

2. (a) $\dfrac{2}{9}[(56^{\frac{3}{2}} - (19)^{\frac{3}{2}}]$ (b) $\dfrac{2}{5}(4\sqrt{2} - 1)$

   (c) $\dfrac{1}{2}\ln\left(\dfrac{1 + 2e}{3}\right)$

3. (a) $\frac{1}{2}e^{x^2} + C$

(b) $\frac{1}{5}\cos^5\theta - \frac{1}{3}\cos^3\theta + C$

(c) $\frac{2}{3}\ln(2 + x^{\frac{3}{2}}) + C$

(d) $\frac{1}{2}\left(\tan^{-1}x + \frac{x}{1+x^2}\right) + C$

(e) $-\frac{1}{2}\sinh^{-1}\left(\frac{1}{2x}\right) + C$

(f) $\frac{2}{5}(x+2)^{\frac{5}{2}} - \frac{8}{3}(x+2)^{\frac{3}{2}} + C$

Exercise 9.4

1. (a) $\frac{e^{2x}}{2}\left(x - \frac{1}{2}\right) + C$

(b) $\frac{1}{2}x\sin 2x + \frac{1}{4}\cos 2x + C$

(c) $-(x^2 + 2x + 2)e^{-x} + C$

(d) $\frac{2}{3}(x+1)^{\frac{3}{2}} - \frac{4}{15}(x+1)^{\frac{5}{2}} + C$

(e) $\frac{x^2}{2}\ln x - \frac{1}{4}x^2 + C$

(f) $\frac{1}{2}e^x(\cos x + \sin x) + C$

2. (a) $-\frac{1}{2}$ (b) $\frac{16}{5}\sqrt{3} - \frac{2}{5}\sqrt{2}$

3. (a) $x\ln x - x + C$
(b) $x\sin^{-1}x - \sqrt{(1-x^2)} + C$

Exercise 9.5

1. (a) 31 (b) 1

Exercise 10.1

1. (a) 5 (b) 1 (c) 0

2. $\frac{32}{3}$

3. (a) $1 - \frac{\pi}{4}$ (b) $\frac{\pi}{2} - \frac{4}{3}$

Exercise 10.2

1. 16π

2. $\frac{\pi a^5}{30}$

3. $\frac{17}{3}\pi$

4. $\frac{\pi^2}{2}$

Exercise 10.3

1.

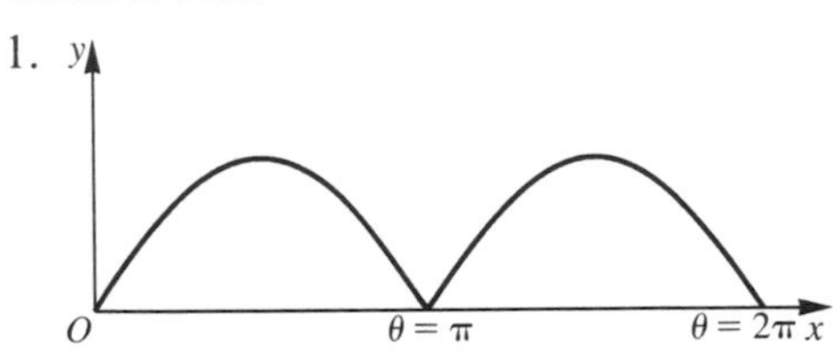

2. $2\pi ah$

3. $\ln(1 + \sqrt{2})$

4. $\left(\frac{15}{4} + \ln 2\right)\pi$

Exercise 10.4

1. (a) $\frac{2}{\pi}$ (b) $\frac{2}{\pi}$ (c) $\frac{7}{3}$

2. $160 - 120e^{-0.5}$

3. (a) $\bar{x} = 1.5$, $\bar{y} = 1.2$
(b) $\bar{x} = 1.6$, $\bar{y} = \frac{16}{7}$
(c) $\bar{x} = \frac{\pi}{2}$ (symmetry), $\bar{y} = \frac{\pi}{8}$

4. (a) $\frac{1}{\sqrt{2}}$ (b) $\frac{1}{\sqrt{2}}$
(c) $\sqrt{\frac{103}{15}}$; $\frac{5}{\sqrt{2}}$

5. (a) $\bar{y} = 0$, $\bar{x} = \frac{4}{3}a$ (b) $\bar{y} = 0$, $\bar{x} = \frac{\pi}{2}$

6. On axis of symmetry, $\frac{1}{4}$ of the way from base to vertex

Exercise 11.1

1. (a) and (d) are scalars, (b) and (c) are vectors

2. (a), (b)

(c), (d)

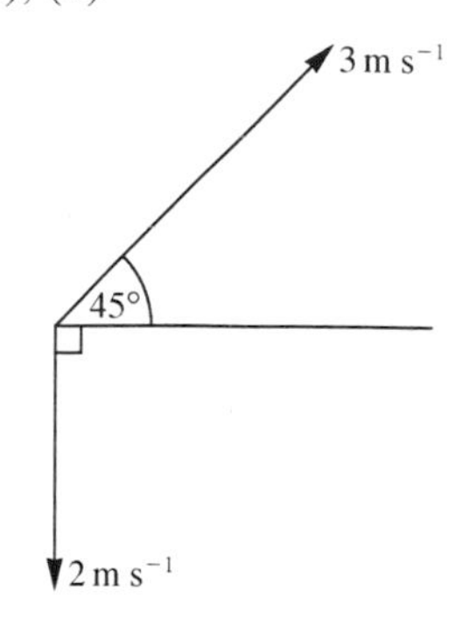

Exercise 11.2

1. 7.28 km per hour, N 60.9° E
2. 9.86 m s^{-2} at 5.82° to the vertical
3. Angle 66.4° to the bank against the tide, 6.5 s
4. 15 km per hour north
5. 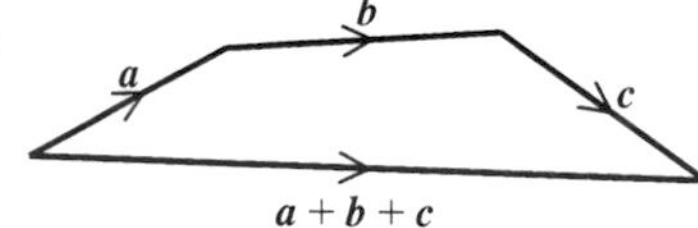

6. 58.3 km per hour from N 59° W

Exercise 11.3

1. (a) $\boldsymbol{a}+\boldsymbol{b}$ (b) $-\frac{1}{2}(\boldsymbol{a}+\boldsymbol{b})$ (c) $\frac{1}{2}(\boldsymbol{b}-\boldsymbol{a})$

2. (a) $-\boldsymbol{c}$ (b) $-\boldsymbol{a}$
(c) $-\boldsymbol{b}$ (d) $\boldsymbol{a}+\boldsymbol{b}+\boldsymbol{c}$
(e) $\boldsymbol{b}+\boldsymbol{c}-\boldsymbol{a}$ (f) $\boldsymbol{c}-\boldsymbol{a}-\boldsymbol{b}$
(g) $\boldsymbol{a}+\boldsymbol{b}$ (h) $\boldsymbol{b}+\boldsymbol{c}$

3. (a) $\boldsymbol{0}$
(b) $\overrightarrow{GH}=\frac{1}{2}\boldsymbol{c}+\frac{1}{2}\boldsymbol{d}$, EF and HG equal and parallel
(c) parallelogram

4. (a) $\boldsymbol{a}+\boldsymbol{b}+\boldsymbol{c}$ (b) $-\boldsymbol{a}$
(c) $\frac{1}{2}\boldsymbol{c}$ (d) $-\frac{1}{2}\boldsymbol{c}$
(e) $\boldsymbol{b}+\frac{1}{2}\boldsymbol{c}$ (f) $\boldsymbol{a}+\boldsymbol{b}+\frac{1}{2}\boldsymbol{c}$

5. (a) $\boldsymbol{c}=-\boldsymbol{a}-\boldsymbol{b}$ (b) $\boldsymbol{a}+\frac{1}{2}\boldsymbol{b}$
(c) $-\boldsymbol{a}-\frac{1}{2}\boldsymbol{c}=\frac{1}{2}(\boldsymbol{b}-\boldsymbol{a})$
(d) $\overrightarrow{AG}=\lambda(\boldsymbol{a}+\frac{1}{2}\boldsymbol{b})$, $\overrightarrow{DG}=\frac{1}{2}\mu(\boldsymbol{b}-\boldsymbol{a})$
(e) $AG=\frac{2}{3}AD$, $BG=\frac{2}{3}BE$
(f) $CG=\frac{2}{3}CF$
(g) Medians are concurrent.

Exercise 11.4

1. (a) $\begin{pmatrix}-1\\3\end{pmatrix}$ (b) $\begin{pmatrix}3\\6\end{pmatrix}$ (c) $\begin{pmatrix}2\\9\end{pmatrix}$
(d) $\sqrt{5}$, $\sqrt{5}$, 5 (e) $\frac{1}{\sqrt{170}}\begin{pmatrix}-1\\13\end{pmatrix}$

2. (a) $\begin{pmatrix}-1\\1\\3\end{pmatrix}$ (b) $\begin{pmatrix}6\\0\\1\end{pmatrix}$ (c) $\begin{pmatrix}8\\-2\\2\end{pmatrix}$
(d) $\sqrt{3}$, $2\sqrt{2}$, $\sqrt{14}$ (e) $\frac{1}{5\sqrt{3}}\begin{pmatrix}5\\-1\\7\end{pmatrix}$

3. $\boldsymbol{a}$: $\frac{1}{\sqrt{5}}, \frac{2}{\sqrt{5}}$; $\boldsymbol{b}$: $\frac{2}{\sqrt{5}}, \frac{-1}{\sqrt{5}}$; $\boldsymbol{c}$: 0, 1

4. $\boldsymbol{a}$: $\frac{1}{\sqrt{3}}, \frac{1}{\sqrt{3}}, \frac{1}{\sqrt{3}}$; $\boldsymbol{b}$: $\frac{1}{\sqrt{2}}, 0, \frac{-1}{\sqrt{2}}$;
$\boldsymbol{c}$: $\frac{3}{\sqrt{14}}, \frac{-1}{\sqrt{14}}, \frac{2}{\sqrt{14}}$

5. (a) $\boldsymbol{r}=\begin{pmatrix}1\\1\end{pmatrix}+\lambda\begin{pmatrix}1\\-4\end{pmatrix}$
(b) $\boldsymbol{r}=\begin{pmatrix}1\\-2\end{pmatrix}+\lambda\begin{pmatrix}-1\\3\end{pmatrix}$
(c) $\boldsymbol{r}=\begin{pmatrix}1\\2\\3\end{pmatrix}+\lambda\begin{pmatrix}1\\-2\\-4\end{pmatrix}$

Exercise 12.1

1. $\boldsymbol{a}$ and $\boldsymbol{b}$

3. $\boldsymbol{b}$ and $\boldsymbol{c}$ 48.2°; $\boldsymbol{c}$ and $\boldsymbol{d}$ 60°; $\boldsymbol{b}$ and $\boldsymbol{d}$ 99.6°

5. 9 units

Exercise 12.2

1. $(2,-7,-3)$, 14.3°

2. $8\sqrt{3}$

3. $\boldsymbol{a}\cdot(\boldsymbol{b}\times\boldsymbol{c})=(\boldsymbol{a}\times\boldsymbol{b})\cdot\boldsymbol{c}=3$;
   $(\boldsymbol{a}\times\boldsymbol{b})\times\boldsymbol{c}=(\boldsymbol{a}\cdot\boldsymbol{c})\boldsymbol{b}-(\boldsymbol{a}\cdot\boldsymbol{b})\boldsymbol{c}=(5,-5,5)$

4. (a) $-2\boldsymbol{i}-2\boldsymbol{k}$ (b) $-\boldsymbol{i}-\boldsymbol{j}-2\boldsymbol{k}$
   (c) $\boldsymbol{i}+\boldsymbol{j}+2\boldsymbol{k}$ (d) $3\boldsymbol{i}-\boldsymbol{j}+4\boldsymbol{k}$
   (e) $\boldsymbol{0}$ (f) $-3\boldsymbol{i}+\boldsymbol{j}-4\boldsymbol{k}$

5. $\boldsymbol{F}=\dfrac{5}{\sqrt{14}}(2,1,3)$, $\boldsymbol{M}=\dfrac{5}{\sqrt{14}}(2,-1,-1)$,
   $|\boldsymbol{M}|=\dfrac{5\sqrt{21}}{7}$ N m

Exercise 12.3

1. $\boldsymbol{r}=\begin{pmatrix}3-\lambda\\ -\lambda\\ 2-\lambda\end{pmatrix}$; $(3\tfrac{1}{2},\tfrac{1}{2},2\tfrac{1}{2})$

2. $\dfrac{5\sqrt{3}}{3}$

3. $\dfrac{1}{\sqrt{14}}\begin{pmatrix}2\\3\\-1\end{pmatrix}$

4. $x+2y-z=2$

5. $\dfrac{1}{\sqrt{6}}$

6. $\dfrac{x+2}{3}=\dfrac{y-6}{-5}=\dfrac{z}{1}$

Exercise 12.4

1. (a) $\begin{pmatrix}2\\1\\4\\3\\6\end{pmatrix}$, $\begin{pmatrix}0\\3\\2\\5\\4\end{pmatrix}$, $\begin{pmatrix}-5\\3\\0\\9\\7\end{pmatrix}$
   (b) $|\boldsymbol{a}|=\sqrt{55}$, $|\boldsymbol{b}|=\sqrt{5}$, $|\boldsymbol{c}|=\sqrt{3}$, $|\boldsymbol{d}|=\sqrt{55}$
   (c) (i) $\boldsymbol{b}$ (ii) $\boldsymbol{a}$, $\boldsymbol{d}$

Exercise 13.1

1. (a) 3×2, 2×4, 2×2, 3×3
   (b) 6, 8, 1, 2
   (c) $\boldsymbol{C}$, $\boldsymbol{D}$; 2, 15

2. $\boldsymbol{C}$ is not square, $\boldsymbol{C}$ is a zero matrix, $\boldsymbol{D}$ is an identity matrix, $\boldsymbol{H}$ and $\boldsymbol{K}$ are diagonal matrices.

3. $\boldsymbol{C}$ and $\boldsymbol{D}$ are symmetric.
   $\boldsymbol{A}$ and $\boldsymbol{H}$ are skew-symmetric.

4. n^2
   (a) n^2-n (b) $\frac{1}{2}(n^2-n)$
   (c) n (d) none

Exercise 13.2

1. $\boldsymbol{A}+\boldsymbol{D}=\begin{bmatrix}1&0&2\\0&1&0\\2&0&1\end{bmatrix}$,

   $\boldsymbol{D}-\boldsymbol{A}=\begin{bmatrix}-1&2&0\\2&-1&2\\0&2&-1\end{bmatrix}$,

   $\boldsymbol{B}^{\mathrm{T}}+\boldsymbol{C}=\begin{bmatrix}2&2\\2&5\\3&6\end{bmatrix}$,

   $\boldsymbol{B}+\boldsymbol{C}^{\mathrm{T}}=\begin{bmatrix}2&2&3\\2&5&6\end{bmatrix}$,

   $4\boldsymbol{A}=\begin{bmatrix}4&-4&4\\-4&4&-4\\4&-4&4\end{bmatrix}$, $\dfrac{1}{2}\boldsymbol{D}=\begin{bmatrix}0&\frac{1}{2}&\frac{1}{2}\\\frac{1}{2}&0&\frac{1}{2}\\\frac{1}{2}&\frac{1}{2}&0\end{bmatrix}$,

   $3\boldsymbol{A}-2\boldsymbol{D}=\begin{bmatrix}3&-5&1\\-5&3&-5\\1&-5&3\end{bmatrix}$

4. (a) $\begin{pmatrix}\dfrac{\sqrt{3}}{2}&\dfrac{1}{2}\\-\dfrac{1}{2}&\dfrac{\sqrt{3}}{2}\end{pmatrix}$ (b) $\begin{pmatrix}\dfrac{1}{\sqrt{2}}&\dfrac{1}{\sqrt{2}}\\-\dfrac{1}{\sqrt{2}}&\dfrac{1}{\sqrt{2}}\end{pmatrix}$

   (c) $\begin{pmatrix}0&1\\-1&1\end{pmatrix}$ (d) $\begin{pmatrix}-1&0\\0&-1\end{pmatrix}$

5. $b=c$ or $a+d=0$

6. $AC = \begin{bmatrix} 4 & 0 \\ 2 & 2 \end{bmatrix}$, $BA = \begin{bmatrix} 2 & 2 & 4 \\ 1 & 0 & 1 \end{bmatrix}$,

$DE = \begin{bmatrix} 3 & -3 & 0 \\ 0 & -1 & 1 \\ 1 & 0 & 3 \end{bmatrix}$, $CD = \begin{bmatrix} 3 & 1 & 0 \\ 0 & -1 & -3 \\ 1 & 0 & 3 \end{bmatrix}$,

$B^2 = \begin{bmatrix} 1 & 2 \\ 0 & 1 \end{bmatrix}$, $D^2 = \begin{bmatrix} 1 & 2 & 1 \\ 1 & 1 & 2 \\ 2 & 1 & 1 \end{bmatrix}$

Exercise 13.3

1. -1, ad, ad, 4, 1, 2, -3
2. $|A| = 50$
4. (a) -27 (b) -88

Exercise 13.4

1. $A^{-1} = \begin{bmatrix} 0 & 1 \\ -1 & 0 \end{bmatrix}$, $B^{-1} = \begin{bmatrix} \cos\theta & -\sin\theta \\ \sin\theta & \cos\theta \end{bmatrix}$,

$D^{-1} = \begin{bmatrix} 1 & -1 \\ 0 & 1 \end{bmatrix}$, $E^{-1} = \dfrac{1}{ad}\begin{bmatrix} d & -b \\ 0 & a \end{bmatrix}$,

$F^{-1} = \dfrac{1}{ad}\begin{bmatrix} d & 0 \\ -c & a \end{bmatrix}$, $G^{-1} = \dfrac{1}{ad}\begin{bmatrix} d & 0 \\ 0 & a \end{bmatrix}$

2. I is orthogonal.

5. $B^{-1} = \dfrac{1}{43}\begin{bmatrix} 2 & -7 & 18 \\ 5 & 4 & 2 \\ 7 & -3 & -23 \end{bmatrix}$,

$D^{-1} = \dfrac{1}{25}\begin{bmatrix} 2 & 19 & 15 \\ 7 & -21 & -10 \\ 1 & -3 & -5 \end{bmatrix}$

Exercise 14.1

1. (a) $2x - 3z = 5$, $x + 3y + 4z = -1$, $4x - 2y + z = 3$

(b) $\begin{bmatrix} 1 & -1 & 2 \\ 1 & 1 & 1 \\ 2 & 0 & 1 \end{bmatrix}\begin{bmatrix} x_1 \\ x_2 \\ x_3 \end{bmatrix} = \begin{bmatrix} 4 \\ 6 \\ 9 \end{bmatrix}$

2. (a) $x = 1.6$, $y = 2.2$ (b) $3x = 2y$
(c) no solutions

3. (a) infinitely many solutions
(b) no solutions
(c) $x = 1$, $y = -2$, $z = 1$

Exercise 14.2

1. (a) $x_1 = 2$, $x_2 = 4$, $x_3 = 6$
(b) $x_1 = 3$, $x_2 = -2$, $x_3 = 1$
2. (a) infinitely many solutions
(b) no solutions
3. $x_3 = 0.143$, $x_2 = 0.428$, $x_1 = 0.715$, exact answers $\frac{1}{7}, \frac{3}{7}, \frac{5}{7}$ (to 3 d.p. 0.143, 0.429, 0.714)
4. $k = 1$
5. Exact solution: $x_1 = 1$, $x_2 = -1$, $x_3 = 2$

Exercise 14.3

1. $x_1 = 4$; $x_2 = 3$; $x_1 = 8$, $x_2 = 1$; ill-conditioned
3. $x_1 = \dfrac{1 - 144a}{1 - 55a}$, $x_2 = \dfrac{1}{1 - 55a}$; 1.88 – 1; -159.2, 100; no solution
4. $x_3 = -7.5$, $x_2 = 381.136$, $x_1 = -3.461\,36$; $x_3 = -8.865\,27$; $x_2 = 450.067$ $x_1 = -8.594\,73$

Exercise 14.4

2. (a) $x_2 = 1$, $x_1 = 0$ (b) $x_2 = 1$, $x_1 = 1$
3. without scaling (a) 8 (b) -0.2
with scaling (a) 1.33 (b) -0.05

Exercise 15.1

1. (a) at least one root in (0,1), in (1,2) and in (2,3)
(b) at least one root in (0,1) and in (2,3)
2. (a) root in $(-3,-4)$ (b) root in (3,4)
(c) root in (0,1) and in (99,100)
(d) no roots
(e) root in $(0,\frac{\pi}{2})$ and roots near $n\pi$, $n = 1,2,3,\ldots$
(f) root in $(\frac{\pi}{2},\pi)$
3. (a) one or three positive roots, no negative roots
(b) one positive root, none or two negative roots

(c) one positive root, none or two negative roots
(d) none or two positive roots, none or two negative roots

4.

	x	0.995	0.995	1.0005	1.005
(a)	$x^8 - 1$	0.019 307	0.003 993	0.004 007	0.040 707
(b)	$x^{\frac{1}{8}} - 1$	0.000 626	0.000 063	0.000 062	0.000 624

Much more accuracy with (b) than (a).

Exercise 15.2

1. (a) (3,3.05), root near 3.0032
(b) (0.45,0.4725), root near 0.461
(c) (4.875,4.9375), root near 4.9206
(d) (2.95,2.975), root near 2.9505
(e) (0.5,0.525), root near 0.5175
(f) (1.75,1.875), root near 1.822

2. (a) (3.003 123 758,3.2)
(b) (0.4,0.461 283 543), $f(x_R) = 0.000\,014\,83$
(c) (4.912 318 43,5)
(d) (2.944 586 259,3), $f(x_L) = -0.000\,054\,823\,4$
(e) (0.514 874 772,0.6), $f(x_L) = -0.000\,586$
(f) (1,1.824 782 655), $f(x_R) = 0.004\,6$

Exercise 15.3

1. (a) 0.57 (b) 1.89

2. $x = x^2$; $x_0 = 0.2$, $x_3 = 0.000\,002\,56$; converges to 0
$x_0 = 0.8$, $x_3 = 0.167\,772\,16$; converges to 0
$x_0 = 1.2$, $x_3 = 4.209\,816\,96$; diverges (to ∞)
$x = x^{\frac{1}{2}}$; $x_0 = 0.2$, $x_3 = 0.817\,765\,433$; converges to 1
$x_0 = 0.8$, $x_3 = 0.972\,492\,472$; converges to 1
$x_0 = 1.2$, $x_3 = 1.023\,051\,875$; converges to 1

3. 1.20, 0.85, 1.75; first and third using $x = \ln(A/x)$, second using $x = Ae^{-x}$

4. Actual roots are $x = 1$ and $x = -2$; 'scale factor' zero.
$x_0 = -3$, $x_3 = -2.097$ (3 d.p.), $x_0 = -1$, $x_3 = -2.012$ (3 d.p.), $x_0 = 0$, $x_3 = 1.012$ (3 d.p.), $x_0 = 2$, $x_3 = 1.000$ (3 d.p.)

Exercise 15.4

1. (a) 0.5671 (b) 1.8955

2. (a) −3.104 (b) 3.104
(c) 0.010 and 99.990
(d) no solution (e) 0.355
(f) 2.554

4. 0.87

Exercise 16.1

1. (a) $\pm 3j$ (b) $1 \pm \sqrt{3}j$

2. $z_1 = \sqrt{13}\angle 2.16^c$, $z_2 = 2\angle(-\pi/2)$, $z_3 = \sqrt{5}\angle(-2.03^c)$, $z_4 = 2\angle 0$

3. $1 \pm \sqrt{3}j$

4. (a) $\sqrt{17}\angle(-0.245^c)$ (b) $3\angle\pi$
(c) $4\angle(-\pi/2)$ (d) $13\angle 2.75^c$

Exercise 16.2

1. (a) $2 + 6j$, $2\sqrt{10}\angle 1.25^c$
(b) $9 + 7j$, $\sqrt{130}\angle 0.66^c$

2. (a) $\frac{2}{5} - \frac{1}{5}j$ (b) $3j$
(c) $-\frac{1}{2} - \frac{5}{2}j$ (d) $-\frac{1}{2} + \frac{1}{2}j$
(e) $1.1 - 0.3j$

3. (a) j (b) j
(c) $\frac{1}{3}j$ All real parts zero.

4. (a) $2 + 2j$ (b) $\frac{3}{2} + \frac{3\sqrt{3}}{2}j$
(c) $-\frac{3}{2} + \frac{3\sqrt{3}}{2}$ (d) $\frac{\sqrt{3}}{2} - \frac{1}{2}j$

5. (a) $2\angle(2\pi/3)$
(b) $2\sqrt{2}\angle(\pi/4)$; $z_1 z_2 = 4\sqrt{2}\angle 11\pi/12$,
$$z_1/z_2 = \frac{1}{\sqrt{2}}\angle 5\pi/12$$

6. $6\angle 5\pi/4$

Exercise 16.3

1. (a)

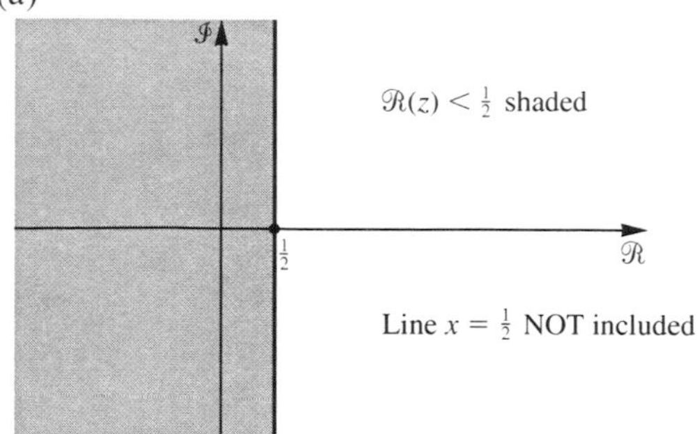

(b)

$\mathcal{R}(z) > -\frac{1}{2}$ shaded

Line $x = -\frac{1}{2}$
NOT included

(c)

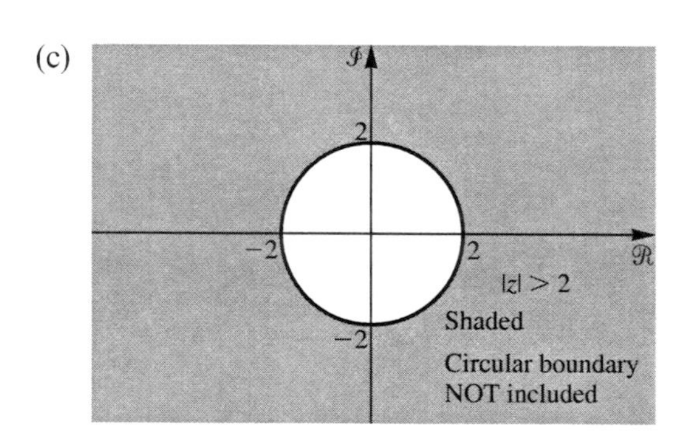

(d)

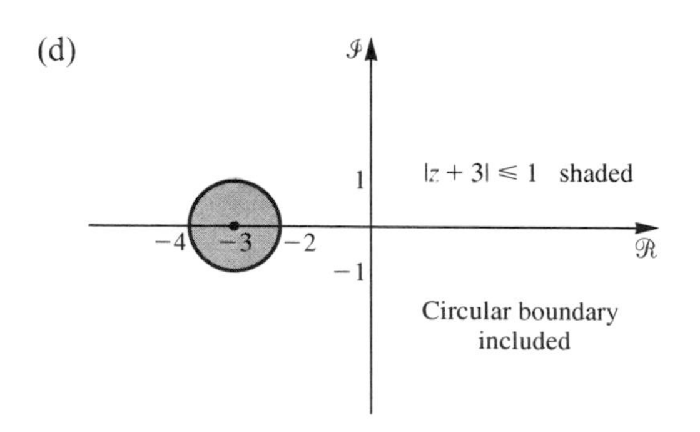

(e)

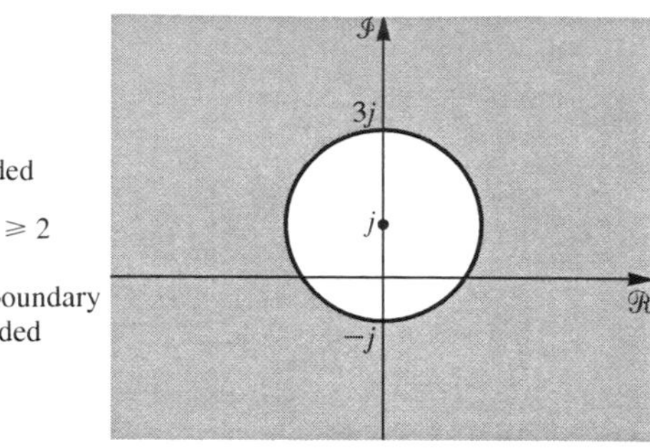

(f)

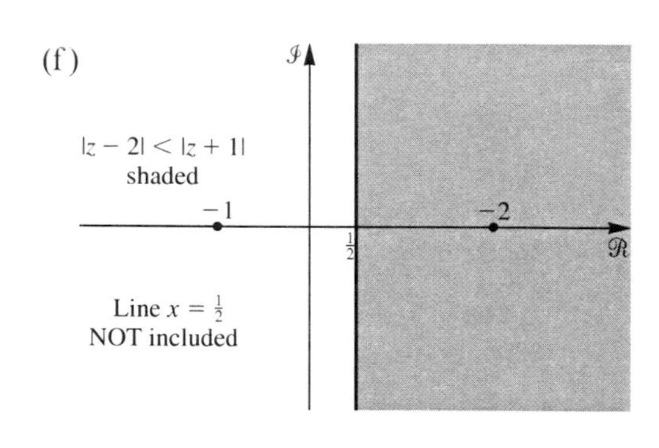

(g)

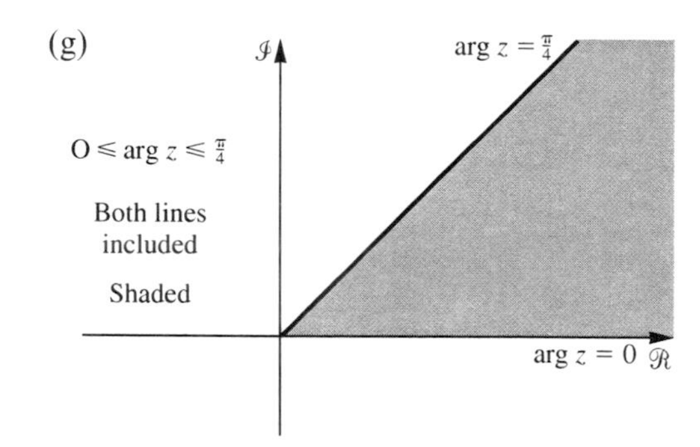

1. (a)

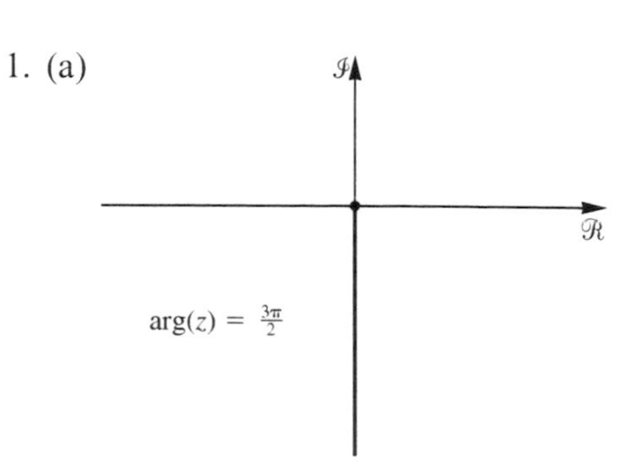

(b)

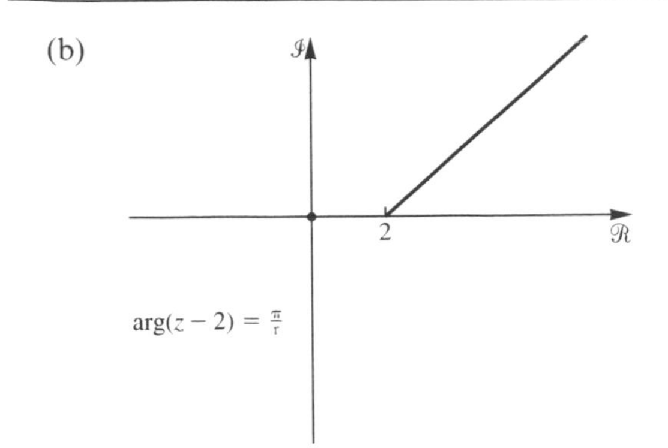

3. (a) $z = \left(1 \pm \frac{1}{\sqrt{2}}\right)(1 + \mathrm{j})$

   (b) $z = \left(\frac{1 \pm \sqrt{7}}{2}\right)(1 + \mathrm{j})$

Exercise 16.4

1. $\cos 5\theta + \sin 5\theta°\mathrm{j}$; $\cos \frac{1}{2}\theta + \sin \frac{1}{2}\theta°\mathrm{j}$
2. (a) $\cos 5\theta + \sin 5\theta°\mathrm{j}$ (b) $\cos 3\theta + \sin 3\theta°\mathrm{j}$
3. $(32)^{\frac{1}{6}}\angle(-15°)$, $(32)^{\frac{1}{6}}\angle(-135)$, $(32)^{\frac{1}{6}}\angle 105°$
5. $2\angle 45°$, $2\angle 135°$, $2\angle -45°\angle -\varnothing <°$ or $\pm\sqrt{2}(1 \pm \mathrm{j})$
6. $1\angle -18°$, $1\angle -90°$, $1\angle -162°$, $1\angle 54°$, $1\angle 126°$

Exercise 17.1

1. (a) $\frac{dy}{dx} = \frac{9x^2 - 2y^2 - y^4}{4x(1 + y^3)}$

   (b) $\frac{dy}{dx} = \frac{x + 1}{2 - y}$ (c) $\frac{dy}{dx} = \frac{2x + 1}{2y + 1}$

   (d) $\frac{dy}{dx} = \frac{-(y + \sin(x + y))}{x + \sin(x + y)}$

   Tangents: $4y = 3x + 1$, $2y = x - 5$, $y = x$

2. $\frac{-2\sqrt{2}b}{a^2}$

3. (a) $(2 - 11x + 4x^2)x(1 - x)^4\mathrm{e}^{-4x}$

   (b) $\frac{(2x\cos x - x\sin x - 2\cos x)}{x^3}\mathrm{e}^{2x}$

   (c) $\frac{(-7 - 2x + 13x^2 - x^3 - 2x^4)(1 + x^3)}{(3 + x)^6}\mathrm{e}^{-2x}$

   (d) $(x + 2x\ln x - 3x^2\ln x)\mathrm{e}^{-3x}$

Exercise 17.2

1. (a) $\frac{1}{3t}$ (b) $\frac{t}{1 + t}$ (c) $-\frac{1}{t^2}$

   (d) $\frac{1}{t}$ (e) $\frac{b}{a}\coth\theta$

2. (a) $\frac{-2}{3t^4}$ (b) $\frac{1}{2(1 + t)^3}$ (c) $\frac{2}{ct^3}$

   (d) $-\frac{1}{2at^3}$ (e) $-\frac{b}{a^2\sinh^3\theta}$

Exercise 17.3

1. (a) $\rho = \frac{1}{2}(1 + 4x^2)^{3/2}$; $\frac{1}{2}\sqrt{17}$, $\frac{1}{2}\sqrt{17}$, $\frac{1}{2}$;

   no greatest value, least at (0,c)

   (b) $\rho = \frac{(a^4y^2 + b^4x^2)^{3/2}}{a^4b^4}$; $-\frac{b^2}{a}$, $-\frac{a^2}{b}$;

   $\frac{(3a^2 + b^2)^{3/2}}{8ab}$; greatest at $x = 0$,

   least where $y = 0$

   (c) $\rho = 1$; circle has constant radius of curvature

2. $\rho = a(t^2 + 1)^{\frac{3}{2}}$

Exercise 17.4

1. $\dot{\boldsymbol{r}} = -\sin t\boldsymbol{i} + \cos t\boldsymbol{j}$, $\ddot{\boldsymbol{r}} = -\cos t\boldsymbol{i} - \sin t\boldsymbol{j}$, $|\dot{\boldsymbol{r}}| = 1$, $|\ddot{\boldsymbol{r}}| = 1$

Exercise 18.1

1. (a) $\mathrm{e}^{-a} - (x - a)\mathrm{e}^{-a}$ (b) $\ln(1 + a) + \frac{(x - a)}{1 + a}$

   (c) $\tan a + \sec^2 a(x - a)$

   (d) $a^2 + 2a(x - a)$

   (e) $a^3 + 3a^2(x - a) + 3a(x - a)^2$

2.

	0	0.1	−0.1	0.5	−0.5	1	−1
(a)	1	0.9	1.1	0.5	1.5	0	2
(b)	0	0.1	−0.1	0.5	−0.5	1	−1
(c)	0	0.1	−0.1	0.5	−0.5	1	−1

3. (a) $e^{-a} - (x-a)e^{-a} + \frac{1}{2}(x-a)^2 e^{-a}$

(b) $\ln(1+a) + \frac{x-a}{1+a} - \frac{(x-a)^2}{2(1+a)^2}$

(c) $\tan a + \sec^2 a(x-a) + \sec^2 a \tan a(x-a)^2$

(d) $a^2 + 2a(x-a) + \frac{(x-a)^2}{a}$

(e) $a^3 + 3a^2(x-a) + 3a(x-a)^2$

4.

	0	0.1	−0.1	0.5	−0.5	1	−1
(a)	1	0.905	1.1105	0.625	1.625	0.5	2.5
(b)	1	0.095	−0.105	0.375	0.625	0.5	−1.5
(c)	0	0.1	−0.1	0.5	−0.5	1	−1

5. (a) $1 - \frac{x^2}{2} + \frac{x^4}{24}$

(b) $1 - x + \frac{1}{2}x^2 - \frac{1}{6}x^3 + \frac{1}{24}x^4$

(c) $(1+x)^4$

Exercise 18.2

1. (a) $|x| < 2$
   (b) $|x| < 1$

2. (a) $1 - \frac{x^2}{8} + \frac{x^4}{384} - \ldots$, all x

(b) $1 + \frac{x}{2} + \frac{x^2}{4} + \frac{x^3}{8} + \ldots$, $|x| < 2$

(c) $-x^2 - \frac{x^4}{2} - \frac{x^6}{3}$, $|x| \leqslant 1$

Exercise 18.3

1. (a) $\frac{x^2}{4} + \frac{x^4}{48} + \frac{x^6}{1440} + \ldots$

(b) $1 + \frac{x^2}{4} + \frac{x^4}{48} + \frac{x^6}{1440} + \ldots$

(c) $-x^2 - \frac{x^4}{2} - \frac{x^6}{3} - \ldots$

2. $1 - x^2 + x^4 - x^6 + \ldots$;

$x - \frac{x^3}{3} + \frac{x^5}{5} - \frac{x^7}{7} + \ldots$

Exercise 18.4

1. (a) $-\frac{1}{2}$ (b) $\frac{1}{2}$ (c) 0 (d) 0
   (e) 1 (f) 4

2. 8

3.

t	1	2	3	4	5	6	7	8
Speed	4.5	10.5	17	13	4.5	0	−4.5	−13
Acceleration	0.75	2.25	1	−3	6.25	−13	22.5	−3

A possible SHM.

Exercise 19.1

1. (a) paraboloid (b) inverted cone
   (c) elliptical cone (d) cone

2. (a) 0, 0, 0.25 (b) 2, 2.25, 3.25

(c) $1, \frac{2}{\sqrt{5}}, \frac{2}{3}$ (d) 1, 0, 0

Exercise 19.2

1.

	$fx(1,2)$	$fx(-1,-1)$	$fx(2,1)$
(a)	2	−2	4
(b)	12	3	12
(c)	6	−3	12
(d)	cos 1	cos 1	cos 2
(e)	$-e^{-1}\cos 2$	$-e\cos 1$	$-e^{-2}\cos 1$

	$fy(1,2)$	$fy(-1,-1)$	$fy(2,1)$
(a)	1	1	1
(b)	4	2	16
(c)	17	−5	18
(d)	$-\sin 2$	$\sin 1$	$-\sin 1$
(e)	$-e^{-1}\sin 2$	$e\sin 1$	$-e^{-2}\sin 1$

2.

	f_{xx}	f_{yy}	f_{xy}
(a)	2	0	0
(b)	$6xy^2$	$2x^3$	$6x^2y$
(c)	$6xy$	$2x+6$	$3x^2+2y$
(d)	$-\sin x$	$-\cos y$	0
(c)	$e^{-x}\cos y$	$-e^{-x}\cos y$	$e^{-x}\sin y$

3. 634 mm s^{-1} decrease

4. (a) $-3x^2y^2a\sin at + 2x^3ya\cos at$

(b) $\frac{6x+2ty}{(x^2+y^2)^{\frac{1}{2}}}$

(c) $\frac{\partial f}{\partial u} = -\sin x e^{y} 2v + \cos x e^{y} .2u,$

$\frac{\partial f}{\partial v} = -\sin x e^{y} .2u - \cos x e^{y} 2u$

5. $\left(\frac{\partial S}{\partial V}\right)_T = \frac{R}{V}; \left(\frac{\partial S}{\partial p}\right)_T = \frac{-R}{p}; \left(\frac{\partial S}{\partial p}\right)_V = \frac{C_V}{p}$

Exercise 19.3

1. (a) $(2x + y^3)\,dx + 3xy^2\,dy$
   (b) $2xy^2z^2\,dx + 2x^2yz^2\,dy + 2x^2y^2z\,dz$
2. ± 0.0146
3. 1% increase
4. $\pm 1\%$

Exercise 20.1

1. (a) 1.724 (3 d.p.); 1.720 (3 d.p.) (correct value 1.718 (3 d.p.))
   (b) 1.1027 (4 d.p.); 1.1016 (4 d.p.) (correct value 1.0986 (4 d.p.))
   (c) 0.2328; 0.203 33 (correct value 0.2)
2. (a) 5 strips: max error = 0.0091; actual error = 0.006 (3 d.p.)
   10 strips: max error = 0.0023; actual error = 0.002 (2 d.p.)
   (b) 5 strips: max error = 0.0533; actual error = 0.0041 (4 d.p.)
   10 strips: max error = 0.0133; actual error = 0.0030 (4 d.p.)
   (c) 5 strips: max error = 0.04; actual error = 0.0328 (4 d.p.)
   10 strips: max error = 0.01; actual error = 0.0033 (4 d.p.)
3. (a) (i) 3 (ii) 7
   (b) (i) 6 (ii) 7
   (c) (i) 5 (ii) 15
4. 58.85

Exercise 20.2

1. (a) 1.7183 (4 d.p.)
   (b) 1.1
   (c) 0.2005 (4 d.p.)
2. (a) max error = 0.000 06; actual error = 0.0000 (4 d.p.)
   (b) max error = 0.0167; actual error = 0.0014 (4 d.p.)
   (c) max error = 0.005; actual error = 0.0005 (4 d.p.)
3. (a) (i) 2 (ii) 2 (b) (i) 4 (ii) 6
   (c) (i) 2 (ii) 4 (n must be even)
4. 59.57

Exercise 20.3

1. (a) $y = 4.8 - 1.93x$
   (b) $y = 5.113 + 0.0285x$
   (c) $y = 12.07 - 2.07x$
2. (a) $xy = ax^2 + b$; xy against x^2
   (b) y against $1/x^2$
   (c) $\ln y = \ln a + b/x$
   (d) $\ln y = \ln a + k \ln x$
3. $y = 1.12x^{1.99}$

Exercise 21.1

1. (a) linear, non-homogeneous, second order, first degree
   (b) non-linear, second order, third degree
   (c) linear, non-homogeneous, first order, first degree
   (d) linear, homogeneous, second order, first degree
   (e) linear, homogeneous, first order, first degree
   (f) linear, non-homogeneous, first order, first degree

Exercise 21.2

1. (a) $\frac{dy}{dx} = 3y^{\frac{2}{3}}$ (b) $\frac{dy}{dx} = -y$
   (c) $\frac{d^2y}{dt^2} = 9y$ (d) $\frac{d^2y}{dt^2} = -4y$
3. (a) $y = 2e^{2t}$ (b) $y = \cos 3t$
   (c) $y = \frac{1}{4}e^{2t} - \frac{1}{4}e^{-2t}$
4. λa is an odd multiple of π when $y = B \sin \lambda x$, B constant

Exercise 21.3

1. (a) $y_4 = 3$ (b) $y_4 = 1.75$
   (c) $y_4 = 6.2208$ (d) $y_4 = 4.3923$
   (e) $y_4 = 4.152$ (f) $y_4 = 2.142$
   (g) $y_4 = 1.061\,106$ (h) $y_4 = 1.5735$

2. (a) analytical answer 3; exact
   (b) analytical answer 2; error -0.25
   (c) analytical answer 6.6766 (4 d.p.); error -0.4588 (4 d.p.)
   (d) analytical answer 4.4755 (4 d.p.); error -0.0826 (4 d.p.)
   (e) analytical answer 5.5413 (4 d.p.); error -1.3893 (4 d.p.)
   (f) analytical answer 2.3319 (4 d.p.); error -0.1891 (4 d.p.)
   (g) analytical answer 1.0833 (4 d.p.); error -0.0222 (4 d.p.)

Exercise 21.4

1. (a) lines $x =$ constant (b) lines $y =$ constant
   (c) ellipses $x^2 + 4y^2 =$ constant

2. (a)

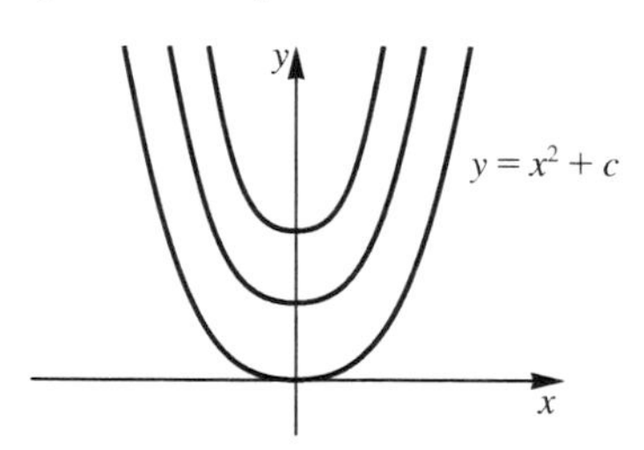

(b)

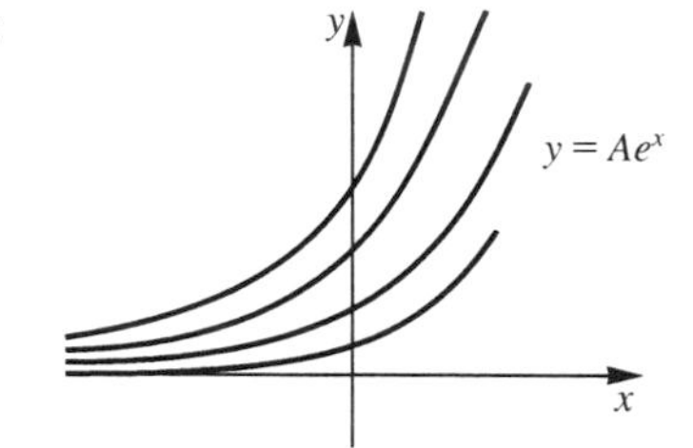

(c)

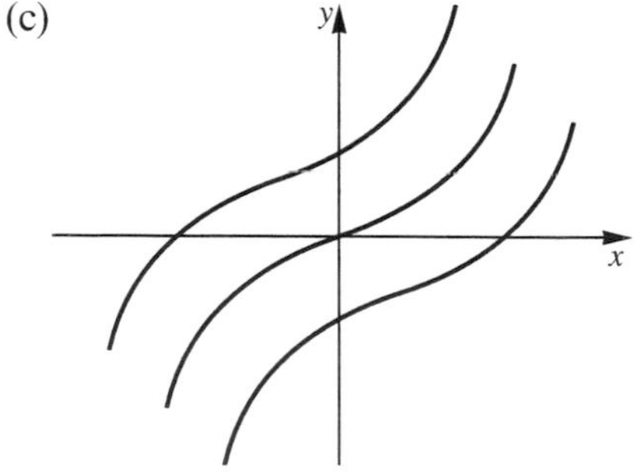

Exercise 22.1

1. (a) separable, $\int (y-2)\,dy = \int (x+3)\,dx$
   (b) separable, $\displaystyle\int \frac{1}{(y-2)}\,dy = \int \frac{1}{(x+3)}\,dx$
   (c) not separable
   (d) not separable
   (e) separable, $\int y\,dy = \int (x-3)\,dx$

2. (a) $\frac{1}{2}y^2 - 2y = \frac{1}{2}x^2 + 3x + C$
   (b) $y - 2 = A(x+3)$
   (e) $\frac{1}{2}y^2 = 3x - x^2 + C$

3. (a) $y = 3e^{-2x}$ (b) $1 - \dfrac{1}{2\omega}\sin 2\omega t$

   (c) $u = \tan\left(-t^2 + \dfrac{\pi}{4}\right)$

4. $v = \left[\dfrac{2gR}{r}(R-r) + v_0^2\right]^{\frac{1}{2}}$

5. $h = h_0 e^{-kt}$; $\dfrac{1}{k}\ln 2$

6. $h = \left(h_0 - \dfrac{1}{2}kt\right)^2$; $\dfrac{1}{k}\left[\ln(2-\sqrt{2}) + \dfrac{1}{2}\ln h_0\right]$

Exercise 22.2

3. (a) $y = e^x(x+1)$

   (b) $y = \left(2 + \dfrac{1}{2}x^2\right)e^{-x^2}$

   (c) $y = x^4 + 3x^3$

4. $i = \dfrac{E}{R}(1 - e^{-Rt/c}) \to \dfrac{E}{R}$; $i = 10(1 - e^{-st})$

   $i = \dfrac{1}{145}(20\sin 60t - 4\cos 60t) + Ce^{-5t}$

Exercise 22.3

1. (a) $y_1 = 2.375$, $y_2 = 2.5$; analytical solution $y = 2 + x - \frac{1}{2}x^2$
   (b) $y_1 = 1.625$, $y_2 = 1.390\,625$; analytical solution $y = 1 + e^{-x}$
   (c) $y_1 = 1.5$, $y_2 = 1$; analytical solution $y = 2 - x$
   (d) $y_1 = 1.0625$, $y_2 = 1.032\,490\,46$;
   analytical solution $y = \dfrac{3e^{2x}+1}{3e^{2x}-1}$

2. (a) $h = 0.5$: $y_1 = 2.375$, $y_2 = 2.5$; $h = 1$: $y_1 = 2.5$
   (b) $h = 0.5$: $y_1 = 1.6068$ (4 d.p.), $y_2 = 1.3681$ (4 d.p.); $h = 1$: $y_1 = 1.375$

Exercise 23.2

1. (a) $y = Ae^{2t} + Be^{-t}$
   (b) $y = (A + Bt)e^{5t}$
   (c) $y = e^{t}(A\cos t + B\sin t)$
   (d) $y = Ae^{\sqrt{2}t} + Be^{-\sqrt{2}t}$
   (e) $v = A\cos\sqrt{2}t + B\sin\sqrt{2}t$
2. (a) $y = Ae^{4t} + Be^{t}$
   (b) $y = (A + Bt)e^{-3t}$
   (c) $y = e^{-2t}(A\cos t + B\sin t)$
   (d) $y = (A + Bt)e^{t}$
   (e) $y = A\cos 3t + B\sin 3t$
   (f) $y = A + Be^{2t}$
   (g) $y = (A + Bt)e^{-Rt/2l}$

Exercise 23.3

1. (a) $\dfrac{1}{3}\sin t$ (b) $3e^{3t}$ (c) $7 + 10t$
   (d) $6t^3 - t^2 - 36t + 2$ (e) $\frac{1}{4}t^2 - \frac{5}{8}t + \frac{21}{32}$
   (f) $\frac{1}{9}e^{t}$ (g) $\frac{1}{6}(\cos t + \sin t)$
2. (a) $\frac{1}{2}e^{-3t}$ (b) $-\frac{1}{2}te^{-3t}$ (c) $\frac{1}{2}t^2e^{-3t}$

Exercise 23.4

1. (a) $i = \dfrac{20}{\sqrt{3}}e^{-50t}\sin 50\sqrt{3}t$; 0
   (b) $i = 6(e^{3t} - e^{-20t})$; 0
   (c) steady state $\cos t + 2\sin t$
   (d) steady state $(70\sin 10t - 90\cos 10t)$
2. (a) $x = 10\cos 10t$ (b) $x = 3\cos t + 4\sin t$
   (c) $x = \dfrac{3}{5}\sin 5t$
3. (a) $x = (1 + 10t)e^{-10t}$
   (b) $x = \left(\dfrac{3\sqrt{5} + 11}{2}\right)e^{(1-\sqrt{5})t/2} - \left(\dfrac{3\sqrt{5} + 9}{2}\right)e^{(1-\sqrt{5})t/2}$
   (c) $x = e^{-4t}\sin 3t$
4. (a) $\dfrac{2001}{2000}(1 + 10t)e^{-10t} - \dfrac{1}{2000}\cos 10t$
   (b) $\dfrac{1}{104}[e^{-4t}(3\cos 3t + 106\sin 3t) - 3\cos 3t + 2\sin 3t]$

Exercise 24.1

1. (a) $\dfrac{24}{s^5}$ (b) $\dfrac{720}{s^7}$
   (c) $\dfrac{3}{s^2+9}$ (d) $\dfrac{1}{s+5}$
   (e) $\dfrac{1}{s-4}$ (f) $\dfrac{9s}{9s^2+1}$
   (g) $\dfrac{2}{s^2-4}$ (h) $\dfrac{1}{s^2-1}$
   (i) $\dfrac{2+4s+4s^2}{s^3}$ (j) $\dfrac{6-4s}{s^2+4}$
   (k) $\dfrac{s}{s^2-9}$
2. (a) $\dfrac{s-4}{(s-4)^2+4}$ (b) $\dfrac{2}{(s+1)^2+4}$
   (c) $\dfrac{2}{(s-2)^3}$ (d) $\dfrac{2+(s-3)}{(s-3)^3}$
   (e) $\dfrac{s+1}{(s+1)^2+4}$ (f) $\dfrac{2}{(s+1)^2-4}$
   (g) $\dfrac{s+1+1}{(s^2+1)^2+1}$ (h) $\dfrac{2s}{(s^2+1)^2}$
   (i) $\dfrac{2s(s^2-27)}{(s^2+9)^3}$ (j) $\dfrac{(s-1)^2-1}{[(s-1)^2+1]^2}$
3. (a) $\dfrac{(s+1)}{s^2}e^{-s}$ (b) $\dfrac{e^{-s}.e}{s-1}$
   (c) $\left(\dfrac{s}{s^2+1}\right)e^{-s}$

Exercise 24.2

3. (a) $sY(s) - 2$
   (b) $s^2Y(s) - 2s - 1$
   (c) $(s^2 + 2s + 5)Y(s) - 2s - 5$
   (d) $(s^2 - 4s + 3)Y(s) - 2s + 7$

Exercise 24.3

1. (a) $5t^3$
   (b) $e^{-2t}\cos 4t$
   (c) $e^{-2t}\sinh 4t$
   (d) $e^{-5t}\cos 2t$
   (e) $e^{-5t}(2\cos 2t - 3\sin 2t)$
   (f) $e^{-0.5t}(5\cos 3t + 2\sin 3t)$
   (g) $(1 + 2t)e^{-3t}$
   (h) $(3\cos t + 2\sin t - 3)e^{-t}$
   (i) $(6\cos 2t - 10\sin 2t)e^{-3t}$
   (j) $(2\cos t + 3\sin t)e^{-t}$

2. $\frac{1}{3}(2e^t + e^{-2t})$

3. (a) $2e^{-t} + 3e^{1.5t}$ (b) $5e^t + e^{2t} - e^{-4t}$
   (c) $\cos 2t + \sin 2t$ (d) $e^t + 2t$
   (e) $\frac{1}{2}t - \frac{1}{2} + 3e^{-t}\sin t$ (f) $e^{-t}\cos t$

Exercise 24.4

2. $\frac{1}{13\Omega}(-10\cos t + 24\sin t + 10e^{-t} + 26te^{-t})$

3. (a) $\frac{1}{s-5}$
   (b) $\frac{1}{s^2 + 4s + 1}$

Index